Tests Concerning Two Population Proportions

p_1 and p_2 are the proportions of the members of two populations that have a certain characteristic. A random sample of size n_1 is chosen from the first population, and an independent random sample of size n_2 is chosen from the second. $\hat{p}_1$ and $\hat{p}_2$ are the proportions of the samples that have the characteristic and $\hat{p}$ is the proportion of the combined samples that has it.

D0087095

H_0	H_1	Test stat TS		
$p_1 = p_2$	$p_1 \neq p_2$	$\dfrac{\hat{p}_1 - \hat{p}_2}{\sqrt{(1/n_1 + 1/n_2)\hat{p}(1 - \hat{p})}}$	Reject H_0 if $\lvert TS \rvert \geq z_{\alpha/2}$	$2P\{Z \geq \lvert v \rvert\}$
$p_1 \leq p_2$	$p_1 > p_2$	$\dfrac{\hat{p}_1 - \hat{p}_2}{\sqrt{(1/n_1 + 1/n_2)\hat{p}(1 - \hat{p})}}$	Reject H_0 if $TS \geq z_\alpha$	$P\{Z \geq v\}$

CHAPTER 11 ANALYSIS OF VARIANCE

One-Factor ANOVA Table

$\bar{X}_i$ and S_i^2, $i = 1, \ldots, m$, are the sample means and sample variances of independent samples of size n from normal populations having means μ_i and a common variance σ^2.

Source of estimator	Estimator of σ^2	Value of test statistic
Between samples	$n\bar{S}^2 = \dfrac{n \sum_{i=1}^{m} (\bar{X}_i - \bar{\bar{X}})^2}{(m-1)}$	$TS = \dfrac{n\bar{S}^2}{\left(\sum_{i=1}^{m} S_i^2\right)/m}$
Within samples	$\left(\sum_{i=1}^{m} S_i^2\right)/m$	

Significance level α test of H_0: all μ_i are equal
  Reject H_0 if $TS \geq F_{m-1, m(n-1), \alpha}$
  Do not reject otherwise

If $TS = v$ then

$$p \text{ value} = P\{F_{m-1, m(n-1)} \geq v\}$$

where $F_{m-1, m(n-1)}$ is an F random variable with $m - 1$ numerator and $m(n - 1)$ denominator degrees of freedom.

Two-factor ANOVA model: For $i = 1, \ldots, m$, $j = 1, \ldots, n$

$$E[X_{ij}] = \mu + \alpha_i + \beta_j$$
$$\sum_{i=1}^{m} \alpha_i = \sum_{j=1}^{n} \beta_j = 0$$

μ is the grand mean, α_i is the deviation from the grand mean due to row i, and β_j is the deviation from the grand mean due to column j. Their estimators are

$$\hat{\mu} = X.. \qquad \hat{\alpha}_i = X_{i\cdot} - X.. \qquad \hat{\beta}_j = X_{\cdot j} - X..$$

Two-Factor ANOVA Table

	Sum of squares	Degrees of freedom
Row	$SS_r = n \sum_{i=1}^{m} (X_{i\cdot} - X..)^2$	$m - 1$
Column	$SS_c = m \sum_{j=1}^{n} (X_{\cdot j} - X..)^2$	$n - 1$
Error	$SS_e = \sum_{i=1}^{m} \sum_{j=1}^{n} (X_{ij} - X_{i\cdot} - X_{\cdot j} + X..)^2$	$N = (n-1)(m-1)$

Null hypothesis	Test statistic	Significance level α test	p value if $TS = v$
No row effect (all $\alpha_i = 0$)	$\dfrac{SS_r/(m-1)}{SS_e/N}$	Reject if $TS \geq F_{m-1, N, \alpha}$	$P\{F_{m-1, N} \geq v\}$
No column effect (all $\beta_j = 0$)	$\dfrac{SS_c/(n-1)}{SS_e/N}$	Reject if $TS \geq F_{n-1, N, \alpha}$	$P\{F_{n-1, N} \geq v\}$

CHAPTER 12 LINEAR REGRESSION

Simple linear regression model: $Y = \alpha + \beta x + e$

Least square estimators: $\hat{\beta} = S_{xY}/S_{xx}$, $\hat{\alpha} = \bar{Y} - \hat{\beta}\bar{x}$

$$S_{xY} = \sum_{i=1}^{n} (x_i - \bar{x})(Y_i - \bar{Y}) = \sum_{1}^{n} x_i Y_i - n\bar{x}\,\bar{Y}$$

$$S_{xx} = \sum_{1}^{n} (x_i - \bar{x})^2 = \sum_{1}^{n} x_i^2 - n\bar{x}^2$$

Estimated regression line: $y = \hat{\alpha} + \hat{\beta}x$

Hypothesis Tests Concerning the Mean μ of a Population

Assumption: Either the distribution is normal or sample size n is large.

H_0	H_1	Test statistic TS	Significance level α test	p value if TS = v
$\mu = \mu_0$	$\mu \neq \mu_0$	$\dfrac{\sqrt{n}(\bar{X} - \mu_0)}{\sigma}$†	Reject H_0 if $\lvert TS \rvert \geq z_{\alpha/2}$	$2P\{Z \geq \lvert v \rvert\}$
$\mu \leq \mu_0$	$\mu > \mu_0$	$\dfrac{\sqrt{n}(\bar{X} - \mu_0)}{\sigma}$†	Reject H_0 if $TS \geq z_\alpha$	$P\{Z \geq v\}$
$\mu = \mu_0$	$\mu \neq \mu_0$	$\dfrac{\sqrt{n}(\bar{X} - \mu_0)}{S}$	Reject H_0 if $\lvert TS \rvert \geq t_{n-1,\alpha/2}$	$2P\{T_{n-1} \geq \lvert v \rvert\}$
$\mu \leq \mu_0$	$\mu > \mu_0$	$\dfrac{\sqrt{n}(\bar{X} - \mu_0)}{S}$	Reject H_0 if $TS \geq t_{n-1,\alpha}$	$P\{T_{n-1} \geq v\}$

†*Assumption:* σ known.

Note: To test H_0: $\mu \geq \mu_0$, multiply data by -1 and use the above.

Hypothesis Tests Concerning p (the proportion of a large population that has a certain characteristic)

X is the number of population members in a sample of size n that have the characteristic. B is a binomial random variable with parameters n and p_0.

H_0	H_1	Test statistic TS	p value if TS = x
$p \leq p_0$	$p > p_0$	X	$P\{B \geq x\}$
$p = p_0$	$p \neq p_0$	X	$2 \text{ Min } \{P\{B \leq x\}, P\{B \geq x\}\}$

CHAPTER 10 HYPOTHESES TESTS CONCERNING TWO POPULATIONS

Tests Concerning the Means of Two Populations When Samples Are Independent

The X sample of size n and the Y sample of size m are independent.

H_0	H_1	Test statistic TS	Assumptions	Significance level α test	p value if TS = v
$\mu_x = \mu_y$	$\mu_x \neq \mu_y$	$\dfrac{\bar{X} - \bar{Y}}{\sqrt{S_x^2/n + S_y^2/m}}$	n, m large	Reject if $\lvert TS \rvert \geq z_{\alpha/2}$	$2P\{Z \geq \lvert v \rvert\}$
$\mu_x \leq \mu_y$	$\mu_x > \mu_y$	$\dfrac{\bar{X} - \bar{Y}}{\sqrt{S_x^2/n + S_y^2/m}}$	n, m large	Reject if $TS \geq z_\alpha$	$P\{Z \geq v\}$
$\mu_x = \mu_y$	$\mu_x \neq \mu_y$	$\dfrac{\bar{X} - \bar{Y}}{\sqrt{S_p^2(1/n + 1/m)}}$	Normal populations $\sigma_x = \sigma_y$	Reject if $\lvert TS \rvert \geq t_{n+m-2,\alpha/2}$	$2P\{T_{n+m-2} \geq \lvert v \rvert\}$
$\mu_x \leq \mu_y$	$\mu_x > \mu_y$	$\dfrac{\bar{X} - \bar{Y}}{\sqrt{S_p^2(1/n + 1/m)}}$	Normal populations $\sigma_x = \sigma_y$	Reject if $TS \geq t_{n+m-2,\alpha}$	$P\{T_{n+m-2} \geq v\}$

$$S_p^2 = \frac{n-1}{n+m-2}S_x^2 + \frac{m-1}{n+m-2}S_y^2 = \text{pooled estimator of } \sigma_x^2 = \sigma_y^2$$

INTRODUCTORY STATISTICS
Sheldon M. Ross

CHAPTER 1   INTRODUCTION TO STATISTICS

Statistics: the art of learning from data

Descriptive statistics: describes and summarizes data

Inferential statistics: draws conclusions from data

Population: collection of elements of interest

Sample: the part of the population from which data is obtained

CHAPTER 2   DESCRIBING DATA SETS

Frequency and relative frequency tables and graphs

Histograms

Stem-and-leaf plots

Scatter plots for paired data

CHAPTER 3   USING STATISTICS TO SUMMARIZE DATA SETS

Sample mean: $\bar{x} = \left(\sum_{i=1}^{n} x_i\right)/n$

Sample median: the middle value

Sample variance: $s^2 = \sum_{i=1}^{n} (x_i - \bar{x})^2/(n - 1)$

Sample standard deviation: $s = \sqrt{s^2}$

Algebraic identity: $\sum_{i=1}^{n} (x_i - \bar{x})^2 = \sum_{i=1}^{n} x_i^2 - n\bar{x}^2$

Empirical rule for normal data sets:

approximately 68% of the data lies within $\bar{x} \pm s$

approximately 95% of the data lies within $\bar{x} \pm 2s$

approximately 99.7% of the data lies within $\bar{x} \pm 3s$

Sample correlation coefficient:

$r = \sum_{i=1}^{n} (x_i - \bar{x})(y_i - \bar{y})/[(n - 1)s_x s_y]$

CHAPTER 4   PROBABILITY

$0 \leq P(A) \leq 1$

$P(S) = 1$, where S is the set of all possible values

$P(A \cup B) = P(A) + P(B)$, when A and B are disjoint

Probability of the complement: $P(A^c) = 1 - P(A)$

Addition rule: $P(A \cup B) = P(A) + P(B) - P(A \cap B)$

Conditional probability: $P(B|A) = P(A \cap B)/P(A)$

Multiplication rule: $P(A \cap B) = P(A)P(B|A)$

Independent events: $P(A \cap B) = P(A)P(B)$

CHAPTER 5   DISCRETE RANDOM VARIABLES

Expected value (or mean): $E[X] = \sum_{i=1}^{n} x_i P\{X = x_i\}$

$E[X + Y] = E[X] + E[Y]$

Variance: $\mathrm{Var}\,(X) = E[(X - E[X])^2] = E[X^2] - (E[X])^2$

Standard deviation: $\mathrm{SD}\,(X) = \sqrt{\mathrm{Var}\,(X)}$

$\mathrm{Var}\,(X + Y) = \mathrm{Var}\,(X) + \mathrm{Var}\,(Y)$ if X and Y are independent

Binomial random variable:

$$P\{X = i\} = \frac{n!}{i!(n - i)!}\, p^i (1 - p)^{n-i}, i = 0, \ldots, n$$

$$E[X] = np \qquad \mathrm{Var}\,(X) = np(1 - p)$$

CHAPTER 6   NORMAL RANDOM VARIABLES

Normal random variable X: characterized by $\mu = E[X]$, $\sigma = \mathrm{SD}\,(X)$

Standard normal random variable Z: normal with $\mu = 0, \sigma = 1$

$P\{|Z| > x\} = 2P\{Z > x\}, x > 0$

$P\{Z < -x\} = P\{Z > x\}$

z_α is such that $P\{Z > z_\alpha\} = \alpha$

If X is normal then $Z = (X - \mu)/\sigma$ is standard normal.

Additive property: If X and Y are independent normals then $X + Y$ is normal with mean $\mu_x + \mu_y$, and variance $\sigma_x^2 + \sigma_y^2$

CHAPTER 7   DISTRIBUTIONS OF SAMPLING STATISTICS

$X_1, \ldots, X_n$ is sample from population: $E[X_i] = \mu$, $\mathrm{Var}\,(X_i) = \sigma^2$

$E[\bar{X}] = \mu$

$\mathrm{Var}\,(\bar{X}) = \sigma^2/n$

Central limit theorem: $\sum_{i=1}^{n} X_i$ is, for large n, approximately normal with mean $n\mu$ and standard deviation $\sigma\sqrt{n}$; equivalently $\sqrt{n}(\bar{X} - \mu)/\sigma$ is approximately standard normal.

Normal approximation to binomial: If $np \geq 5, n(1 - p) \geq 5$ then $[\mathrm{Bin}\,(n, p) - np]/\sqrt{np(1 - p)}$ is approximately standard normal.

CHAPTER 8   ESTIMATION

$\bar{X}$ is the estimator of the population mean μ.

$\hat{p}$, the proportion of the sample that has a certain property, estimates p, the population proportion having this property.

S^2 estimates σ^2, and S estimates σ.

$100(1 - \alpha)$ confidence interval estimator for μ:

data normal or n large, σ known: $\bar{X} \pm z_{\alpha/2}\sigma/\sqrt{n}$

data normal, σ unknown: $\bar{X} \pm t_{n-1,\alpha/2}S/\sqrt{n}$

$100(1 - \alpha)$ confidence interval for p: $\hat{p} \pm z_{\alpha/2}\sqrt{\hat{p}(1 - \hat{p})/n}$

CHAPTER 9   TESTING STATISTICAL HYPOTHESES

H_0 = null hypothesis: hypothesis that is to be tested

Significance level α: the (largest possible) probability of rejecting H_0 when it is true

p value: the smallest significance level at which H_0 would be rejected

Chapter 12 (Cont.)

Error term e is normal with mean 0 and variance σ^2. Estimator of σ^2 is $SS_R/(n-2)$, $SS_R = \sum_i (Y_i - \hat\alpha - \hat\beta x_i)^2 = (S_{xx}S_{YY} - S_{xY}^2)/S_{xx}$

To test H_0: $\beta = 0$. Use TS $= \sqrt{(n-2)S_{xx}/SS_R}\, \hat\beta$

Significance level γ test is to reject H_0 if $|TS| \geq t_{n-2,\,\gamma/2}$.

If TS $= v$, p value $= 2P\{T_{n-2} \geq v\}$

$100(1-\gamma)$ confidence prediction interval for response at input x_0

$$\hat\alpha + \hat\beta x_0 \pm t_{n-2,\gamma/2}\sqrt{(1 + 1/n + (x_0 - \bar x)^2/S_{xx})SS_R/(n-2)}$$

Coefficient of determination: $R^2 = 1 - SS_R/S_{YY}$ is the proportion of the variation in the response variables that is explained by the different input values. Its square root is the absolute value of the sample correlation coefficient.

Multiple linear regression model:
$Y = \beta_0 + \beta_1 x_1 + \cdots + \beta_k x_k + e$

CHAPTER 13   CHI-SQUARED GOODNESS OF FIT TESTS

P_i is the proportion of population with value i, $i = 1, \ldots, k$. To test H_0: $P_i = p_i$, $i = 1, \ldots, k$, take a sample of size n. Let N_i be the number equal to i, $e_i = np_i$, TS $= \sum_{i=1}^{k}(N_i - e_i)^2/e_i$. Significance level α test rejects H_0 if TS $\geq \chi^2_{k-1,\alpha}$.

If TS $= v$, then p value $= P\{\chi^2_{k-1} \geq v\}$.

Suppose each member of a population has an X and a Y characteristic. Assume r possible X and s possible Y characteristics. To test for independence of the characteristics of a randomly chosen member, choose a sample of size n.

N_{ij} = number with X characteristic i and Y characteristic j
N_i = number with X characteristic i
M_j = number with Y characteristic j $\hat e_{ij} = N_i M_j/n$

If $\sum_i \sum_j (N_{ij} - \hat e_{ij})^2/\hat e_{ij} \geq \chi^2_{(r-1)(s-1),\alpha}$ then the hypothesis of independence is rejected at significance level α.

CHAPTER 14   NONPARAMETRIC HYPOTHESES

Let η = median of population. The *sign* test of

H_0: $\eta = m$ against H_1: $\eta \neq m$

takes a sample of size n. If i are less than m, then

p value $= 2$ Min $(P\{N \leq i\}, P\{N \geq i\})$

where N is a binomial $(n, 1/2)$ random variable.

The *signed rank* test is used to test the hypothesis that a population distribution is symmetric about 0. It ranks the data in terms of absolute value. TS is the sum of the ranks of the negative values. If TS $= t$, then

p value $= 2$ Min $(P\{TS \leq t\}, P\{TS \geq t\})$

TS is approximately normal with mean $n(n+1)/4$ and variance $n(n+1)(2n+1)/24$.

To test equality of two population distributions, draw random samples of sizes n and m and rank the $n + m$ data values. The *rank sum* test uses TS = sum of ranks of first sample. It rejects H_0 if TS is either significantly large or significantly small. If TS $= t$, then

p value $= 2$ Min $(P\{TS \leq T\}, P\{TS \geq t\})$

TS is approximately normal with mean $n(n + m + 1)/2$ and variance $nm(n + m + 1)/12$.

To test the hypothesis that a sequence of 0s and 1s is random, use the *runs* test by counting R, the number of runs. Reject randomness when R is either too small or too large to be explained by chance. Use the result that when H_0 is true, R is approximately normal with mean $1 + 2nm/(n+m)$ and variance

$$\frac{2nm(2nm - n - m)}{(n+m)^2(n+m-1)}$$

CHAPTER 15   QUALITY CONTROL

Control chart limits $\mu \pm 3\sigma/\sqrt{n}$ n = subgroup size

Area under the Standard Normal Curve to the Left of x

x	.00	.01	.02	.03	.04	.05	.06	.07	.08	.09
.0	.5000	.5040	.5080	.5120	.5160	.5199	.5239	.5279	.5319	.5359
.1	.5398	.5438	.5478	.5517	.5557	.5596	.5636	.5675	.5714	.5733
.2	.5793	.5832	.5871	.5910	.5948	.5987	.6026	.6064	.6103	.6141
.3	.6179	.6217	.6255	.6293	.6331	.6368	.6406	.6443	.6480	.6517
.4	.6554	.6591	.6628	.6664	.6700	.6736	.6772	.6808	.6844	.6879
.5	.6915	.6950	.6985	.7019	.7054	.7088	.7123	.7157	.7190	.7224
.6	.7257	.7291	.7324	.7357	.7389	.7422	.7454	.7486	.7517	.7549
.7	.7580	.7611	.7642	.7673	.7704	.7734	.7764	.7794	.7823	.7852
.8	.7881	.7910	.7939	.7967	.7995	.8023	.8051	.8078	.8106	.8133
.9	.8159	.8186	.8212	.8238	.8264	.8289	.8315	.8340	.8365	.8389
1.0	.8413	.8438	.8461	.8485	.8508	.8531	.8554	.8577	.8599	.8621
1.1	.8643	.8665	.8686	.8708	.8729	.8749	.8770	.8790	.8810	.8830
1.2	.8849	.8869	.8888	.8907	.8925	.8944	.8962	.8980	.8997	.9015
1.3	.9032	.9049	.9066	.9082	.9099	.9115	.9131	.9147	.9162	.9177
1.4	.9192	.9207	.9222	.9236	.9251	.9265	.9279	.9292	.9306	.9319
1.5	.9332	.9345	.9357	.9370	.9382	.9394	.9406	.9418	.9429	.9441
1.6	.9452	.9463	.9474	.9484	.9495	.9505	.9515	.9525	.9535	.9545
1.7	.9554	.9564	.9573	.9582	.9591	.9599	.9608	.9616	.9625	.9633
1.8	.9641	.9649	.9656	.9664	.9671	.9678	.9686	.9693	.9699	.9706
1.9	.9713	.9719	.9726	.9732	.9738	.9744	.9750	.9756	.9761	.9767
2.0	.9772	.9778	.9783	.9788	.9793	.9798	.9803	.9808	.9812	.9817
2.1	.9821	.9826	.9830	.9834	.9838	.9842	.9846	.9850	.9854	.9857
2.2	.9861	.9864	.9868	.9871	.9875	.9878	.9881	.9884	.9887	.9890
2.3	.9893	.9896	.9898	.9901	.9904	.9906	.9909	.9911	.9913	.9916
2.4	.9918	.9920	.9922	.9925	.9927	.9929	.9931	.9932	.9934	.9936
2.5	.9938	.9940	.9941	.9943	.9945	.9946	.9948	.9949	.9951	.9952
2.6	.9953	.9955	.9956	.9957	.9959	.9960	.9961	.9962	.9963	.9964
2.7	.9965	.9966	.9967	.9968	.9969	.9970	.9971	.9972	.9973	.9974
2.8	.9974	.9975	.9976	.9977	.9977	.9978	.9979	.9979	.9980	.9981
2.9	.9981	.9982	.9982	.9983	.9984	.9984	.9985	.9985	.9986	.9986
3.0	.9987	.9987	.9987	.9988	.9988	.9989	.9989	.9989	.9990	.9990
3.1	.9990	.9991	.9991	.9991	.9992	.9992	.9992	.9992	.9993	.9993
3.2	.9993	.9993	.9994	.9994	.9994	.9994	.9994	.9995	.9995	.9995
3.3	.9995	.9995	.9995	.9996	.9996	.9996	.9996	.9996	.9996	.9997
3.4	.9997	.9997	.9997	.9997	.9997	.9997	.9997	.9997	.9997	.9998

INTRODUCTORY STATISTICS

INTRODUCTORY
STATISTICS

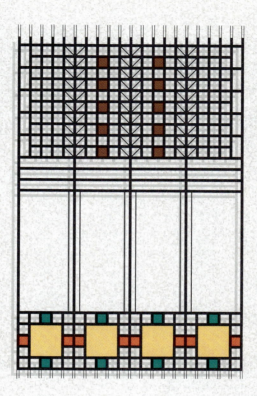

Sheldon M. Ross

University of California, Berkeley

The McGraw-Hill Companies, Inc.
New York St. Louis San Francisco Auckland Bogotá Caracas
Lisbon London Madrid Mexico City Milan Montreal
New Delhi San Juan Singapore Sydney Tokyo Toronto

TO MY FAMILY

This book was set in Times Roman by GTS Graphics.
The editors were Jack Shira, Maggie Lanzillo, and David A. Damstra;
the designer was Robin Hoffmann;
cover photo by Patrick Barta;
the production supervisor was Friederich W. Schulte.
The photo editor was Kathy Bendo;
the photo researcher was Debra P. Hershkowitz.
Von Hoffmann Press, Inc., was printer and binder.

Introductory Statistics

This book is printed on acid-free paper.

1 2 3 4 5 6 7 8 9 0   VNH VNH 9 0 9 8 7 6 5

P/N 053932-4

Library of Congress Cataloging-in-Publication Data
Ross, Sheldon M.
    Introductory statistics / Sheldon M. Ross.
      p.   cm. — (McGraw-Hill series in probability and statistics)
    Includes index.
    ISBN 0-07-912244-2 (IBM set). — ISBN 0-07-912245-0 (MAC set)
    1. Mathematical statistics.   I. Title.   II. Series.
  QA276.R684   1996
  519.5—dc20   95-19670

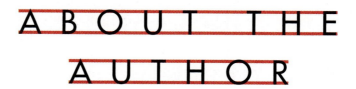

ABOUT THE AUTHOR

Sheldon M. Ross is a professor in the Department of Industrial Engineering and Operations Research at the University of California at Berkeley. He received his Ph.D. in statistics at Stanford University in 1968 and has been at Berkeley ever since. He has published many technical articles and textbooks in the areas of statistics and applied probability. Among his texts are *A First Course in Probability* (4th edition), *Introduction to Probability Models* (5th edition), and *Introduction to Probability and Statistics for Engineers and Scientists.*

Professor Ross is the founding and continuing editor of the journal *Probability in the Engineering and Informational Sciences.* He is a fellow of the Institute of Mathematical Statistics, and a recipient of the Humboldt U.S. Senior Scientist award.

CONTENTS

APPENDIXES — 751

PREFACE

*Statistical thinking will one day be as necessary
for efficient citizenship as the ability to read and write.*
H. G. Wells (1866–1946)

In today's complicated world, very few issues are clear cut and without controversy. In order to understand and form an opinion about an issue, one must usually gather information, or data. To learn from data, one must know something about statistics, which is the art of learning from data.

This introductory statistics text is written for college level students in any field of study. It can be used in a quarter, semester, or full year course. Its only prerequisite is high school algebra. Our goal in writing it is to present statistical concepts and techniques in a manner that will teach students not only how and when to utilize the statistical procedures developed, but also to understand why these procedures should be used. As a result we have made a great effort to explain the ideas behind the statistical concepts and techniques presented. Concepts are motivated, illustrated, and explained in a way that attempts to increase one's intuition. It is only when a student develops a feel or intuition for statistics that she or he is really on the path toward making sense of data.

To illustrate the diverse applications of statistics, and to offer students different perspectives about the use of statistics, we have provided a wide variety of text examples and problems to be worked by students. Most refer to real world issues, such as gun control, stock price models, health issues, driving age limits, school admission ages, public policy issues, gender issues, use of helmets, sports, disputed authorship, scientific fraud, Vitamin C, and many others. Many of them use data that are not only real but are themselves of interest. The exercises have been posed in a clear and concise manner, and include many thought provoking problems that emphasize thinking and problem solving skills. In addition, some of the problems are designed to be open ended and can be used as starting points for term projects.

SOME SPECIAL FEATURES OF THE TEXT

Introduction   The first numbered section of each chapter is an introduction which poses a realistic statistical situation to help students gain perspective on what they will encounter in the chapter.

Statistics in Perspective   Statistics in Perspective highlights are placed throughout the book to illustrate real-world application of statistical techniques and concepts. These perspectives are designed to help students analyze and interpret data, as well as to utilize proper statistical techniques and methodology.

Real Data   Throughout the text discussions, examples, perspective highlights, and exercises, real data sets are used to enhance the students' understanding of the material. These data sets provide information for the study of current issues in a variety of disciplines, such as health, medicine, sports, business, and education.

Historical Perspectives   These enrichment sections profile prominent statisticians and historical events, giving students an understanding of how the discipline of statistics has evolved.

Problems/Review Problems   This text includes hundreds of exercises which are placed at the end of each section within a chapter, as well as more comprehensive review exercises at the end of each chapter. Many of these problems utilize real data and are designed to assess the students' conceptual as well as computational understanding of the material. Selected problems are open-ended and offer excellent opportunity for extended discussion, group activities or student projects.

Summary/Key Terms   An end-of-chapter summary provides a detailed review of important concepts and formulas covered in the chapter. Key terms and their definitions are listed which serve as a working glossary within each chapter.

Formula Card   Important tables and formulas which students refer to and utilize often are included in a detachable table/formula card. This card can be used as a quick reference when doing homework or studying for an exam.

Minitab Laboratories/Computer Applications   Extended laboratory sessions for Minitab are provided at the end of most chapters. These laboratories, written by Lloyd R. Jaisingh of Morehead State University, include detailed instructions and extended examples for using Minitab and allow students to explore the topics covered in the corresponding chapter. In addition, exercises are placed at the end of each laboratory which can be used with Minitab or any other statistical software package.

Data/Program Disk   A disk is provided within each text which includes major data sets from the text and exercise material, in addition to programs that can be used to solve basic statistical computation problems. Please refer to Appendix E for a listing of these programs and easy-to-use instructions.

SUPPLEMENTS

Instructor's Resource Manual   The Instructor's Resource Manual contains detailed solutions to the even-numbered end-of-section and end-of-chapter problems.

Print Test Bank/Computerized Test Bank   The Print Test Bank provides instructors with over 1,000 test questions to choose from. A computerized version of this print test bank is also available in both IBM and Macintosh formats.

Student Solutions Manual   The Student Solutions Manual contains detailed solutions to the odd-numbered end-of-section and end-of-chapter exercises.

McGraw-Hill's Statistics Discovery Series: A Guide to Learning Statistics   This supplement is intended to help students enhance their understanding of introductory statistics. Each section of this study guide contains study objectives, an overview of the topics covered, key terms and definitions, worked-out examples, helpful hints to the student, and new exercises and their solutions.

McGraw-Hill's Statistics Discovery Series: A Guide to Minitab   This supplement helps the student gain a better understanding of statistics through the use of the statistics software Minitab. Worked-out examples and new exercises for use with Minitab are presented, along with a data disk containing data sets ready for use with Minitab. The supplement contains command information for DOS, Windows, and Macintosh platforms, and is packaged with either an IBM or a Macintosh disk.

McGraw-Hill's Statistics Discovery Series: A Guide to TI Graphing Calculators for Statistics   This supplement contains instructions for the student using a graphic calculator in an introductory statistics course through worked-out examples and new exercises. Appendixes for Texas Instruments 81, 82, and 85 graphic calculators are also included.

Against All Odds Videotapes   Videotapes depicting the use of statistics in our world, produced by the Annenberg/CPB Project, are available.

Mystat   This student version of the statistical software program Systat is available through McGraw-Hill.

THE TEXT

In chapter 1 we introduce the subject matter of statistics and present its two branches. The first of these, called descriptive statistics, is concerned with the collection, description, and summarization of data. The second branch, called inferential statistics, deals with the drawing of conclusions from data.

Chapters 2 and 3 are concerned with descriptive statistics. In Chapter 2 we discuss tabular and graphical methods of presenting a set of data. We see that an effective presentation of a data set can often reveal certain of its essential features. Chapter 3 shows how to summarize certain of the features of a data set.

In order to be able to draw conclusions from data it is necessary to have some understanding of what they represent. For instance, it is often assumed that the data constitute a "random sample from some population." In order to understand exactly what this and similar phrases signify, it is necessary to have some understanding of probability, and that is the subject of Chapter 4. The study of probability is often a troublesome issue in an introductory statistics class because many students find it a difficult subject. As a result, certain textbooks have chosen to downplay its importance and present it in a rather cursory style. We have chosen a different approach, and have attempted to concentrate on its essential features, and present them in a clear and easily understood manner. Thus, while avoiding the topic of counting principles, we have briefly but carefully dealt with the concepts of the events of an experiment, the properties of the probabilities that are assigned to the events, and the idea of conditional probability and independence. Our study of probability is continued in chapter 5 where discrete random variables are introduced and in chapter 6, which deals with the normal and other continuous random variables.

Chapter 7 is concerned with the probability distributions of sampling statistics. In this chapter we learn why the normal distribution is of such importance in statistics.

Chapter 8 deals with the problem of using data to estimate certain parameters of interest. For instance, we might want to estimate the proportion of people that are presently in favor of congressional term limits. Two types of estimators are studied. The first of these estimates the quantity of interest with a single number (for instance, it might estimate that 52 percent of the voting population favors term limits). The second type provides an estimator in the form of an interval of values (for instance, it might estimate that between 49 and 55 percent of the voting population favors term limits).

Chapter 9 introduces the important topic of statistical hypothesis testing, which is concerned with using data to test the plausibility of a specified hypothesis. For instance, such a test might reject the hypothesis that over 60 percent of the voting population favors term limits. The concept of p-value, which measures the degree of plausibility of the hypothesis after the data has been observed, is introduced.

Whereas the tests in chapter 9 deal with a single population, the ones in chapter 10 relate to two separate populations. For instance, we might be interested in testing whether the proportions of men and of women that favor term limits are the same.

Probably the most widely used statistical inference technique is that of the analysis of variance, and this is introduced in chapter 11. This technique allows us to test inferences about parameters that are affected by many different factors. Both one- and two-factor analysis of variance problems are considered in this chapter.

In Chapter 12 we learn about linear regression and how it can be used to relate the value of one variable (say the height of a man) to that of another (the height of his father). The concept of regression to the mean is discussed and the regression fallacy is introduced and carefully explained. We also learn about the relation between regression and correlation. Also, in an optional section, we use regression to the mean along with the central limit theorem to present a simple, original argument to explain why biological data sets often appear to be normally distributed.

In Chapter 13 we present goodness of fit tests, which can be used to test whether a proposed model is consistent with data. This chapter also considers populations classified according to two characteristics and shows how to test whether the characteristics of a randomly chosen member of the population are independent.

Chapter 14 deals with nonparametric hypothesis test, which are tests that can be used in situations where the ones of earlier chapters are inappropriate. Chapter 15 introduces the subject matter of quality control, a key statistical technique in manufacturing and production processes.

ACKNOWLEDGMENTS

We would like to thank Margaret Lin, Erol Pekoz, and the following reviewers of this text for their many helpful comments: William H. Beyer, University of Akron; Patricia Buchanan, Pennsylvania State University; Michael Eurgubian, Santa Rosa Junior College; Larry Griffey, Florida Community College, Jacksonville; James E. Holstein, University of Missouri; James Householder, Humboldt State University; Robert Lacher, South Dakota State University; Jacinta Mann, Seton Hill College; C. J. Park, San Diego State University; Ronald Pierce, Eastern Kentucky University; Lawrence Riddle, Agnes Scott College; Gaspard T. Rizzuto, University of Southwestern Louisiana; Jim Robison-Cox, Montana State University; Walter Rosencrantz, University of Massachusetts, Amherst; Bruce Sisko, Belleville Area College; Glen Swindle, University of California, Santa Barbara; Paul Vetrano, Santa Rose Junior College; Joseph J. Walker, Georgia State University; Deborah White, College of the Redwoods; and Cathleen Zucco, LeMoyne College.

Sheldon M. Ross

INTRODUCTORY STATISTICS

CHAPTER 1

Introduction to Statistics

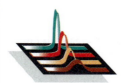

This chapter introduces the subject matter of statistics, the art of learning from data. It describes the two branches of statistics, descriptive and inferential. The idea of learning about a population by sampling and studying certain of its members is discussed. Some history is presented.

1.1   INTRODUCTION

Is it better for children to start school at a younger or older age? This is certainly a question of interest to many parents as well as to people who set public policy. How can we answer it?

It is reasonable to start by thinking about the above question, relating it to your own experiences, and talking it over with friends. However, if you want to convince others and obtain a consensus, it is then necessary to gather some objective information. For instance, in many states, achievement tests are given to children at the end of their first year in school. The children's results on these tests can be obtained and then analyzed to see whether there appears to be a connection between children's ages at school entrance and their scores on the test. In fact, such studies have been done, and they have generally concluded that older student entrants have, as a group, fared better than younger entrants. However, it has also been noted that the reason for this may just be that those students who entered at an older age would be older at the time of the examination, and this by itself may be what is responsible for their higher scores. For instance, suppose parents did not send their 6-year-olds to school but rather waited an additional year. Then, since these children will probably learn a great deal at home in that year, they will probably score higher when they take the test at the end of their first year of school than they would have if they had started school at age 6.

A recent study (see Table 1.1) has attempted to improve upon earlier work by examining the effect of children's age upon entering school on the eventual number of years of school completed. These authors argue that the total number of years spent in school is a better measure of school success than is a score on an achievement test

| Table 1.1 | **Total Years in School Related to Starting Age** |

	Younger half of children		Older half of children	
Year	Average age on starting school	Average number of years completed	Average age on starting school	Average number of years completed
1946	6.38	13.84	6.62	13.67
1947	6.34	13.80	6.59	13.86
1948	6.31	13.78	6.56	13.79
1949	6.29	13.77	6.54	13.78
1950	6.24	13.68	6.53	13.68
1951	6.18	13.63	6.45	13.65
1952	6.08	13.49	6.37	13.53

Source: J. Angrist and A. Krueger, "The Effect of Age at School Entry on Educational Attainment: An Application of Instrumental Variables with Moments from Two Samples," *Journal of the American Statistical Association,* vol. 87, no. 18, 1992, pp. 328–336

taken in an early grade. Using 1960 and 1980 census data, they concluded that the age at which a child enters school has very little effect on the total number of years that child spends in school. Table 1.1 is an abridgment of one presented in their work. The table indicates that for children beginning school in 1949, the younger half (whose average entrance age was 6.29 years) spent an average of 13.77 years, and the older half an average of 13.78 years, in school.

Note that we have not presented the preceding in order to make the case that the ages at which children enter school do not affect their performance in school. Rather we are using it to indicate the modern approach to learning about a complicated question. Namely, one must collect relevant information, or *data,* and these data must then be described and analyzed. Such is the subject matter of statistics.

1.2   THE NATURE OF STATISTICS

It has become a truism in today's world that in order to learn about something, you must first collect data. For instance, the first step in learning about such things as

1. The present state of the economy

2. The percentage of the voting public who favors a certain proposition

3. The average miles per gallon of a newly developed automobile

4. The efficacy of a new drug

5. The usefulness of a new way of teaching reading to children in elementary school

is to collect relevant data.

Definition

Statistics is the art of learning from data. It is concerned with the collection of data, its subsequent description, and its analysis, which often leads to the drawing of conclusions.

1.2.1   Data Collection

Sometimes a statistical analysis begins with a given set of data; for instance, the government regularly collects and publicizes data about such quantities as the unemployment rate and the gross domestic product. Statistics would then be used to describe, summarize, and analyze these data.

In other situations, data are not yet available, and statistics can be utilized to design an appropriate experiment to generate data. The experiment chosen should depend on the use that one wants to make of the data. For instance, if a cholesterol-lowering drug has just been developed and its efficacy needs to be determined, volunteers will be recruited and their cholesterol levels noted. They will then be given the drug for some period, and their levels will be taken again. However, it would be an ineffective experiment if *all* the volunteers were given the drug. For if this were so, then even if the cholesterol levels of all the volunteers were significantly reduced, we would not be justified in concluding that the improvements were due to the drug used and not to some other possibility. For instance, it is a well-documented fact that any medication received by a patient, whether or not it is directly related to that patient's suffering, will often lead to an improvement in the patient's condition. This is the *placebo effect,* which is not as surprising as it might seem at first, since a patient's belief that she or he is being effectively treated often leads to a reduction in stress, which can result in an improved state of health. In addition, there might have been other—usually unknown—factors that played a role in the reduction of cholesterol levels. Perhaps the weather was unusually warm (or cold), causing the volunteers to spend more or less time outdoors than usual, and this was a factor. Thus, we see that the experiment which calls for giving the drug to all the volunteers is not well designed for generating data from which we can learn about the efficacy of that drug.

A better experiment is one that tries to neutralize all other possible causes of the change of cholesterol level except the drug. The accepted way of accomplishing this is to divide the volunteers into two groups; then one group receives the drug, and the other group receives a tablet (known as a *placebo*) that looks and tastes like the drug but has no physiological effect. The volunteers should not know whether they are receiving the true drug or the placebo, and indeed it is best if the medical people overseeing the experiment also do not know, so their own biases will not play a role. In addition, we want the division of the volunteers into the two groups to be done in such a manner so that neither of the groups is favored in that it tends to have the "better" patients. The accepted best approach for arranging this is to break up the volunteers "at random," where by this term we mean that the breakup is done in such a manner that all possible choices of people in the group receiving the drug are equally likely. The group that does not receive any treatment (that is, the volunteers that receive a placebo) is called the *control* group.

At the end of the experiment, the data should be described. For instance, the before and after cholesterol levels of each volunteer should be presented, and the experimenter should note whether the volunteer received the drug or the placebo. In addition, summary measures such as the average reduction in cholesterol of members of the control group and members of the drug group should be determined.

Definition

The part of statistics concerned with the description and summarization of data is called *descriptive statistics.*

1.2.2   Inferential Statistics and Probability Models

When the experiment is completed and the data are described and summarized, we hope to be able to draw a conclusion about the efficacy of the drug. For instance, can we conclude that it is effective in reducing blood cholesterol levels?

Definition
The part of statistics concerned with the drawing of conclusions from data is called *inferential statistics*.

To be able to draw a conclusion from the data, we must take into account the possibility of chance. For instance, suppose that the average reduction in cholesterol is lower for the group receiving the drug than for the control group. Can we conclude that this result is due to the drug? Or is it possible that the drug is really ineffective and that the improvement was just a chance occurrence? For instance, the fact that a coin comes up heads 7 times in 10 flips does not necessarily mean that the coin is more likely to come up heads than tails in future flips. Indeed, it could be a perfectly ordinary coin that, by chance, just happened to land heads 7 times out of the total of 10 flips. (On the other hand, if the coin had landed heads 47 times out of 50 flips, then we would be quite certain that it was not an ordinary coin.)

To be able to draw logical conclusions from data, it is usually necessary to make some assumptions about the chances (or *probabilities*) of obtaining the different data values. The totality of these assumptions is referred to as a *probability model* for the data.

Sometimes the nature of the data suggests the form of the probability model that is assumed. For instance, suppose the data consist of the responses of a selected group of individuals to a question about whether they are in favor of President Clinton's welfare reform proposal. Provided that this group was *randomly* selected, it is reasonable to suppose that each individual queried was in favor of the proposal with probability p, where p represents the unknown proportion of all citizens in favor of the proposal. The resultant data can then be used to make inferences about p.

In other situations, the appropriate probability model for a given data set will not be readily apparent. However, a careful description and presentation of the data sometimes enable us to infer a reasonable model, which we can then try to verify with the use of additional data.

Since the basis of statistical inference is the formulation of a probability model to describe the data, an understanding of statistical inference requires some knowledge of the theory of probability. In other words, statistical inference starts with the assumption that important aspects of the phenomenon under study can be described in terms of probabilities, and then it draws conclusions by using data to make inferences about these probabilities.

1.3 POPULATIONS AND SAMPLES

In statistics, we are interested in obtaining information about a total collection of elements, which we will refer to as the *population*. The population is often too large for us to examine each of its members. For instance, we might have all the residents of a given state, or all the television sets produced in the last year by a particular manufacturer, or all the households in a given community. In such cases, we try to learn about the population by choosing and then examining a subgroup of its elements. This subgroup of a population is called a *sample*.

Definition
The total collection of all the elements that we are interested in is called a *population*.   A subgroup of the population that will be studied in detail is called a *sample*.

In order for the sample to be informative about the total population, it must be, in some sense, representative of that population. For instance, suppose that we are interested in learning about the age distribution of people residing in a given city, and we obtain the ages of the first 100 people to enter the town library. If the average age of these 100 people is 46.2 years, are we justified in concluding that this is approximately the average age of the entire population? Probably not, for we could certainly argue that the sample chosen in this case is not representative of the total population because usually more young students and senior citizens use the library than do working-age citizens. Note that *representative* does not mean that the age distribution of people in the sample is exactly that of the total population, but rather that the sample was chosen in such a way that all parts of the population had an equal chance to be included in the sample.

In certain situations, such as the library illustration, we are presented with a sample and must then decide whether this sample is reasonably representative of the entire population. In practice, a given sample generally cannot be considered to be representative of a population unless that sample has been chosen in a random manner. This is because any specific nonrandom rule for selecting a sample often results in one that is inherently biased toward some data values as opposed to others.

Definition
A sample of k members of a population is said to be a *random sample*, sometimes called a *simple random sample*, if the members are chosen in such a way that all possible choices of the k members are equally likely.

Thus, although it may seem paradoxical, we are most likely to obtain a representative sample by choosing its members in a totally random fashion without any prior considerations of the elements that will be chosen. In other words, we need not attempt to deliberately choose the sample so that it contains, for instance, the same gender percentage and the same percentage of people in each profession as found in the general population. Rather, we should just leave it up to "chance" to obtain roughly the correct percentages. The actual mechanics of choosing a random sample involve the use of random numbers and will be presented in App. C.

Once a random sample is chosen, we can use statistical inference to draw conclusions about the entire population by studying the elements of the sample.

*1.3.1 Stratified Random Sampling

A more sophisticated approach to sampling than simple random sampling is the *stratified random sampling* approach. This approach, which requires more initial information about the population than does simple random sampling, can be explained as follows. Consider a high school that contains 300 students in the first-year class, 500 in the second-year class, and 600 each in the third- and fourth-year classes. Suppose that in order to learn about the students' feelings concerning a military draft for 18-year-olds, an in-depth interview of 100 students will be done. Rather than randomly choosing 100 people from the 2000 students, in a stratified sample one calculates how many to choose from each class. Since the proportion of students who are first-year is 300/2000 = .15, in a stratified sample the percentage is the same and thus there are 100 × .15 = 15 first-year students in the sample. Similarly, one selects 100 × .25 = 25 second-year students and 100 × .30 = 30 third-year and 30 fourth-year students. Then one selects students from each class at random.

In other words, in this type of sample, first the population is *stratified* into subpopulations, and then the correct number of elements is randomly chosen from each of the subpopulations. As a result, the proportions of the sample members that belong to each of the subpopulations are exactly the same as the proportions for the total population. Stratification is particularly effective for learning about the "average" member of the entire population when there are inherent differences between the subpopulations with respect to the question of interest. For instance, in the above survey, the upper-grade students, being older, would be more immediately affected by a military draft than the lower-grade students. Thus, each class might have inherently different feelings about the draft, and stratification would be effective in learning about the feelings of the average student.

* The asterisk signifies optional material not used in the sequel.

1.4 A BRIEF HISTORY OF STATISTICS

A systematic collection of data on the population and the economy was begun in the Italian city-states of Venice and Florence during the Renaissance. The term *statistics,* derived from the word *state,* was used to refer to a collection of facts of interest to the state. The idea of collecting data spread from Italy to the other countries of western Europe. Indeed, by the first half of the 16th century, it was common for European governments to require parishes to register births, marriages, and deaths. Because of poor public health conditions this last statistic was of particular interest.

The high mortality rate in Europe before the 19th century was due mainly to epidemic diseases, wars, and famines. Among epidemics the worst were the plagues. Starting with the Black Plague in 1348, plagues recurred frequently for nearly 400 years. In 1562, as a way to alert the King's court to consider moving to the countryside, the city of London began to publish weekly bills of mortality. Initially these mortality bills listed the places of death and whether a death had resulted from plague. Beginning in 1625, the bills were expanded to include all causes of death.

In 1662 the English tradesman John Graunt published a book entitled *Natural and Political Observations Made upon the Bills of Mortality.* Table 1.2, which notes the total number of deaths in England and the number due to the plague for five different plague years, is taken from this book.

Graunt used the London bills of mortality to estimate the city's population. For instance, to estimate the population of London in 1660, Graunt surveyed households in certain London parishes (or neighborhoods) and discovered that, on average, there were approximately 3 deaths for every 88 people. Dividing by 3 shows that, on average, there was roughly 1 death for every 88/3 people. Since the London bills cited 13,200 deaths in London for that year, Graunt estimated the London population to be about

$$13,200 \cdot \frac{88}{3} = 387,200$$

Table 1.2   **Total Deaths in England**

Year	Burials	Plague deaths
1592	25,886	11,503
1593	17,844	10,662
1603	37,294	30,561
1625	51,758	35,417
1636	23,359	10,400

Graunt used this estimate to project a figure for all England. In his book he noted that these figures would be of interest to the rulers of the country, as indicators of both the number of men who could be drafted into an army and the number who could be taxed.

Graunt also used the London bills of mortality—and some intelligent guesswork as to what diseases killed whom and at what age—to infer ages at death. (Recall that the bills of mortality listed only causes and places of death, not the ages of those dying.) Graunt then used this information to compute tables giving the proportion of the population that dies at various ages. Table 1.3 is one of Graunt's mortality tables. It states, for instance, that of 100 births, 36 people will die before reaching age 6, 24 will die between the age of 6 and 15, and so on.

Graunt's estimates of the ages at which people were dying were of great interest to those in the business of selling annuities. Annuities are the opposite of life insurance in that one pays in a lump sum as an investment and then receives regular payments for as long as one lives.

Graunt's work on mortality tables inspired further work by Edmund Halley in 1693. Halley, the discoverer of the comet bearing his name (and also the man who was most responsible, by both his encouragement and his financial support, for the publication of Isaac Newton's famous *Principia Mathematica*), used tables of mortality to compute the odds that a person of any age would live to any other particular age. Halley was influential in convincing the insurers of the time that an annual life insurance premium should depend on the age of the person being insured.

Following Graunt and Halley, the collection of data steadily increased throughout the remainder of the 17th century and on into the 18th century. For instance, the city of Paris began collecting bills of mortality in 1667; and by 1730 it had become common practice throughout Europe to record ages at death.

Table 1.3	Graunt's Mortality Table

Age at death	Deaths per 100 births
0–6	36
6–16	24
16–26	15
26–36	9
36–46	6
46–56	4
56–66	3
66–76	2
≥ 76	1

Note: The categories go up to, but do not include, the right-hand value. For instance, 0–6 means ages 0 through 5 years.

The term *statistics,* which was used until the 18th century as a shorthand for the descriptive science of states, in the 19th century became increasingly identified with numbers. By the 1830s the term was almost universally regarded in Britain and France as being synonymous with the *numerical science* of society. This change in meaning was caused by the large availability of census records and other tabulations that began to be systematically collected and published by the governments of western Europe and the United States beginning around 1800.

Throughout the 19th century, although probability theory had been developed by such mathematicians as Jacob Bernoulli, Karl Friedrich Gauss, and Pierre Simon Laplace, its use in studying statistical findings was almost nonexistent, as most social statisticians at the time were content to let the data speak for themselves. In particular, at that time statisticians were not interested in drawing inferences about individuals, but rather were concerned with the society as a whole. Thus, they were not concerned with sampling but rather tried to obtain censuses of the entire population. As a result, probabilistic inference from samples to a population was almost unknown in 19th-century social statistics.

It was not until the late 1800s that statistics became concerned with inferring conclusions from numerical data. The movement began with Francis Galton's work on analyzing hereditary genius through the uses of what we would now call regression and correlation analysis (see Chap. 12) and obtained much of its impetus from the work of Karl Pearson. Pearson, who developed the chi-squared goodness-of-fit test (see Chap. 13), was the first director of the Galton laboratory, endowed by Francis Galton in 1904. There Pearson originated a research program aimed at developing new methods of using statistics in inference. His laboratory invited advanced students from science and industry to learn statistical methods that could then be applied in their fields. One of his earliest visiting researchers was W. S. Gosset, a chemist by training, who showed his devotion to Pearson by publishing his own works under the name *Student.* (A famous story has it that Gosset was afraid to publish under his own name for fear that his employers, the Guinness brewery, would be unhappy to discover that one of its chemists was doing research in statistics.) Gosset is famous for his development of the t test (see Chap. 9).

Two of the most important areas of applied statistics in the early 20th century were population biology and agriculture. This was due to the interest of Pearson and others at his laboratory and to the remarkable accomplishments of the English scientist Ronald A. Fisher. The theory of inference developed by these pioneers, including among others Karl Pearson's son Egon and the Polish-born mathematical statistician Jerzy Neyman, was general enough to deal with a wide range of quantitative and practical problems. As a result, after the early years of this century, a rapidly increasing number of people in science, business, and government began to regard statistics as a tool able to provide quantitative solutions to scientific and practical problems.

Nowadays the ideas of statistics are everywhere. Descriptive statistics are featured in every newspaper and magazine. Statistical inference has become indispensable to public health and medical research, to marketing and quality control, to education, to accounting, to economics, to meteorological forecasting, to polling and surveys, to sports, to insurance, to gambling, and to all research that makes any claim to being scientific. Statistics has indeed become ingrained in our intellectual heritage.

Problems

1. This problem refers to Table 1.1.
   (a) In which year was there the largest difference between the average number of years of school completed by the younger and older starters?
   (b) Were there more years in which the average number of years completed by the younger starting group exceeded that of the older group, or the opposite?

2. A social scientist, studying whether people in the United States had become more isolated from one another in the 1980s, has gathered these data concerning membership totals in the Boy Scouts and Girl Scouts.

	Number of members (in thousands)								
	1975	**1980**	**1981**	**1982**	**1983**	**1984**	**1985**	**1986**	**1987**
Boy Scouts	5318	4318	4355	4542	4689	4755	4845	5171	5347
Girl Scouts	3234	2784	2829	2819	2888	2871	2802	2917	2947

Source: Boy Scouts of America and Girl Scouts of America, annual reports.

   (a) Does it appear from the above data that fewer people were becoming scouts as the 1980s progressed?
   (b) Were the Boy Scouts or Girl Scouts more stable in terms of yearly membership?
   (c) Which of the two groups is increasing in size?

3. The following data concern the number of passenger cars produced and sold in the United States from 1985 to 1990.

Year	Number of cars (in thousands)
1985	7337
1986	6869
1987	6487
1988	6437
1989	6181
1990	5502

Source: Motor Vehicle Manufacturers Association of the United States.

   (a) How would you describe the sales progression from 1985 to 1990?
   (b) Compare the total number of cars sold from 1985 to 1987 with the number sold from 1988 to 1990.
   (c) Would you agree that sales fell roughly 10 percent from the previous year's number in each of the years from 1986 to 1990?

4. A medical researcher, trying to establish the efficacy of a new drug, has begun testing the drug along with a placebo. To make sure that the two groups of volunteer patients—those receiving the drug and those receiving a placebo—are as nearly alike as possible, the researcher has decided not to rely on chance but rather to carefully scrutinize the volunteers and then choose the groupings himself. Is this approach advisable? Why or why not?

5. Explain why it is important that a researcher who is trying to learn about the usefulness of a new drug not know which patients are receiving the new drug and which are receiving a placebo.

6. An election will be held next week, and by polling a sample of the voting population, we are trying to predict whether the Republican or Democratic candidate will prevail. Which of the following methods of selection will yield a representative sample?
   (a) Poll all people of voting age attending a college basketball game.
   (b) Poll all people of voting age leaving a fancy midtown restaurant.
   (c) Obtain a copy of the voter registration list, randomly choose 100 names, and question them.
   (d) Use the results of a television call-in poll, in which the station asked its viewers to call and tell their choice.
   (e) Choose names from the telephone directory and call these people.

7. The approach used in Prob. 6e led to a disastrous prediction in the 1936 Presidential election in which Franklin Roosevelt defeated Alfred Landon by a landslide. A Landon victory had been predicted by the *Literary Digest*. The magazine based its prediction on the preferences of a sample of voters chosen from lists of automobile and telephone owners.
   (a) Why do you think the *Literary Digest*'s prediction was so far off?
   (b) Has anything changed between 1936 and now that would make you believe that the approach used by the *Literary Digest* would work better today?

8. A researcher is trying to discover the average age at death for people in the United States today. To obtain data, the obituary columns of *The New York Times* are read for 30 days, and the ages at death of people in the United States are noted. Do you think this approach will lead to a representative sample?

9. If, in Prob. 8, the average age at death of those recorded is 82.4 years, what conclusion could you draw?

10. To determine the proportion of people in your town who are smokers, it has been decided to poll people at one of the following local spots:
    (a) The pool hall   (b) The bowling alley
    (c) The shopping mall   (d) The library

Which of these potential polling places would most likely result in a reasonable approximation to the desired proportion? Why?

11. A university plans on conducting a survey of its recent graduates to determine information on their yearly salaries. It randomly selected 200 recent graduates and sent them questionnaires dealing with their present jobs. Of these 200, however, only 86 questionnaires were returned. Suppose that the average of the yearly salaries reported was $75,000.
    (a) Would the university be correct in thinking that $75,000 was a good approximation to the average salary level of all its graduates? Explain the reasoning behind your answer.
    (b) If your answer to (a) is no, can you think of any set of conditions relating to the group that returns questionnaires for which $75,000 would be a good approximation?

12. An article reported that a survey of clothing worn by pedestrians killed at night in traffic accidents revealed that about 80 percent of the victims were wearing dark-colored clothing and 20 percent were wearing light-colored clothing. The conclusion drawn in the article was that it is safer to wear light-colored clothing at night.
    (a) Is this conclusion justified? Explain.
    (b) If your answer to (a) is no, what other information would be needed before a final conclusion could be drawn?

13. Critique Graunt's method for estimating the population of London. What implicit assumption is he making?

14. The London bills of mortality listed 12,246 deaths in 1658. Supposing that a survey of London parishes showed that roughly 2 percent of the population died that year, use Graunt's method to estimate London's population in 1658.

15. Suppose you were a seller of annuities in 1662 when Graunt's book was published. Explain how you would make use of his data on the ages at which people were dying.

16. Based on Table 1.2, which of the five plague years appears to have been the most severe? Explain your reasoning.

17. Based on Graunt's mortality table:
    (a) What proportion of babies survived to age 6?
    (b) What proportion survived to age 46?
    (c) What proportion died between the ages of 6 and 36?

18. Why do you think that the study of statistics is important in your field? How do you expect to utilize it in your future work?

The Changing Definition of Statistics

Statistics has then for its object that of presenting a faithful representation of a state at a determined epoch. (Quetelet, 1849)

Statistics are the only tools by which an opening can be cut through the formidable thicket of difficulties that bars the path of those who pursue the Science of man. (Galton, 1889)

Statistics may be regarded (i) as the study of populations, (ii) as the study of variation, and (iii) as the study of methods of the reduction of data. (Fisher, 1925)

Statistics is a scientific discipline concerned with collection, analysis, and interpretation of data obtained from observation or experiment. The subject has a coherent structure based on the theory of Probability and includes many different procedures which contribute to research and development throughout the whole of Science and Technology. (E. Pearson, 1936)

Statistics is the name for that science and art which deals with uncertain inferences—which uses numbers to find out something about nature and experience. (Weaver, 1952)

Statistics has become known in the 20th century as the mathematical tool for analyzing experimental and observational data. (Porter, 1986)

Statistics is the art of learning from data. (Ross, 1995)

Key Terms

Statistics: The art of learning from data.

Descriptive statistics: The part of statistics that deals with the description and summarization of data.

Inferential statistics: The part of statistics that is concerned with drawing conclusions from data.

Probability model: The mathematical assumptions relating to the likelihood of different data values.

Population: A collection of elements of interest.

Sample: A subgroup of the population that is to be studied.

Random sample of size k: A sample chosen in such a manner that all subgroups of size k are equally likely to be selected.

Stratified random sample: A sample obtained by dividing the population into distinct subpopulations and then choosing random samples from each subpopulation.

Introduction to
the Minitab Software

Note:

1. The labs presented in this text will assume that you are using the Windows version (release 10) of the Minitab* software. The commands presented will vary slightly if you are using the earlier Windows version (release 9).

2. The procedures presented in these labs may not be the only way to achieve the end results, so feel free to try other procedures as you gain more experience with the software. Also, whenever graphs are presented, only the Minitab graphics features are used.

3. We will assume that the Windows software and Minitab (release 10) are already installed on the computer that you are using.

To start Minitab for Windows, turn on your computer, if necessary, and display the Program Manager window in Windows. Double-click on the Minitab 10 for Windows group icon to open the Minitab group window. The Minitab group window is shown in Fig. M1.1.

Figure M1.1

*Minitab is a registered trademark of Minitab Inc., 3081 Enterprise Drive, State College, PA 16801.

Note: If the software is located on a network server, you may need to consult with the technical support person or the lab attendant as to how you may access the software.

Double-click on the Minitab for Windows program-item icon. The welcome screen appears briefly, and then the main Minitab window will be displayed. This is shown in Fig. M1.2.

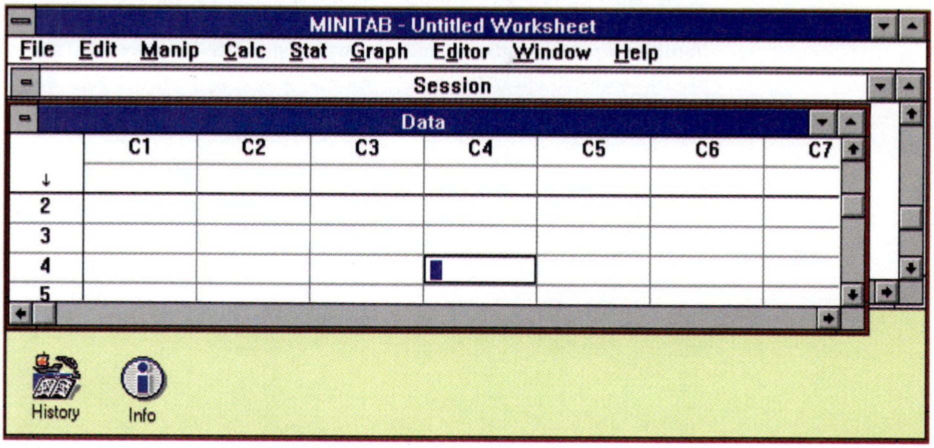

Figure M1.2

In Fig. M1.2, the *menu bar* (with File, Edit, etc.) appears at the top of the window. Each menu lists groups of commands that you will need to use the program. In addition, Fig. M1.2 lists two open windows, the Session window and the Data window. Observe that the Data window title bar is highlighted, indicating that this is the active window. To make the Session window the active window, all you need to do is to click anywhere on it. The Data window or worksheet is where the data will be entered, and it consists of columns and rows. The intersection of a row and a column constitutes a cell. The active cell in Fig. M1.2 is at the intersection of row 4 and column C4. The Session window is where Minitab commands could be entered or where computations and results will be displayed. That is, the Session window keeps a record of your entire session. The History and the Info windows are minimized to icons and are shown at the bottom of the Minitab window. The History window lists all the commands issued during a session, and the Info window contains a summary of all columns and constants in the current active worksheet. Observe that the title for the main Minitab window is *Minitab— Untitled Worksheet.* This worksheet title will change depending on the name of the worksheet on which you are working.

Observe at the top left-hand corner of the Data window the symbol ↓. This means that if you enter a value in a cell and press the Enter key, the cell vertically below will become active so you can then enter a value in that cell. If you click on the ↓ symbol, it will change to →; so if you press the Enter key, the next horizontal cell will become active. To rename a column, you can type the name in the box along the ↓ or → row. That is, you can type the name in the box just below C1, C2, etc.

In these labs you will observe sequences such as Calc→Random Data→Integer. This means that you have to first click on Calc on the menu bar, then click on Random data in the Calc menu, and then click on Integer in the Random data menu.

You can print any current window by selecting File→Print Window if your computer is connected to a printer. To open a saved worksheet, select File→Open Worksheet and follow the instructions on the screen. To exit Minitab, select File→Exit and your session will be terminated.

You should not be afraid to use the Help menu when in doubt.

Note: This is just a very brief introduction to Minitab for Windows. You need to consult with your instructor, lab attendant, or the Minitab manual for more detailed information. However, as you proceed with the labs that are included in the text, you will certainly become more familiar with the software and you will appreciate its power and the ease with which you can do the required analyses.

CHAPTER 2

Describing Data Sets

CHAPTER 2

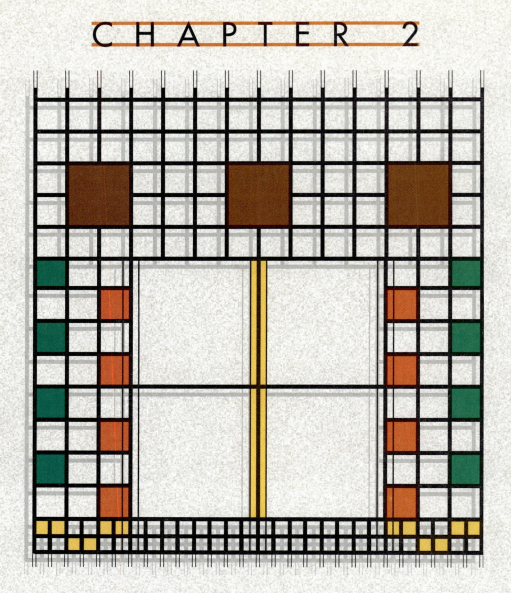

Describing Data Sets

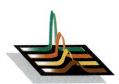

n this chapter we learn methods for presenting and describing sets of data. We introduce different types of tables and graphs, which enable us to easily see key features of a data set.

2.1   INTRODUCTION

It is very important that the numerical findings of any study be presented clearly and concisely, and in a manner that enables one to quickly obtain a feel for the essential characteristics of the data. This is particularly needed when the set of data is large, as is frequently the case in surveys or controlled experiments. Indeed, an effective presentation of the data often quickly reveals important features such as their range, degree of symmetry, how concentrated or spread out they are, where they are concentrated, and so on. In this chapter we will be concerned with techniques, both tabular and graphic, for presenting data sets.

Frequency tables and frequency graphs are presented in Sec. 2.2. These include a variety of tables and graphs—line graphs, bar graphs, and polygon graphs—that are useful for describing data sets having a relatively small number of distinct values. As the number of distinct values becomes too large for these forms to be effective, it is useful to break up the data into disjoint classes and consider the number of data values that falls in each class. This is done in Sec. 2.3, where we study the histogram, a bar graph that results from graphing class frequencies. A variation of the histogram, called a stem-and-leaf plot, which uses the actual data values to represent the size of a class, is studied in Sec. 2.4. In Sec. 2.5 we consider the situation where the data consist of paired values, such as the population and the crime rate of various cities, and introduce the scatter diagram as an effective way of presenting such data. Some historical comments are presented in Sec. 2.6.

2.2   FREQUENCY TABLES AND GRAPHS

The following data represent the number of days of sick leave taken by each of 50 workers of a given company over the last 6 weeks:

2, 2, 0, 0, 5, 8, 3, 4, 1, 0, 0, 7, 1, 7, 1, 5, 4, 0, 4, 0, 1, 8, 9, 7, 0,
1, 7, 2, 5, 5, 4, 3, 3, 0, 0, 2, 5, 1, 3, 0, 1, 0, 2, 4, 5, 0, 5, 7, 5, 1

Since this data set contains only a relatively small number of distinct, or different, values, it is convenient to represent it in a *frequency table* which presents each distinct value along with its frequency of occurrence. Table 2.1 is a frequency table of the above data. In Table 2.1 the frequency column represents the number of occurrences of each distinct value in the data set. Note that the sum of all the frequencies is 50, the total number of data observations.

Example 2.1   Use Table 2.1 to answer the following questions:

(a) How many workers had at least 1 day of sick leave?

(b) How many workers had between 3 and 5 days of sick leave?

(c) How many workers had more than 5 days of sick leave?

Table 2.1	**A Frequency Table of Sick Leave Data**

Value	Frequency	Value	Frequency
0	12	5	8
1	8	6	0
2	5	7	5
3	4	8	2
4	5	9	1

Solution

(a) Since 12 of the 50 workers had no days of sick leave, the answer is
$50 - 12 = 38$.

(b) The answer is the sum of the frequencies for values 3, 4, and 5, that is,
$4 + 5 + 8 = 17$.

(c) The answer is the sum of the frequencies for the values 6, 7, 8, and 9. Therefore, the answer is $0 + 5 + 2 + 1 = 8$.

2.2.1 Line Graphs, Bar Graphs, and Frequency Polygons

Data from a frequency table can be graphically pictured by a *line graph* which plots the successive values on the horizontal axis and indicates the corresponding frequency by the height of a vertical line. A line graph for the data of Table 2.1 is shown in Fig. 2.1.

Sometimes the frequencies are represented not by lines but rather by bars having some thickness. These graphs, called *bar graphs,* are often utilized. Figure 2.2 presents a bar graph for the data of Table 2.1.

Another type of graph used to represent a frequency table is the *frequency polygon,* which plots the frequencies of the different data values and then connects the plotted points with straight lines. Figure 2.3 presents the frequency polygon of the data of Table 2.1.

A set of data is said to be *symmetric* about the value x_0 if the frequencies of the values $x_0 - c$ and $x_0 + c$ are the same for all c. That is, for every constant c, there are just as many data points that are c less than x_0 as there are that are c greater than x_0. The data set presented in the frequency table, Table 2.2, is symmetric about the value $x_0 = 3$.

Data that are "close to" being symmetric are said to be *approximately symmetric.* The easiest way to determine whether a data set is approximately symmetric is to represent it graphically. Figure 2.4 presents three bar graphs: one of a symmetric data set, one of an approximately symmetric data set, and one of a data set that exhibits no symmetry.

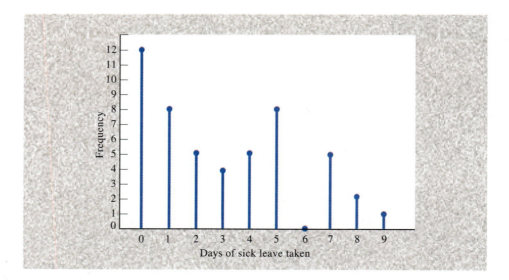

Figure 2.1
A line graph.

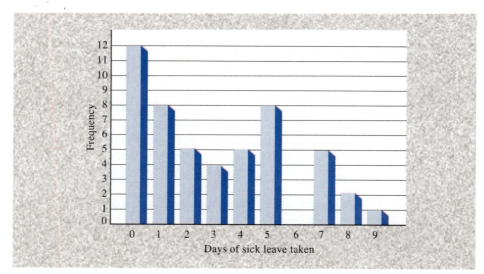

Figure 2.2
A bar graph.

Table 2.2	**Frequency Table of a Symmetric Data Set**

Value	Frequency	Value	Frequency
0	1	4	2
2	2	6	1
3	3		

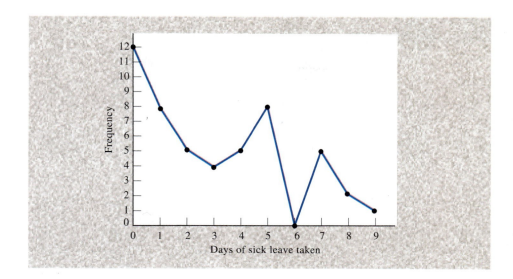

Figure 2.3
A frequency
polygon.

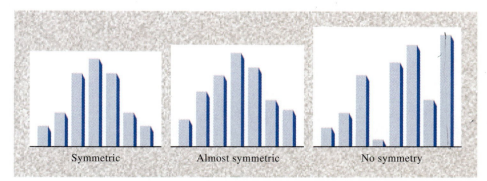

Figure 2.4
Bar graphs and
symmetry.

2.2.2   Relative Frequency Graphs

It is sometimes convenient to consider and plot the *relative* rather than the absolute frequencies of the data values. If f represents the frequency of occurrence of some data value x, then the *relative frequency* f/n can be plotted versus x, where n represents the total number of observations in the data set. For the data of Table 2.1, $n = 50$ and so the relative frequencies are as given in Table 2.3. Note that whereas the sum of the frequency column should be the total number of observations in the data set, the sum of the relative frequency column should be 1.

A polygon plot of these relative frequencies is presented in Fig. 2.5. A plot of the relative frequencies looks exactly like a plot of the absolute frequencies, except that the labels on the vertical axis are the old labels divided by the total number of observations in the data set.

Table 2.3

Relative Frequencies, $n = 50$, of Sick Leave Data

Value x	Frequency f	Relative frequency f/n
0	12	$\dfrac{12}{50} = .24$
1	8	$\dfrac{8}{50} = .16$
2	5	$\dfrac{5}{50} = .10$
3	4	$\dfrac{4}{50} = .08$
4	5	$\dfrac{5}{50} = .10$
5	8	$\dfrac{8}{50} = .16$
6	0	$\dfrac{0}{50} = .00$
7	5	$\dfrac{5}{50} = .10$
8	2	$\dfrac{2}{50} = .04$
9	1	$\dfrac{1}{50} = .02$

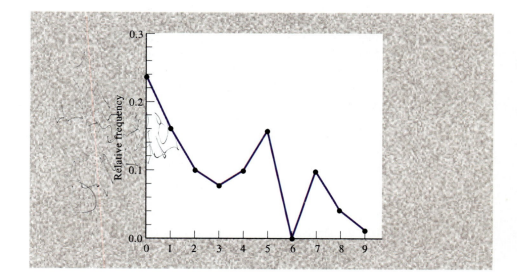

Figure 2.5
A relative frequency polygon.

To construct a relative frequency table from a data set
Arrange the data set in increasing order of values. Determine the distinct values and how often they occur. List these distinct values alongside their frequencies f and their relative frequencies f/n, where n is the total number of observations in the data set.

Example 2.2   The Masters Golf Tournament is played each year at the Augusta, Georgia, National Golf Club. To discover what type of score it takes to win this tournament, we have gathered all the winning scores from 1955 to 1991.

The Masters Golf Tournament Winners

Year	Winner	Score	Year	Winner	Score
1955	Cary Middlecoff	279	1974	Gary Player	278
1956	Jack Burke	289	1975	Jack Nicklaus	276
1957	Doug Ford	283	1976	Ray Floyd	271
1958	Arnold Palmer	284	1977	Tom Watson	276
1959	Art Wall, Jr.	284	1978	Gary Player	277
1960	Arnold Palmer	282	1979	Fuzzy Zoeller	280
1961	Gary Player	280	1980	Severiano Ballesteros	275
1962	Arnold Palmer	280	1981	Tom Watson	280
1963	Jack Nicklaus	286	1982	Craig Stadler	284
1964	Arnold Palmer	276	1983	Severiano Ballesteros	280
1965	Jack Nicklaus	271	1984	Ben Crenshaw	277
1966	Jack Nicklaus	288	1985	Bernhard Langer	282
1967	Gay Brewer, Jr.	280	1986	Jack Nicklaus	279
1968	Bob Goalby	277	1987	Larry Mize	285
1969	George Archer	281	1988	Sandy Lyle	281
1970	Billy Casper	279	1989	Nick Faldo	283
1971	Charles Coody	279	1990	Nick Faldo	278
1972	Jack Nicklaus	286	1991	Ian Woosnam	277
1973	Tommy Aaron	283			

(a) Arrange the data set of winning scores in a relative frequency table.

(b) Plot these data in a relative frequency bar graph.

Solution

(a) The 37 winning scores range from a low of 271 to a high of 289. This is the relative frequency table:

Winning score	Frequency f	Relative frequency $f/37$
271	2	.054
275	1	.027
276	3	.081
277	4	.108
278	2	.054
279	4	.108
280	6	.162
281	2	.054
282	2	.054
283	3	.081
284	3	.081
285	1	.027
286	2	.054
288	1	.027
289	1	.027

(b) The following is a relative frequency bar graph of the preceding data.

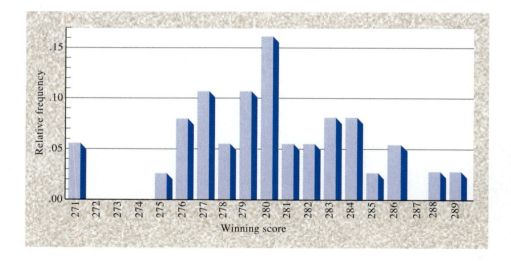

Relative frequency bar graph for winning scores.

2.2.3 Pie Charts

A *pie chart* is often used to plot relative frequencies when the data are nonnumeric. A circle is constructed and then is sliced up into distinct sectors, one for each different data value. The area of each sector, which is meant to represent the relative frequency of the value that the sector represents, is determined as follows. If the relative frequency of the data value is f/n, then the area of the sector is the fraction f/n of the

| Table 2.4 | **Murder Weapons** |

Type of weapon	Percentage of murders caused by this weapon
Handgun	52
Knife	18
Shotgun	7
Rifle	4
Personal weapon	6
Other	13

[handwritten margin notes: 96.5 - 97 [96.5 - 97] Don't use 97 ...]

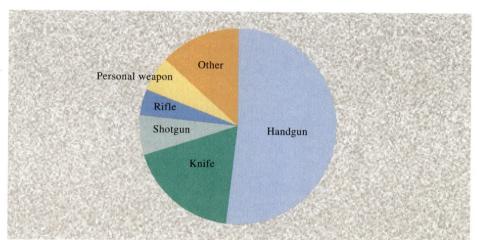

Figure 2.6
A pie chart.

total area of the circle. For instance, the data in Table 2.4 give the relative frequencies of types of weapons used in murders in a large midwestern city in 1985. These data are represented in a pie chart in Fig. 2.6.

If a data value has relative frequency f/n, then its sector can be obtained by setting the angle at which the lines of the sector meet equal to 360 f/n degrees. For instance, in Fig. 2.6, the angle of the lines forming the knife sector is 360(.18) = 64.8°.

Problems

1. The following data represent the sizes of 30 families that reside in a small town in Guatemala.

   5, 13, 9, 12, 7, 4, 8, 6, 6, 10, 7, 11, 10, 8, 15, 8, 6, 9, 12, 10, 7, 11, 10, 8, 12, 9, 7, 10, 7, 8

(a) Construct a frequency table for these data.
(b) Using a line graph, plot the data.
(c) Plot the data as a frequency polygon.

2. The following frequency table relates the weekly sales of bicycles at a given store over a 42-week period.

Value	0	0	2	3	4	5	6	7
Frequency	3	6	7	10	8	5	2	1

(a) In how many weeks were at least 2 bikes sold?
(b) In how many weeks were at least 5 bikes sold?
(c) In how many weeks were an even number of bikes sold?

3. Draw a line graph for the data of Prob. 2. Is this data set symmetric? Approximately symmetric?

4. Label each of the following data sets as symmetric, approximately symmetric, or not at all symmetric.

A: 6, 0, 2, 1, 8, 3, 5
B: 4, 0, 4, 0, 2, 1, 3, 2
C: 1, 1, 0, 1, 0, 3, 3, 2, 2, 2
D: 9, 9, 1, 2, 3, 9, 8, 4, 5

5. The following table lists all the values but only some of the frequencies for a symmetric data set. Fill in the missing numbers.

Value	Frequency
10	8
20	
30	7
40	
50	3
60	

6. The following are the scores of 32 students who took a statistics test.

55, 70, 80, 75, 90, 80, 60, 100, 95, 70, 75, 85, 80, 80, 70, 95, 100, 80, 85, 70, 85, 90, 80, 75, 85, 70, 90, 60, 80, 70, 85, 80

Represent this data set in a frequency table, and then draw a bar graph.

7. Draw a relative frequency table for the data of Prob. 1. Plot these relative frequencies in a line graph.

8. The following data represent the time to tumor progression, measured in months, for 65 patients having a particular type of brain tumor called *glioblastoma*:

6, 5, 37, 10, 22, 9, 2, 16, 3, 3, 11, 9, 5, 14, 11, 3, 1, 4, 6, 2, 7, 3, 7, 5, 4, 8, 2, 7, 13, 16, 15, 9, 4, 4, 2, 3, 9, 5, 11, 3, 7, 5, 9, 3, 8, 9, 4, 10, 3, 2, 7, 6, 9, 3, 5, 4, 6, 4, 14, 3, 12, 6, 8, 12, 7

(a) Make up a relative frequency table for this data set.
(b) Plot the relative frequencies in a frequency polygon.
(c) Is this data set approximately symmetric?

9. The following relative frequency table is obtained from a data set of the number of emergency appendectomies performed each month at a certain hospital.

Value	0	1	2	3	4	5	6	7
Relative frequency	.05	.08	.12	.14	.16	.20	.15	.10

(a) What proportion of months has fewer than 2 emergency appendectomies?
(b) What proportion of months has more than 5?
(c) Is this data set symmetric?

10. Relative frequency tables and plots are particularly useful when we want to compare different sets of data. The following two data sets relate the number of months from diagnosis to death of AIDS patients for samples of male and female AIDS sufferers.

Males	15	13	16	10	8	20	14	19	9	12	16	18	20	12	14	14
Females	8	12	10	8	14	12	13	11	9	8	9	10	14	9	10	

Plot these two data sets together in a relative frequency polygon. Use a different color for each set. What conclusion can you draw about which data set tends to have larger values?

11. Using the data of Example 2.2, determine the proportion of winning scores in the Masters Golf Tournament that is
(a) Below 280
(b) 282 or higher
(c) Between 278 and 284 inclusive

The following table gives the average number of days in each month that various cities have at least .01 inch of precipitation. Problems 12 through 14 refer to it.

Average Number of Days with Precipitation of .01 Inch or More

State	City	Length of record (yr.)	Jan.	Feb.	Mar.	Apr.	May	June	July	Aug.	Sept.	Oct.	Nov.	Dec.	Annual
AL	Mobile	46	11	10	11	7	8	11	16	14	10	6	8	10	123
AK	Juneau	43	18	17	18	17	17	16	17	18	20	24	19	21	220
AZ	Phoenix	48	4	4	4	2	1	1	4	5	3	3	3	4	36
AR	Little Rock	45	9	9	10	10	10	8	8	7	7	7	8	9	103
CA	Los Angeles	52	6	6	6	3	1	1	1	0	1	2	4	5	36
	Sacramento	48	10	9	9	5	3	1	0	0	1	3	7	9	58
	San Diego	47	7	6	7	5	2	1	0	1	1	3	5	6	43
	San Francisco	60	11	10	10	6	3	1	0	0	1	4	7	10	62
CO	Denver	53	6	6	9	9	11	9	9	9	6	5	5	5	89
CT	Hartford	33	11	10	11	11	12	11	10	10	9	8	11	12	127
DE	Wilmington	40	11	10	11	11	11	10	9	9	8	8	10	10	117
DC	Washington	46	10	9	11	10	11	10	10	9	8	7	8	9	111
FL	Jacksonville	46	8	8	8	6	8	12	15	14	13	9	6	8	116
	Miami	45	6	6	6	6	10	15	16	17	17	14	9	7	129
GA	Atlanta	53	11	10	11	9	9	10	12	9	8	6	8	10	115
HI	Honolulu	38	10	9	9	9	7	6	8	6	7	9	9	10	100
ID	Boise	48	12	10	10	8	8	6	2	3	4	6	10	11	91
IL	Chicago	29	11	10	12	12	11	10	10	9	10	9	10	12	127
	Peoria	48	9	8	11	12	11	10	9	8	9	8	9	10	114
IN	Indianapolis	48	12	10	13	12	12	10	9	9	8	8	10	12	125
IA	Des Moines	48	7	7	10	11	11	11	9	9	9	8	7	8	107
KS	Wichita	34	6	5	8	8	11	9	7	8	8	6	5	6	86
KY	Louisville	40	11	11	13	12	12	10	11	8	8	8	10	11	125
LA	New Orleans	39	10	9	9	7	8	11	15	13	10	6	7	10	114
ME	Portland	47	11	10	11	12	13	11	10	9	8	9	12	12	128
MD	Baltimore	37	10	9	11	11	11	9	9	10	7	7	9	9	113
MA	Boston	36	12	10	12	11	12	11	9	10	9	9	11	12	126
MI	Detroit	29	13	11	13	12	11	11	9	9	10	9	12	14	135
	Sault Ste. Marie	46	19	15	13	11	11	12	10	11	13	13	17	20	165
MN	Duluth	46	12	10	11	10	12	13	11	11	12	10	11	12	134
	Minneapolis-St. Paul	49	9	7	10	10	11	12	10	10	10	8	8	9	115
MS	Jackson	24	11	9	10	8	10	8	10	10	8	6	8	10	109
MO	Kansas City	15	7	7	11	11	11	11	7	9	8	8	8	8	107
	St. Louis	30	8	8	11	11	11	10	8	8	8	8	10	9	111
MT	Great Falls	50	9	8	9	9	12	12	7	8	7	6	7	8	101
NE	Omaha	51	6	7	9	10	12	11	9	9	9	7	5	6	98
NV	Reno	45	6	6	6	4	4	3	2	2	2	3	5	6	51
NH	Concord	46	11	10	11	12	12	11	10	10	9	9	11	11	125
NJ	Atlantic City	44	11	10	11	11	10	9	9	9	8	7	9	10	112
NM	Albuquerque	48	4	4	5	3	4	4	9	9	6	5	3	4	61
NY	Albany	41	12	10	12	12	13	11	10	10	10	9	12	12	134
	Buffalo	44	20	17	16	14	12	10	10	11	11	12	16	20	169
	New York	118	11	10	11	11	11	10	10	10	8	8	9	10	121
NC	Charlotte	48	10	10	11	9	10	10	11	9	7	7	8	10	111
	Raleigh	43	10	10	10	9	10	9	11	10	8	7	8	9	111
ND	Bismarck	48	8	7	8	8	10	12	9	9	7	6	6	8	97
OH	Cincinnati	40	12	11	13	13	11	11	10	9	8	8	11	12	129
	Cleveland	46	16	14	15	14	13	11	10	10	10	11	14	16	156
	Columbus	48	13	12	14	13	13	11	11	9	8	9	11	13	137
OK	Oklahoma City	48	5	6	7	8	10	9	6	6	7	6	5	5	82
OR	Portland	47	18	16	17	14	12	9	4	5	8	13	18	19	152

Average Number of Days with Precipitation of .01 Inch or More (Continued)

State	City	Length of record (yr.)	Jan.	Feb.	Mar.	Apr.	May	June	July	Aug.	Sept.	Oct.	Nov.	Dec.	Annual
PA	Philadelphia	47	11	9	11	11	11	10	9	9	8	8	9	10	117
	Pittsburgh	35	16	14	16	14	12	12	11	10	9	11	13	17	154
RI	Providence	34	11	10	12	11	11	9	9	10	8	8	11	12	124
SC	Columbia	40	10	10	11	8	9	9	12	11	8	6	7	9	109
SD	Sioux Falls	42	6	6	9	9	10	11	9	9	8	6	6	6	97
TN	Memphis	37	10	9	11	10	9	8	9	8	7	6	9	10	106
	Nashville	46	11	11	12	11	11	9	10	9	8	7	10	11	119
TX	Dallas-Fort Worth	34	7	7	7	8	9	6	5	5	7	6	6	6	78
	El Paso	48	4	3	2	2	2	4	8	8	5	4	3	4	48
	Houston	18	10	8	9	7	9	9	9	10	10	8	9	9	106
UT	Salt Lake City	59	10	9	10	9	8	5	5	6	5	6	8	9	91
VT	Burlington	44	14	12	13	12	14	13	12	12	12	12	14	15	154
VA	Norfolk	39	10	10	11	10	10	9	11	10	8	8	8	9	114
	Richmond	50	10	9	11	9	11	9	11	10	8	7	8	9	113
WA	Seattle	43	19	16	17	14	10	9	5	6	9	13	18	20	156
	Spokane	40	14	12	11	9	9	8	4	5	6	8	12	15	113
WV	Charleston	40	16	14	15	14	13	11	13	11	9	10	12	14	151
WI	Milwaukee	47	11	10	12	12	12	11	10	9	9	9	10	11	125
WY	Cheyenne	52	6	6	9	10	12	11	11	10	7	6	6	5	99
PR	San Juan	32	16	13	12	13	17	16	19	18	17	17	18	19	195

Source: U.S. National Oceanic and Atmospheric Administration, *Comparative Climatic Data.*

12. Construct a relative frequency table for the average number of rainy days in January for the different cities. Then plot the data in a relative frequency polygon.

13. Using only the data relating to the first 12 cities listed, construct a frequency table for the average number of rainy days in either November or December.

14. Using only the data relating to the first 24 cities, construct relative frequency tables for the month of June and separately for the month of December. Then plot these two sets of data together in a relative frequency polygon.

15. The following table gives the number of deaths on British roads in 1987 for individuals in various classifications.

Classification	Number of deaths
Pedestrians	1699
Bicyclists	280
Motorcyclists	650
Automobile drivers	1327

Express this data set in a pie chart.

16. The following data, taken from *The New York Times,* represent the percentage of items, by total weight, in the garbage of New York City in 1990. Represent this in a pie chart.

Organic material (food, yard waste, lumber, etc.)	37.3
Paper	30.8
Bulk (furniture, refrigerators, etc.)	10.9
Plastic	8.5
Glass	5
Metal	4
Inorganic	2.2
Aluminum	0.9
Hazardous waste	0.4

2.3   GROUPED DATA AND HISTOGRAMS

As seen in Sec. 2.2, using a line or a bar graph to plot the frequencies of data values is often an effective way of portraying a data set. However, for some data sets the number of distinct values is too large to utilize this approach. Instead, in such cases, we divide the values into groupings, or *class intervals,* and then plot the number of data values falling in each class interval. The number of class intervals chosen should be a tradeoff between (1) choosing too few classes at a cost of losing too much information about the actual data values in a class and (2) choosing too many classes, which will result in the frequencies of each class being too small for a pattern to be discernible. Although 5 to 10 class intervals are typical, the appropriate number is a subjective choice, and of course, you can try different numbers of class intervals to see which of the resulting charts appears to be most revealing about the data. It is common, although not essential, to choose class intervals of equal length.

The endpoints of a class interval are called the *class boundaries.* We will adopt the *left-end inclusion convention,* which stipulates that a class interval contains its left-end but not its right-end boundary point. Thus, for instance, the class interval 20–30 contains all values that are both greater than *or equal to* 20 and less than 30.

The data in Table 2.5 represent the blood cholesterol levels of 40 first-year students at a particular college. As a prelude to determining class size frequencies, it is useful to rearrange the data in increasing order. This give the 40 values of Table 2.6.

Since the data range from a minimum value of 171 to a maximum of 227, the left-end boundary of the first class interval must be less than or equal to 171, and the right-end boundary of the final class interval must be greater than 227. One choice would be to have the first class interval be 170 to 180. This will result in six class intervals. A frequency table giving the frequency (as well as the relative frequency) of data values falling in each class interval is seen in Table 2.7.

Note: Because of the left-end inclusion convention, the values of 200 were placed in the class interval of 200 to 210, not in the interval of 190 to 200.

Table 2.5   **Blood Cholesterol Levels**

213	174	193	196	220	183	194	200
192	200	200	199	178	183	188	193
187	181	193	205	196	211	202	213
216	206	195	191	171	194	184	191
221	212	221	204	204	191	183	227

Table 2.6   **Blood Cholesterol Levels in Increasing Order**

171, 174, 178, 181, 183, 183, 183, 184, 187, 188, 191, 191, 191, 192, 193, 193, 193, 194, 194, 195, 196, 196, 199, 200, 200, 200, 202, 204, 204, 205, 206, 211, 212, 213, 213, 216, 220, 221, 221, 227

Table 2.7   **Frequency Table of Blood Cholesterol Levels**

Class intervals	Frequency	Relative frequency
170–180	3	$\frac{3}{40} = .075$
180–190	7	$\frac{7}{40} = .175$
190–200	13	$\frac{13}{40} = .325$
200–210	8	$\frac{8}{40} = .20$
210–220	5	$\frac{5}{40} = .125$
220–230	4	$\frac{4}{40} = .10$

A bar graph plot of the data, with the bars placed adjacent to each other, is called a *histogram*. The vertical axis of a histogram can represent either the class frequency or the relative class frequency. In the former case, the histogram is called a *frequency histogram* and in the latter a *relative frequency histogram*. Figure 2.7 presents a frequency histogram of the data of Table 2.7.

It is important to recognize that a class frequency table or a histogram based on that table does not contain all the information in the original data set. These two representations note only the *number* of data values in each class and not the actual data values themselves. Thus, whereas such tables and charts are useful for illustrating data, the original raw data set should *always* be saved.

To construct a histogram from a data set

1. Arrange the data in increasing order.

2. Choose class intervals so that all data points are covered.

3. Construct a frequency table.

4. Draw adjacent bars having heights determined by the frequencies in step 3.

The importance of a histogram is that it enables us to organize and present data graphically so as to draw attention to certain important features of the data. For instance, a histogram can often indicate

1. How symmetric the data are

2. How spread out the data are

3. Whether there are intervals having high levels of data concentration

4. Whether there are gaps in the data

5. Whether some data values are far apart from others

For instance, the histogram presented in Fig. 2.7 indicates that the frequencies of the successive classes first increase and then decrease, reaching a maximum in the class having limits of 190 to 200. The histograms of Fig. 2.8 give valuable information about the data sets they represent. The data set whose histogram is on the left side of Fig. 2.8(a) is symmetric, whereas the one on the right side is not. The data set represented on the left side of Fig. 2.8(b) is fairly evenly spread out, whereas the one for the right side is more concentrated. The data set represented by the left side of Fig. 2.8(c) has a gap, whereas the one represented on the right side has certain values far apart from the rest.

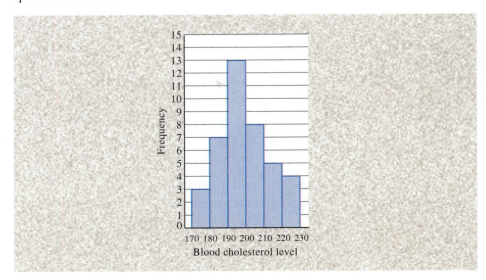

Figure 2.7
Frequency histogram for the data of Table 2.7.

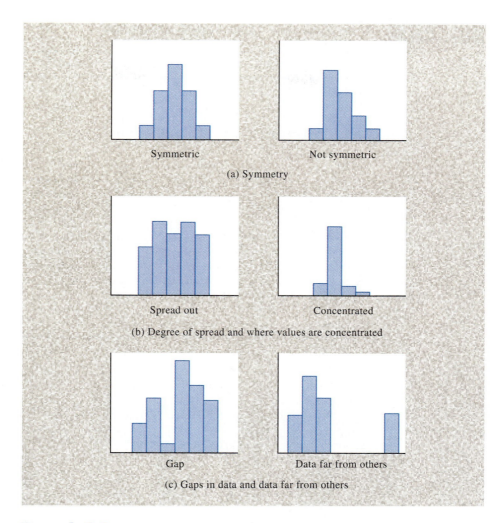

Figure 2.8
Characteristics
of data detected
by histograms.

Example 2.3   Table 2.8 (page 40) gives the 1989 birth rates (per 1000 population) in each of the 50 states of the United States. Plot these data in a histogram.

Solution Since the data range from a low value of 12.4 to a high of 21.9, let us use class intervals of length 1.5, starting at the value 12. With these class intervals, we obtain the following frequency table.

Class intervals	Frequency	Class intervals	Frequency
12.0–13.5	2	18.0–19.5	2
13.5–15.0	15	19.5–21.0	0
15.0–16.5	22	21.0–22.5	2
16.5–18.0	7		

A histogram plot of these data is presented in Fig. 2.9.

Table 2.8	Birth Rates per 1000 Population, 1989

State	Rate	State	Rate	State	Rate
Alabama	14.2	Louisiana	15.7	Ohio	14.9
Alaska	21.9	Maine	13.8	Oklahoma	14.4
Arizona	19.0	Maryland	14.4	Oregon	15.5
Arkansas	14.5	Massachusetts	16.3	Pennsylvania	14.1
California	19.2	Michigan	15.4	Rhode Island	15.3
Colorado	15.9	Minnesota	15.3	South Carolina	15.7
Connecticut	14.7	Mississippi	16.1	South Dakota	15.4
Delaware	17.1	Missouri	15.5	Tennessee	15.5
Florida	15.2	Montana	14.1	Texas	17.7
Georgia	17.1	Nebraska	15.1	Utah	21.2
Hawaii	17.6	Nevada	16.5	Vermont	14.0
Idaho	15.2	New Hampshire	16.2	Virginia	15.3
Illinois	16.0	New Jersey	15.1	Washington	15.4
Indiana	14.8	New Mexico	17.9	West Virginia	12.4
Iowa	13.1	New York	16.2	Wisconsin	14.8
Kansas	14.2	North Carolina	15.6	Wyoming	13.7
Kentucky	14.1	North Dakota	16.5		

Source: Department of Health and Human Services.

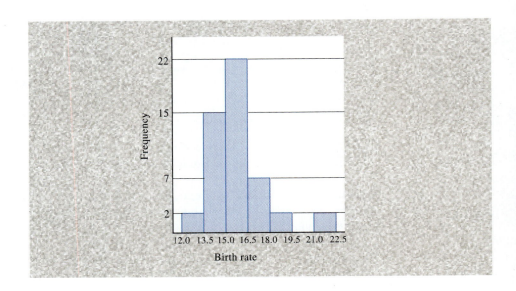

Figure 2.9
A histogram for
birth rates in
the 50 states.

A histogram is, in essence, a bar chart that graphs the frequencies, or relative frequencies, of data falling into different class intervals. These class frequencies can also be represented graphically by a frequency (or relative frequency) polygon. Each class interval is represented by a value, usually taken to be the midpoint of that interval. A plot is made of these values versus the frequencies of the class intervals they represent. These plotted points are then connected by straight lines to yield the frequency polygon. Such graphs are particularly useful for comparing data sets, since the different frequency polygons can be plotted on the same chart.

Example 2.4   The data of Table 2.9 represent class frequencies for the systolic blood pressure of two groups of male industrial workers: those aged 30 to 40 and those aged 50 to 60.

It is difficult to directly compare the blood pressures for the two age groups since the total number of workers in each group is different. To remove this difficulty, we can compute and graph the *relative* frequencies of each of the classes. That is, we divide all the frequencies relating to workers aged 30 to 39 by 2540 (the number of such workers) and all the frequencies relating to workers aged 50 to 59 by 731. This results in Table 2.10.

Table 2.9   **Class Frequencies of Systolic Blood Pressure of Two Groups of Male Workers**

	Number of workers	
Blood pressure	**Aged 30–40**	**Aged 50–60**
Less than 90	3	1
90–100	17	2
100–110	118	23
110–120	460	57
120–130	768	122
130–140	675	149
140–150	312	167
150–160	120	73
160–170	45	62
170–180	18	35
180–190	3	20
190–200	1	9
200–210		3
210–220		5
220–230		2
230–240		1
Total	2540	731

Table 2.10	**Relative Class Frequencies of Blood Pressures**

	Percentage of workers	
Blood pressure	**Aged 30–40**	**Aged 50–60**
Less than 90	.12	.14
90–100	.67	.27
100–110	4.65	3.15
110–120	18.11	7.80
120–130	30.24	16.69
130–140	26.57	20.38
140–150	12.28	22.84
150–160	4.72	9.99
160–170	1.77	8.48
170–180	.71	4.79
180–190	.12	2.74
190–200	.04	1.23
200–210		.41
210–220		.68
220–230		.27
230–240		.14
Total	100.00	100.00

Figure 2.10 graphs the relative frequency polygons for both age groups. Having both frequency polygons on the same graph makes it easy to compare the two data sets. For instance, it appears that the blood pressures of the older group are more spread out among larger values than are those of the younger group.

Problems

1. The following data set represents the scores on intelligence quotient (IQ) examinations of 40 sixth-grade students at a particular school:

114, 122, 103, 118, 99, 105, 134, 125, 117, 106, 109, 104, 111, 127, 133, 111, 117, 103, 120, 98, 100, 130, 141, 119, 128, 106, 109, 115, 113, 121, 100, 130, 125, 117, 119, 113, 104, 108, 110, 102

   (a) Present this data set in a frequency histogram.
   (b) Which class interval contains the greatest number of data values?
   (c) Is there a roughly equal number of data in each class interval?
   (d) Does the histogram appear to be approximately symmetric? If so, about which interval is it approximately symmetric?

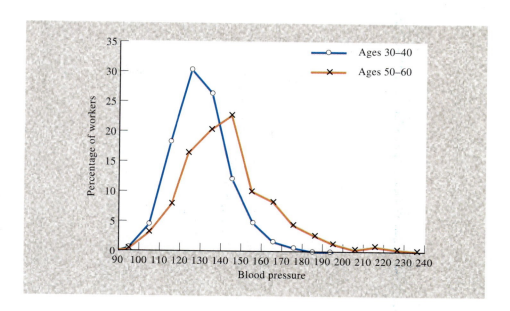

Figure 2.10
Relative
frequency
polygons.

2. The following data represent the daily high temperature (in degrees Celsius) on July 4 in San Francisco over a sequence of 30 years:

class intervals 2½

> 22.8, 26.2, 31.7, 31.1, 26.9, 28.0, 29.4, 28.8, 26.7, 27.4, 28.2, 30.3, 29.5,
> 28.9, 27.5, 28.3, 24.1, 25.3, 28.5, 27.7, 24.4, 29.2, 30.3, 33.7, 27.5, 29.3,
> 30.2, 28.5, 32.2, 33.7

(a) Present this data set in a frequency histogram.
(b) What would you say is a "typical" July 4 temperature in San Francisco?
(c) What other conclusions can be drawn from the histogram?

3. The following data (in thousands of dollars) represent the gross adjusted annual income for a sample of taxpayers.

> 47, 55, 18, 24, 27, 41, 50, 38, 33, 29, 15, 77, 64, 22, 19, 35, 39, 41, 67,
> 55, 121, 77, 80, 34, 41, 48, 60, 30, 22, 28, 84, 55, 26, 105, 62, 30, 17, 23,
> 31, 28, 56, 64, 88, 104, 115, 39, 25, 18, 21, 30, 57, 40, 38, 29, 19, 46, 40,
> 49, 72, 70, 37, 39, 18, 22, 29, 52, 94, 86, 23, 36

(a) Graph this data set in a frequency histogram having 5 class intervals.
(b) Graph this data set in a frequency histogram having 10 class intervals.
(c) Which histogram do you think is more informative? Why?

4. A set of 200 data points was broken up into 8 classes each of size (in the units of the data) 3, and the frequency of values in each class was determined. A frequency table was then constructed. However, some of the entries of this table were lost. Suppose that the part of the frequency table that remains is as follows:

Class interval	Frequency	Relative frequency
		.05
	14	
	18	
15–18	38	
		.10
	42	
	11	

Fill in the missing numbers and draw a relative frequency histogram.

5. The following is the ozone concentration (measured in parts per 100 million) of air in the downtown Los Angeles area during 25 consecutive summer days in 1984:

   6.2, 9.1, 2.4, 3.6, 1.9, 1.7, 4.5, 4.2, 3.3, 5.1, 6.0, 1.8, 2.3, 4.9, 3.7, 3.8, 5.5, 6.4, 8.6, 9.3, 7.7, 5.4, 7.2, 4.9, 6.2

   (a) Construct a frequency histogram for this data set having 3 to 5 as a class interval.
   (b) Construct a frequency histogram for this data set having 2 to 3 as a class interval.
   (c) Which frequency histogram do you find more informative?

6. The following is the 1985 meat production, in thousands of metric tons, for 11 different countries.

Country	Production	Country	Production
Argentina	3,077	Mexico	2,289
Australia	2,086	Poland	2,192
Brazil	3,045	United Kingdom	2,419
France	3,733	United States	17,874
Italy	2,462	West Germany	4,845
Japan	2,087		

Source: *Statistical Abstract of the United States, 1990.*

   (a) Represent the above in a frequency histogram.
   (b) A data value that is far removed from the others is called an *outlier*. Is there an outlier in the above data?

7. Consider the blood cholesterol levels of the first 100 students in the data set presented in App. A. Divide these students by gender groupings, and construct a class relative frequency table for each. Plot, on the same chart, separate class relative frequency polygons for the female and male students. Can any conclusions be drawn about the relationship between gender and cholesterol level?

8. To aid in analyzing the amount of property taxes paid by residents of different states, we can utilize the following table, which gives the 1989 per capita property tax (in dollars) in all 50 states and the District of Columbia. Use the table to construct a frequency histogram.

Per Capita Property Tax by State, 1989

State	Property tax per capita	State	Property tax per capita
Alabama	148	Nebraska	698
Alaska	1257	Nevada	386
Arizona	600	New Hampshire	1058
Arkansas	211	New Jersey	1056
California	543	New Mexico	180
Colorado	635	New York	929
Connecticut	1003	North Carolina	322
Delaware	276	North Dakota	441
Florida	549	Ohio	476
Georgia	444	Oklahoma	267
Hawaii	335	Oregon	795
Idaho	386	Pennsylvania	467
Illinois	657	Rhode Island	737
Indiana	480	South Carolina	353
Iowa	624	South Dakota	544
Kansas	617	Tennessee	296
Kentucky	228	Texas	613
Louisiana	241	Utah	420
Maine	666	Vermont	729
Maryland	546	Virginia	541
Massachusetts	743	Washington	542
Michigan	764	West Virginia	237
Minnesota	659	Wisconsin	710
Mississippi	291	Wyoming	870
Missouri	318	District of Columbia	1117
Montana	679		

Source: *Tax Features,* May–June 1991.

The following table provides data concerning the death rates by different causes in various countries. Use it to answer Probs. 9 through 12.

Death Rates (per 100,000 Population) by Selected Causes

Country	Year	Ischemic heart disease	Cerebro-vascular disease	Malignant Neoplasm of—			Bronchitis,† emphysema, asthma	Chronic liver disease and cirrhosis	Motor vehicle traffic accidents	Suicide
				Lung, trachea, bronchus	Stomach	Female breast*				
United States	1984	218.1	60.1	52.7	6.0	31.9	8.3	12.9	18.1	12.4
Australia	1985	230.9	95.6	41.0	10.1	30.0	16.9	8.7	17.9	11.8
Austria	1986	155.1	133.2	34.3	20.7	31.6	22.3	26.6	17.3	26.7
Belgium	1984	120.6	95.0	55.9	14.7	36.8	22.6	12.4	18.7	22.7
Bulgaria	1985	245.9	254.5	30.6	24.2	21.5	28.6	16.2	12.0	15.6
Canada	1985	200.6	57.5	50.6	9.0	34.5	9.7	10.1	15.3	12.8
Czechoslovakia	1985	289.4	194.3	51.3	22.4	27.3	33.8	19.6	10.9	19.6
Denmark	1985	243.8	73.4	52.2	10.9	39.7	37.1	12.2	14.0	26.9
Finland	1986	259.8	105.0	36.4	17.3	23.9	19.8	8.8	11.9	26.3
France	1985	76.0	79.7	32.2	10.8	27.1	11.7	22.9	17.3	21.8
Hungary	1986	240.1	186.5	55.0	25.9	31.2	43.8	42.1	15.7	44.1
Italy	1983	128.9	121.9	42.1	23.9	28.9	30.9	31.5	16.6	7.3
Japan	1986	41.9	112.8	24.9	40.7	8.1	12.2	14.4	10.1	21.1
Netherlands	1985	164.6	71.1	56.3	15.6	38.2	17.8	5.5	9.2	11.2
New Zealand	1985	250.5	98.4	42.0	11.2	37.7	25.8	4.8	20.9	10.8
Norway	1985	208.5	88.6	26.3	14.4	25.9	18.2	6.9	9.7	14.2
Poland	1986	109.4	75.3	47.2	24.2	21.1	33.4	12.0	14.3	13.8
Portugal	1986	76.6	216.4	18.7	26.5	22.6	17.8	30.0	24.4	9.4
Spain	1981	79.0	133.9	26.0	19.7	19.0	19.1	23.3	15.4	4.8
Sweden	1985	244.7	73.0	23.2	12.5	26.0	14.3	6.4	9.2	17.2
Switzerland	1986	112.0	65.6	36.6	12.0	36.6	17.5	10.4	14.8	21.4
United Kingdom:										
England, Wales	1985	247.6	104.5	57.2	15.2	41.9	24.2	4.8	8.8	8.6
Scotland	1986	288.0	128.4	68.7	14.9	41.2	14.8	7.3	11.1	10.9
West Germany	1986	159.5	100.4	34.6	18.3	32.6	26.1	19.3	12.5	17.0

* Data for female population only.   † Chronic and unspecified.

Source: World Health Organization, Geneva, Switzerland, *World Health Statistics,* annual.

9. Construct a relative frequency histogram, using the death rates due to ischemic heart disease in the different countries as the data.

10. Construct a relative frequency histogram of the death rates from bronchitis, emphysema, and asthma.

11. Construct a frequency histogram of the total death rate from all the listed causes.

12. Roughly speaking, state whether the United States is in the low end, middle range, or high end of the death rates due to

(a) Ischemic heart disease

(b) Bronchitis, emphysema, and asthma

(c) All listed causes

13. Using the table preceding Prob. 12 in Sec. 2.2, construct a histogram for the average yearly number of rainy days for the cities listed.

14. Consider the following table.

Age of driver, yrs	Percentage of all drivers	Percentage of all drivers in fatal accidents
15–20	9	18
20–25	13	21
25–30	13	14
30–35	11	11
35–40	9	7
40–45	8	6
45–50	8	5
50–55	7	5
55–60	6	4
60–65	6	3
65–70	4	2
70–75	3	2
Over 75	3	2

By the left-end convention, 13 percent of all drivers are at least 25 but less than 30 years old, and 11 percent of drivers killed in car accidents are at least 30 but less than 35 years old.

(a) Draw a relative frequency histogram for the age breakdown of drivers.

(b) Draw a relative frequency histogram for the age breakdown of those drivers who are killed in car accidents.

(c) Which age group accounts for the largest number of fatal accidents?

(d) Which age group should be charged the highest insurance premiums? Explain your reasoning.

15. A cumulative relative frequency table gives, for an increasing sequence of values, the percentage of data values that are less than that value. It can be constructed from a relative frequency table by simply adding the relative frequencies in a cumulative fashion. The table below is the beginning of such a table for the two data sets shown in Table 2.9. It says, for instance, that 5.44 percent of men aged 30 to 40 years have blood pressures below 110, as opposed to only 3.56 percent of those aged 50 to 60 years.

A Cumulative Relative Frequency Table for Data Sets of Table 2.9

	Percentage of workers	
Blood pressure less than	**Aged 30–40**	**Aged 50–60**
90	.12	.14
100	.79	.41
110	5.44	3.56
120		
130		
•		
•		
•		
240	100	100

(a) Explain why the cumulative relative frequency for the last class must be 100.
(b) Complete the table.
(c) What does the table tell you about the two data sets? (That is, which one tends to have smaller values?)
(d) Graph, on the same chart, cumulative relative frequency polygons for the above data. Such graphs are called *ogives* (pronounced "oh jives").

2.4 STEM-AND-LEAF PLOTS

A very efficient way of displaying a small to moderate-size data set is to utilize a *stem-and-leaf plot*. These plots are obtained by dividing each data value into two parts—its stem and its leaf. For instance, if the data are all two-digit numbers, then we could let the stem of a data value be the tens digit and the leaf be the ones digit. That is, the value 84 is expressed as

Stem Leaf
8 | 4

and the two data values 84 and 87 are expressed as

Stem Leaf
8 | 4, 7

Example 2.5   Table 2.11 presents the property tax per $1000 of personal income for each of the 50 states and the District of Columbia. The data are for 1989.

| Table 2.11 | **Property Tax per $1000 of Personal Income by State, 1989** |

State	Tax ($)	State	Tax ($)
Alabama	12	Nebraska	47
Alaska	66	Nevada	23
Arizona	41	New Hampshire	56
Arkansas	17	New Jersey	48
California	30	New Mexico	15
Colorado	39	New York	48
Connecticut	44	North Carolina	23
Delaware	16	North Dakota	34
Florida	34	Ohio	31
Georgia	30	Oklahoma	20
Hawaii	20	Oregon	54
Idaho	31	Pennsylvania	29
Illinois	38	Rhode Island	44
Indiana	32	South Carolina	28
Iowa	43	South Dakota	43
Kansas	39	Tennessee	22
Kentucky	18	Texas	42
Louisiana	20	Utah	35
Maine	45	Vermont	48
Maryland	28	Virginia	31
Massachusetts	36	Washington	34
Michigan	46	West Virginia	20
Minnesota	40	Wisconsin	46
Mississippi	26	Wyoming	63
Missouri	21	District of Columbia	54
Montana	53		

Source: *Tax Features,* May–June 1991

The data presented in Table 2.11 can be represented in the following stem-and-leaf plot.

```
1 | 2, 5, 6, 7, 8
2 | 0, 0, 0, 0, 1, 2, 3, 3, 6, 8, 8, 9
3 | 0, 0, 1, 1, 1, 2, 4, 4, 4, 5, 6, 8, 9, 9
4 | 0, 1, 2, 3, 3, 4, 4, 5, 6, 6, 7, 8, 8, 8
5 | 3, 4, 4, 6
6 | 3, 6
```

Note that the values of the leaves are put in the plot in increasing order.

The choice of stems should always be made so that the resultant stem-and-leaf plot is informative about the data. For instance, consider Example 2.6.

Example 2.6   The following data represent the proportion of the public elementary school students that is classified as minority in each of 18 cities.

55.2, 47.8, 44.6, 64.2, 61.4, 36.6, 28.2, 57.4, 41.3, 44.6, 55.2, 39.6, 40.9, 52.2, 63.3, 34.5, 30.8, 45.3

If we let the stem denote the tens digit and the leaf represent the remainder of the value, then the stem-and-leaf plot of the above is as follows:

```
2 | 8.2
3 | 0.8, 4.5, 6.6, 9.6
4 | 0.9, 1.3, 4.6, 4.6, 5.3, 7.8
5 | 2.2, 5.2, 5.2, 7.4
6 | 1.4, 3.3, 4.2
```

We could have let the stem denote the integer part and the leaf the decimal part of the value, so that the value 28.2 would be represented as

```
28 | .2
```

However, this would have resulted in too many stems (with too few leaves each) to clearly illustrate the data set.

Example 2.7   The following stem-and-leaf plot represents the weights of 80 attendees at a sporting convention. The stem represents the tens digit, and the leaves are the ones digit.

```
10 | 2, 3, 3, 4, 7                    (5)
11 | 0, 1, 2, 2, 3, 6, 9             (7)
12 | 1, 2, 4, 4, 6, 6, 6, 7, 9       (9)
13 | 1, 2, 2, 5, 5, 6, 6, 8, 9       (9)
14 | 0, 4, 6, 7, 7, 9, 9             (7)
15 | 1, 1, 5, 6, 6, 6, 7             (7)
16 | 0, 1, 1, 1, 2, 4, 5, 6, 8, 8   (10)
17 | 1, 1, 3, 5, 6, 6, 6             (7)
18 | 1, 2, 2, 5, 5, 6, 6, 9         (8)
19 | 0, 0, 1, 2, 4, 5                (6)
20 | 9, 9                            (2)
21 | 7                               (1)
22 | 1                               (1)
23 |                                 (0)
24 | 9                               (1)
```

The numbers in parentheses on the right represent the numbers of values in each stem class. These summary numbers are often useful. They tell us, for instance, that there are 10 values having stem 16; that is, 10 individuals have weights between 160 and 169. Note that a stem without any leaves (such as stem value 23 above) indicates that there are no occurrences in that class.

It is clear from this plot that almost all the data values are between 100 and 200, and the spread is fairly uniform throughout this region, with the exception of fewer values in the intervals between 100 and 110 and between 190 and 200.

Stem-and-leaf plots are quite useful in showing all the data values in a clear representation which can be the first step in describing, summarizing, and learning from the data. It is most helpful in moderate-size data sets. (If the size of the data set were very large, then from a practical point of view, the values of all the leaves might be too overwhelming and a stem-and-leaf plot might not be any more informative than a histogram.) Physically this plot looks like a histogram turned on its side, with the additional plus that it presents the original within-group data values. These within-group values can be quite valuable to help you discover patterns in the data, such as that all the data values are multiples of some common value, or find out which values occur most frequently within a stem group.

Sometimes a stem-and-leaf plot appears to have too many leaves per stem line and as a result looks cluttered. One possible solution is to double the number of stems by having two stem lines for each stem value. On the top stem line in the pair we could include all leaves having values 0 through 4, and on the bottom stem line all leaves having values 5 through 9. For instance, suppose one line of a stem-and-leaf plot is as follows:

6|0, 0, 1, 2, 2, 3, 4, 4, 4, 4, 5, 5, 6, 6, 7, 7, 7, 7, 8, 9, 9

This could be broken into two lines as follows:

6|0, 0, 1, 2, 2, 3, 4, 4, 4, 4
6|5, 5, 6, 6, 7, 7, 7, 7, 8, 9, 9

Problems

1. For the following data, draw stem-and-leaf plots having (a) 4 stems and (b) 8 stems.

   124, 129, 118, 135, 114, 139, 127, 141, 111, 144, 133, 127, 122, 119, 132, 137, 146, 122, 119, 115, 125, 132, 118, 126, 134, 147, 122, 119, 116, 125, 128, 130, 127, 135, 122, 141

2. The following data are the U.S. marriage and divorce rates (per 1000 population) in each year from 1967 to 1990.

Year	Marriage rate	Divorce rate
1967	9.7	2.6
1968	10.4	2.9
1969	10.6	3.2
1970	10.6	3.5
1971	10.6	3.7
1972	11.0	4.1
1973	10.9	4.4
1974	10.5	4.6
1975	10.1	4.9
1976	10.0	5.0
1977	10.1	5.0
1978	10.5	5.2
1979	10.6	5.4
1980	10.6	5.2
1981	10.6	5.3
1982	10.8	5.1
1983	10.5	5.0
1984	10.5	4.9
1985	10.2	5.0
1986	10.0	4.8
1987	9.9	4.8
1988	9.7	4.8
1989	9.7	4.7
1990	9.8	4.7

Source: Department of Health and Human Services, National Center for Health Statistics.

(a) Draw a stem-and-leaf plot for the marriage rates in the 24 years.

(b) Draw a stem-and-leaf plot for the divorce rates in the 24 years.

3. The following are the ages, to the nearest year, of 43 patients admitted to the emergency ward of a certain adult hospital:

23, 18, 31, 79, 44, 51, 24, 19, 17, 25, 27, 19, 44, 61, 22, 18, 14, 17, 29, 31, 22, 17, 15, 40, 55, 16, 17, 19, 20, 32, 20, 45, 53, 27, 16, 19, 22, 20, 18, 30, 20, 33, 21

Draw a stem-and-leaf plot for this data set. Use this plot to determine the 5-year interval of ages that contains the largest number of data points.

4. A psychologist recorded the following 48 reaction times (in seconds) to a certain stimulus.

   1.1 2.1 0.4 3.3 1.5 1.3 3.2 2.0 1.7 0.6 0.9 1.6 2.2 2.6 1.8 0.9 2.5 3.0 0.7
   1.3 1.8 2.9 2.6 1.8 3.1 2.6 1.5 1.2 2.5 2.8 0.7 2.3 0.6 1.8 1.1 2.9 3.2 2.8
   1.2 2.4 0.5 0.7 2.4 1.6 1.3 2.8 2.1 1.5

   (a) Construct a stem-and-leaf plot for these data.
   (b) Construct a second stem-and-leaf plot, using additional stems.
   (c) Which one seems more informative?
   (d) Suppose a newspaper article stated, "The typical reaction time was _____ seconds." Fill in your guess as to the missing word.

5. The following data represent New York City's daily revenue from parking meters (in units of $1000) during 30 days in 1980:

   108, 77, 58, 88, 65, 52, 104, 75, 80, 83, 74, 68, 94, 97, 83, 71, 78, 83, 90, 79, 84, 81, 68, 57, 59, 32, 75, 93, 100, 88

   (a) Represent this data set in a stem-and-leaf plot.
   (b) Do any of the data values seem "suspicious"? Why?

6. The volatility of a stock is an important property in the theory of stock options pricing. It is an indication of how much change there tends to be in the day-to-day price of the stock. A volatility of 0 means that the price of the stock always remains the same. The higher the volatility, the more the stock's price tends to change. The following is a list of the volatility of 32 companies whose stock is traded on the American Stock Exchange:

   .26, .31, .45, .30, .26, .17, .33, .32, .37, .38, .35, .28, .37, .35, .29, .20, .33, .19, .31, .26, .24, .50, .22, .33, .51, .44, .63, .30, .28, .48, .42, .37

   (a) Represent these data in a stem-and-leaf plot.
   (b) What is the largest data value?
   (c) What is the smallest data value?
   (d) What is a "typical" data value?

7. The following table gives the scores of the first 25 Super Bowl games in professional football. Use it to construct a stem-and-leaf plot of
   (a) The winning scores
   (b) The losing scores
   (c) The amounts by which the winning teams outscored the losing teams

Super Bowls I–XXV

Game	Date	Winner	Loser
XXV	Jan. 27, 1991	Giants (NFC) 20	Buffalo (AFC) 19
XXIV	Jan. 28, 1990	San Francisco (NFC) 55	Denver (AFC) 10
XXIII	Jan. 22, 1989	San Francisco (NFC) 20	Cincinnati (AFC) 16
XXII	Jan. 31, 1988	Washington (NFC) 42	Denver (AFC) 10
XXI	Jan. 25, 1987	Giants (NFC) 39	Denver (AFC) 20
XX	Jan. 26, 1986	Chicago (NFC) 46	New England (AFC) 10
XIX	Jan. 20, 1985	San Francisco (NFC) 38	Miami (AFC) 16
XVIII	Jan. 22, 1984	Los Angeles Raiders (AFC) 38	Washington (NFC) 9
XVII	Jan. 30, 1983	Washington (NFC) 27	Miami (AFC) 17
XVI	Jan. 24, 1982	San Francisco (NFC) 26	Cincinnati (AFC) 21
XV	Jan. 25, 1981	Oakland (AFC) 27	Philadelphia (NFC)10
XIV	Jan. 20, 1980	Pittsburgh (AFC) 31	Los Angeles (NFC) 19
XIII	Jan. 21, 1979	Pittsburgh (AFC) 35	Dallas (NFC) 31
XII	Jan. 15, 1978	Dallas (NFC) 27	Denver (AFC) 10
XI	Jan. 9, 1977	Oakland (AFC) 32	Minnesota (NFC) 14
X	Jan. 18, 1976	Pittsburgh (AFC) 21	Dallas (NFC) 17
IX	Jan. 12, 1975	Pittsburgh (AFC) 16	Minnesota (NFC) 6
VIII	Jan. 13, 1974	Miami (AFC) 24	Minnesota (NFC) 7
VII	Jan. 14, 1973	Miami (AFC) 14	Washington (NFC) 7
VI	Jan. 16, 1972	Dallas (NFC) 24	Miami (AFC) 3
V	Jan. 17, 1971	Baltimore (AFC) 16	Dallas (NFC) 13
IV	Jan. 11, 1970	Kansas City (AFL) 23	Minnesota (NFL) 7
III	Jan. 12, 1969	New York (AFL) 16	Baltimore (NFL) 7
II	Jan. 14, 1968	Green Bay (NFL) 33	Oakland (AFL) 14
I	Jan. 15, 1967	Green Bay (NFL) 35	Kansas City (AFL) 10

8. Consider the following stem-and-leaf plot and histogram concerning the same set of data.

2 | 1, 1, 4, 7	2–3 | x, x, x, x
3 | 0, 0, 3, 3, 6, 9, 9, 9	3–4 | x, x, x, x, x, x, x, x
4 | 2, 2, 5, 8, 8, 8	4–5 | x, x, x, x, x, x
5 | 1, 1, 7, 7	5–6 | x, x, x, x
6 | 3, 3, 3, 6	6–7 | x, x, x, x
7 | 2, 2, 5, 5, 5, 8	7–8 | x, x, x, x, x, x

What can you conclude from the stem-and-leaf plot that would not have been apparent from the histogram?

9. Use the data represented in the stem-and-leaf plot in Prob. 8 to answer the following questions.

(a) How many data values are in the 40s?

(b) What percentage of values is greater than 50?

(c) What percentage of values has the ones digit equal to 1?

10. The following table gives both the average amount spent per person on hospital care, doctors' services, and prescription drugs in each state in 1991 and the percentage of per capita income that this amount represents.

	Per capita health spending	Percentage of per capita income		Per capita health spending	Percentage of per capita income
Mass.	$2402	10.5	Wis.	1762	9.8
N.Y.	2134	9.5	Neb.	1754	9.8
Penn.	2105	11.0	Tex.	1742	10.1
N.D.	2083	12.9	Ind.	1738	10.1
Conn.	2080	8.0	S.D.	1736	10.6
Fla.	2037	10.8	N.H.	1735	8.3
Del.	2028	10.0	Kan.	1727	9.3
Tenn.	2009	12.3	Va.	1722	8.6
Mo.	1983	11.1	Ky.	1713	11.0
Minn.	1969	10.3	Ariz.	1712	10.4
La.	1960	12.9	Wash.	1691	8.7
Md.	1918	8.7	Ark.	1673	11.3
R.I.	1916	10.2	Iowa	1657	9.5
Calif.	1914	9.1	N.C.	1631	9.8
Hawaii	1889	8.9	Me.	1592	9.2
N.J.	1887	7.4	N.M.	1582	10.7
Ga.	1884	10.8	Ore.	1573	8.9
U.S.	1877	11.5	S.C.	1558	10.1
Ohio	1859	10.4	Okla.	1553	9.8
Mich.	1856	9.9	Vt.	1454	8.2
Ill.	1840	8.8	Miss.	1440	10.8
Ala.	1833	11.8	Mont.	1440	9.0
W. Va.	1832	12.9	Utah	1434	9.9
Colo.	1807	9.3	Wyo.	1301	7.6
Alaska	1801	8.2	Idaho	1234	8.0
Nev.	1771	9.2			

Source: Department of Health and Human Services, Census estimates.

(a) Represent the amounts spent per capita in a histogram.

(b) Represent the percentages of per capita income spent on health care in a stem-and-leaf plot.

11. A useful way of comparing two data sets is to put their stem-and-leaf plots side by side. The following represents the scores of students in two different schools on a standard examination. In both schools 24 students took the examination.

School A		School B
Leaves	**Stem**	**Leaves**
0	5	3, 5, 7
8, 5	6	2, 5, 8, 9, 9
9, 7, 4, 2, 0	7	3, 6, 7, 8, 8, 9
9, 8, 8, 7, 7, 6, 5, 3	8	0, 2, 3, 5, 6, 6
8, 8, 6, 6, 5, 5, 3, 0	9	0, 1, 5
	10	0

(a) Which school had the "high scorer"?
(b) Which school had the "low scorer"?
(c) Which school did better on the examination?
(d) Combine the two schools, and draw a stem-and-leaf plot for all 48 values.

12. The following data show the sentences (in months) of criminals convicted of assault in a certain county in 1975 with those of criminals convicted in the same county in 1985.
    1975 33, 44, 40, 38, 52, 19, 29, 31, 38, 32, 41, 45, 47, 50, 35, 39, 34, 32
    1985 24, 27, 30, 32, 22, 25, 30, 26, 22, 31, 33, 27, 29, 36, 27, 28
    (a) Draw side-by-side stem-and-leaf plots for these data sets.
    (b) What conclusions can you draw about sentences in 1975 versus those in 1985?

2.5 SETS OF PAIRED DATA

Sometimes a data set consists of pairs of values that have some relationship to each other. Each member of the data set is thought of as having an x value and a y value. We often express the ith pair by the notation (x_i, y_i), $i = 1, \ldots, n$. For instance, in the data set presented in Table 2.12, x_i represents the score on an intelligence quotient (IQ) test, and y_i represents the annual salary (to the nearest $1000) of the ith chosen worker in a sample of 30 workers from a particular company. In this section, we show how to effectively display data sets of paired values.

One approach to representing such a data set is to first consider each part of the paired data separately and then plot the relevant histograms or stem-and-leaf plots for each. For instance, Figs. 2.11 and 2.12 are stem-and-leaf plots of, respectively, the IQ test scores and the annual salaries for the data presented in Table 2.12.

However, although Figs. 2.11 and 2.12 tell us a great deal about the individual IQ scores and worker salaries, they tell us nothing about the relationship between these two variables. Thus, for instance, by themselves they would not be useful in helping us learn whether higher IQ scores tend to go along with higher income at this company. To learn about how the data relate to such questions, it is necessary to consider the paired values of each data point simultaneously.

Table 2.12	Worker i	IQ score x_i	Annual salary y_i (in units of $1000)	Worker i	IQ score x_i	Annual salary y_i (in units of $1000)
	1	110	68	16	84	19
	2	107	30	17	83	16
	3	83	13	18	112	52
	4	87	24	19	80	11
	5	117	40	20	91	13
	6	104	22	21	113	29
	7	110	25	22	124	71
	8	118	62	23	79	19
	9	116	45	24	116	43
	10	94	70	25	113	44
	11	93	15	26	94	17
	12	101	22	27	95	15
	13	93	18	28	104	30
	14	76	20	29	115	63
	15	91	14	30	90	16

A useful way of portraying a data set of paired values is to plot the data on a two-dimensional rectangular plot with the x axis representing the x value of the data and the y axis representing the y value. Such a plot is called a *scatter diagram*. Figure 2.13 presents a scatter diagram for the data of Table 2.12.

It is clear from Fig. 2.13 that higher incomes appear to go along with higher scores on the IQ test. That is, while not every worker with a high IQ score receives a larger salary than another worker with a lower score (compare worker 5 with worker 29), it appears to be generally true.

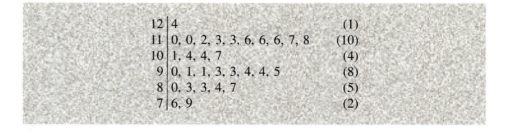

Figure 2.11
Stem-and-leaf plot for IQ scores.

```
12 | 4                                (1)
11 | 0, 0, 2, 3, 3, 6, 6, 6, 7, 8     (10)
10 | 1, 4, 4, 7                       (4)
 9 | 0, 1, 1, 3, 3, 4, 4, 5           (8)
 8 | 0, 3, 3, 4, 7                    (5)
 7 | 6, 9                             (2)
```

Figure 2.12
Stem-and-leaf plot for annual salaries (in $1000).

```
 7 | 0, 1                               (2)
 6 | 2, 3, 8                            (3)
 5 | 2                                  (1)
 4 | 0, 3, 4, 5                         (4)
 3 | 0, 0                               (2)
 2 | 0, 2, 2, 4, 5, 9                   (6)
 1 | 1, 3, 3, 4, 5, 5, 6, 6, 7, 8, 9, 9 (12)
```

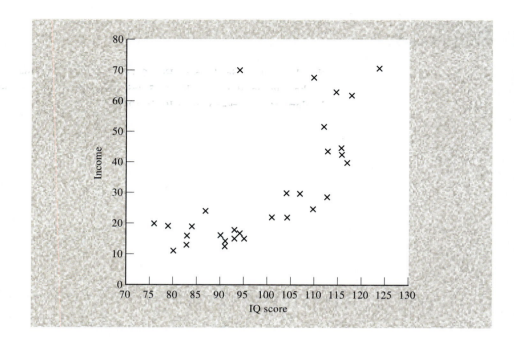

Figure 2.13
Scatter diagram
of IQ versus
income data.

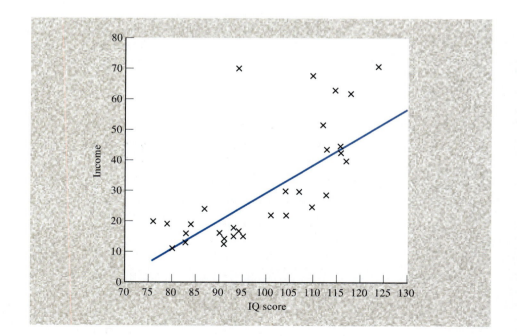

Figure 2.14
Scatter diagram
for IQ versus
income: fitting
a straight line
by eye.

The scatter diagram of Fig. 2.13 also appears to have some predictive uses. For instance, suppose we wanted to predict the salary of a worker, similar to the ones considered above, whose IQ test score is 120. One way to do this is to "fit by eye" a line to the data set, as is done in Fig. 2.14. Since the y value on the line corresponding to the x value of 120 is about 45, this seems like a reasonable prediction for the annual salary of a worker whose IQ is 120.

In addition to displaying joint patterns of two variables and guiding predictions, a scatter diagram is useful in detecting *outliers,* which are data points that do not appear to follow the pattern of the other data points. [For example, the point (94, 70) in Fig. 2.13 does not appear to follow the general trend.] Having noted the outliers, we can then decide whether the data pair is meaningful or is caused by an error in data collection.

Problems

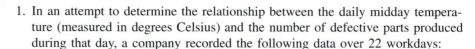

1. In an attempt to determine the relationship between the daily midday temperature (measured in degrees Celsius) and the number of defective parts produced during that day, a company recorded the following data over 22 workdays:

Temperature	Number of defective parts	Temperature	Number of defective parts
24.2	25	24.8	23
22.7	31	20.6	20
30.5	36	25.1	25
28.6	33	21.4	25
25.5	19	23.7	23
32.0	24	23.9	27
28.6	27	25.2	30
26.5	25	27.4	33
25.3	16	28.3	32
26.0	14	28.8	35
24.4	22	26.6	24

(a) Draw a scatter diagram.
(b) What can you conclude from the scatter diagram?
(c) If tomorrow's midday temperature reading were 24.0, what would your best guess be as to the number of defective parts produced?

2. The following data yield the registration rate, equal to the proportion of the voting-age population who were registered to vote, and the actual turnout rate, equal to the proportion of the voting-age population who actually voted, for 27 cities. (The data are from 1960 records.)

City	Registration rate	Turnout rate
Des Moines, IA	92.6	85.2
Lansing, MI	91.9	72.4
Spokane, WA	89.4	67.0
Peoria, IL	87.4	64.9
Portland, OR	85.8	74.1
Pasadena, CA	83.2	69.2
Tulsa, OK	82.4	69.4
Camden, NJ	81.3	69.0
Youngstown, OH	81.0	71.6
Omaha, NE	79.8	66.8
Syracuse, NY	79.3	72.3
Philadelphia, PA	77.6	69.8
Los Angeles, CA	77.0	64.2
Trenton, NJ	75.8	63.8
Boston, MA	74.0	63.3
Columbus, OH	72.4	63.1
Winston-Salem, NC	71.2	50.5
Bridgeport, CT	70.6	67.5
St. Louis, MO	68.5	62.0
San Francisco, CA	68.0	64.4
New York, NY	65.7	58.8
Wichita, KS	62.2	43.0
Honolulu, HI	60.0	54.7
Nashville, TN	55.9	38.0
Forth Worth, TX	48.4	23.9
San Antonio, TX	42.6	31.4
Atlanta, GA	33.8	25.6

(a) Draw a scatter diagram to represent these data.

(b) What can be concluded from the diagram?

(c) Using the scatter diagram, estimate the fraction of registered voters who actually voted in 1960.

3. The following data relate the number of tons of fertilizer produced by a company over 10 years and the yearly production cost (adjusted for inflation) per ton.

Year	Number of tons (in thousands)	Dollar cost per ton	Year	Number of tons (in thousands)	Dollar cost per ton
1980	3.2	41.0	1985	5.5	31.4
1981	4.1	39.1	1986	8.2	27.6
1982	5.2	34.8	1987	7.3	29.4
1983	4.4	33.0	1988	9.1	25.0
1984	6.6	29.2	1989	9.0	25.6

(a) Draw a scatter diagram to represent these data.

(b) What conclusions can you draw from an inspection of this diagram?

(c) If 10 tons will be produced in 1990, what is your estimate, based on the scatter diagram, of the cost per ton?

4. The following data relate the number of violent crimes (defined as murder, rape, robbery, and aggravated assault) at a sampling of college campuses in 1985 and 1989. The data listed are in terms of number of crimes per 1000 students.

School	1985	1989
University of California, Berkeley	1.88	1.83
University of California, Los Angeles	1.21	1.43
University of California, Davis	0.32	0.58
Colorado State (Fort Collins)	0.48	0.94
University of Florida (Gainesville)	0.83	0.94
University of Minnesota (Minneapolis)	0.26	0.39
University of Iowa (Iowa City)	0.19	0.39
Penn State	0.24	0.37
Northeastern (Boston)	0.41	1.54
Boston College	0.54	3.15

(a) Draw a scatter diagram relating the crime rate in 1985 to that in 1989 for the 10 campuses.

(b) Do high rates in 1985 appear to "go" with high rates in 1989?

5. The following data relate the attention span (in minutes) to a score on an IQ examination of 18 preschool-age children.

Attention span	IQ score	Attention span	IQ score	Attention span	IQ score
2.0	82	6.3	105	5.5	118
3.0	88	5.4	108	3.6	128
4.4	86	6.6	112	5.4	128
5.2	94	7.0	116	3.8	130
4.9	90	6.5	122	2.7	140
6.1	99	7.2	110	2.2	142

(a) Draw a scatter diagram.

(b) Give a plausible inference concerning the relation of attention span to IQ score.

6. The following data relate prime lending rates and the corresponding inflation rate during 8 years in the 1970s.

Inflation rate	Prime lending rate	Inflation rate	Prime lending rate
3.3	5.2	5.8	6.8
6.2	8.0	6.5	6.9
11.0	10.8	7.6	9.0
9.1	7.9		

(a) Draw a scatter diagram.

(b) Fit a straight line drawn "by hand" to the data pairs.

(c) Using your straight line, predict the prime lending rate in a year whose inflation rate is 7.2 percent.

7. The following data are the 1979 and 1985 per capita incomes for residents of 12 different U.S cities.

City	Per capita income 1979	Per capita income 1985	City	Per capita income 1979	Per capita income 1985
New York	7,271	11,188	San Diego	8,016	11,766
Los Angeles	8,415	12,084	Dallas	8,610	12,816
Chicago	6,933	9,642	San Antonio	5,758	8,499
Houston	8,817	12,115	Phoenix	7,552	11,363
Philadelphia	6,053	8,807	Honolulu	7,912	11,434
Detroit	6,215	8,852	Baltimore	5,877	8,647

Source: U.S. Bureau of the Census, 1980 Census of Population.

(a) Represent these data in a scatter diagram.

(b) The 1979 per capita income of residents of San Jose was $8382. Give a rough prediction of their annual 1985 per capita income.

(c) Repeat part (b) for Kansas City, whose 1979 per capita income was $7480.

8. Use the table on death rates from different causes in various countries, presented prior to Prob. 9 of Sec. 2.3, to draw a scatter diagram relating death rates due to ischemic heart disease and due to stomach cancer. What conclusions, if any, can be drawn from the diagram?

9. Use the table on death rates from different causes in various countries, presented prior to Prob. 9 of Sec. 2.3, to draw a scatter diagram relating death rates due to motor traffic accidents and due to suicide. What conclusions, if any, can be drawn from the diagram?

10. Problem 7 of Sec. 2.4 gives the scores of the first 25 Super Bowl football games. For each game, let y denote the score of the winning team, and let x denote the number of points by which that team won. Draw a scatter diagram relating x and y. Do high values of one tend to go with high values of the other?

2.6 SOME HISTORICAL COMMENTS

Probably the first recorded instance of statistical graphics—that is, the representation of data by tables or graphs—was Sir Edmund Halley's graphical analysis of barometric pressure as a function of altitude, published in 1686. Using the rectangular coordinate system introduced by the French scientist René Descartes in his study of analytic geometry, Halley plotted a scatter diagram and was then able to fit a curve to the plotted data.

In spite of Halley's demonstrated success with graphical plotting, almost all the applied scientists until the latter part of the 18th century emphasized tables rather than graphs in presenting their data. Indeed, it was not until 1786, when William Playfair invented the bar graph to represent a frequency table, that graphs began to be regularly employed. In 1801 Playfair invented the pie chart and a short time later originated the use of histograms to display data.

The use of graphs to represent continuous data—that is, data in which all the values are distinct—did not regularly appear until the 1830s. In 1833 the Frenchman A. M. Guerry applied the bar chart form to continuous crime data, by first breaking up the data into classes, to produce a histogram. Systematic development of the histogram was carried out by the Belgian statistician and social scientist Adolphe Quetelet about 1846. Quetelet and his students demonstrated the usefulness of graphical analysis in their development of the social sciences. In doing so, Quetelet popularized the practice, widely followed today, of initiating a research study by first gathering and presenting numerical data. Indeed, along with the additional steps of summarizing the data and then utilizing the methods of statistical inference to draw conclusions, this has become the accepted paradigm for research in all fields connected with the social sciences. It has also become an important technique in other fields, such as medical research (the testing of new drugs and therapies), as well as in such traditionally nonnumerical fields as literature (in deciding authorship) and history (particularly as developed by the French historian F. Braudel).

The term *histogram* was first used by Karl Pearson in his 1895 lectures on statistical graphics. The stem-and-leaf plot, which is a variant of the histogram, was introduced by the U.S. statistician John Tukey in 1970. In the words of Tukey, "whereas a histogram uses a nonquantitative mark to indicate a data value, clearly the best type of mark is a digit."

(Princeton University Libraries)

John Tukey

Key Terms

Frequency: The number of times that a given value occurs in a data set.

Frequency table: A table that presents, for a given set of data, each distinct data value along with its frequency.

Line graph: A graph of a frequency table. The abscissa specifies a data value, and the frequency of occurrence of that value is indicated by the height of a vertical line.

Bar chart (or **bar graph**): Similar to a line graph, except now the frequency of a data value is indicated by the height of a bar.

Frequency polygon: A plot of the distinct data values and their frequencies that connects the plotted points by straight lines.

Symmetric data set: A data set is symmetric about a given value x_0 if the frequencies of the data values $x_0 - c$ and $x_0 + c$ are the same for all values of c.

Relative frequency: The frequency of a data value divided by the number of pieces of data in the set.

Pie chart: A chart that indicates relative frequencies by slicing up a circle into distinct sectors.

Histogram: A graph in which the data are divided into class intervals, whose frequencies are shown in a bar graph.

Relative frequency histogram: A histogram that plots relative frequencies for each data value in the set.

Stem-and-leaf plot: Similar to a histogram except that the frequency is indicated by stringing together the last digits (the leaves) of the data.

Scatter diagram: A two-dimensional plot of a data set of paired values.

Summary

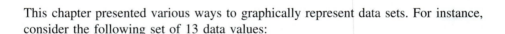

This chapter presented various ways to graphically represent data sets. For instance, consider the following set of 13 data values:

$$1, 2, 3, 1, 4, 2, 6, 2, 4, 3, 5, 4, 2$$

These values can be represented in a *frequency table,* which lists each value and the number of times it occurs in the data, as follows:

A Frequency Table

Value	Frequency	Value	Frequency
1	2	4	3
2	4	5	1
3	2	6	1

The data also can be graphically pictured by either a *line graph* or a *bar chart* (facing page). Sometimes the frequencies of the different data values are plotted on a graph, and then the resulting points are connected by straight lines. This gives rise to a *frequency polygon* (facing page).

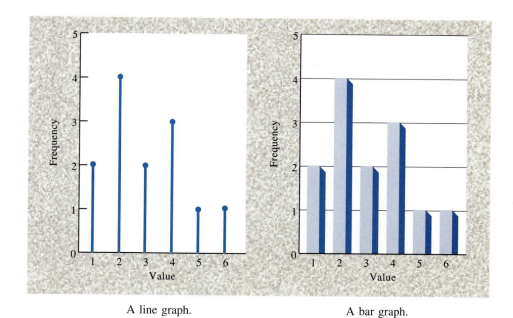

A line graph.

A bar graph.

A frequency
polygon.

When there are a large number of data values, often we break them up into
class intervals. A bar chart plot relating each class interval to the number of data
values falling in the interval is called a *histogram*. The y axis of this plot can represent either the class frequency (that is, the number of data values in the interval) or

the proportion of all the data that lies in the class. In the former case we call the plot a *frequency histogram* and in the latter case a *relative frequency histogram*.

Consider this data set:

41, 38, 44, 47, 33, 35, 55, 52, 41, 66, 64, 50, 49, 56, 55, 48, 52, 63, 59, 57, 75, 63, 38, 37

Using the five class intervals

30–40, 40–50, 50–60, 60–70, 70–80

along with the left-end inclusion convention (which signifies that the interval contains all points greater than or equal to its left-end member and less than its right-end member), we have the following histogram to represent this data set.

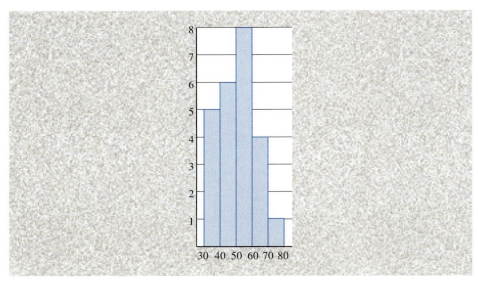

A histogram.

Data sets can also be graphically displayed in a *stem-and-leaf plot*. The following stem-and-leaf plot is for the preceding data set.

```
7 | 5
6 | 3, 3, 4, 6
5 | 0, 2, 2, 5, 5, 6, 7, 9
4 | 1, 1, 4, 7, 8, 9
3 | 3, 5, 7, 8, 8
```

A stem-and-leaf plot.

Often data come in pairs. That is, for each element of the data set there is an x value and a y value. A plot of the x and y values is called a *scatter diagram*. A scatter diagram can be quite useful in ascertaining such things as whether high x values appear to go along with high y values, or whether high x values tend to go along with low y values, or whether there appears to be no particular association between the x and y values of a pair.

The following data set of pairs

i	1	2	3	4	5	6	7	8
x_i	8	12	7	15	5	12	10	22
y_i	14	10	17	9	13	8	12	6

is represented in the following scatter diagram. The diagram indicates that high values of one member of the pair appear to be generally associated with low values of the other member of the pair.

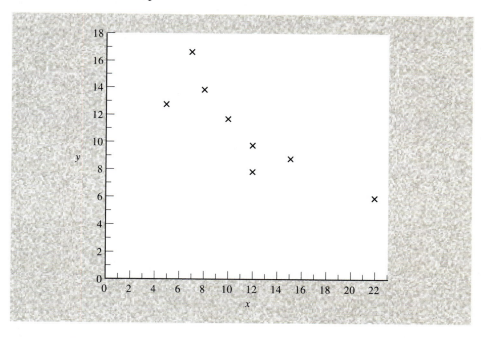

A scatter diagram.

Using these graphical tools, often we can communicate pertinent features of a data set at a glance. As a result, we can learn things about the data that are not immediately evident in the raw numbers themselves. The choice of which display to use depends on such things as the size of the data set, the type of data, and the number of distinct values.

Review Problems

1. The following data are the blood types of 50 volunteers at a blood plasma donation clinic:

O A O AB A A O O B A O A AB B O O O A B A A O A A O B A O
AB A O O A B A A A O B O O A O A B O AB A O B

(a) Represent these data in a frequency table.

(b) Represent them in a relative frequency table.

(c) Represent them in a pie chart.

2. The following is a sample of prices, rounded to the nearest cent, charged per gallon of standard gasoline in the San Francisco Bay area in May 1991.

121, 119, 117, 121, 120, 120, 118, 124, 123, 139, 120, 115, 117, 121, 123, 120, 123, 118, 117, 122, 122, 119

(a) Construct a frequency histogram for this data set.

(b) Construct a frequency polygon.

(c) Construct a stem-and-leaf plot.

(d) Does any data value seem out of the ordinary? If so, explain why.

3. The following frequency table presents the number of female suicides that took place in eight German states over 14 years.

Number of suicides per year	0	1	2	3	4	5	6	7	8	9	10
Frequency	9	19	17	20	15	11	8	2	3	5	3

Thus, for instance, there were a total of 20 cases in which states had 3 suicides in a year.

(a) How many suicides were reported over the 14 years?

(b) Represent the above data in a histogram.

4. The following table gives the 1991 crime rate (per 100,000 population) in each state. Use it to construct a

(a) Frequency histogram of the total violent crime rates in the states

(b) Relative frequency histogram of the total property crime rates in the states

(c) Stem-and-leaf plot of the murder rates

(d) Stem-and-leaf plot of the burglary rates

		1991								
		Violent crime					**Property crime**			
Region, Division, and State	Total	Total	Murder	Forcible rape	Robbery	Aggravated assault	Total	Burglary	Larceny —theft	Motor vehicle theft
United States	5,898	758	9.8	42	273	433	5,140	1,252	3,229	659
Northeast	5,155	752	8.4	29	352	363	4,403	1,010	2,598	795
New England	4,950	532	4.1	30	159	338	4,419	1,103	2,600	716
Maine	3,768	132	1.2	22	23	86	3,636	903	2,570	163
New Hampshire	3,448	119	3.6	30	33	53	3,329	735	2,373	220
Vermont	3,955	117	2.1	31	12	72	3,838	1,020	2,674	144
Massachusetts	5,332	736	4.2	32	195	505	4,586	1,167	2,501	919
Rhode Island	5,039	462	3.7	31	123	304	4,577	1,127	2,656	794
Connecticut	5,364	540	5.7	29	224	280	4,824	1,191	2,838	796
Middle Atlantic	5,227	829	9.9	29	419	372	4,398	978	2,598	823
New York	6,245	1,164	14.2	28	622	499	5,081	1,132	2,944	1,004
New Jersey	5,431	635	5.2	29	293	307	4,797	1,016	2,855	926
Pennsylvania	3,559	450	6.3	29	194	221	3,109	720	1,907	482

Region, Division, and State	Total	1991 Violent crime					1991 Property crime			
		Total	Murder	Forcible rape	Robbery	Aggravated assault	Total	Burglary	Larceny —theft	Motor vehicle theft
Midwest	5,257	631	7.8	45	223	355	4,626	1,037	3,082	507
East north central	5,482	704	8.9	50	263	383	4,777	1,056	3,151	570
Ohio	5,033	562	7.2	53	215	287	4,471	1,055	2,916	500
Indiana	4,818	505	7.5	41	116	340	4,312	977	2,871	465
Illinois	6,132	1,039	11.3	40	456	532	5,093	1,120	3,318	655
Michigan	6,138	803	10.8	79	243	470	5,335	1,186	3,469	680
Wisconsin	4,466	277	4.8	25	119	128	4,189	752	3,001	436
West north central	4,722	457	5.4	34	129	288	4,265	991	2,918	356
Minnesota	4,496	316	3.0	40	98	175	4,180	854	2,963	363
Iowa	4,134	303	2.0	21	45	235	3,831	832	2,828	171
Missouri	5,416	763	10.5	34	251	467	4,653	1,253	2,841	558
North Dakota	2,794	65	1.1	18	8	38	2,729	373	2,229	127
South Dakota	3,079	182	1.7	40	19	122	2,897	590	2,192	115
Nebraska	4,354	335	3.3	28	54	249	4,020	727	3,080	213
Kansas	5,534	500	6.1	45	138	310	5,035	1,307	3,377	351
South	6,417	798	12.1	45	252	489	5,618	1,498	3,518	603
South Atlantic	6,585	851	11.4	44	286	510	5,734	1,508	3,665	561
Delaware	5,869	714	5.4	86	215	408	5,155	1,128	3,652	375
Maryland	6,209	956	11.7	46	407	492	5,253	1,158	3,365	731
District of Columbia	10,768	2,453	80.6	36	1,216	1,121	8,315	2,074	4,880	1,360
Virginia	4,607	373	9.3	30	138	196	4,234	783	3,113	339
West Virginia	2,663	191	6.2	23	43	119	2,472	667	1,631	175
North Carolina	5,889	658	11.4	35	178	434	5,230	1,692	3,239	299
South Carolina	6,179	973	11.3	59	171	731	5,207	1,455	3,365	387
Georgia	6,493	738	12.8	42	268	415	5,755	1,515	3,629	611
Florida	8,547	1,184	9.4	52	400	723	7,363	2,006	4,573	784
East south central	4,687	631	10.4	41	149	430	4,056	1,196	2,465	395
Kentucky	3,358	438	6.8	35	83	313	2,920	797	1,909	215
Tennessee	5,367	726	11.0	46	213	456	4,641	1,365	2,662	614
Alabama	5,366	844	11.5	36	153	644	4,521	1,269	2,889	363
Mississippi	4,221	389	12.8	46	116	214	3,832	1,332	2,213	286
West south central	7,118	806	14.2	50	254	488	6,312	1,653	3,871	788
Arkansas	5,175	593	11.1	45	136	402	4,582	1,227	3,014	341
Louisiana	6,425	951	16.9	41	279	614	5,473	1,412	3,489	573
Oklahoma	5,669	584	7.2	51	129	397	5,085	1,478	3,050	557
Texas	7,819	840	15.3	53	286	485	6,979	1,802	4,232	944
West	6,478	841	9.6	46	287	498	5,637	1,324	3,522	791
Mountain	6,125	544	6.5	44	122	371	5,581	1,247	3,843	491
Montana	3,648	140	2.6	20	19	99	3,508	524	2,778	206
Idaho	4,196	290	1.8	29	21	239	3,905	826	2,901	178
Wyoming	4,389	310	3.3	26	17	264	4,079	692	3,232	155
Colorado	6,074	559	5.9	47	107	399	5,515	1,158	3,930	426
New Mexico	6,679	835	10.5	52	120	652	5,845	1,723	3,775	346
Arizona	7,406	671	7.8	42	166	455	6,735	1,607	4,266	861
Utah	5,608	287	2.9	46	55	183	5,321	840	4,240	241
Nevada	6,299	677	11.8	66	312	287	5,622	1,404	3,565	652
Pacific	6,602	945	10.7	47	345	542	5,656	1,351	3,409	896
Washington	6,304	523	4.2	70	146	303	5,781	1,235	4,102	444
Oregon	5,755	506	4.6	53	150	298	5,249	1,176	3,598	474
California	6,773	1,090	12.7	42	411	624	5,683	1,398	3,246	1,039
Alaska	5,702	614	7.4	92	113	402	5,088	979	3,575	534
Hawaii	5,970	242	4.0	33	87	118	5,729	1,234	4,158	336

Source: U.S. Federal Bureau of Investigation, *Crime in the United States*, annual.

5. Construct a frequency table for a data set of 10 values that is symmetric and has (a) 5 distinct values and (b) 4 distinct values. (c) About what values are the data sets in parts (a) and (b) symmetric?

6. The following are the estimated oil reserves, in billions of barrels, for four regions in the western hemisphere.

   United States   38.7

   South America   22.6

   Canada   8.8

   Mexico   60.0

   Represent the preceding in a pie chart.

7. The following tables give the 1985 and 1990 crude steel production, in millions of metric tons, for a variety of countries.

	Crude steel production	
Country	**1985**	**1990**
Argentina	2.8	3.6
Australia	6.3	6.7
Brazil	20.5	20.5
Canada	13.5	12.2
France	18.8	19.3
Italy	23.9	25.5
Spain	14.7	12.8
United States	80.1	88.6
United Kingdom	15.7	17.8

   Source: Statistical Office of the United Nations, *Statistical Yearbook* and *Industrial Statistics Yearbook.*

   (a) Represent the 1985 and 1990 data in back-to-back stem-and-leaf plots.

   (b) Draw a scatter diagram to relate 1985 and 1990 data.

8. The following data refer to the ages (to the nearest year) at which patients died at a large inner-city (nonbirthing) hospital:

   1, 1, 1, 1, 3, 3, 4, 9, 17, 18, 19, 20, 20, 22, 24, 26, 28, 34, 45, 52, 56, 59, 63, 66, 68, 68, 69, 70, 74, 77, 81, 90

   (a) Represent this data set in a histogram.

   (b) Represent this data set in a frequency polygon.

   (c) Represent it in a cumulative frequency polygon.

   (d) Represent it in a stem-and-leaf plot.

Problems 9 to 11 refer to the last 50 student entries in App. A.

9. (a) Draw a histogram of the weights of these students.
   (b) Comment on this histogram.

10. Draw a scatter diagram relating weight and cholesterol level. Comment on what the scatter diagram indicates.

11. Draw a scatter diagram relating weight and blood pressure. What does this diagram indicate?

Problems 12 and 13 refer to the following table concerning the mathematics and verbal SAT scores of a graduating class of high school seniors.

Student	Verbal score	Mathematics score	Student	Verbal score	Mathematics score
1	520	505	8	620	576
2	605	575	9	604	622
3	528	672	10	720	704
4	720	780	11	490	458
5	630	606	12	524	552
6	504	488	13	646	665
7	530	475	14	690	550

12. Draw side-by-side stem-and-leaf plots of the student scores on the mathematics and verbal SAT examinations. Did the students, as a group, perform better on one examination? If so, which one?

13. Draw a scatter diagram of student scores on the two examinations. Do high scores on one tend to go along with high scores on the other?

14. The following table gives information about the age of the population in both the United States and Mexico.

Age, years	Proportion of population (percent)	
	Mexico	United States
0– 9	32.5	17.5
10–19	24	20
20–29	14.5	14.5
30–39	11	12
40–49	7.5	12.5
50–59	4.5	10.5
60–69	3.5	7
70–79	1.5	4
Over 80	1	2

(a) What percentage of the Mexican population is less than 30 years old?

(b) What percentage of the U.S. population is less than 30 years old?

(c) Draw two relative frequency polygons on the same graph. Use different colors for Mexican and for U.S. data.

(d) In general, how do the age distributions compare for the two countries?

15. The following data relate to the normal monthly and annual precipitation (in inches) for various cities.

Normal Monthly and Annual Precipitation in Selected Cities

State	City	Jan.	Feb.	Mar.	Apr.	May	June	July	Aug.	Sept.	Oct.	Nov.	Dec.	Annual
AL	Mobile	4.59	4.91	6.48	5.35	5.46	5.07	7.74	6.75	6.56	2.62	3.67	5.44	64.64
AK	Juneau	3.69	3.74	3.34	2.92	3.41	2.98	4.13	5.02	6.40	7.71	5.15	4.66	53.15
AZ	Phoenix	.73	.59	.81	.27	.14	.17	.74	1.02	.64	.63	.54	.83	7.11
AR	Little Rock	3.91	3.83	4.69	5.41	5.29	3.67	3.63	3.07	4.26	2.84	4.37	4.23	49.20
CA	Los Angeles	3.06	2.49	1.76	.93	.14	.04	.01	.10	.15	.26	1.52	1.62	12.08
	Sacramento	4.03	2.88	2.06	1.31	.33	.11	.05	.07	.27	.86	2.23	2.90	17.10
	San Diego	2.11	1.43	1.60	.78	.24	.06	.01	.11	.19	.33	1.10	1.36	9.32
	San Francisco	4.65	3.23	2.64	1.53	.32	.11	.03	.05	.19	1.06	2.35	3.55	19.71
CO	Denver	.51	.69	1.21	1.81	2.47	1.58	1.93	1.53	1.23	.98	.82	.55	15.31
CT	Hartford	3.53	3.19	4.15	4.02	3.37	3.38	3.09	4.00	3.94	3.51	4.05	4.16	44.39
DE	Wilmington	3.11	2.99	3.87	3.39	3.23	3.51	3.90	4.03	3.59	2.89	3.33	3.54	41.38
DC	Washington	2.76	2.62	3.46	2.93	3.48	3.35	3.88	4.40	3.22	2.90	2.82	3.18	39.00
FL	Jacksonville	3.07	3.48	3.72	3.32	4.91	5.37	6.54	7.15	7.26	3.41	1.94	2.59	52.76
	Miami	2.08	2.05	1.89	3.07	6.53	9.15	5.98	7.02	8.07	7.14	2.71	1.86	57.55
GA	Atlanta	4.91	4.43	5.91	4.43	4.02	3.41	4.73	3.41	3.17	2.53	3.43	4.23	48.61
HI	Honolulu	3.79	2.72	3.48	1.49	1.21	.49	.54	.60	.62	1.88	3.22	3.43	23.47
ID	Boise	1.64	1.07	1.03	1.19	1.21	.95	.26	.40	.58	.75	1.29	1.34	11.71
IL	Chicago	1.60	1.31	2.59	3.66	3.15	4.08	3.63	3.53	3.35	2.28	2.06	2.10	33.34
	Peoria	1.60	1.41	2.86	3.81	3.84	3.88	3.99	3.39	3.63	2.51	1.96	2.01	34.89
IN	Indianapolis	2.65	2.46	3.61	3.68	3.66	3.99	4.32	3.46	2.74	2.51	3.04	3.00	39.12
IA	Des Moines	1.01	1.12	2.20	3.21	3.96	4.18	3.22	4.11	3.09	2.16	1.52	1.05	30.83
KS	Wichita	.68	.85	2.01	2.30	3.91	4.06	3.62	2.80	3.45	2.47	1.47	.99	28.61
KY	Louisville	3.38	3.23	4.73	4.11	4.15	3.60	4.10	3.31	3.35	2.63	3.49	3.48	43.56
LA	New Orleans	4.97	5.23	4.73	4.50	5.07	4.63	6.73	6.02	5.87	2.66	4.06	5.27	59.74

Source: U.S. National Oceanic and Atmospheric Administration, *Climatography of the United States,* September 1982.

(a) Represent the normal precipitation amounts for April in a stem-and-leaf plot.

(b) Represent the annual amounts in a histogram.

(c) Draw a scatter diagram relating the April amount to the annual amount.

16. A data value that is far away from the other values is called an *outlier*. In the following data sets, specify which, if any, of the data values are outliers.

(a) 14, 22, 17, 5, 18, 22, 10, -17, 25, 28, 33, 12

(b) 5, 2, 13, 16, 9, 12, 7, 10, 54, 22, 18, 15, 12

(c) 18, 52, 14, 20, 24, 27, 43, 17, 25, 28, 3, 22, 6

17. The amount of bananas consumed per capita in the United States each year from 1980 to 1987 is as follows.

Year	Per capita banana consumption (pounds)
1980	58.5
1981	59.0
1982	59.8
1983	60.0
1984	64.6
1985	63.7
1986	66.6
1987	71.5

Source: U.S. Department of Agriculture, *Food Consumption, Prices and Expenditures*

(a) Does banana consumption appear to be increasing as time passes?
(b) Represent the above in a scatter diagram.
(c) Predict the per capita banana consumption in 1988.

The following table gives the suicide rate per 100,000 population for selected countries, by sex and age group. Problems 18 to 20 are based on it.

Suicide Rates for Selected Countries, by Sex and Age Group

Sex and age	United States, 1984	Australia, 1985	Austria, 1986	Canada, 1985	Denmark, 1985	France, 1985	Italy, 1983	Japan, 1986	Netherlands, 1985	Poland, 1986	Sweden, 1985	United Kingdom, 1985	West Germany, 1986
MALE													
Total	19.7	18.2	42.1	20.5	35.1	33.1	11.0	27.8	14.6	22.0	25.0	12.1	26.6
15–24 yrs. old	20.5	24.0	31.0	25.2	17.0	17.0	5.2	14.1	10.6	17.5	14.3	8.2	17.7
25–34 yrs. old	24.9	26.6	48.6	27.0	38.7	35.2	8.4	25.1	17.6	29.3	32.0	15.3	25.3
35–44 yrs. old	22.6	22.5	53.3	24.7	38.5	36.5	9.1	31.6	16.1	33.5	29.0	16.3	28.5
45–54 yrs. old	23.7	21.5	54.2	26.4	56.6	45.4	14.8	51.0	20.0	36.7	39.3	17.1	35.6
55–64 yrs. old	27.2	22.0	52.5	26.5	55.3	48.0	18.4	44.8	21.0	35.9	32.7	18.1	36.7
65–74 yrs. old	33.5	24.8	72.8	28.5	57.7	61.4	29.7	43.9	26.1	30.6	36.2	16.9	44.7
75 yrs. old and over	49.1	27.4	106.5	28.4	83.4	120.5	47.9	78.8	41.0	29.3	45.3	22.3	72.8
FEMALE													
Total	5.4	5.1	15.8	5.4	20.6	12.7	4.3	14.9	8.1	4.4	11.5	5.7	12.0
15–24 yrs. old	4.4	4.9	9.7	4.0	8.1	4.7	1.3	8.0	3.1	2.7	7.6	1.8	5.3
25–34 yrs. old	6.1	4.7	14.3	6.6	14.3	10.6	3.3	11.6	9.3	5.3	13.2	4.4	9.1
35–44 yrs. old	7.7	6.1	18.5	8.0	28.6	14.6	3.7	12.8	9.9	5.8	15.5	5.4	10.5
45–54 yrs. old	9.2	8.7	20.2	9.0	34.0	17.7	5.6	18.4	12.5	8.4	14.8	9.2	14.8
55–64 yrs. old	8.5	8.3	18.9	8.0	41.5	20.8	8.3	20.2	14.6	7.6	13.8	10.5	16.5
65–74 yrs. old	7.3	7.6	29.0	7.8	34.1	26.8	10.1	33.0	15.5	6.6	20.1	11.9	23.6
75 yrs. old and over	6.0	9.7	31.5	5.3	24.6	27.5	11.0	59.1	10.8	5.5	14.0	10.1	24.8

Source: World Health Organization, *World Health Statistics*.

18. (a) Represent the suicide rates for 15- to 24-year-old males in a stem-and-leaf plot.
    (b) Represent the suicide rates for 15- to 24-year-old females in a stem-and-leaf plot.
    (c) Draw a scatter diagram relating the suicide rates in each of the countries for 15- to 24-year-old females and males.

19. Draw a scatter diagram relating for the United States and Japan the male and female suicide rates in the different age groups specified. (For instance, the x value of one pair will consist of the suicide rate of U.S. 15- to 24-year-old males, and the y value will be the rate for Japanese males in this age group.)

20. Draw a scatter diagram relating for the United States and France the female and male suicide rates in the different age groups specified.

MINITAB LAB

Describing Data Sets

Purpose

Use Minitab to do the following:

1. Construct frequency tables.

2. Construct line graphs, bar graphs, frequency polygons, relative frequency graphs, and pie charts.

3. Construct histograms for grouped data.

4. Construct stem-and-leaf plots.

5. Construct scatter diagrams.

Procedures

First, load the Minitab (Windows version) software as described in the *Chap. 1 Minitab Lab.*

1. CONSTRUCTING FREQUENCY TABLES

Example 1   Use Minitab to construct frequency, cumulative frequency, relative frequency, and cumulative relative frequency tables for the data relating to the average number of rainy days in March from the table for Probs. 12 and 13 in Sec. 2.2 of your text.

To create such tables in Minitab, type in the data set in C1 and rename it MARCH. *You can rename a column by typing the name in the box that lies between the column number (for example, C1) and row 1.* See the Chap. 1 Minitab lab. Select Stat→Tables→Tally, and the Tally dialog box will be displayed. Make the appropriate selections as shown in Fig. M2.1. To select MARCH for the Variables text box, click on C1 MARCH to highlight it and then click on the Select button in the dialog box; or you can just click on the Variables text box and type in the word *MARCH*. To select Counts, Percents, etc., use the mouse and click on the boxes until an × is displayed.

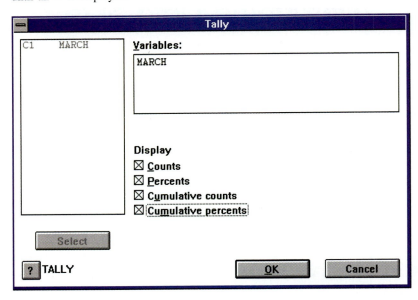

Figure M2.1

Click on the OK button in the dialog box and the distributions will be displayed in the Session window shown in Fig. M2.2. The first column lists the values of the variable MARCH, the second column lists the frequency counts, the third column lists the cumulative frequency counts, the fourth column lists the relative frequencies as a percentage, and the last column lists the cumulative relative frequencies as a percentage for the values of the variable.

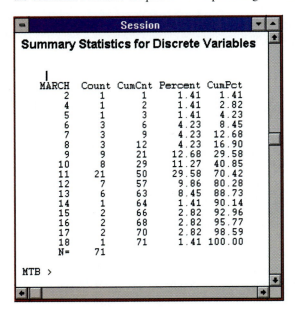

Figure M2.2

Note   If your computer is connected to a printer with the appropriate printer driver, you can print any *active* window by selecting File→Print Window.

To display this set of information in the Data window so that you can construct appropriate graphs and charts, in the Session window type in the following:

MTB>Tally C1;
SUBC>All;
SUBC>Store C2-C6.

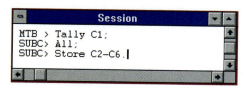

Figure M2.3

Here MTB and SUBC are Minitab prompts. The Session window entries are shown in Fig. M2.3. You should press the Enter key after each line is typed in. The five columns of information in Fig. M2.2 will be displayed in the Data window (C2 to C6). Use the above procedure to rename the columns. For this presentation, C2 was renamed *VALUES*, C3 was renamed *f* for frequency, C4 was renamed *CUMf* for cumulative frequency, C5 was renamed *RELf* for relative frequency, and C6 was renamed *CUMRELf* for cumulative relative frequency.

2. CONSTRUCTING LINE AND BAR GRAPHS, FREQUENCY POLYGONS, RELATIVE FREQUENCY GRAPHS, AND PIE CHARTS

Example 1 (Continued)   To construct a *line graph* for the March precipitation data, select Graph→Plot. In the Graph variable dialog box, select C3 (f) for Y and C2 (VALUES) for X. In the Data display drop-down Display box, select Project. This sequence of commands will draw a line graph for you with the frequencies along the *Y* axis and the data values along the *X* axis. To make it more presentable, select Edit attributes and enter the appropriate Line type, color, and size. Select Annotation→Title to type in the title, and select Frame→Axes or Frame→Tick to darken the axes or to use an appropriate scale along the axes. The line graph for the March data is shown in Fig. M2.4.

To construct a *bar graph (chart)* for the March precipitation data, follow the above procedure in constructing a line graph and select a wider line size in the Edit attributes box. The bar chart is shown in Fig. M2.5.

To construct a *frequency polygon* for the March precipitation data, follow the above procedure in constructing a line graph. Select Connect in the Data display drop-down Display box, and select Polygon in the Annotation box. The frequency polygon is shown in Fig. M2.6.

To construct *relative frequency* graphs, use column C5 (RELf) instead of column C3 (f) in the above procedures in the construction of the line graph, bar chart, and frequency polygon.

To construct a *pie chart* for the March precipitation data, select Graph→Pie Chart. In the *Chart data in* box, select C3 (f). In the *Title* box you can type in an appropriate title, and in the Options box you can select *Add lines connecting labels to slices*. The pie chart for the March precipitation data is shown in Fig. M2.7.

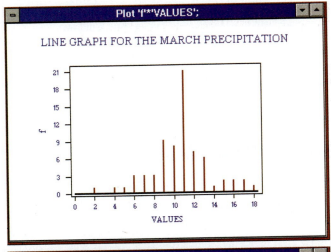

Figure M2.4

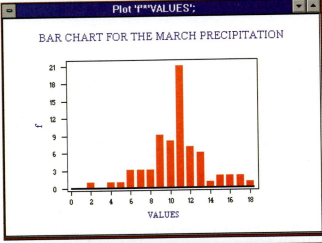

Figure M2.5

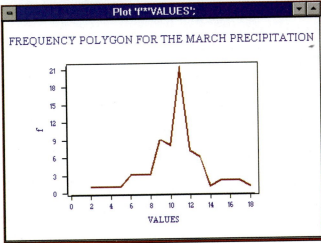

Figure M2.6

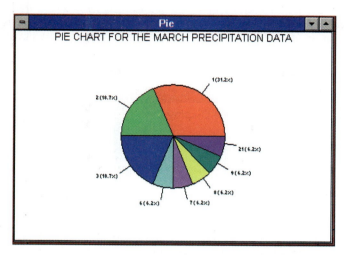

Figure M2.7

3. CONSTRUCTING HISTOGRAMS FOR GROUPED DATA

Example 1 (Continued) To construct a *histogram* for the March precipitation data, select Graph→Histogram. In the Graph variables box, select VALUES; and select Bar in the Data display box. Select Options; and in the Histogram options box, select Frequency for the type of histogram. For the type of intervals, select Cut Point. This selection will place the *class boundaries* along the horizontal axis. Under Definition of Intervals you can select any of the options. For Fig. M2.8 the Midpoint/Cut Point option was chosen with 0:20/5. That is, the values along the X axis ranged from 0 to 20 with class intervals of size 5.

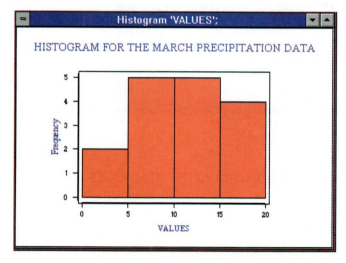

Figure M2.8

4. CONSTRUCTING STEM-AND-LEAF-PLOTS

Example 1 (Continued) To construct *stem-and-leaf plots* for the March precipitation data, select Graph→Character Graphs→Stem-and-Leaf. In the Variables text box that appears, select VALUES. If you select OK, a stem-and-leaf display will be shown in the Session window. You can select different lengths of intervals for the display from the Increment box. For Fig. M2.9 an increment of 4 was used, and for Fig. M2.10 an increment of 2 was used.

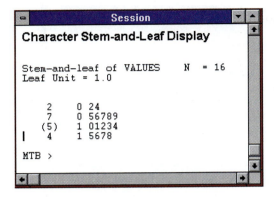

Figure M2.9

Figure M2.10

5. CONSTRUCTING SCATTER DIAGRAMS

Example 2   Use Minitab to present a scatter diagram for the information given in Prob. 3 in Sec. 2.5 that relates the number of tons of fertilizer produced by a company and the yearly production cost. Let the dollar cost be the *dependent variable (Y)* and the number of tons be the *independent variable (X)*.

First, enter the values in the Data window, and name the columns appropriately. Here the *Y* variable is named *$cost,* and the *X* variable is named *tons.* Select Graph→Plot. In the Graph variables box in the Plot box, select the appropriate columns for *X* and *Y*. In the Data display box, select symbol in the Display drop-down box, and select a symbol to use for the plotted points. In this example, the plus symbol (+) was used. Select Edit attributes to select the type, color, and size of the symbol. Select Annotation→Title to type in an appropriate title. The plot of $cost versus tons is shown in Fig. M2.11.

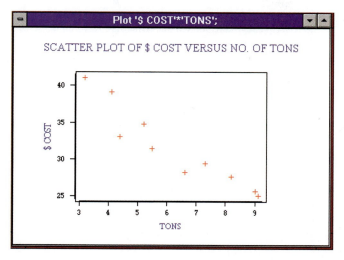

Figure M2.11

To fit a line to the scatter diagram, you can use the "fit by eye" method as in your text. To use Minitab to fit a line to the plot, select Stat→Regression→Fitted Line Plot. In the Fitline box, just select the appropriate columns for the variables X and Y. You can type in a title if you wish, but ignore all the other prompts at this time. The other prompts deal with concepts that are discussed in Chap. 12 of the text. A fitted line is shown in Fig. M2.12 for the data in Example 2. You can now use this line to predict dollar cost per ton of fertilizer for a given number of tons within the range of the number of tons of fertilizer. Observe from the plot that as the number of tons increases, the cost per ton is decreasing.

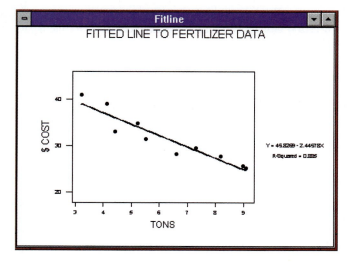

Figure M2.12

Computer Exercises

1. For the table given in Chap. 2 Review Prob. 4 for the 1991 crime rates, use Minitab or any other statistical software to do the following:
   (a) Construct a frequency, cumulative frequency, relative frequency, and cumulative relative frequency table for violent crime due to forcible rape.
   (b) Construct a line graph, bar graph, frequency polygon, relative frequency polygon, and pie chart for violent crime due to forcible rape.
   (c) Construct a histogram using five classes for violent crime due to forcible rape.
   (d) Construct a stem-and-leaf plot for violent crime due to forcible rape.
   (e) Present a report with your displays and discuss any observations that you make from these displays.

2. Use the information given on registration rate and turnout rate in Sec. 2.5, Prob. 2, to do the following:
   (a) Repeat Exercise 1 for the individual variables of registration rate and turnout rate.
   (b) Construct a scatter diagram with turnout rate being the *dependent variable Y* and registration rate being the *independent variable X.*
   (c) Fit a line to the plot and discuss any observations from it.
   (d) Use the fitted line to help predict the estimated turnout rate for five different registration rate values within the range of the observed registration rates.
   (e) Present a report with your displays, and discuss any observation that you make from these displays.

3. Collect data (from the *Wall Street Journal,* etc.) for two variables that have integer values. Use Minitab or any other statistical software, and repeat 2(a) through 2(e) for the observed variables.

4. Using the information given on property tax per capita in Sec. 2.3, Prob. 8, with Minitab or any other statistical software, repeat 1(a) through 1(e).

5. For the information given on the average number of days with precipitation of .01 inch or more in Sec. 2.2, Probs. 12 and 13, use Minitab or any other statistical software to repeat 2(a) through 2(e) for the precipitation values for the months of August and September.

CHAPTER 3

Using Statistics to
Summarize Data Sets

Using Statistics to Summarize Data Sets

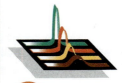

I do hate averages. There is no greater mistake than to call arithmetic an exact science. There are permutations and aberrations discernible to minds entirely noble like mine; subtle variations which ordinary accountants fail to discover, hidden laws of numbers which it requires a mind like mine to perceive. For instance if you average numbers from the bottom up and then again from the top down, the result is always different.
A letter to the *Mathematical Gazette* (a 19th-century British mathematical journal)

The way to make sense out of raw data is to compare and contrast, to understand differences.
Gregory Bateson (in *Steps to an Ecology of the Mind*)

Our objective in this chapter is to develop measures that can be used to summarize a data set. These measures, formally called *statistics,* are quantities whose values are determined by the data. We study the sample mean, sample median, and sample mode. These are all statistics which measure the center or middle value of a data set. Statistics that indicate the amount of variation in the data set are also considered. We learn about what it means for a data set to be normal, and we present an empirical rule concerning such sets. We also consider data sets consisting of paired values, and we present a statistic that measures the degree to which a scatter diagram of paired values can be approximated by a straight line.

3.1   INTRODUCTION

Modern-day experiments often track certain characteristics of thousands of individuals over time. For instance, in an attempt to learn about the health consequences of certain common practices, the medical statisticians R. Doll and A. B. Hill sent questionnaires in 1951 to all doctors in the United Kingdom and received 40,000 replies. Their questionnaire dealt with age, eating habits, exercise habits, and smoking habits. These doctors were then monitored for 10 years, and the causes of death of those who died were determined. As one can imagine, this study resulted in huge sets of data. For instance, even if we just focus on one component of the study at a single moment of time, such as the doctors' ages in 1951, the resulting data set of 40,000 values is vast. To obtain a feel for such a large data set, it is often necessary to summarize it by some suitably chosen measures. In this chapter, we introduce different statistics that can be used to summarize certain features of data sets.

To begin, suppose that we have in our possession sample data from some underlying population. Now, whereas in Chap. 2 we showed how to describe and portray data sets in their entirety, here we will be concerned with determining certain summary measures about the data. These summary measures are called *statistics,* where by a statistic we mean any numerical quantity whose value is determined by the data.

Definition
Numerical quantities computed from a data set are called *statistics.*

We will be concerned with statistics that describe the central tendency of the data set; that is, they describe the center of the set of data values. Three different statistics for describing this—the sample mean, sample median, and sample mode—will be presented in Secs. 3.2, 3.3, and 3.4, respectively. Once we have some idea of the center of a data set, the question naturally arises as to how much *variation* there is. That is, are most of the values close to the center, or do they vary widely about the center? In Sec. 3.5 we will discuss the sample variance and sample standard deviation, which are statistics designed to measure such variation.

In Sec. 3.6 we introduce the concept of a normal data set, which is a data set having a bell-shaped histogram. For data sets that are close to being normal, we present a rule which can be used to approximate the proportion of the data that is within a specified number of sample standard deviations from the sample mean.

In the first six sections of this chapter, we concern ourselves with data sets where each datum is a single value. However, in Sec. 3.7, we deal with paired data. That is, each data point will consist of an x value and a y value. For instance, the x value might represent the average number of cigarettes that an individual smoked per day, and the y value could be the age at which that individual died. We introduce a statistic called the *sample correlation coefficient* whose value indicates the degree to which data points having large *x* values also have large *y* values and correspondingly those having small *x* values also have small *y* values.

> The Doll-Hill study yielded the result that only about 1 in 1000 nonsmoking doctors died of lung cancer. For heavy smokers the figure was 1 in 8. In addition, death rates from heart attacks were 50 percent higher for smokers.

3.2 SAMPLE MEAN

Suppose we have a sample of *n* data points whose values we designate by $x_1, x_2, \ldots, x_n$. One statistic for indicating the center of this data set is the *sample mean*, defined to equal the arithmetic average of the data values.

Definition

The *sample mean*, which we designate by $\bar{x}$ (pronounced "*x* bar"), is defined by

$$\bar{x} = \frac{\sum_{i=1}^{n} x_i}{n} = \frac{x_1 + x_2 + \cdots + x_n}{n}$$

Example 3.1   The U.S. per capita consumption of soft drinks in the years 1982 to 1986 was* (in gallons)

     26.9, 26.9, 27.2, 29.1, 30.3

For instance, in 1986 the soft drink consumption was 30.3 gallons per person. Find the simple mean of this set of data.

Solution   The sample mean $\bar{x}$ is the average of the five data values. That is,

$$\bar{x} = \frac{26.9 + 26.9 + 27.2 + 29.1 + 30.3}{5} = \frac{140.4}{5} = 28.08$$

*Source: U.S. Department of Agriculture, *Food Consumption, Prices, and Expenditures.*

Note from this example that whereas the sample mean is the average of all the data values, it need not itself be one of them.

Consider again the data set $x_1, x_2, \ldots, x_n$. If each data value is increased by a constant amount c, then this causes the sample mean also to be increased by c. Mathematically, we can express this by saying that if

$$y_i = x_i + c \qquad \text{for } i = 1, \ldots, n$$

then

$$\bar{y} = \bar{x} + c$$

where $\bar{y}$ and $\bar{x}$ are the sample means of the y_i and the x_i, respectively. Therefore, when it is convenient, we can compute $\bar{x}$ by first adding c to all the data values, then computing the sample mean $\bar{y}$ of the new data, and finally subtracting c from $\bar{y}$ to obtain $\bar{x}$. Since it is sometimes a lot easier to work with the transformed rather than the original data, this can greatly simplify the computation of $\bar{x}$. Our next example illustrates this point.

Example 3.2   The winning scores in the U.S. Masters Golf Tournament in the years from 1981 to 1990 were as follows:

280, 284, 280, 277, 282, 279, 285, 281, 283, 278

Find the sample mean of these winning scores.

Solution Rather than directly adding the preceding numbers, first we subtract 280 from (that is, add $c = -280$ to) each one to obtain the following transformed data:

0, 4, 0, −3, 2, −1, 5, 1, 3, −2

The sample mean of these transformed data, call it $\bar{y}$, is

$$\bar{y} = \frac{0 + 4 + 0 - 3 + 2 - 1 + 5 + 1 + 3 - 2}{10} = \frac{9}{10}$$

Adding 280 to $\bar{y}$ shows that the sample mean of the original data is

$$\bar{x} = 280.9$$

If each data value is multiplied by c, then so is the sample mean. That is, if

$$y_i = cx_i \qquad i = 1, \ldots, n$$

then

$$\bar{y} = c\bar{x}$$

For instance, suppose that the sample mean of the height of a collection of individuals is 5.0 feet. Suppose that we now want to change the unit of measurement from feet to inches. Then since each new data value is the old value multiplied by 12, it follows that the sample mean of the new data is $12 \cdot 5 = 60$. That is, the sample mean is 60 inches.

Our next example considers the computation of the sample mean when the data are arranged in a frequency table.

Example 3.3 The number of suits sold daily by a women's boutique on the past 6 days has been arranged in the following frequency table:

Value	Frequency
3	2
4	1
5	3

What is the sample mean?

Solution Since the original data set consists of the 6 values

3, 3, 4, 5, 5, 5

it follows that the sample mean is

$$\bar{x} = \frac{3 + 3 + 4 + 5 + 5 + 5}{6}$$

$$= \frac{3 \times 2 + 4 \times 1 + 5 \times 3}{6}$$

$$= \frac{25}{6}$$

That is, the sample mean of the number of suits sold daily is 4.25.

In Example 3.3 we have seen that when the data are arranged in a frequency table, the sample mean can be expressed as the sum of the products of the distinct values and their frequencies, all divided by the size of the data set. This result holds in general. To see this, suppose the data are given in a frequency table which lists k distinct values $x_1, x_2, \ldots, x_k$ with respective frequencies $f_1, f_2, \ldots, f_k$. It follows that the data set consists of n observations, where $n = \sum_{i=1}^{k} f_i$ and where the value x_i appears f_i times for $i = 1, 2, \ldots, k$. Hence, the sample mean for this data set is

$$\bar{x} = \frac{x_1 + \cdots + x_1 + x_2 + \cdots + x_2 + \cdots + x_k + \cdots + x_k}{n}$$

$$= \frac{f_1 x_1 + f_2 x_2 + \cdots + f_k x_k}{n}$$

(3.1)

Now, if $w_1, w_2, \ldots, w_k$ are nonnegative numbers that sum to 1, then

$$w_1 x_1 + w_2 x_2 + \cdots + w_k x_k$$

is said to be a *weighted average* of the values $x_1, x_2, \ldots, x_k$ with w_i being the weight of x_i. For instance, suppose that $k = 2$. Now, if $w_1 = w_2 = 1/2$, then the weighted average

$$w_1 x_1 + w_2 x_2 = \frac{1}{2} x_1 + \frac{1}{2} x_2$$

is just the ordinary average of x_1 and x_2. On the other hand, if $w_1 = 2/3$ and $w_2 = 1/3$, then the weighted average

$$w_1 x_1 + w_2 x_2 = \frac{2}{3} x_1 + \frac{1}{3} x_2$$

gives twice as much weight to x_1 as it does to x_2.

By writing Eq. (3.1) as

$$\bar{x} = \frac{f_1}{n} x_1 + \frac{f_2}{n} x_2 + \cdots + \frac{f_k}{n} x_k$$

we see that the sample mean $\bar{x}$ is a weighted average of the set of distinct values. The weight given to the value x_i is f_i/n, the proportion of the data values that is equal to x_i. Thus, for instance, in Example 3.3 we could have written that

$$\bar{x} = \frac{2}{6} \times 3 + \frac{1}{6} \times 4 + \frac{3}{6} \times 5 = \frac{25}{6}$$

Example 3.4   In a paper entitled "The Effects of Helmet Use on the Severity of Head Injuries in Motorcycle Accidents" (published in the *Journal of the American Statistical Association,* 1992, pp. 48–56), A. Weiss analyzed a sample of 770 similar motorcycle accidents that occurred in the Los Angeles area in 1976 and 1977. Each accident was classified according to the severity of the head injury suffered by the motorcycle operator. The classification used was as follows:

Classification of accident	Interpretation
0	No head injury
1	Minor head injury
2	Moderate head injury
3	Severe, not life-threatening
4	Severe and life-threatening
5	Critical, survival uncertain at time of accident
6	Fatal

In 331 of the accidents the operator wore a helmet, whereas in the other 439 accidents the operator did not. The following are frequency tables giving the severities of the accidents that occurred when the operator was wearing and was not wearing a helmet.

Classification	Frequency of driver with helmet	Frequency of driver without helmet
0	248	227
1	58	135
2	11	33
3	3	14
4	2	3
5	8	21
6	1	6
	331	439

Find the sample mean of the head severity classifications for both those operators who wore helmets and those who did not.

Solution The sample mean for those wearing helmets is

$$\bar{x} = \frac{0 \cdot 248 + 1 \cdot 58 + 2 \cdot 11 + 3 \cdot 3 + 4 \cdot 2 + 5 \cdot 8 + 6 \cdot 1}{331}$$

$$= \frac{143}{331} = .432$$

The sample mean for those who did not wear a helmet is

$$\bar{x} = \frac{0 \cdot 227 + 1 \cdot 135 + 2 \cdot 33 + 3 \cdot 14 + 4 \cdot 3 + 5 \cdot 21 + 6 \cdot 6}{439}$$

$$= \frac{396}{439} = .902$$

Therefore, the data indicate that those cyclists who were wearing a helmet suffered, on average, less severe head injuries than those who were not wearing a helmet.

3.2.1   Deviations

Again suppose that sample data consist of the n values $x_1, \ldots, x_n$ and that $\bar{x} = \sum_{i=1}^{n} x_i/n$ is the sample mean. The differences between each of the data values and the sample mean are called *deviations*.

Definition

The *deviations* are the differences between the data values and the sample mean. The value of the ith deviation is $x_i - \bar{x}$.

A useful identity is that the sum of all the deviations must equal 0. That is,

$$\sum_{i=1}^{n} (x_i - \bar{x}) = 0$$

That the above equality is true is seen by the following argument.

$$\sum_{i=1}^{n} (x_i - \bar{x}) = \sum_{i=1}^{n} x_i - \sum_{i=1}^{n} \bar{x}$$
$$= n\bar{x} - n\bar{x}$$
$$= 0$$

The above equality states that the sum of the positive deviations from the sample mean must exactly balance the sum of the negative deviations. In physical terms, this means that if n weights of equal mass are placed on a (weightless) rod at the points $x_i, i = 1, \ldots, n$, then $\bar{x}$ is the point at which the rod will be in balance. This balancing point is called the *center of gravity* (see Fig. 3.1).

Figure 3.1
The center of gravity of 0, 1, 2, 6, 10, 11 is $(0 + 1 + 2 + 6 + 10 + 11)/6$ $= 30/6 = 5$.

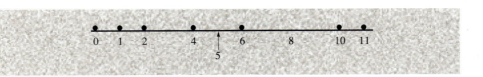

Historical Perspective

In the early days of sea voyages it was quite common for large portions of a ship's cargo to be either lost or damaged due to storms. To handle this potential loss, there was a standard agreement that all those having merchandise aboard the ship would contribute to pay for the value of all lost or damaged goods. The amount of money that each of them was called upon to pay was known as *havaria*, and from this Latin word derives our present word *average*. [Typically, if there were n shippers having damages $x_1, \ldots, x_n$, then the total loss was $x_1 + \cdots + x_n$ and the havaria for each was $(x_1 + \cdots + x_n)/n$.]

Example 3.5   Given the data of Example 3.1, the deviations from the sample mean of 28.08 gallons of soft drink consumed per capita are

$$x_1 - \bar{x} = 26.9 - 28.08 = -1.18$$
$$x_2 - \bar{x} = 26.9 - 28.08 = -1.18$$
$$x_3 - \bar{x} = 27.2 - 28.08 = -.88$$
$$x_4 - \bar{x} = 29.1 - 28.08 = 1.02$$
$$x_5 - \bar{x} = 30.3 - 28.08 = 2.22$$

As a check, we note that the sum of the deviations is

$$-1.18 - 1.18 - .88 + 1.02 + 2.22 = 0$$

Problems

1. The following data represent the scores on a statistics examination of a sample of students:

   87, 63, 91, 72, 80, 77, 93, 69, 75, 79, 70, 83, 94, 75, 88

   What is the sample mean?

2. The following data (from U.S. Department of Agriculture, *Food Consumption, Prices, and Expenditures*) give the U.S. per capita consumption (in gallons) of milk in a sample of years.

Year	1970	1975	1982	1985	1987
Per capita consumption	31.2	29.5	26.4	26.4	25.9

   Find the sample mean of the above data.

3. The following data give the annual average number of inches of precipitation and the average number of days of precipitation in a sample of cities.

City	Average amount of precipitation	Average number of days
Albany, NY	35.74	134
Baltimore, MD	31.50	83
Casper, WY	11.43	95
Denver, CO	15.31	88
Fargo, ND	19.59	100
Houston, TX	44.76	105
Knoxville, TN	47.29	127
Los Angeles, CA	12.08	36
Miami, FL	57.55	129
New Orleans, LA	59.74	114
Pittsburgh, PA	36.30	154
San Antonio, TX	29.13	81
Wichita, KS	28.61	85

Source: National Oceanic and Atmospheric Administration.

(a) Find the sample mean of the average number of inches of precipitation.
(b) Find the sample mean of the average number of days of precipitation.

4. Consider five numbers. Suppose the mean of the first four numbers is 14.
(a) If the fifth number is 24, what is the mean of all five numbers?
(b) If the mean of all five numbers is 24, what is the fifth number?

5. The following data, taken from the *1993 Statistical Abstract of the United States*, give the number of law enforcement officers who were killed in the United States in each year from 1979 to 1990. Find the sample mean of these data.

   164, 165, 157, 164, 152, 147, 148, 131, 147, 155, 145, 132

6. Suppose that the sample mean of a set of 10 data points is $\bar{x} = 20$.
(a) If it is discovered that a data point having value 15 was incorrectly read as having value 13, what should be the revised value of $\bar{x}$?
(b) Suppose there is an additional data point whose value is 22. Will this increase or decrease the value of $\bar{x}$?
(c) Using the original data [and not the revised data in part (a)], what is the new value of $\bar{x}$ in part (b)?

7. The following table lists the yearly number of reported cases of tetanus within the United States in a sample of years. Find the sample mean.

Year	1970	1975	1980	1982	1984	1985	1987
Number of cases	148	102	95	88	74	83	48

Source: U.S. Center for Disease Control, *Summary of Notifiable Diseases, Morbidity and Mortality,* September 1988

8. The following stem-and-leaf plot portrays the most recent 15 league bowling scores of the author of this text. Compute the sample mean.

```
18 | 2, 4, 7
17 | 0
16 | 1, 9
15 | 2, 2, 4, 8, 8
14 |
13 | 2, 1, 5, 5
```

9. Find the sample mean for this data set:

   1, 2, 4, 7, 10, 12

   Now find the sample means for the data sets

   3, 6, 12, 21, 30, 36

   and

   6, 7, 9, 12, 15, 17

10. Suppose that $\bar{x}$ is the sample mean of the data set consisting of the data x_1, ..., x_n. If the data are transformed according to the formula

    $$y_i = ax_i + b \qquad i = 1, \ldots, n$$

    what is the sample mean of the data set $y_1, \ldots, y_n$? (In the above equation, a and b are given constants.)

11. The following data give the total number of fires (in units of 1000 fires) reported to fire departments in the years from 1980 to 1986.

Year	1980	1981	1982	1983	1984	1985	1986
Number	2988	2893	2538	2327	2343	2371	2271

Source: "A Look at Fire Loss in the United States during 1986," Fire Journal, September 1987.

Thus, for instance, 2,988,000 fires were reported in 1980. Find the sample mean of the above data set.

12. The data set below specifies the total number of passenger cars, made both in and outside the United States, that were sold in the United States over a sample of years. The data are in units of 1000 cars. Find the sample mean of the number of cars sold annually in these years.

Year	1975	1980	1983	1985	1987	1988
Number sold	8640	8979	9182	11,042	10,278	10,626

Source: *Statistical Abstract of the United States, 1990.*

13. One-half the values of a sample are equal to 10, and the other half are equal to 20. What is the sample mean?

14. The following is a frequency table of the ages of a sample of members of a symphony for young adults.

Age value	Frequency
16	9
17	12
18	15
19	10
20	8

Find the sample mean of the above ages.

15. Half the values of a sample are equal to 10, one-sixth are equal to 20, and one-third are equal to 30. What is the sample mean?

16. There are two entrances to a parking lot. Student 1 counts the daily number of cars that pass through entrance 1, and student 2 does the same for entrance 2. Over 30 days, the data of student 1 yielded a sample mean of 122, and the data of student 2 yielded a sample mean of 160. Over these 30 days, what was the daily average number of cars that entered the parking lot?

17. A company runs two manufacturing plants. A sample of 30 engineers at plant 1 yielded a sample mean salary of $33,600. A sample of 20 engineers at plant 2 yielded a sample mean salary of $42,400. What is the sample mean salary for all 50 engineers?

18. Suppose that we have two distinct samples of sizes n_1 and n_2. If the sample mean of the first sample is $\bar{x}_1$ and that of the second is $\bar{x}_2$, what is the sample mean of the combined sample of size $n_1 + n_2$?

19. Find the deviations for each of the three data sets of Prob. 9, and verify your answers by showing that in each case the sum of the deviations is 0.

20. Calculate the deviations for the data of Prob. 14 and check that they sum to 0.

3.3 SAMPLE MEDIAN

The following data represent the number of weeks after completion of a learn-to-drive course that it took a sample of seven people to obtain a driver's license:

2, 110, 5, 7, 6, 7, 3

The sample mean of this data set is $\bar{x} = 140/7 = 20$; and so six of the seven data values are quite a bit less than the sample mean, and the seventh is much greater. This points out a weakness of the sample mean as an indicator of the center of a data set— namely, its value is greatly affected by extreme data values.

A statistic which is also used to indicate the center of a data set but which is not affected by extreme values is the *sample median,* defined as the middle value when the data are ranked in order from smallest to largest. We will let *m* denote the sample median.

Definition
Order the data values from smallest to largest. If the number of data values is odd, then the *sample median* is the middle value in the ordered list; if it is even, then the *sample median* is the average of the two middle values.

It follows from the above definition that if there are three data values, then the sample median is the second smallest value; and if there are four, then it is the average of the second and the third smallest values.

Example 3.6 The following data represent the number of weeks it took seven individuals to obtain their driver's licenses. Find the sample median.

2, 110, 5, 7, 6, 7, 3

Solution First arrange the data in increasing order.

2, 3, 5, 6, 7, 7, 110

Since the sample size is 7, it follows that the sample median is the fourth-smallest value. That is, the sample median number of weeks it took to obtain a driver's license is $m = 6$ weeks.

Example 3.7 The following data represent the number of days it took 6 individuals to quit smoking after completing a course designed for this purpose.

1, 2, 3, 5, 8, 100

What is the sample median?

Solution Since the sample size is 6, the sample median is the average of the two middle values; thus,

$$m = \frac{3+5}{2} = 4$$

That is, the sample median is 4 days.

In general, for a data set of n values, the sample median is the $(n + 1)/2$ smallest value when n is odd and is the average of the $n/2$ smallest value and the $n/2 + 1$ smallest value when n is even.

The sample mean and sample median are both useful statistics for describing the central tendency of a data set. The sample mean, being the arithmetic average, makes use of all the data values. The sample median, which makes use of only one or two middle values, is not affected by extreme values.

Example 3.8 The following data give the names of the National Basketball Association (NBA) individual scoring champions and their season scoring averages in each of the seasons from 1953 to 1991.

Season	Player, team	Average
1953–54	Neil Johnston, Philadelphia Warriors	24.4
1954–55	Neil Johnston, Philadelphia Warriors	22.7
1955–56	Bob Pettit, St. Louis Hawks	25.7
1956–57	Paul Arizin, Philadelphia Warriors	25.6
1957–58	George Yardley, Detroit Pistons	27.8
1958–59	Bob Pettit, St. Louis Hawks	29.2
1959–60	Wilt Chamberlain, Philadelphia Warriors	37.6
1960–61	Wilt Chamberlain, Philadelphia Warriors	38.4
1961–62	Wilt Chamberlain, Philadelphia Warriors	50.4
1962–63	Wilt Chamberlain, San Francisco Warriors	44.8
1963–64	Wilt Chamberlain, San Francisco Warriors	36.9
1964–65	Wilt Chamberlain, San Francisco Warriors–Phila. 76ers	34.7
1965–66	Wilt Chamberlain, Philadelphia 76ers	33.5
1966–67	Rick Barry, San Francisco Warriors	35.6

Season	Player, team	Average
1967–68	Dave Bing, Detroit Pistons	27.1
1968–69	Elvin Hayes, San Diego Rockets	28.4
1969–70	Jerry West, Los Angeles Lakers	31.2
1970–71	Lew Alcindor, Milwaukee Bucks	31.7
1971–72	Kareem Abdul-Jabbar, Milwaukee Bucks	34.8
1972–73	Nate Archibald, Kansas City–Omaha Kings	34.0
1973–74	Bob McAdoo, Buffalo Braves	30.8
1974–75	Bob McAdoo, Buffalo Braves	34.5
1975–76	Bob McAdoo, Buffalo Braves	31.1
1976–77	Pete Maravich, New Orleans Jazz	31.1
1977–78	George Gervin, San Antonio Spurs	27.2
1978–79	George Gervin, San Antonio Spurs	29.6
1979–80	George Gervin, San Antonio Spurs	33.1
1980–81	Adrian Dantley, Utah Jazz	30.7
1981–82	George Gervin, San Antonio Spurs	32.3
1982–83	Alex English, Denver Nuggets	28.4
1983–84	Adrian Dantley, Utah Jazz	30.6
1984–85	Bernard King, New York Knicks	32.9
1985–86	Dominique Wilkins, Atlanta Hawks	30.3
1986–87	Michael Jordan, Chicago Bulls	37.1
1987–88	Michael Jordan, Chicago Bulls	35.0
1988–89	Michael Jordan, Chicago Bulls	32.5
1989–90	Michael Jordan, Chicago Bulls	33.6
1990–91	Michael Jordan, Chicago Bulls	31.5
1991–92	Michael Jordan, Chicago Bulls	30.1

(a) Find the sample median of the scoring averages.

(b) Find the sample mean of the scoring averages.

Eliminate the seasons starting with 1961 and 1962, when Wilt Chamberlain averaged 50.4 and 44.8 points per game, respectively, and find the

(c) Sample median

(d) Sample mean

Solution

(a) Since there are 39 data values, the sample median is the 20th smallest. There are 11 values in the 20s, and thus the sample median is the 9th smallest value when we eliminate all values under 30. Ordering these remaining values gives

$$30.1, 30.3, 30.6, 30.7, 30.8, 31.1, 31.1, 31.2, 31.5, \ldots$$

Therefore, the sample median is

$$m = 31.5$$

(b) The sum of all 39 values is 1253.8, and so the sample mean is

$$\bar{x} = \frac{1253.8}{39} \approx 32.15$$

(c) If we eliminate the two years specified, then the sample median is the 19th smallest of the remaining 37 values. By the ordering given in (b) which starts at the 12th smallest, we obtain that the sample median is now

$$m = 31.2$$

(d) Eliminating the two years reduces the sum of all the data values to

$$1253.8 - 50.4 - 44.8 = 1158.6$$

Thus, the sample mean is now

$$\bar{x} = \frac{1158.6}{37} = 31.31$$

Thus, we see that eliminating the two largest values in the data set has only a relatively small effect on the sample median, reducing it from 31.50 to 31.20, but a larger effect on the sample mean, reducing it from 32.15 to 31.31.

For data sets that are roughly symmetric about their central values, the sample mean and sample median will have values close to each other. For instance, the data

4, 6, 8, 8, 9, 12, 15, 17, 19, 20, 22

are roughly symmetric about the value 12, which is the sample median. The sample mean is $\bar{x} = 140/11 = 12.73$, which is close to 12.

The question as to which of the two summarizing statistics is the more informative depends on what you are interested in learning from the data set. For instance, if a city government has a flat-rate income tax and is trying to figure out how much income it can expect, then it would be more interested in the sample mean of the income of its citizens than in the sample median (why is this?). On the other hand, if the city government were planning to construct some middle-income housing and were interested in the proportion of its citizens who would be able to afford such housing, then the sample median might be more informative (why is this?).

Although it is interesting to consider whether the sample mean or sample median is more informative in a particular situation, note that we need never restrict ourselves to a knowledge of just one of these quantities. They are both important, and thus both should always be computed when a data set is summarized.

Problems

1. The following are the total yardages of a sample of 12 municipal golf courses.

    7040, 6620, 6050, 6300, 7170, 5990, 6330, 6780, 6540, 6690, 6200, 6830

   (a) Find the sample median.
   (b) Find the sample mean.

2. (a) Determine the sample median of the data set

      14, 22, 8, 19, 15, 7, 8, 13, 20, 22, 24, 25, 11, 9, 14

   (b) Increase each value in (a) by 5, and find the new sample median.
   (c) Multiply each value in (a) by 3, and find the new sample median.

3. If the median of the data set x_i, $i = 1, \ldots, n$, is 10, what is the median of the data set $2x_i + 3$, $i = 1, \ldots, n$?

4. The following are the speeds of 40 cars as measured by a radar device on a city street.

      22, 26, 31, 38, 27, 29, 33, 40, 36, 27, 25, 42, 28, 19, 28, 26, 33, 26, 37, 22, 31, 30, 44, 29, 25, 17, 46, 28, 31, 29, 40, 38, 26, 43, 45, 21, 29, 36, 33, 30

   Find the sample median.

5. The following presents the male and female suicide rates per 100,000 population for a variety of countries.

Suicide rates per 100,000 population					
Sex	**United States**	**Australia**	**Austria**	**Canada**	**Denmark**	**France**
Female	5.4	5.1	15.8	5.4	20.6	12.7
Male	19.7	18.2	42.1	20.5	35.1	33.1

Sex	**Italy**	**Japan**	**Netherlands**	**Poland**	**Sweden**	**U.K.**	**W. Germany**
Female	4.3	14.9	8.1	4.4	11.5	5.7	12.0
Male	11.0	27.8	14.6	22.0	25.0	12.1	26.6

Source: World Health Organization, *World Health Statistics.*

(a) Find the sample median of the male suicide rates.
(b) Find the sample median of the female suicide rates.
(c) Find the sample mean of the male suicide rates.
(d) Find the sample mean of the female suicide rates.

6. Find the sample median of the average annual number of days of precipitation in the cities noted in Prob. 3 of Sec. 3.2.

7. Find the sample median of the average annual number of inches of precipitation in the cities noted in Prob. 3 of Sec. 3.2.

8. Use the data presented in Prob. 8 of Sec. 2.3 to find the sample median of the per capita property tax in each of the 50 states and the District of Columbia.

9. Use the table on death rates preceding Prob. 9 of Sec. 2.3 to find the sample median of the death rates due to
   (a) Cerebrovascular disease
   (b) Suicide
   (c) Ischemic heart disease

10. The sample median of 10 distinct values is 5. What can you say about the new sample median if
    (a) An additional datum whose value is 7 is added to the data set?
    (b) Two additional data values—3 and 42—are added to the data set?

11. The histogram in the figure on the facing page describes the annual rainfall, in inches, over the last 34 years in a certain western city. Since the raw data are not recoverable from a histogram, we cannot use them to exactly compute the value of the sample mean and sample median. Still, based on this histogram, what is the largest possible value of
    (a) The sample mean?
    (b) The sample median?

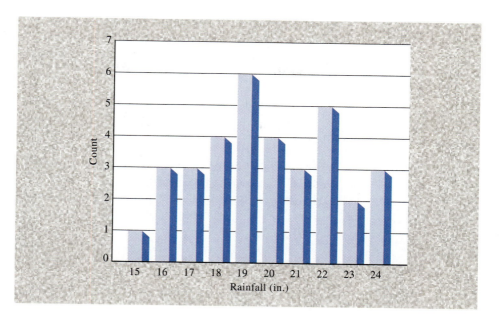

Annual rainfall.

What is the smallest possible value of
(c) The sample mean?
(d) The sample median?
(e) The actual data follow:

> 15.2, 16.1, 16.5, 16.7, 17.2, 17.5, 17.7, 18.3, 18.6, 18.8, 18.9, 19.1,
> 19.2, 19.2, 19.6, 19.8, 19.9, 20.2, 20.3, 20.3, 20.8, 21.1, 21.4, 21.7,
> 22.2, 22.5, 22.5, 22.7, 22.9, 23.3, 23.6, 24.1, 24.5, 24.9

Determine the sample mean and sample median and see that they are consistent with your previous answers.

12. The following were the high/low temperatures (in degrees Fahrenheit) in various cities on July 4, 1993.

City	High/low temperature on July 4, 1993
Atlanta	96/75
Boise	75/53
Cleveland	90/68
Jacksonville	95/75
Norfolk	89/73
Providence	89/68
Rochester	85/59
Seattle	68/55
Toledo	93/71
Wilmington	95/71

Source: *The New York Times*, July 5, 1993.

(a) Find the sample median of the high temperatures.

(b) Find the sample median of the low temperatures.

(c) Find the sample median of the differences between the high and low temperatures.

13. Use the data from Example 3.8 to find the sample median and sample mean of the leading NBA scoring averages when the 1961–1962 NBA season is removed from the data set.

14. In the following situations, which do you think is a more informative statistic, the sample mean or sample median?

(a) In order to decide whether to discontinue a bus service from Rochester to New York City, an executive studies the number of riders on a sample of days.

(b) To determine how present-day college-bound students compare with those of earlier years, a sample of entrance examination scores from several years is consulted.

(c) A lawyer representing a defendant in a jury trial is studying the IQ scores of the jurors who were selected.

(d) You purchased your home 6 years ago in a small suburban community for $105,000, which was both the mean and the median price for all homes sold that year in that community. However, in the last couple of years some new, more expensive homes have been built. To get an idea of the present value of your home, you study recent sales prices of homes in your community.

15. Women make up the following percentages of the workforce in the 14 occupations listed.

Occupation	Percentage women	Occupation	Percentage women
Corporate executives	36.8	Doctors	17.6
Nurses	94.3	Lawyers	18.0
Sales supervisors	30.5	Elementary school teachers	85.2
Sales workers	68.6	Postal clerks	43.5
Firefighters	1.9	Police workers	10.9
Cleaning jobs	41.5	Construction supervisors	1.6
Construction workers	2.8	Truck drivers	2.1

For these percentages find

(a) The sample mean

(b) The sample median

It also turns out that women make up 44.4 percent of the total workforce for these occupations. Is this consistent with your answers in (a) and (b)? Explain!

16. Using data concerning the first 30 students in App. A, find the sample median and the sample mean for
    (a) Weight
    (b) Cholesterol
    (c) Blood pressure

17. The following lists the yearly number of reported cases in the United States of leprosy for a sample of years.

Year	1970	1975	1980	1981	1982	1983	1984	1985	1986	1987
Cases	129	162	223	256	250	259	290	361	270	238

Source: U.S. Center for Disease Control, *Summary of Notifiable Diseases, United States, Morbidity and Mortality.*

    (a) Find the sample median.
    (b) Find the sample mean.

18. Using the data from Example 3.4, compute the sample medians of the severity of head injuries suffered by motorcycle operators who were wearing and who were not wearing helmets.

3.3.1 Sample Percentiles

The sample median is a special type of statistic known as a *sample 100p percentile,* where p is any fraction between 0 and 1. Loosely speaking, a sample $100p$ percentile is the value such that $100p$ percent of the data values are less than it and $100(1 - p)$ percent of the values are greater than it.

Definition

The *sample 100p percentile* is that data value having the property that at least $100p$ percent of the data are less than or equal to it and at least $100(1 - p)$ percent of the data values are greater than or equal to it. If two data values satisfy this condition, then the sample $100p$ percentile is the arithmetic average of these values.

Note that the sample median is the sample 50th percentile. That is, it is the sample $100p$ percentile when $p = .50$.

Suppose the data from a sample of size n are arranged in increasing order from smallest to largest. To determine the sample $100p$ percentile, we must determine the data value such that

1. At least np of the data values are less than or equal to it.

2. At least $n(1 - p)$ of the data values are greater than or equal to it.

Now if np is not an integer, then the only data value satisfying these requirements is the one whose position is the smallest integer greater than np. For instance, suppose we want the sample 90th percentile from a sample of size $n = 12$. Since $p = .9$, we have $np = 10.8$ and $n(1 - p) = 1.2$. Thus, we require those data values for which

1. At least 10.8 values are less than or equal to it (and so the data value must be in position 11 or higher).

2. At least 1.2 values are greater than or equal to it (and so it must be in position 11 or lower).

Clearly, the only data value that satisfies both requirements is the one that is in position 11, and thus this is the sample 90th percentile.

On the other hand, if np *is* an integer, then both the value in position np and the value in position $np + 1$ satisfy requirements (1) and (2); and so the sample $100p$ percentile value would be the average of these two data values. For instance, suppose we wanted the sample 95th percentile from a data set of $n = 20$ values. Then both the 19th and the 20th values (that is, the two largest values) will be greater than or equal to at least $np = 20(.95) = 19$ of the values and less than or equal to at least $n(1 - p) = 1$ value. The 95th percentile is thus the average of the 19th and 20th largest values.

Summing up, we have shown the following:

To find the sample $100p$ percentile of a data set of size n

1. Arrange the data in increasing order.

2. If np is not an integer, determine the smallest integer greater than np. The data value in that position is the sample $100p$ percentile.

3. If np is an integer, then the average of the values in positions np and $np + 1$ is the sample $100p$ percentile.

Example 3.9   Which data value is the sample 90th percentile when the sample size is (a) 8, (b) 16, and (c) 100?

Solution

(a) Since $.9 \times 8 = 7.2$, which is not an integer, it follows that if the data are arranged from smallest to largest, then the sample 90th percentile value would be the 8th smallest value (that is, the largest value).

(b) Since .9 × 16 = 14.4, which is not an integer, it follows that the sample 90th percentile would be the 15th smallest value.

(c) Since .9 × 100 = 90 is an integer, the sample 90th percentile value is the average of the 90th and the 91st values when the data are arranged from smallest to largest.

Example 3.10   Table 3.1 lists the 50 largest U.S. based industrial corporations in 1990, based on yearly sales. The companies along with their 1990 sales figures (in millions of dollars) are listed. Using the data on the yearly sales figures, find the

(a) Sample 90th percentile

(b) Sample 20th percentile

Solution

(a) Since the sample size is 50 and 50 × .9 = 45, the sample 90th percentile is the average of the 45th and 46th smallest sales figures. Equivalently, it is the average of the 6th and 5th largest values. Hence, the sample 90th percentile is

Table 3.1   **Largest Industrial Corporations, 1990**

Corporation	1990 sales figures (in millions of dollars)	Corporation	1990 sales figures (in millions of dollars)
General Motors	126,017.0	Tenneco	14,893.0
Exxon	105,885.0	Phillips Petroleum	14,032.0
Ford Motor	98,274.7	RJR Nabisco Holdings	13,879.0
International Business Machines	69,018.0	Hewlett-Packard	13,233.0
Mobil	58,770.0	Digital Equipment	13,084.5
General Electric	58,414.0	Minnesota Mining & Mfg.	13,021.0
Philip Morris	44,323.0	International Paper	12,960.0
Texaco	41,235.0	Westinghouse Electric	12,915.0
E. I. du Pont de Nemours	39,839.0	Georgia-Pacific	12,665.0
Chevron	39,262.0	Rockwell International	12,442.5
Chrysler	30,868.0	Allied-Signal	12,396.0
Amoco	28,277.0	Sun	11,909.0
Boeing	27,595.0	Sara Lee	11,652.0
Shell Oil	24,423.0	Caterpillar	11,540.0
Procter & Gamble	24,376.0	Goodyear Tire & Rubber	11,453.1
Occidental Petroleum	21,947.0	Johnson & Johnson	11,232.0
United Technologies	21,783.2	Motorola	10,885.0
Dow Chemical	20,005.0	Aluminum Co. of America	10,865.1
USX	19,462.0	Anheuser-Busch	10,750.6
Eastman Kodak	19,075.0	Unocal	10,740.0
Atlantic Richfield	18,819.0	Bristol-Myers Squibb	10,509.0
Xerox	18,382.0	Coca-Cola	10,406.3
Pepsico	17,802.7	General Dynamics	10,182.0
McDonnell Douglas	16,351.0	Unisys	10,111.3
Conagra	15,517.7	Lockheed	9,977.0

Source: Fortune magazine, 1991.

$$\frac{58,770.0 + 58,414.0}{2} = 58,592$$

That is, the sample 90th percentile of the sales figures is $58,592,000,000.

(b) Since $50 \times .2 = 10$, the sample 20th percentile is the average of the 10th and 11th smallest values on the list. Hence,

$$\text{Sample 20th percentile} = \frac{11,232 + 11,453.1}{2} = 11,342.55$$

That is, the sample 20th percentile of the sales figures is $11,342,550,000.

The sample 25th percentile, 50th percentile, and 75th percentile are known as the *quartiles.*

Definition

The sample 25th percentile is called the *first quartile.* The sample 50th percentile is called the *median,* or the *second quartile.* The sample 75th percentile is called the *third quartile.*

The quartiles break up a data set into four parts with about 25 percent of the data values being less than the first quartile, about 25 percent being between the first and second quartiles, about 25 percent being between the second and third quartiles, and about 25 percent being larger than the third quartile.

Example 3.11 Find the sample quartiles for the following 18 data values, which represent the ordered values of a sample of scores from a league bowling tournament:

122, 126, 133, 140, 145, 145, 149, 150, 157, 162, 166, 175, 177, 177, 183, 188, 199, 212

Solution Since $.25 \times 18 = 4.5$, the sample 25th percentile is the fifth smallest value, which is 145.

Since $.50 \times 18 = 9$, the second quartile (or sample median) is the average of the 9th and 10th smallest values and so is

$$\frac{157 + 162}{2} = 159.5$$

Since $.75 \times 18 = 13.5$, the third quartile is the 14th smallest value, which is 177.

3.3.1

Problems

1. Seventy-five values are arranged in increasing order. How would you determine the sample
   (a) 80th percentile
   (b) 60th percentile
   (c) 30th percentile

   of this data set?

2. The following table, taken from *Fortune* magazine, lists the 1990 sales figures in millions of dollars for the 25 largest retailing companies in the United States.

25 Largest Retailing Companies	Sales
Sears Roebuck	$55,971.7
Wal-Mart Stores	32,601.6
Kmart	32,080.0
American Stores	22,155.5
Kroger	20,261.0
J.C. Penney	17,410.0
Safeway	14,873.6
Dayton Hudson	14,739.0
Great Atlantic & Pacific Tea	11,164.2
May Department Stores	11,027.0
Woolworth	9,789.0
Winn-Dixie	9,744.5
Melville	8,686.8
Albertson's	8,218.6
Southland	8,037.1
R.H. Macy	7,266.8
McDonald's	6,639.6
Supermarkets General Holdings	6,126.0
Walgreen	6,063.0
Publix Super Markets	5,820.7
Food Lion	5,584.5
Toys "R" Us	5,521.2
Price	5,428.8
Vons	5,333.9
The Limited	5,253.5

In terms of these sales figures, find the
(a) Sample 90th percentile
(b) Sample 80th percentile
(c) Sample median
(d) Sample 10th percentile

3. Consider a data set of n values 1, 2, 3, . . . , n. Find the value of the sample 95th percentile when
   (a) $n = 100$
   (b) $n = 101$

The following table gives the number of physicians and of dentists per 100,000 population for 12 midwestern states in 1986. Problems 4 and 5 are based on it.

State	Physician's rate	Dentist's rate
Ohio	188	56
Indiana	146	48
Illinois	206	61
Michigan	177	64
Wisconsin	177	70
Minnesota	207	70
Iowa	141	60
Missouri	186	55
North Dakota	157	55
South Dakota	129	54
Nebraska	162	71
Kansas	166	52

Source: American Medical Association, *Physician Characteristics and Distribution in the U.S.*

4. For the physician's rates per 100,000 population, find the
   (a) Sample 40th percentile
   (b) Sample 60th percentile
   (c) Sample 80th percentile

5. For the dentist's rates per 100,000 population, find the
   (a) Sample 90th percentile
   (b) Sample 50th percentile
   (c) Sample 10th percentile

6. Suppose the sample $100p$ percentile of a set of data is 120. If we add 30 to each data value, what is the new value of the sample $100p$ percentile?

7. Suppose the sample $100p$ percentile of a set of data is 230. If we multiply each data value by a positive constant c, what is the new value of the sample $100p$ percentile?

8. Find the sample 90th percentile of this data set:

   75, 33, 55, 21, 46, 98, 103, 88, 35, 22, 29, 73, 37, 101, 121, 144, 133, 52, 54, 63, 21, 7

9. Consider data concerning the first 50 students listed in App. A. Find (i) the sample 90th percentile and (ii) the sample 95th percentile for
   (a) Weight
   (b) Blood cholesterol
   (c) Blood pressure

10. The following table lists the 1990 average per capita income by states (including the District of Columbia).

State	Per capita income, 1990	State	Per capita income, 1990
Alabama	$14,826	Montana	$15,110
Alaska	21,761	Nebraska	17,221
Arizona	16,297	Nevada	19,416
Arkansas	14,218	New Hampshire	20,789
California	20,795	New Jersey	24,968
Colorado	18,794	New Mexico	14,228
Connecticut	25,358	New York	21,975
Delaware	20,039	North Carolina	16,203
D.C.	24,181	North Dakota	15,255
Florida	18,586	Ohio	17,473
Georgia	16,944	Oklahoma	15,444
Hawaii	20,254	Oregon	17,156
Idaho	15,160	Pennsylvania	18,672
Illinois	20,303	Rhode Island	18,841
Indiana	16,864	South Carolina	15,099
Iowa	17,249	South Dakota	15,872
Kansas	17,986	Tennessee	15,798
Kentucky	14,929	Texas	16,759
Louisiana	14,391	Utah	14,083
Maine	17,200	Vermont	17,436
Maryland	21,864	Virginia	19,746
Massachusetts	22,642	Washington	18,858
Michigan	18,346	West Virginia	13,747
Minnesota	18,731	Wisconsin	17,503
Mississippi	12,735	Wyoming	16,398
Missouri	17,497	United States	18,685

Source: U.S. Department of Commerce, Bureau of Economic Analysis, *Survey of Current Business.*

Find the sample quartiles of this data set.

11. The following are the quartiles of a large data set:

    First quartile   = 35
    Second quartile = 47
    Third quartile   = 66

(a) Give an interval in which approximately 50 percent of the data lie.
(b) Give a value which is greater than approximately 50 percent of the data.
(c) Give a value such that approximately 25 percent of the data values are greater than it.

12. A symmetric data set has its median equal to 40 and its third quartile equal to 55. What is the value of the first quartile?

3.4 Sample Mode

Another indicator of central tendency is the *sample mode,* which is the data value that occurs most frequently in the data set.

Example 3.12   The following are the sizes of the last 8 dresses sold at a women's boutique:

8, 10, 6, 4, 10, 12, 14, 10

What is the sample mode?

Solution The sample mode is 10, since the value of 10 occurs most frequently.

If no single value occurs most frequently, then all the values that occur at the highest frequency are called *modal values.* In such a situation we say that there is no unique value of the sample mode.

Example 3.13   The ages of 6 children at a day care center are

2, 5, 3, 5, 2, 4

What are the modal values of this data set?

Solution Since the ages 2 and 5 both occur most frequently, both 2 and 5 are modal values.

It is easy to pick out the modal value from a frequency table, since it is just that value having the largest frequency.

Example 3.14   The following frequency table gives the values obtained in 30 throws of a die.

Value	Frequency
1	6
2	4
3	5
4	8
5	3
6	4

For these data, find the

(a) Sample mode

(b) Sample median

(c) Sample mean

Solution

(a) Since the value 4 appears with the highest frequency, the sample mode is 4.

(b) Since there are 30 data values, the sample median is the average of the 15th and 16th smallest values. Since the 15th smallest value is 3 and the 16th smallest is 4, the sample median is 3.5.

(c) The sample mean is

$$\bar{x} = \frac{1 \cdot 6 + 2 \cdot 4 + 3 \cdot 5 + 4 \cdot 8 + 5 \cdot 3 + 6 \cdot 4}{30} = \frac{100}{30} \approx 3.333$$

Problems

1. Match each statement in the left-hand column with the correct data set from the right-hand column.
   1. Sample mode is 9   *A:* 5, 7, 8, 10, 13, 14
   2. Sample mean is 9   *B:* 1, 2, 5, 9, 9, 15
   3. Sample median is 9   *C:* 1, 2, 9, 12, 12, 18

2. Using the data from Example 2.2, find the sample mode of the winning Masters Golf Tournament scores.

3. Using data concerning the first 100 students in App. A, find the sample mode for
   (a) Weight
   (b) Blood pressure
   (c) Cholesterol

4. Suppose you want to guess the salary of a bank vice president whom you have just met. If you want to have the greatest chance of being correct to the nearest $1000, would you rather know the sample mean, sample median, or sample mode of the salaries of bank vice presidents?

5. Construct a data set for which the sample mean is 10, the sample median is 8, and the sample mode is 6.

6. If the sample mode of the data x_i, $i = 1, \ldots, n$, is equal to 10, what is the sample mode of the data $y_i = 2x_i + 5$, $i = 1, \ldots, n$?

7. Joggers use a quarter-mile track around an athletic field. In a sample of 17 joggers, 1 did 2 loops, 4 did 4 loops, 5 did 6 loops, 6 did 8 loops, and 1 did 12 loops.
   (a) What is the sample mode of the number of loops run by these joggers?
   (b) What is the sample mode of the distances run by these joggers?

3.5 SAMPLE VARIANCE AND SAMPLE STANDARD DEVIATION

Whereas so far we have talked about statistics that measure the central tendency of a data set, we have not yet considered ones that measure its spread or variability. For instance, although data sets A and B below have the same sample mean and sample median, there is clearly more spread in the values of B than in those of A.

$$A: 1, 2, 5, 6, 6 \qquad B: -40, 0, 5, 20, 35$$

One way of measuring the variability of a data set is to consider the deviations of the data values from a central value. The most commonly used central value for this purpose is the sample mean. If the data values are $x_1, \ldots, x_n$ and the sample mean is $\bar{x} = \left(\sum_{i=1}^{n} x_i \right)/n$ then the deviation of the value x_i from the sample mean is $x_i - \bar{x}$, $i = 1, \ldots, n$.

One might suppose that a natural measure of the variability of a set of data would be the average of the deviations from the mean. However, as we have shown in Sec. 3.2, $\sum_{i=1}^{n} (x_i - \bar{x}) = 0$. That is, the sum of the deviations from the sample mean is always equal to 0, and thus the average of the deviations from the sample mean must also be 0. However, after some additional reflection it should be clear that we really do not want to allow the positive and the negative deviations to cancel. Instead, we should be concerned about the individual deviations without regard to their signs. This can be accomplished either by considering the absolute values of the deviations or, as turns out to be more useful, by considering their squares.

The sample variance is a measure of the "average" of the squared deviations from the sample mean. However, for technical reasons (which will become clear in Chap. 8) this "average" divides the sum of the n squared deviations by the quantity $n - 1$, rather than by the usual value n.

Definition

The *sample variance*, call it s^2, of the data set $x_i, \ldots, x_n$ having sample mean $\bar{x} = \left(\sum_{i=1}^{n} x_i \right)/n$, is defined by

$$s^2 = \frac{\sum_{i=1}^{n} (x_i - \bar{x})^2}{n - 1}$$

Example 3.15 Find the sample variance of data set A.

Solution It is determined as follows:

x_i	1	2	5	6	6
$\bar{x}$	4	4	4	4	4
$x_i - \bar{x}$	-3	-2	1	2	2
$(x_i - \bar{x})^2$	9	4	1	4	4

Hence, for data set A,

$$s^2 = \frac{9 + 4 + 1 + 4 + 4}{4} = 5.5$$

Example 3.16 Find the sample variance for data set B.

Solution The sample mean for data set B is also $\bar{x} = 4$. Therefore, for this set, we have

x_i	-40	0	5	20	35
$x_i - \bar{x}$	-44	-4	1	16	31
$(x_i - \bar{x})^2$	1936	16	1	256	961

Thus,

$$s^2 = \frac{3170}{4} = 792.5$$

The following algebraic identity is useful for computing the sample variance by hand:

An Algebraic Identity

$$\sum_{i=1}^{n} (x_i - \bar{x})^2 = \sum_{i=1}^{n} x_i^2 - n\bar{x}^2 \tag{3.2}$$

Example 3.17   Check that the above identity holds for data set A.

Solution Since $n = 5$ and $\bar{x} = 4$,

$$\sum_{i=1}^{5} x_i^2 - n\bar{x}^2 = 1 + 4 + 25 + 36 + 36 - 5(16) = 102 - 80 = 22$$

From Example 3.15,

$$\sum_{i=1}^{5} (x_i - \bar{x})^2 = 9 + 4 + 1 + 4 + 4 = 22$$

and so the identity checks out.

Suppose that we add a constant c to each of the data values $x_1, \ldots, x_n$ to obtain the new data set $y_1, \ldots, y_n$, where

$$y_i = x_i + c$$

To see how this affects the value of the sample variance, recall from Sec. 3.2 that

$$\bar{y} = \bar{x} + c$$

and so

$$y_i - \bar{y} = x_i + c - (\bar{x} + c) = x_i - \bar{x}$$

That is, the y deviations are equal to the x deviations, and therefore their sums of squares are equal. Thus, we have shown the following useful result.

The sample variance remains unchanged when a constant is added to each data value.

The preceding result can often be used in conjunction with the algebraic identity (3.2) to greatly reduce the time it takes to compute the sample variance.

Example 3.18 The following data give the number of law enforcement officers killed in the United States in each of the years from 1979 to 1988.

Year	1979	1980	1981	1982	1983	1984	1985	1986	1987	1988
Number killed	164	165	157	164	152	147	148	131	147	155

Source: *Statistical Abstract of the United States, 1990.*

Find the sample variance of the number killed in these years.

Solution Rather than working directly with the above data, let us subtract the value 150 from each one. (That is, we are adding $c = -150$ to each data value.) This results in the new data set

$$14, 15, 7, 14, 2, -3, -2, -19, -3, 5$$

Its sample mean is

$$\bar{y} = \frac{14 + 15 + 7 + 14 + 2 - 3 - 2 - 19 - 3 + 5}{10} = 3.0$$

The sum of the squares of the new data is

$$\sum_{i=1}^{10} y_i^2 = 14^2 + 15^2 + 7^2 + 14^2 + 2^2 + 3^2 + 2^2 + 19^2 + 3^2 + 5^2 = 1078$$

Therefore, using the algebraic identity (3.2) shows that

$$\sum_{i=1}^{10} (y_i - \bar{y})^2 = 1078 - 10(9) = 988$$

Hence, the sample variance of the revised data, which is equal to the sample variance of the original data, is

$$s^2 = \frac{988}{9} \approx 109.78$$

The positive square root of the sample variance is called the *sample standard deviation*.

Definition

The quantity s, defined by

$$s = \sqrt{\frac{\sum_{i=1}^{n} (x_i - \bar{x})^2}{n - 1}}$$

is called the *sample standard deviation*.

The sample standard deviation is measured in the same units as the original data. That is, for instance, if the data are in feet, then the sample variance will be expressed in units of square feet and the sample standard deviation in units of feet.

If each data value x_i, $i = 1, \ldots, n$, is multiplied by a constant c to obtain the new data set

$$y_i = cx_i \qquad i = 1, \ldots, n$$

then the sample variance of the y data is the sample variance of the x data multiplied by c^2. That is,

$$s_y^2 = c^2 s_x^2$$

where s_y^2 and s_x^2 are the sample variances of the new and old data sets, respectively. Taking the square root of both sides of the preceding equation shows that the standard deviation of the y data is equal to the absolute value of c times the standard deviation of the x data, or

$$s_y = |c| s_x$$

Example 3.19   The following data give the monthly high prices for sales of a seat on the New York Stock Exchange for the first 6 months of 1990.

Month	High sale price
January	$400,000
February	360,000
March	360,000
April	410,000
May	410,000
June	430,000

Source: New York Stock Exchange.

Find the sample standard deviation of the monthly high price.

Solution Letting the unit of measurement be $100,000 (or, equivalently, by multiplying each data value by 1/100,000) transforms the data to

$$4, 3.6, 3.6, 4.1, 4.1, 4.3$$

To compute the sample standard deviation of the above data, it is easiest (if you are working by hand) to subtract 4 from each value to obtain the new data set

$$0, -.4, -.4, .1, .1, .3$$

The sample mean of these values is

$$\bar{y} = \frac{-.3}{6} = -.05$$

The sum of the squares is

$$\sum_{i=1}^{6} y_i^2 = 0 + .16 + .16 + .01 + .01 + .09 = .43$$

Using identity (3.2) shows that the sample variance is

$$s_y^2 = \frac{.43 - 6(.05)^2}{5} = .083$$

Hence, the sample standard deviation is

$$s_y = \sqrt{.083} \approx .289$$

Therefore, the sample standard deviation of the original data set (before we divided by 100,000) is

$$s_x = 100,000(.289) = 28,900$$

That is, the sample standard deviation of the data consisting of the monthly high sale prices of a seat on the New York Stock Exchange is $28,900. ◾ ◾

The computation of s or s^2 may easily be performed either by hand or by using a calculator. Also, Program 3-1 on the enclosed diskette will compute the sample mean, sample variance, and sample standard deviation for a given set of data.

Example 3.20 To see how to use Program 3-1, suppose we wanted to compute the sample variance of data set A. To do so, we run Program 3-1, which yields the following:

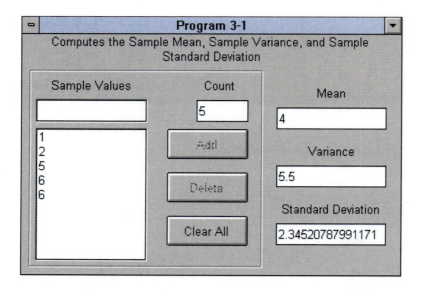

The *range* of a data set is another measure of its variability, defined as the difference between the largest and smallest values. For instance, the range of the data set

A: 1, 2, 5, 6, 6

is 6 − 1 = 5. For data set

B: −40, 0, 5, 20, 35

the range is 35 − (−40) = 75.

Another indicator of the variability of a data set is the interquartile range, which is equal to the third minus the first quartile. That is, roughly speaking, the interquartile range is the length of the interval in which the middle half of the data values lie.

Example 3.21 The Miller Analogies Test is a standardized test that is taken by a variety of students applying to graduate and professional schools. Table 3.2 presents some of the percentile scores on this examination for students, classified according to the graduate fields they are entering. For instance, Table 3.2 states that the median grade of students in the physical sciences is 68, whereas it is 49 for those applying to law school.

| Table 3.2 | **Selected Percentiles on the Miller Analogies Test for Five Categories of Students** |

Percentile	**Physical sciences**	**Medical school**	**Social sciences**	**Languages and literature**	**Law school**
99	93	92	90	87	84
90	88	78	82	80	73
75	80	71	74	73	60
50	68	57	61	59	49
25	55	45	49	43	37

Determine the interquartile ranges of the scores of students in the five specified categories.

Solution Since the interquartile range is the difference between the 75th and the 25th sample percentiles, it follows that its value is

$80 - 55 = 25$ for scores of physical science students
$71 - 45 = 26$ for scores of medical school students
$74 - 49 = 25$ for scores of social science students
$73 - 43 = 33$ for scores of language and literature students
$60 - 37 = 23$ for scores of physical science students

Problems

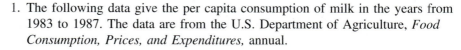

1. The following data give the per capita consumption of milk in the years from 1983 to 1987. The data are from the U.S. Department of Agriculture, *Food Consumption, Prices, and Expenditures,* annual.

Year	Amount (in gallons per capita)
1983	26.3
1984	26.2
1985	26.4
1986	26.3
1987	25.9

Find the sample mean and the sample variance of this set.

2. You are given these data sets:

   A: 66, 68, 71, 72, 72, 75   *B:*   2, 5, 9, 10, 10, 16

(a) Which one appears to have the larger sample variance?

(b) Determine the sample variance of data set *A*.

(c) Determine the sample variance of data set *B*.

3. The Masters Golf Tournament and the U.S. Open are the two most prestigious golf tournaments in the United States. The Masters is always played on the Augusta National golf course whereas the U.S. Open is played on different courses in different years. As a result, one might expect the sample variance of the winning scores in the U.S. Open to be higher than that of the winning scores in the Masters. To check whether this is so, we have collected the winning scores in both tournaments for 1981 to 1990.

	Winning score									
Tournament	1981	1982	1983	1984	1985	1986	1987	1988	1989	1990
U.S. Open	273	282	280	276	279	279	277	278	278	280
Masters	280	284	280	277	282	279	285	281	283	278

(a) Compute the sample variance of the winning scores in the U.S. Open tournament.

(b) Compute the sample variance of the winning scores in the Masters tournament.

The following table presents the number of physicians and dentists in the United States per 100,000 population in a sample of years from 1960 to 1986. Problems 4 and 5 are based on this table.

	Number per 100,000 population						
	1960	1970	1975	1980	1982	1983	1986
Physicians	151	168	187	211	222	228	246
Dentists	47	47	50	54	55	56	57

Source: U.S. Department of Health and Human Services.

4. Determine the sample variance of the yearly number of doctors per 100,000 population.

5. Determine the sample variance of the yearly number of dentists per 100,000 population.

6. An individual needing automobile insurance requested quotes from 10 different insurers for identical coverage and received the following values (amounts are annual premiums in dollars).

720, 880, 630, 590, 1140, 908, 677, 720, 1260, 800

Find

(a) The sample mean

(b) The sample median

(c) The sample standard deviation

The following table gives the number of reported cases in the United States of different diseases from the years 1980 to 1987. Problems 7, 8, and 9 refer to this table.

Disease	1980	1981	1982	1983	1984	1985	1986	1987
AIDS	(NA)	199	744	2,117	4,445	8,249	13,166	21,070
Amebiasis	5,271	6,632	7,304	6,658	5,252	4,433	3,532	3,123
Aseptic meningitis	8,028	9,547	9,680	12,696	8,326	10,619	11,374	11,487
Botulism	89	103	97	133	123	122	109	82
Brucellosis (undulant fever)	183	185	173	200	131	153	106	129
Chickenpox (1000)	190.9	200.8	167.4	177.5	222.0	178.2	183.2	213.2
Diphtheria	3	5	2	5	1	3	—	3
Encephalitis:								
Primary infectious	1,362	1,492	1,464	1,761	1,257	1,376	1,302	1,418
Post infectious	40	43	36	34	108	161	124	121
Hepatitis: B (serum) (1000)	19.0	21.2	22.2	24.3	26.1	26.6	26.1	25.9
A (infectious) (1000)	29.1	25.8	23.4	21.5	22.0	23.2	23.4	25.3
Unspecified (1000)	11.9	11.0	8.6	7.1	5.5	5.5	3.9	3.1
Non-A, non-B (1000)			2.6	3.5	3.9	4.2	3.6	3.0
Legionellosis			654	852	750	830	948	1,038
Leprosy	223	256	250	259	290	361	270	238
Leptospirosis	85	82	100	61	40	57	41	43
Malaria	2,062	1,388	1,056	813	1,007	1,049	1,123	944
Measles (1000)	13.5	3.1	1.7	1.5	2.6	2.8	6.3	3.7
Meningococcal infections	2,840	3,525	3,056	2,736	2,746	2,479	2,594	2,930
Mumps (1000)	8.6	4.9	5.3	3.4	3.0	3.0	7.8	12.8
Pertussis (1000)	1.7	1.2	1.9	2.5	2.3	3.6	4.2	2.8
Plague	18	13	19	40	31	17	10	12
Poliomyelitis, acute	9	6	8	15	8	7	8	—
Psittacosis	124	136	152	142	172	119	224	98
Rabies, animal	6,421	7,118	6,212	5,878	5,567	5,565	5,504	4,658
Rabies, human	—	2	—	2	3	1	—	1
Rheumatic fever, acute	432	264	137	88	117	90	147	141
Rubella (1000)	3.9	2.1	2.3	1.0	1.0	.6	.6	.3
Salmonellosis (1000)	33.7	40.0	40.9	44.3	40.9	65.3	50.0	50.9
Shigellosis (1000)	19.0	19.9	18.1	19.7	17.4	17.1	17.1	23.9
Tetanus	95	72	88	91	74	83	64	48
Toxic-shock syndrome				502	482	384	412	372
Trichinosis	131	206	115	45	68	61	39	40
Tuberculosis (1000)	27.7	27.4	25.5	23.8	22.3	22.2	22.8	22.5
Tularemia	234	288	275	310	291	177	170	214
Typhoid fever	510	584	425	507	390	402	362	400
Typhus fever:								
Flea-borne (endemic-murine)	81	81	58	62	53	37	67	49
Tick-borne (Rocky Mt. spotted fever)	1,163	1,192	976	1,126	838	714	760	604
Venereal diseases (civilian cases):								
Gonorrhea (1000)	1,004	991	961	900	879	911	901	781
Syphilis (1000)	69	73	76	75	70	68	68	87
Other (1000)	1.0	1.2	1.6	1.2	1.0	2.3	4.2	5.3

Source: U.S. Centers for Disease Control, *Summary of Notifiable Diseases, Morbidity and Mortality, 1988.*

7. Find the sample variance of the yearly number of cases of typhoid fever.

8. Find the sample variance of the yearly number of cases of leprosy.

9. Find the sample variance of the yearly number of cases of mumps.

10. If s^2 is the sample variance of the data x_i, $i = 1, \ldots, n$, what is the sample variance of the data $ax_i + b$, $i = 1, \ldots, n$, when a and b are given constants?

11. Compute the sample variance and sample standard deviation of the following data sets:
    (a) 1, 2, 3, 4, 5
    (b) 6, 7, 8, 9, 10
    (c) 11, 12, 13, 14, 15
    (d) 2, 4, 6, 8, 10
    (e) 10, 20, 30, 40, 50

12. On the U.S. side of the U.S.-Canada border, temperatures are measured in degrees Fahrenheit whereas on the Canadian side they are measured in degrees Celsius (also called Centigrade). Suppose that during the month of January the sample mean of the temperatures, as recorded on the U.S. side of the border, was 40°F with a sample variance of 12.

    Use the formula for converting a Fahrenheit temperature to a Celsius temperature

    $$C = \frac{5}{9}(F - 32)$$

    to find
    (a) The sample mean recorded by the Canadians
    (b) The sample variance recorded by the Canadians

13. Compute the sample mean and sample variance of the systolic blood pressures of the first 50 students of the data set of App. A. Now do the same with the last 50 students of this data set. Compare your answers. Comment on the results of this comparison. Do you find it surprising?

14. If s is the sample standard deviation of the data x_i, $i = 1, \ldots, n$, what is the sample standard deviation of $ax_i + b$, $i = 1, \ldots, n$? In this problem, a and b are given constants.

15. The following table gives the number of motorcycle retail sales in the United States for 8 different years. Use it to find the sample standard deviation of the number of motorcycle sales in the 8 years.

Year	1975	1980	1983	1984	1985	1986	1987	1988
Motorcycle sales (in thousands)	940	1070	1185	1305	1260	1045	935	710

Source: Motorcycle Industry Council.

16. Find the sample standard deviation of the data set given by the following frequency table:

Value	Frequency	Value	Frequency
3	1	5	3
4	2	6	2

17. The following data represent the acidity of 40 successive rainfalls in the state of Minnesota. The acidity is measured on a pH scale which varies from 1 (very acidic) to 7 (neutral).

  3.71, 4.23, 4.16, 2.98, 3.23, 4.67, 3.99, 5.04, 4.55, 3.24, 2.80, 3.44, 3.27, 2.66, 2.95, 4.70, 5.12, 3.77, 3.12, 2.38, 4.57, 3.88, 2.97, 3.70, 2.53, 2.67, 4.12, 4.80, 3.55, 3.86, 2.51, 3.33, 3.85, 2.35, 3.12, 4.39, 5.09, 3.38, 2.73, 3.07

  (a) Find the sample standard deviation.   (c) Find the interquartile range.
  (b) Find the range.

18. Consider the following two data sets.

  A: 4.5, 0, 5.1, 5.0, 10, 5.2   *B:*   0.4, 0.1, 9, 0, 10, 9.5

  (a) Determine the range for each data set.
  (b) Determine the sample standard deviation for each data set.
  (c) Determine the interquartile range for each data set.

3.6   NORMAL DATA SETS AND THE EMPIRICAL RULE

Many of the large data sets one encounters in practice have histograms that are similar in shape. These histograms are often symmetric about their point of highest frequency and then decrease on both sides of this point in a bell-shaped fashion. Such data sets are said to be *normal*, and their histograms are called *normal histograms*.

Definition

A data set is said to be *normal* if a histogram describing it has the following properties:

1. It is highest at the middle interval.

2. Moving from the middle interval in either direction, the height decreases in such a way that the entire histogram is bell-shaped.

3. The histogram is symmetric about its middle interval.

Figure 3.2 shows the histogram of a normal data set.

If the histogram of a data set is close to being a normal histogram, then we say that the data set is *approximately normal*. For instance, the histogram given in Fig. 3.3 is from an approximately normal data set whereas the ones presented in Fig. 3.4 and 3.5 are not (since each is too nonsymmetric). Any data set that is not approximately symmetric about its sample median is said to be *skewed*. It is called *skewed to the right* if it has a long tail to the right and *skewed to the left* if it has a long tail to the left. Thus the data set presented in Fig. 3.4 is skewed to the left, and the one of Fig. 3.5 is skewed to the right.

It follows from the symmetry of the normal histogram that a data set that is approximately normal will have its sample mean and sample median approximately equal.

Suppose that $\bar{x}$ and s are the sample mean and sample standard deviation, respectively, of an approximately normal data set. The following rule, known as the *empirical rule,* specifies the approximate proportions of the data observations that are within s, $2s$, and $3s$ of the sample mean $\bar{x}$.

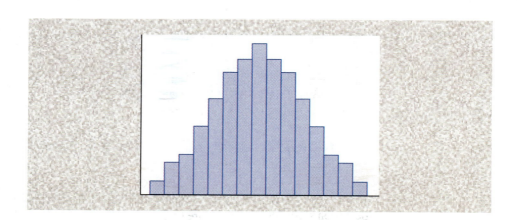

Figure 3.2
Histogram of a normal data set.

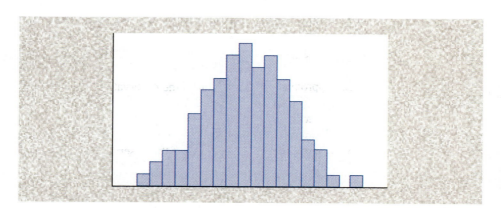

Figure 3.3
Histogram of an approximately normal data set.

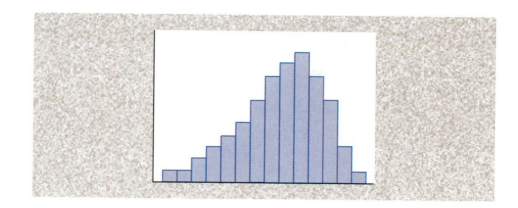

Figure 3.4
Histogram of a data set skewed to the left.

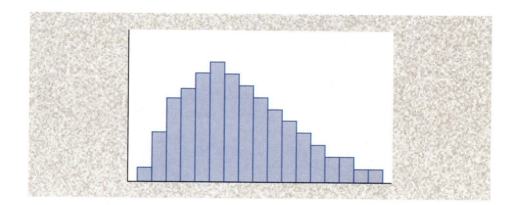

Figure 3.5
Histogram of a data set skewed to the right.

Empirical rule

If a data set is approximately normal with sample mean $\bar{x}$ and sample standard deviation s, then the following are true.

1. Approximately 68 percent of the observations lie within

   $$\bar{x} \pm s$$

2. Approximately 95 percent of the observations lie within

   $$\bar{x} \pm 2s$$

3. Approximately 99.7 percent of the observations lie within

   $$\bar{x} \pm 3s$$

Example 3.22   The scores of 25 students on a history examination are listed on the following stem-and-leaf plot.

```
9 | 0, 0, 4
8 | 3, 4, 4, 6, 6, 9
7 | 0, 0, 3, 5, 5, 8, 9
6 | 2, 2, 4, 5, 7
5 | 0, 3, 5, 8
```

By standing this figure on its side (or, equivalently, by turning the textbook), we can see that the corresponding histogram is approximately normal. Use it to assess the empirical rule.

Solution   A calculation yields that the sample mean and sample standard deviation of the data are

$$\bar{x} = 73.68 \quad \text{and} \quad s = 12.80$$

The empirical rule states that approximately 68 percent of the data values are between $\bar{x} - s = 60.88$ and $\bar{x} + s = 86.48$. Since 17 of the observations actually fall within 60.88 and 86.48, the actual percentage is $100(17/25) = 68$ percent. Similarly, the empirical rule states that approximately 95 percent of the data are between $\bar{x} - 2s = 48.08$ and $\bar{x} + 2s = 96.28$ whereas, in actuality, 100 percent of the data fall in this range.

A data set that is obtained by sampling from a population that is itself made up of subpopulations of different types is usually not normal. Rather, the histogram from such a data set often appears to resemble a combining, or superposition, of normal histograms and thus will often have more than one local peak or hump. Because the histogram will be higher at these local peaks than at their neighboring values, these peaks are similar to modes. A data set whose histogram has two local peaks is said to be *bimodal*. The data set represented in Fig. 3.6 is bimodal.

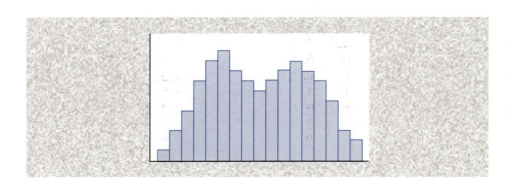

Figure 3.6
Histogram of a bimodal data set.

Since a stem-and-leaf plot can be regarded as a histogram lying on its side, it is useful in showing us whether a data set is approximately normal.

Example 3.23 The following is the stem-and-leaf plot of the weights of 200 members of a health club.

```
24 | 9
23 |
22 | 1
21 | 7
20 | 2, 2, 5, 5, 6, 9, 9, 9
19 | 0, 0, 0, 0, 0, 1, 1, 2, 4, 4, 5, 8
18 | 0, 1, 1, 2, 2, 2, 3, 4, 4, 4, 5, 5, 5, 6, 6, 6, 6, 7, 9, 9, 9
17 | 1, 1, 1, 2, 3, 3, 4, 4, 4, 5, 5, 6, 6, 6, 6, 7, 7, 7, 7, 9
16 | 0, 0, 1, 1, 1, 1, 2, 4, 5, 5, 6, 6, 8, 8, 8, 8
15 | 0, 1, 1, 1, 1, 1, 1, 5, 5, 5, 5, 6, 6, 6, 7, 7, 8, 9
14 | 0, 0, 0, 1, 2, 3, 4, 5, 6, 7, 7, 7, 8, 9, 9
13 | 0, 0, 0, 1, 1, 1, 2, 2, 2, 2, 2, 3, 3, 4, 5, 5, 6, 6, 6, 6, 7, 7, 8, 8, 8, 9, 9, 9
12 | 1, 1, 1, 2, 2, 2, 3, 4, 4, 5, 5, 6, 6, 6, 6, 6, 6, 6, 6, 7, 7, 7, 7, 8, 8, 9, 9, 9
11 | 0, 1, 1, 2, 2, 2, 2, 3, 3, 4, 4, 5, 5, 6, 9, 9
10 | 0, 2, 3, 3, 3, 4, 4, 5, 7, 7, 8
 9 | 0, 0, 9
 8 | 6
```

By standing it on its side, we see that its histogram does not appear to be approximately normal. However, it is important to note that these data consist of the weights of all members of the health club, both female and male. Since these are clearly separate populations with regard to weight, it makes sense to consider the data for each gender separately. We will now do so.

It turns out that these 200 data values are the weights of 97 women and 103 men. Separating the data for women and men results in the stem-and-leaf plots in Figs. 3.7 and 3.8.

```
16 | 0, 5
15 | 0, 1, 1, 1, 5
14 | 0, 0, 1, 2, 3, 4, 6, 7, 9
13 | 0, 0, 1, 1, 2, 2, 2, 2, 3, 4, 5, 5, 6, 6, 6, 6, 7, 8, 8, 8, 9, 9, 9
12 | 1, 1, 1, 2, 2, 2, 3, 4, 4, 5, 5, 6, 6, 6, 6, 6, 6, 6, 6, 6, 7, 7, 7, 7, 8, 8, 9, 9
11 | 0, 0, 1, 1, 2, 2, 2, 2, 3, 3, 4, 4, 5, 5, 6, 9, 9
10 | 2, 3, 3, 3, 4, 4, 5, 7, 7, 8
 9 | 0, 0, 9
 8 | 6
```

Figure 3.7
Weights of 97 female health club members.

```
24 | 9
23 |
22 | 1
21 | 7
20 | 2, 2, 5, 5, 6, 9, 9, 9
19 | 0, 0, 0, 0, 0, 1, 1, 2, 4, 4, 5, 8
18 | 0, 1, 1, 2, 2, 2, 3, 4, 4, 4, 5, 5, 5, 6, 6, 6, 6, 7, 9, 9, 9
17 | 1, 1, 1, 2, 3, 3, 4, 4, 4, 5, 5, 6, 6, 6, 6, 7, 7, 7, 7, 9
16 | 0, 1, 1, 1, 1, 2, 4, 5, 6, 6, 8, 8, 8, 8
15 | 1, 1, 1, 5, 5, 5, 6, 6, 6, 7, 7, 8, 9
14 | 0, 5, 7, 7, 8, 9
13 | 0, 1, 2, 3, 7
12 | 9
```

Figure 3.8
Weights of 103
male health
club members.

As we can see from Figs. 3.7 and 3.8, the separated data for each sex appear to be approximately normal. Let us calculate $\bar{x}_w$, s_w, $\bar{x}_m$, and s_m, the sample mean and sample standard deviation for, respectively, the women and the men. This calculation yields

$$\bar{x}_w = 125.70 \qquad \bar{x}_m = 174.69$$
$$s_w = 15.58 \qquad s_m = 21.23$$

A further corroboration of the approximate normality of the two sets of separated data is provided by noting the similar values in each set of the sample mean and sample median. The sample median of the women's weights is the 49th smallest data value, which equals 126, whereas for the men's data the sample median is the 52nd smallest data value, which equals 174. These are quite close to the two sample means whose values are 125.7 and 174.69.

Given the values of the sample mean and sample standard deviation, it follows from the empirical rule that approximately 68 percent of the women will weigh between 110.1 and 141.3 and approximately 95 percent of the men will weigh between 132.2 and 217.2. The actual percentages from Figs. 3.7 and 3.8 are

$$100 \times \frac{68}{97} = 70.1 \qquad \text{and} \qquad 100 \times \frac{101}{103} = 98.1$$

Problems

1. The daily numbers of animals treated at a certain veterinarian clinic over a 24-day period are as follows:

   22, 17, 19, 31, 28, 29, 21, 33, 36, 24, 15, 28, 25, 28, 22, 27, 33, 19, 25, 28, 26, 20, 30, 32

Historical Perspective

(North Wind Picture Archives)

Adolphe Quetelet

Quetelet and How the Normal Curve Uncovered Fraud

The Belgian social scientist and statistician Adolphe Quetelet was a great believer in the hypothesis that most data sets relating to human measurements are normal. In one study he measured the chests of 5738 Scottish soldiers, plotted the resulting data in a histogram, and concluded that it was normal.

In a later study Quetelet used the shape of the normal histogram to uncover evidence of fraud in regard to draft conscripts to the French army. He studied data concerning the heights of a huge sample of 100,000 conscripts. Plotting the data in a histogram—with class intervals of 1 inch—he found that, with the exception of three class intervals around 62 inches, the data appeared to be normal. In particular, there were fewer values in the interval from 62 to 63 inches and slightly more in the intervals from 60 to 61 and from 61 to 62 inches than would have occurred with a perfect normal fit of the data. Trying to figure out why the normal curve did not fit as well as he had supposed it would, Quetelet discovered that 62 inches was the minimum height required for soldiers in the French army. Based on this and his confidence in the widespread applicability of normal data, Quetelet concluded that some conscripts whose heights were slightly above 62 inches

were "bending their knees" to appear shorter so as to avoid the draft.

For 50 years following Quetelet, that is, roughly from 1840 to 1890, it was widely believed that most data sets from homogeneous populations (that is, data that were not obviously a mixture of different populations) would appear to be normal if the sample size were sufficiently large. Whereas present-day statisticians have become somewhat skeptical about this claim, it is quite common for a data set to appear to come from a normal population. This phenomenon, which often appears in data sets originating in either the biological or the physical sciences, is partially explained by a mathematical result known as the *central limit theorem*. Indeed, the central limit theorem (studied in Chap. 7) will in itself explain why many data sets originating in the physical sciences are approximately normal. To explain why biometric data (that is, data generated by studies in biology) often appear to be normal, we will use what was originally an empirical observation noted by Francis Galton, but which nowadays has a sound scientific explanation, called *regression to the mean*. Regression to the mean, in conjunction with the central limit theorem and the passing of many generations, will yield our explanation as to why a biometric data set is often normal. The explanation will be presented in Chap. 12.

(a) Plot these data in a histogram.
(b) Find the sample mean.
(c) Find the sample median.
(d) Is this data set approximately normal?

2. The following data give the injury rates per 100,000 worker-hours for a sample of 20 semiconductor firms.

   1.4, 2.4, 3.7, 3.1, 2.0, 1.9, 2.5, 2.8, 2.2, 1.7, 3.1, 4.0, 2.2, 1.8, 2.6, 3.6, 2.9, 3.3, 2.0, 2.4

   (a) Plot the data in a histogram.
   (b) Is the data set roughly symmetric?
   (c) If the answer to (b) is no, is it skewed to the left or to the right?
   (d) If the answer to (b) is yes, is it approximately normal?

The following table gives the 1988 and 1989 birth rates per 1000 population for all 50 states and the District of Columbia. Problems 3 and 4 refer to it.

Birth Rates

State	1989 rate	1988 rate	State	1989 rate	1988 rate
Alabama	14.2	14.8	Montana	14.1	14.5
Alaska	21.9	21.4	Nebraska	15.1	14.9
Arizona	19.0	18.8	Nevada	16.5	17.1
Arkansas	14.5	14.6	New Hampshire	16.2	16.0
California	19.2	18.8	New Jersey	15.1	15.3
Colorado	15.9	16.2	New Mexico	17.9	17.9
Connecticut	14.7	14.9	New York	16.2	15.7
Delaware	17.1	15.8	North Carolina	15.6	15.0
D.C.	37.3	17.1	North Dakota	16.5	15.1
Florida	15.2	14.9	Ohio	14.9	14.8
Georgia	17.1	16.7	Oklahoma	14.4	14.6
Hawaii	17.6	17.3	Oregon	15.5	14.5
Idaho	15.2	15.7	Pennsylvania	14.1	13.8
Illinois	16.0	15.9	Rhode Island	15.3	14.3
Indiana	14.8	14.7	South Carolina	15.7	15.9
Iowa	13.1	13.5	South Dakota	15.4	15.7
Kansas	14.2	15.5	Tennessee	15.5	14.4
Kentucky	14.1	13.7	Texas	17.7	18.0
Louisiana	15.7	16.8	Utah	21.2	21.3
Maine	13.8	14.3	Vermont	14.0	14.6
Maryland	14.4	16.4	Virginia	15.3	15.5
Massachusetts	16.3	15.0	Washington	15.4	15.6
Michigan	15.4	15.1	West Virginia	12.1	11.6
Minnesota	15.3	15.5	Wisconsin	14.8	14.6
Mississippi	16.1	16.1	Wyoming	13.7	15.0
Missouri	15.5	14.9	**Total**	**16.2**	**15.9**

Source: Department of Health and Human Services, National Center for Health Statistics.

3. This problem refers to the data for 1988.
   (a) Plot the data in a stem-and-leaf plot.
   (b) Is the data set approximately normal?
   (c) Compute the sample mean.
   (d) Compute the sample median.

4. Repeat Prob. 3, but use the data for 1989.

5. The following is a sample of sales prices of homes in a middle-class California community. The data are in thousands of dollars.

   166, 82, 175, 181, 169, 177, 180, 185, 159, 164, 170, 149, 188, 173, 170, 164, 158, 177, 173, 175, 190, 172

   (a) Find the sample mean.
   (b) Find the sample median.
   (c) Plot the data in a histogram.
   (d) Is this data set approximately normal?

6. The following data give the age at inauguration of all 42 presidents of the United States:

President	Age at inauguration	President	Age at inauguration
1. Washington	57	22. Cleveland	47
2. J. Adams	61	23. B. Harrison	55
3. Jefferson	57	24. Cleveland	55
4. Madison	57	25. McKinley	54
5. Monroe	58	26. T. Roosevelt	42
6. J. Q. Adams	57	27. Taft	51
7. Jackson	61	28. Wilson	56
8. Van Buren	54	29. Harding	55
9. W. Harrison	68	30. Coolidge	51
10. Tyler	51	31. Hoover	54
11. Polk	49	32. F. Roosevelt	51
12. Taylor	64	33. Truman	60
13. Fillmore	50	34. Eisenhower	62
14. Pierce	48	35. Kennedy	43
15. Buchanan	65	36. L. Johnson	55
16. Lincoln	52	37. Nixon	56
17. A. Johnson	56	38. Ford	61
18. Grant	46	39. Carter	52
19. Hayes	54	40. Reagan	69
20. Garfield	49	41. Bush	64
21. Arthur	50	42. Clinton	46

   (a) Find the sample mean and sample standard deviation of this data set.
   (b) Draw a histogram for the above data.
   (c) Do the data appear to be approximately normal?
   (d) If the answer to (c) is yes, give an interval that you would expect to contain approximately 95 percent of the data observations.
   (e) What percentage of the data lies in the interval given in part (d)?

7. For the data on the weights of female health club members presented in Fig. 3.7, the sample mean and sample standard deviation were computed to be 125.70 and 15.58, respectively. Based on the shape of Fig. 3.7 and these values, approximate the proportion of the women whose weight is between 94.54 and 156.86 pounds. What is the actual proportion?

8. A sample of 36 male coronary patients yielded the following data concerning the ages at which they suffered their first heart attacks.

$$
\begin{array}{r|l}
7 & 1, 2, 4, 5 \\
6 & 0, 1, 2, 2, 3, 4, 5, 7 \\
5 & 0, 1, 2, 3, 3, 4, 4, 4, 5, 6, 7, 8, 9 \\
4 & 1, 2, 2, 3, 4, 5, 7, 8, 9 \\
3 & 7, 9
\end{array}
$$

   (a) Determine $\bar{x}$ and s.
   (b) From the shape of the stem-and-leaf plot, what percentage of data values would you expect to be between $\bar{x} - s$ and $\bar{x} + s$? Between $\bar{x} - 2s$ and $\bar{x} + 2s$?
   (c) Find the actual percentages for the intervals given in (b).

9. If the histogram is skewed to the right, which statistic will be larger—the sample mean or the sample median? (*Hint:* If you are not certain, construct a data set that is skewed to the right and then calculate the sample mean and sample median.)

10. The following data are the ages of a sample of 36 victims of violent crime in a large eastern city:

      25, 16, 14, 22, 17, 20, 15, 18, 33, 52, 70, 38, 18, 13, 22, 27, 19, 23, 33, 15, 13, 62, 21, 57, 66, 16, 24, 22, 31, 17, 20, 14, 26, 30, 18, 25

   (a) Determine the sample mean.   27
   (b) Find the sample median.
   (c) Determine the sample standard deviation.   15.44
   (d) Does this data set appear to be approximately normal?

(e) What proportion of the data lies within 1 sample standard deviation of the sample mean?

(f) Compare your answer in (e) to the approximation provided by the empirical rule.

The following table lists the 1985 per capita income for the 50 largest U.S. cities. Problems 11 to 13 refer to it.

1985 Per Capita Money Income for 50 Largest Cities

Cities ranked by population	Income	Cities ranked by population	Income
New York	11,188	Seattle	12,919
Los Angeles	12,084	Nashville-Davidson	11,253
Chicago	9,642	Austin	11,633
Houston	12,115	Oklahoma City	11,527
Philadelphia	8,807	Kansas City	11,153
Detroit	8,852	Fort Worth	10,515
San Diego	11,766	St. Louis	8,799
Dallas	12,816	Atlanta	10,341
San Antonio	8,499	Long Beach	11,729
Phoenix	11,363	Portland	10,770
Honolulu	11,434	Pittsburgh	9,998
Baltimore	8,647	Miami	8,904
San Francisco	13,575	Tulsa	12,670
Indianapolis	10,836	Cincinnati	10,247
San Jose	12,583	Albuquerque	11,133
Memphis	9,362	Tucson	9,430
Washington	13,530	Oakland	10,911
Jacksonville	10,466	Minneapolis	12,302
Milwaukee	9,765	Charlotte	12,259
Boston	10,774	Omaha	12,886
Columbus	9,909	Toledo	10,050
New Orleans	8,975	Virginia Beach	12,039
Cleveland	8,018	Buffalo	8,840
Denver	12,490	Sacramento	10,627
El Paso	7,670	Newark	6,494

Source: U.S. Bureau of the Census, *Current Population Reports,* no. 86.

11. This problem concerns the cities on the left side of the preceding table.
    (a) Plot the data in a histogram.
    (b) Compute the sample mean.
    (c) Compute the sample median.
    (d) Compute the sample variance.

(e) Are the data approximately normal?

(f) Use the empirical rule to give an interval which should contain approximately 68 percent of the observations.

(g) Use the empirical rule to give an interval which should contain approximately 95 percent of the observations.

(h) Determine the actual proportion of observations in the interval specified in (f).

(i) Determine the actual proportion of observations in the interval specified in (g).

12. Repeat Prob. 11, this time using the data on the right side of the table.

13. Repeat Prob. 11, this time using all the data in the table.

3.7 SAMPLE CORRELATION COEFFICIENT

Consider the data set of paired values $(x_1, y_1), (x_2, y_2), \ldots, (x_n, y_n)$. In this section we will present a statistic, called the *sample correlation coefficient,* that measures the degree to which larger x values go with larger y values and smaller x values go with smaller y values.

The data in Table 3.3 represent the average daily number of cigarettes smoked (the x variable) and the number of free radicals (the y variable), in a suitable unit, found in the lungs of 10 smokers. (A free radical is a single atom of oxygen. It is believed to be potentially harmful because it is highly reactive and has a strong tendency to combine with other atoms within the body.) Figure 3.9 shows the scatter diagram for these data.

From an examination of Fig. 3.9 we see that when the number of cigarettes is high, there tends to be a large number of free radicals, and when the number of cig-

Table 3.3	Person	Number of cigarettes smoked	Free radicals
	1	18	202
	2	32	644
	3	25	411
	4	60	755
	5	12	144
	6	25	302
	7	50	512
	8	15	223
	9	22	183
	10	30	375

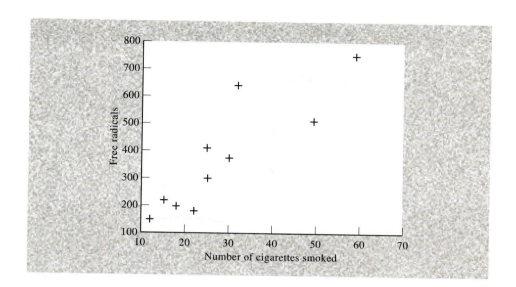

Figure 3.9
Cigarettes
smoked versus
number of free
radicals.

arettes smoked is low, there tends to be a small number of free radicals. In this case, we say that there is a *positive correlation* between these two variables.

We are also interested in determining the strength of the relationship between a pair of variables in which large values of one variable tend to be associated with small values of the other. For instance, the data of Table 3.4 represent the years of schooling (variable x) and the resting pulse rate in beats per minute (variable y) of 10 individuals. A scatter diagram of this data is presented in Fig. 3.10. From Fig. 3.10 we see that higher numbers of years of schooling tend to be associated with lower resting pulse rates and that lower numbers of years of schooling tend to be associated with the higher resting pulse rates. This is an example of a *negative correlation*.

To obtain a statistic which can be used to measure the association between the individual values of a paired set, suppose the data set consists of the paired values (x_i, y_i), $i = 1, \ldots, n$. Let $\bar{x}$ and $\bar{y}$ denote the sample mean of the x values and the sample mean of the y values, respectively. For data pair i, consider $x_i - \bar{x}$ the deviation of its x value from the sample mean and $y_i - \bar{y}$ the deviation of its y value from the sample mean. Now if x_i is a large x value, then it will be larger than the average value

Table 3.4　**Pulse Rate and Years of School Completed**

					Person					
	1	2	3	4	5	6	7	8	9	10
Years of school	12	16	13	18	19	12	18	19	12	14
Pulse rate	73	67	74	63	73	84	60	62	76	71

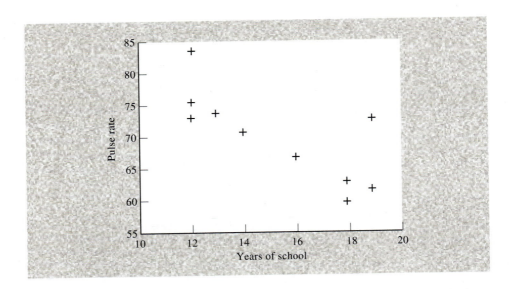

Figure 3.10
Scatter diagram
of years in
school and
pulse rate.

of all the x's, and so the deviation $x_i - \bar{x}$ will be a positive value. Similarly, when x_i is a small x value, then the deviation $x_i - \bar{x}$ will be a negative value. Since the same statements are true about the y deviations, we can conclude the following:

> When large values of the x variable tend to be associated with large values of the y variable and small values of the x variable tend to be associated with small values of the y variable, then the signs, either positive or negative, of $x_i - \bar{x}$ and $y_i - \bar{y}$ will tend to be the same.

Now, if $x_i - \bar{x}$ and $y_i - \bar{y}$ both have the same sign (either positive or negative), then their product $(x_i - \bar{x})(y_i - \bar{y})$ will be positive. Thus, it follows that when large x values tend to be associated with large y values and small x values are associated with small y values, then $\sum_{i=1}^{n} (x_i - \bar{x})(y_i - \bar{y})$ will tend to be a large positive number.

The same logic as above also implies that when large values of one of the variables tend to go along with small values of the other, then the signs of $x_i - \bar{x}$ and $y_i - \bar{y}$ will be opposite, and so $\sum_{i=1}^{n} (x_i - \bar{x})(y_i - \bar{y})$ will be a large negative number.

To determine what it means for $\sum_{i=1}^{n} (x_i - \bar{x})(y_i - \bar{y})$ to be "large," we standardize this sum first by dividing by $n - 1$ and then by dividing by the product of the two sample standard deviations. The resulting statistic is called the *sample correlation coefficient*.

Definition

Let s_x and s_y denote, respectively, the sample standard deviations of the x values and the y values. The *sample correlation coefficient*, call it r, of the data pairs (x_i, y_i), $i = 1, \ldots, n$, is defined by

$$r = \frac{\displaystyle\sum_{i=1}^{n} (x_i - \bar{x})(y_i - \bar{y})}{(n-1)s_x s_y}$$

$$= \frac{\displaystyle\sum_{i=1}^{n} (x_i - \bar{x})(y_i - \bar{y})}{\sqrt{\displaystyle\sum_{i=1}^{n} (x_i - \bar{x})^2 \sum_{i=1}^{n} (y_i - \bar{y})^2}}$$

When $r > 0$, we say that the sample data pairs are *positively correlated;* and when $r < 0$, we say that they are *negatively correlated.*

We now list some of the properties of the sample correlation coefficient.

1. The sample correlation coefficient r is always between -1 and $+1$.

2. The sample correlation coefficient r will equal $+1$ if, for some constant a,

$$y_i = a + bx_i \qquad \text{for } i = 1, \ldots, n$$

where b is a positive constant.

3. The sample correlation coefficient r will equal -1 if, for some constant a,

$$y_i = a + bx_i \qquad \text{for } i = 1, \ldots, n$$

where b is a negative constant.

4. If r is the sample correlation coefficient for the data $x_i, y_i, i = 1, \ldots, n$, then for any constants a, b, c, d, it is also the sample correlation coefficient for the data

$$a + bx_i, c + dy_i \qquad i = 1, \ldots, n$$

provided that b and d have the same sign (that is, provided that $bd \geq 0$).

Property 1 says that the sample correlation coefficient r is always between -1 and $+1$. Property 2 says that r will equal $+1$ when there is a straight-line (also called a *linear*) relation between the paired data which is such that large y values are attached to large x values. Property 3 says that r will equal -1 when the relation is linear and large y values are attached to small x values. Property 4 states that the value of r is unchanged when a constant is added to each of the x variables (or to each of the y variables) or when each x variable (or each y variable) is multiplied by a positive constant. This property implies that r does not depend on the dimensions chosen to measure the data. For instance, the sample correlation coefficient between a person's height and weight does not depend on whether the height is measured in feet or in inches or whether the weight is measured in pounds or kilograms. Also if one of the values in the pair is temperature, then the sample correlation coefficient is the same whether it is measured in degrees Fahrenheit or Celsius.

For computational purposes, the following is a convenient formula for the sample correlation coefficient.

Computational formula for r

$$r = \frac{\sum_{i=1}^{n} x_i y_i - n\bar{x}\bar{y}}{\sqrt{\left(\sum_{i=1}^{n} x_i^2 - n\bar{x}^2\right)\left(\sum_{i=1}^{n} y_i^2 - n\bar{y}^2\right)}}$$

Example 3.24 The following table gives the U.S. per capita consumption of whole milk x and of low-fat milk in three different years.

	Per capita consumption (gallons)		
	1980	**1984**	**1987**
Whole milk x	17.1	14.7	12.8
Low-fat milk (y)	10.6	11.5	13.2

Source: U.S. Department of Agriculture, *Food Consumption, Prices, and Expenditures.*

Find the sample correlation coefficient r for the above data.

Solution To make the computation easier, let us first subtract 12.8 from each of the x values and 10.6 from each of the y values. This gives the new set of data pairs:

	i		
	1	**2**	**3**
x_i	4.3	1.9	0
y_i	0	.9	2.6

Now,

$$\bar{x} = \frac{4.3 + 1.9 + 0}{3} = 2.0667$$

$$\bar{y} = \frac{0 + .9 + 2.6}{3} = 1.1667$$

$$\sum_{i=1}^{3} x_i y_i = (1.9)(.9) = 1.71$$

$$\sum_{i=1}^{3} x_i^2 = (4.3)^2 + (1.9)^2 = 22.10$$

$$\sum_{i=1}^{3} y_i^2 = (.9)^2 + (2.6)^2 = 7.57$$

Thus,

$$r = \frac{1.71 - 3(2.0667)(1.1667)}{\sqrt{[22.10 - 3(2.0667)^2][7.57 - 3(1.1667)^2]}} = -.97$$

Therefore, our three data pairs exhibit a very strong negative correlation between consumption of whole and of low-fat milk.

For small data sets such as in Example 3.24, the sample correlation coefficient can be easily obtained by using a hand calculator. However, for large data sets this computation can become tedious. To relieve this, Program 3-2 on the enclosed diskette or any statistical package can be utilized.

Example 3.25   Compute the sample correlation coefficient of the data of Table 3.3, which relates the number of cigarettes smoked to the number of free radicals found in a person's lungs.

Solution You can solve this example by running Program 3-2. The number of pairs is 10. The pairs are as follows:

18, 202
32, 644
25, 411
60, 755
12, 144
25, 302
50, 512
15, 223
22, 183
30, 375

Program 3-2 computes the sample correlation coefficient as .8759639.

The large value of the sample correlation coefficient indicates a strong positive correlation between the number of cigarettes a person smokes and the number of free radicals in that person's lungs.

Example 3.26 Compute the sample correlation coefficient of the data of Table 3.4, which relates a person's resting pulse rate to the number of years of school completed.

Solution Program 3-2 can also be used to solve this example. Again, the number of pairs is 10. The pairs are as follows:

12, 73
16, 67
13, 74
18, 63
19, 73
12, 84
18, 60
19, 62
12, 76
14, 71

Program 3-2 computes the sample correlation coefficient as $-.7638035$.

The large negative value of the sample correlation coefficient indicates that, for the data set considered, a high pulse rate tends to be associated with a small number of years spent in school and a low pulse rate tends to be associated with a large number of years spent in school.

The absolute value of the sample correlation coefficient r (that is, $|r|$—its value without regard to its sign) is a measure of the strength of the linear relationship between the x and the y values of a data pair. A value of $|r|$ equal to 1 means that there is a perfect linear relation; that is, a straight line can pass through all the data points (x_i, y_i), $i = 1, \ldots, n$. A value of $|r|$ of about .8 means that the linear relation is relatively strong; although there is no straight line that passes through all the data points, there is one that is "close" to them all. A value of $|r|$ around .3 means that the linear relation is relatively weak. The sign of r gives the direction of the relation. It is positive when the linear relation is such that smaller y values tend to go with smaller x values and larger y values with larger x values (and so a straight-line approximation points upward); and it is negative when larger y values tend to go with smaller x values and smaller y values with larger x values (and so a straight-line approximation points downward). Figure 3.11 displays scatter diagrams for data sets with various values of r.

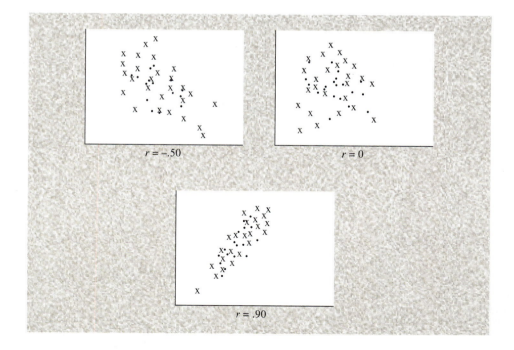

Figure 3.11
Sample correlation coefficients.

Historical Perspective

(Bettmann)

Francis Galton

The development of the concept and utility of the sample correlation coefficient involved the efforts of four of the great men of statistics. The original concept was due to Francis Galton, who was trying to study the laws of inheritance from a quantitative point of view. As such, he wanted to be able to quantify the degree to which characteristics of an offspring relate to those of its parents. This led him to name and define a form of the sample correlation coefficient that differs somewhat from the one presently in use. Although originally it was meant to be used to assess the hereditary influence of a parent on an offspring, Galton later realized that the sample correlation coefficient presented a method of assessing the interrelation between any two variables.

Although Francis Galton was the founder of the field of biometrics—the quantitative study of biology—its ac-

knowledged leader, at least after 1900, was Karl Pearson. After the Royal Society of London passed a resolution in 1900 stating that it would no longer accept papers that applied mathematics to the study of biology, Pearson, with financial assistance from Galton, founded the statistical journal *Biometrika,* which still flourishes today. The form of the sample correlation coefficient that is presently in use (and that we have presented) is due to Karl Pearson and was originally called *Pearson's product-moment correlation coefficient.*

The probabilities associated with the possible values of the sample correlation coefficient r were discovered, in the case where the data pairs come from a normal population, by William Gosset. There were, however, some technical errors in his derivations, and these were subsequently corrected in a paper by Ronald Fisher.

Problems

1. Explain why the sample correlation coefficient for the data pairs

   (121, 360), (242, 362), (363, 364)

   is the same as that for the pairs

   (1, 0), (2, 2), (3, 4)

   which is the same as that for the pairs

   (1, 0), (2, 1), (3, 2)

2. Compute the sample correlation coefficient for the data pairs in Prob. 1.

Correlation Measures Association, Not Causation

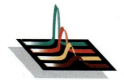

The results of Example 3.26 indicated a strong negative correlation between an individual's years of education and that individual's resting pulse rate. However, this does not imply that additional years of school will directly reduce one's pulse rate. That is, whereas additional years of school tend to be associated with a lower resting pulse rate, this does not mean that it is a direct *cause* of it. Often the explanation for such an association lies with an unexpressed factor that is related to both variables under consideration. In this instance, it may be that a person who has spent additional time in school is more aware of the latest findings in the area of health and thus may be more aware of the importance of exercise and good nutrition; or perhaps it is not knowledge that is making the difference but rather that people who have had more education tend to end up in jobs that allow them more time for exercise and good nutrition. Probably the strong negative correlation between years in school and resting pulse rate results from a combination of these as well as other underlying factors.

3. The following data represent the IQ scores of 10 mothers and their eldest daughters.

Mother's IQ	Daughter's IQ
135	121
127	131
124	112
120	115
115	99
112	118
104	106
96	89
94	92
85	90

(a) Draw a scatter diagram.
(b) Guess at the value of the sample correlation coefficient r.
(c) Compute r.
(d) What conclusions can you draw about the relationship between the mother's and daughter's IQs?

4. The following is a sampling of 10 recently released first-time federal prisoners. The data give their crime, their sentence, and the actual time that they served.

Number	Crime	Sentence (months)	Time served (months)
1	Drug abuse	44	24
2	Forgery	30	12
3	Drug abuse	52	26
4	Kidnapping	240	96
5	Income tax fraud	18	12
6	Drug abuse	60	28
7	Robbery	120	52
8	Embezzlement	24	14
9	Robbery	60	35
10	Robbery	96	49

Draw a scatter diagram of the sentence time versus time actually served. Compute the sample correlation coefficient. What does this say about the relationship between the length of a sentence and the time actually served?

5. Using the data of Prob. 4, determine the sample correlation coefficient of the sentence time and the proportion of that time actually served. What does this say about the relationship between the length of a sentence and the proportion of this time that is actually served?

6. The following data refer to the number of adults in prison and on parole in 12 midwestern states. The data are in thousands of adults.

State	In prison	On parole
Illinois	18.63	11.42
Indiana	9.90	2.80
Iowa	2.83	1.97
Kansas	4.73	2.28
Michigan	17.80	6.64
Minnesota	2.34	1.36
Missouri	9.92	4.53
Nebraska	1.81	.36
North Dakota	.42	.17
Ohio	20.86	6.51
South Dakota	1.05	.42
Wisconsin	5.44	3.85

(a) Draw a scatter diagram.
(b) Determine the sample correlation coefficient between the number of adults in state prison and on parole in that state.
(c) Fill in the missing word. States having a large prison population tend to have a(n) _____ number of individuals on parole.

7. The following data relate the number of criminal cases filed in various U.S. cities to the percentage of those cases that result in a plea of guilty.

City	Percentage of cases resulting in a guilty plea	Number of cases filed
San Diego, CA	73	11,534
Dallas, TX	72	14,784
Portland, OR	62	3,892
Chicago, IL	41	35,528
Denver, CO	68	3,772
Philadelphia, PA	26	13,796
Lansing, MI	68	1,358
St. Louis, MO	63	3,649
Davenport, IA	60	1,312
Tallahassee, FL	50	2,879
Salt Lake City, UT	61	2,745

Determine the sample correlation coefficient between the number of cases filed and the percentage of guilty pleas. What can you say about the degree of association between these two variables for these data?

8. The following data relate the U.S. per capita consumption of whole milk and of low-fat milk in each of the years from 1980 to 1987, excluding 1981. (Some of these data were used in Example 3.24.)

	Consumption (gallons)						
	1980	**1982**	**1983**	**1984**	**1985**	**1986**	**1987**
Whole milk	17.1	15.6	15.2	14.7	14.3	13.4	12.8
Low-fat milk	10.6	10.8	11.1	11.1	12.1	12.8	13.2

Source: *Food Consumption, Prices, and Expenditures.*

Compute the sample correlation coefficient of whole and low-fat milk consumption for the years specified.

9. The following data give the per capita money income, in dollars, for 12 U.S. cities in the years 1979 and 1985.

City	1979 Income	1985 Income	City	1979 Income	1985 Income
New York	7,271	11,188	Detroit	6,215	8,852
Baltimore	5,877	8,647	Memphis	6,466	9,362
Denver	8,553	12,490	Milwaukee	7,029	9,765
Austin	7,368	11,633	St. Louis	5,877	8,799
Cincinnati	6,874	10,247	Charlotte	7,952	12,259
Omaha	7,714	12,886	Buffalo	5,929	8,840

Compute the sample correlation coefficient between the 1979 and 1985 per capita incomes in these cities.

10. The following data give the numbers of physicians and dentists, per 100,000 population, in the United States for six different years.

	1980	1981	1982	1983	1985	1986
Physicians	211	217	222	228	237	246
Dentists	54	54	55	56	57	57

Source: Health Resources Statistics, annual.

(a) Show that the number of physicians and the number of dentists are positively correlated for these years.

(b) Do you think that a large value of one of these variables by itself causes a large value of the other? If not, how would you explain the reason for the positive correlation?

The following table gives the death rates by selected causes in different countries. It will be used in Probs. 11 to 13.

Death Rates per 100,000 Population by Selected Causes and Countries

Country	Year	Ischemic heart disease	Cerebro-vascular disease	Malignant neoplasm of— Lung, tra-chea, bron-chus	Stom-ach	Female breast	Bronchi-tis, emphy-sema, asthma	Chronic liver disease and cirrho-sis
United States	1984	218.1	60.1	52.7	6.0	31.9	8.3	12.9
Australia	1985	230.9	95.6	41.0	10.1	30.0	16.9	8.7
Austria	1986	155.1	133.2	34.3	20.7	31.6	22.3	26.6
Belgium	1984	120.6	95.0	55.9	14.7	36.8	22.6	12.4
Bulgaria	1985	245.9	254.5	30.6	24.2	21.5	28.6	16.2
Canada	1985	200.6	57.5	50.6	9.0	34.5	9.7	10.1
Czechoslovakia	1985	289.4	194.3	51.3	22.4	27.3	33.8	19.6
Denmark	1985	243.8	73.4	52.2	10.9	39.7	37.1	12.2
Finland	1986	259.8	105.0	36.4	17.3	23.9	19.8	8.8
France	1985	76.0	79.7	32.2	10.8	27.1	11.7	22.9
Hungary	1986	240.1	186.5	55.0	25.9	31.2	43.8	42.1
Italy	1983	128.9	121.9	42.1	23.9	28.9	30.9	31.5
Japan	1986	41.9	112.8	24.9	40.7	8.1	12.2	14.4
Netherlands	1985	164.6	71.1	56.3	15.6	38.2	17.8	5.5
New Zealand	1985	250.5	98.4	42.0	11.2	37.7	25.8	4.8
Norway	1985	208.5	88.6	26.3	14.4	25.9	18.2	6.9
Poland	1986	109.4	75.3	47.2	24.2	21.1	33.4	12.0
Portugal	1986	76.6	216.4	18.7	26.5	22.6	17.8	30.0
Spain	1981	79.0	133.9	26.0	19.7	19.0	19.1	23.3
Sweden	1985	244.7	73.0	23.2	12.5	26.0	14.3	6.4
Switzerland	1986	112.0	65.6	36.6	12.0	36.6	17.5	10.4
United Kingdom:								
England and Wales	1985	247.6	104.5	57.2	15.2	41.9	24.2	4.8
Scotland	1986	288.0	128.4	68.7	14.9	41.2	14.8	7.3
West Germany	1986	159.5	100.4	34.6	18.3	32.6	26.1	19.3

Source: World Health Organization, *World Health Statistics.*

In doing Probs. 11 to 13, use all the data if you are using either Program 3-2 or a statistical package. If you are using a hand calculator, use only the data relating to the first seven countries.

11. Find the sample correlation coefficient between the death rates of ischemic heart disease and of chronic liver disease.

12. Find the sample correlation coefficient between the death rates of stomach cancer and of female breast cancer.

13. Find the sample correlation coefficient between the death rates of lung cancer and of bronchitis, emphysema, and asthma.

14. In a well-publicized experiment, a University of Pittsburgh researcher enlisted the cooperation of public school teachers in Boston in obtaining a baby tooth from each of their pupils. These teeth were then sawed open and analyzed for lead content. The lead content of each tooth was plotted against the pupil's IQ test score. A strong negative correlation resulted between the amount of lead in the tooth and the IQ score. Newspapers headlined this result as "proof" that lead ingestion results in decreased scholastic aptitude.
    (a) Does this conclusion necessarily follow?
    (b) Offer some other possible explanations.

15. A recent study has found a strong positive correlation between the cholesterol levels of young adults and the amounts of time they spend watching television.
    (a) Would you have expected such a result? Why?
    (b) Do you think that watching television causes higher cholesterol levels?
    (c) Do you think that having a high cholesterol level makes a young adult more likely to watch television?
    (d) How would you explain the results of the study?

16. An analysis relating the number of points scored and fouls committed by basketball players in the Pacific Ten conference has established a strong positive correlation between these two variables. The analyst has gone on record as claiming that this verifies the hypothesis that offensive-minded basketball players tend to be very aggressive and so tend to commit a large number of fouls. Can you think of a simpler explanation for the positive correlation? (*Hint:* Think in terms of the average number of minutes per game that a player is on the court.)

17. A *New England Journal of Medicine* study published in October 1993 found that people who have guns in their homes for protection are 3 times more likely to be murdered than those with no guns in the home. Does this prove that an individual's chance of being murdered is increased when he or she purchases a gun to keep at home? Explain your answer.

Key Terms

Statistic: A numerical quantity whose value is determined by the data.

Sample mean: The arithmetic average of the values in a data set.

Deviation: The difference between the individual data values and the sample mean. If x_i is the ith data value and $\bar{x}$ is the sample mean, then $x_i - \bar{x}$ is called the ith *deviation.*

Sample median: The middle value of an ordered set of data. For a data set of n values, the sample median is the $(n + 1)/2$ smallest value when n is odd and the average of the $n/2$ and $n/2 + 1$ smallest values when n is even.

Sample 100p percentile: That data value such that at least $100p$ percent of the data are less than or equal to it and at least $100(1 - p)$ percent of the data are greater than or equal to it. If two data values satisfy this criterion, then it is the average of them.

First quartile: The sample 25th percentile.

Second quartile: The sample 50th percentile, which is also the sample median.

Third quartile: The sample 75th percentile.

Sample mode: The data value that occurs most frequently in a data set.

Sample variance: The statistic s^2, defined by

$$s^2 = \frac{\sum\limits_{i=1}^{n} (x_i - \bar{x})^2}{n - 1}$$

It measures the average of the squared deviations.

Sample standard deviation: The positive square root of the sample variance.

Range: The largest minus the smallest data value.

Interquartile range: The third quartile minus the first quartile.

Normal data set: One whose histogram is symmetric about its middle interval and decreases on both sides of the middle in a bell-shaped manner.

Skewed data set: One whose histogram is not symmetric about its middle interval. It is said to be skewed to the right if it has a long tail to the right and skewed to the left if it has a long tail to the left.

Bimodal data set: One whose histogram has two local peaks or humps.

Sample correlation coefficient: For the set of paired values x_i, y_i, $i = 1, \ldots, n$, it is defined by

$$r = \frac{\sum\limits_{i=1}^{n} (x_i - \bar{x})(y_i - \bar{y})}{(n - 1)s_x s_y}$$

where $\bar{x}$ and s_x are, respectively, the sample mean and sample standard deviation of the x values, and similarly for $\bar{y}$ and s_y. A value of r near $+1$ indicates that larger x values tend to be paired with larger y values and smaller x values tend to be paired with smaller y values. A value near -1 indicates that larger x values tend to be paired with smaller y values and smaller x values tend to be paired with larger y values.

Summary

We have seen three different statistics which describe the center of a data set: the sample mean, sample median, and sample mode.

The sample mean of the data $x_1 \ldots, x_n$ is defined by

$$\bar{x} = \frac{\sum\limits_{i=1}^{n} x_i}{n}$$

and is a measure of the center of the data.

If the data are specified by the frequency table

Value	Frequency
x_1	f_1
x_2	f_2
.	.
.	.
.	.
x_k	f_k

then the sample mean of the $n = \sum\limits_{i=1}^{k} f_i$ data values can be expressed as

$$\bar{x} = \frac{\sum\limits_{i=1}^{n} f_i x_i}{n}$$

A useful identity is

$$\sum_{i=1}^{n} (x_i - \bar{x}) = 0$$

The sample median is the middle value when the data are arranged from smallest to largest. If there are an even number of data points, then it is the average of the two middle values. It is also a measure of the center of the data set.

The sample mode is that value in the data set which occurs most frequently.

Suppose a data set of size n is arranged from smallest to largest. If np is not an integer, then the sample $100p$ percentile is the value whose position is the smallest integer larger than np. If np is an integer, then the sample $100p$ percentile is the average of the values in positions np and $np + 1$.

The sample 25th percentile is the first quartile. The sample 50th percentile (which is equal to the sample median) is called the *second quartile,* and the sample 75th percentile is called the *third quartile.*

The sample variance s^2 is a measure of the spread in the data and is defined by

$$s^2 = \frac{\sum_{i=1}^{n} (x_i - \bar{x})^2}{n - 1}$$

where n is the size of the set. Its square root s is called the *sample standard deviation,* and it is measured in the same units as the data.

The identity

$$\sum_{i=1}^{n} (x_i - \bar{x})^2 = \sum_{i=1}^{n} x_i^2 - n\bar{x}^2$$

is useful for computing the sample variance by using pencil and paper or a hand calculator.

Program 3-1 will compute the sample mean, sample variance, and sample standard deviation of any set of data.

Another statistic which describes the spread of the data is the *range,* the difference between the largest and smallest data values.

Normal data sets will have their sample mean and sample median approximately equal. Their histograms are symmetric about the middle interval and exhibit a bell shape.

The sample correlation coefficient r measures the degree of association between two variables. Its value is between -1 and $+1$. A value of r near $+1$ indicates that when one of the variables is large, the other one also tends to be large and when one of them is small, the other also tends to be small. A value of r near -1 indicates that when one of the variables is large, the other one tends to be small.

A large value of $|r|$ indicates a strong association between the two variables. Association does not imply causation.

Review Problems

1. Construct a data set that is symmetric about 0 and contains
   (a) Four distinct values
   (b) Five distinct values
   (c) In both cases, compute the sample mean and sample median.

2. The following stem-and-leaf plot records the diastolic blood pressure of a sample of 30 men.

   ```
   9 | 3, 5, 8,
   8 | 6, 7, 8, 9, 9, 9
   7 | 0, 1, 2, 2, 4, 5, 5, 6, 7, 8
   6 | 0, 1, 2, 2, 3, 4, 5, 5
   5 | 4, 6, 8
   ```

   (a) Compute the sample mean $\bar{x}$.
   (b) Compute the sample median.
   (c) Compute the sample mode.
   (d) Compute the sample standard deviation s.
   (e) Do the data appear to be approximately normal?
   (f) What proportion of the data values lies between $\bar{x} + 2s$ and $\bar{x} - 2s$?
   (g) Compare the answer in part (f) to the one prescribed by the empirical rule.

3. The following data are the median ages of residents in each of the 50 states of the United States.

29.3	27.7	30.4	31.1	28.5
32.1	28.0	31.3	26.6	25.8
25.9	33.0	31.5	30.0	28.4
24.9	31.6	26.6	25.4	29.2
29.3	27.9	31.8	31.5	30.3
28.5	29.3	26.6	31.2	32.1
31.4	30.1	27.0	28.5	27.6
28.9	29.4	30.5	31.2	29.4
29.3	30.1	28.8	27.9	30.4
32.3	30.4	25.8	27.1	26.9

(a) Find the median of these ages.
(b) Is this necessarily the median age of all people in the United States? Explain.
(c) Find the quartiles.
(d) Find the sample 90th percentile.

4. Use Table 3.2 (shown in Example 3.21) to fill in the answers.
   (a) To have one's score be among the top 10 percent of all physical science students, it must be at least _____.
   (b) To have one's score be among the top 25 percent of all social science students, it must be at least _____.
   (c) To have one's score be among the bottom 50 percent of all medical students, it must be less than or equal to _____.
   (d) To have one's score be among the middle 50 percent of all law school students, it must be between _____ and _____.

5. The winning scores in the Masters Golf Tournament from 1955 to 1991 are provided in Example 2.2. Is this data set approximately normal?

6. The following data represent the birth weights at an inner-city hospital in a large eastern city.

   2.4, 3.3, 4.1, 5.0, 5.1, 5.2, 5.6, 5.8, 5.9, 5.9, 6.0, 6.1, 6.2, 6.3, 6.3, 6.4, 6.4, 6.5, 6.7, 6.8, 7.2, 7.4, 7.5, 7.5, 7.6, 7.6, 7.7, 7.8, 7.8, 7.9, 7.9, 8.3, 8.5, 8.8, 9.2, 9.7, 9.8, 9.9, 10.0, 10.3, 10.5

   (a) Plot this in a stem-and-leaf diagram.
   (b) Find the sample mean $\bar{x}$.
   (c) Find the sample median.
   (d) Find the sample standard deviation s.
   (e) What proportion of the data lies within $\bar{x} \pm 2s$?
   (f) Do the data appear to be approximately normal?
   (g) If your answer to (f) is yes, what would you have estimated, based on your answers to (b) and (d), for (e)?

*7. Let a and b be constants. Show that if $y_i = a + bx_i$ for $i = 1, \ldots, n$, then r, the sample correlation coefficient of the data pairs x_i, y_i, $i = 1, \ldots, n$, is given by
(a) $r = 1$ when $b > 0$
(b) $r = -1$ when $b < 0$

(*Hint:* Use the definition of r, not its computational formula.)

8. The following data are taken from the book *Researches on the Probability of Criminal and Civil Verdicts,* published in 1837 by the French mathematician and probabilist Simeon Poisson. The book emphasized legal applications of probability. The data refer to the number of people accused and convicted of crimes in France from 1825 to 1830.

Year	Number accused	Number convicted
1825	6652	4037
1826	6988	4348
1827	6929	4236
1828	7396	4551
1829	7373	4475
1830	6962	4130

(a) Determine the sample mean and sample median of the number accused.
(b) Determine the sample mean and sample median of the number convicted.
(c) Determine the sample standard deviation of the number accused.
(d) Determine the sample standard deviation of the number convicted.
(e) Would you expect the number accused and the number convicted to have a positive or a negative sample correlation coefficient?
(f) Determine the sample correlation coefficient of the number of accused and number of convicted.
(g) Determine the sample correlation coefficient between the number accused and the percentage of these who are convicted.
(h) Draw scatter diagrams for parts (f) and (g).
(i) Guess at the value of the sample correlation coefficient between the number of convicted and the percentage convicted.
(j) Draw a scatter diagram for the variables in (i).
(k) Determine the sample correlation coefficient for the variables in part (i).

9. Recent studies have been inconclusive about the connection between coffee consumption and coronary heart disease. If a study indicated that consumers of large amounts of coffee appeared to have a greater chance of suffering heart attacks than did drinkers of moderate amounts or drinkers of no coffee at all, would this necessarily "prove" that excessive coffee drinking leads to an increased risk of heart attack? What other explanations are possible?

10. Recent studies have indicated that death rates for married middle-aged people appear to be lower than for single, middle-aged people. Does this mean that marriage tends to increase one's life span? What other explanations are possible?

11. A June 9, 1994, article in *The New York Times* noted a study showing that years with low inflation rates tend to be years with high average-productivity increases. The article claimed that this supported the Federal Reserve Board's claim that a low rate of inflation tends to result in an increase in productivity. Do you think the study provides strong evidence for this claim? Explain your answer.

Using Statistics to Summarize Data

Purpose

Use Minitab to do the following:

1. Compute the *sample statistics*—mean, median, percentiles, mode, variance, standard deviation, and correlation coefficient.

2. Investigate the *empirical rule.*

Procedures

First, load the Minitab (Windows version) software as described in the Chap. 1 Minitab Lab.

1. COMPUTING SAMPLE STATISTICS

Example 1 Use Minitab to compute the sample mean for the mother's IQ scores and the daughter's IQ scores in Prob. 3, Sec. 3.7.

To compute the sample mean in Minitab, type in the sample data set in C1 (named *MIQ*) and C2 (named *DIQ*). Select Calc→Column Statistics→Mean. Use the mouse to select the check box for the Mean by clicking on it; and in the Input variable text box, use the mouse to select C1 DIQ. The Column statistics dialog box is shown in Fig. M3.1 with the appropriate selections. Click on the OK button, and the column mean of 111.20 will be displayed in the Session window. This is the mean for this sample of IQ scores for the mothers. Repeat for C2 MIQ to obtain a mean of 107.30 for the daughters. Note that if you had selected the text box Store result in, you would need to specify a single constant (say, K1) in which the mean would have been stored. To display the stored value in the Session window, you need to type at the MTB> prompt Print K1 and press the Enter key.

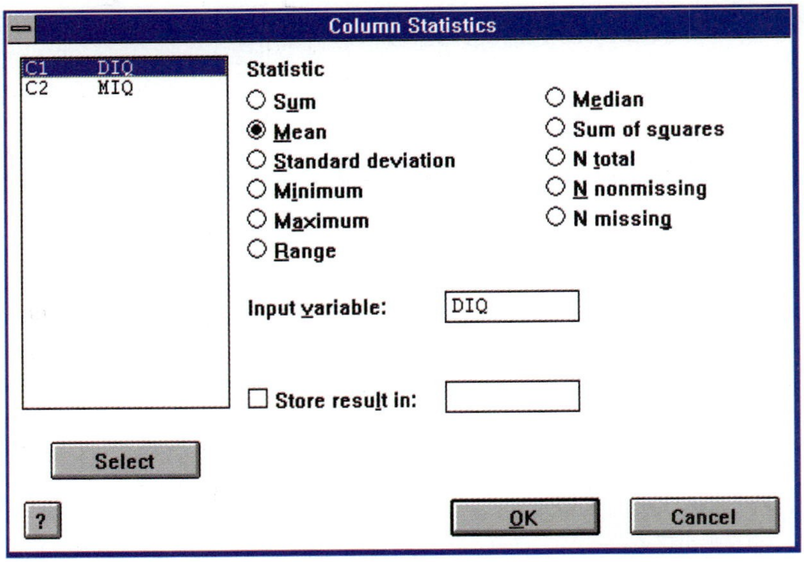

Figure M3.1

Use Minitab to compute the sample median for the mother's IQ scores and the daughter's IQ scores in Prob. 3, Sec. 3.7.

Repeat the process, but select Median in the Column statistics dialog box. The results that should be displayed in the Session window are 113.5 for the variable DIQ and 109 for the variable MIQ.

Recall, to find the sample median, you need to sort the data set (ascending or descending) and since both variables here (MIQ and DIQ) have an even set of 10 values, the sample median will be the average of the fifth and sixth numbers in the ordered set. To sort the values of the variables, select Manip→Sort. In the Sort dialog box, select C1 DIQ for the Sort column(s) box; and in the box Store sorted column(s) in, type in C3. In the box Sort by column, select DIQ. That is, you are sorting the values of the variable DIQ by itself and storing the sorted values in C3. Since the Descending box was not selected, the values will be sorted in an ascending order. The Sort dialog box is shown in Fig. M3.2. Repeat for the variable MIQ. From the sorted values, the median for the variable DIQ will be $(112 + 115)/2 = 113.5$, and for the variable MIQ it will be $(106 + 112)/2 = 109$. These were the values obtained above.

Use Minitab to compute the sample standard deviation and sample variance for the mother's IQ scores and the daughter's IQ scores in Prob. 3, Sec. 3.7.

In Fig. M3.1 select Standard deviation in the Column statistics dialog box to compute the standard deviations for variables MIQ and DIQ. These values will be displayed in the Session window if they are not stored as constants. Verify that the sample standard deviations for the variables MIQ and DIQ are, respectively, 16.116 and 14.469. The *sample variance* for a variable is found by just squaring the sample standard deviation if it is known. Thus the sample variance for the mother's IQ is $(16.116)^2 = 259.726$, and the variance for the daughter's IQ scores is $(14.469)^2 = 209.352$. *Note:* If you had stored the sample standard deviations as constants, you could use Minitab to square the values. For example, if 16.116 was stored as K1, in the Session window at the MTB> prompt, type in Let K2=K1**2 and press the Enter key. At the next prompt, type in Print K2, and a value of 259.726 will be displayed. Alternatively, you can use the sequence Calc→Mathematical Expressions and define K2 in that dialog box.

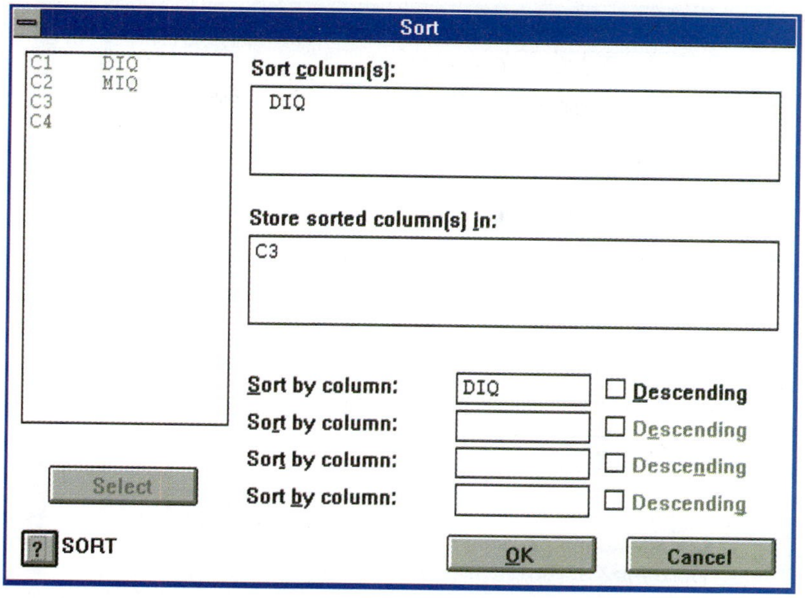

Figure M3.2

Use Minitab to compute the sample mode for the mother's IQ scores and the daughter's IQ scores in Prob. 3, Sec. 3.7.

To determine the sample mode, we need to construct a frequency distribution for the variables MIQ and DIQ as was done in the lab for Chap. 2 by using the sequence Stat→Tables→ Tally. Construct these tables and observe the frequency (COUNTS) column. In both cases, the frequency counts are 1 for each value of the variables, thus there is *no unique value* for the sample mode.

Use Minitab to compute the sample correlation coefficient between the mother's IQ scores and the daughter's IQ scores in Prob. 3, Sec. 3.7.

To compute the correlation coefficient (Pearson) between the variables DIQ and MIQ, select Stat→Basic Statistics→Correlation. In the Correlation dialog box, select the variables DIQ and MIQ for the Variables text box. This is shown in Fig. M3.3. Select the OK button, and the sample correlation coefficient will be computed and displayed in the Session window. The computed sample correlation coefficient is .862. That is, there seems to be a moderate to high positive correlation between IQ scores for the mothers and their daughters.

Use Minitab to *simultaneously* compute the sample mean, sample median, and sample standard deviation for variables DIQ and MIQ.

Select Stat→Basic Statistics→Descriptive Statistics. In the Descriptive statistics dialog box, select the variables for the Variables text box, as shown in Fig. M3.4. Select the OK button, and these statistics will be computed and displayed in the Session window. These values are shown in Fig. M3.5. Note that there are other statistics that have been computed by this sequence of commands. Ignore them for the time being.

Use Minitab to help to compute the sample 70th and 45th percentiles for the variable MIQ.

First, sort the variable MIQ. Save the sorted (ordered) values in C3, and follow the procedure in your text to determine the percentiles. In this case $n = 10$, $p = .7$, and $np = 7$. Thus the sample 70th percentile is the average of the seventh and eighth values, or

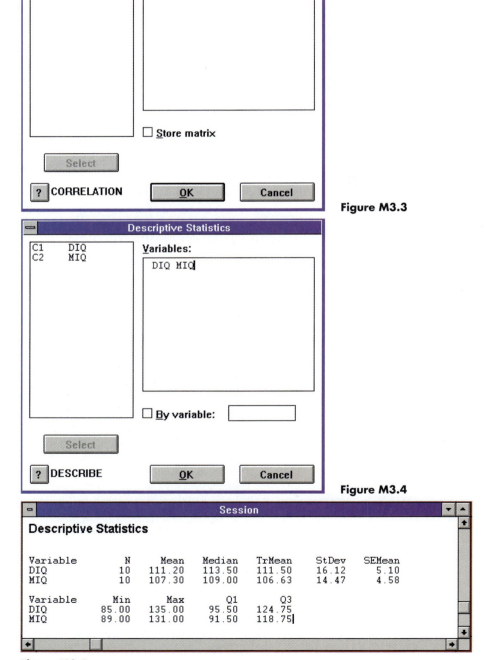

Figure M3.3

Figure M3.4

Figure M3.5

$(115 + 118)/2 = 116.5$. For the sample 45th percentile, $np = 4.5$; thus the 45 percentile will be located at the fifth position in the ordered set. This value is 106.

Use Minitab to find the first, second, and third quartiles for variables DIQ and MIQ.

These values are always computed whenever you use the Descriptive statistics procedure, as in Fig. M3.4. Figure M3.5 displays these results. The first quartile is denoted by Q_1, the third quartile is denoted by Q_3, and the second quartile is the median.

Example 2   Use Minitab to compute the sample mean for the frequency table in Prob. 14, Sec. 3.2.

First enter the variable AGE in C1 and the frequency f in C2. In the Session window type:

    MTB>LET C3 = SUM(C1*C2)/SUM(C2)
    MTB>PRINT C3

Here, C3 is equivalent to Eq. (3.1) of the text. Minitab responds with a value of 17.9259, correct to four decimal places. Alternatively, you could have used the sequence Calc→Mathematical Expressions and defined C3 in that dialog box.

2. INVESTIGATING THE EMPIRICAL RULE

The empirical rule will be investigated by use of simulation. Recall that the rule was associated with *normal* data sets. We can generate normal data sets by selecting Calc→Random Data→Normal. In the Normal distribution dialog box, generate 500 rows of values and store in column C1 (GVALUES), as shown in Fig. M3.6. Observe that the mean for the distribution is 0 and the standard deviation is 1. You can select any value for the mean and any *positive* number for the standard deviation. When the OK button is selected, 500 values will be generated and placed in C1.

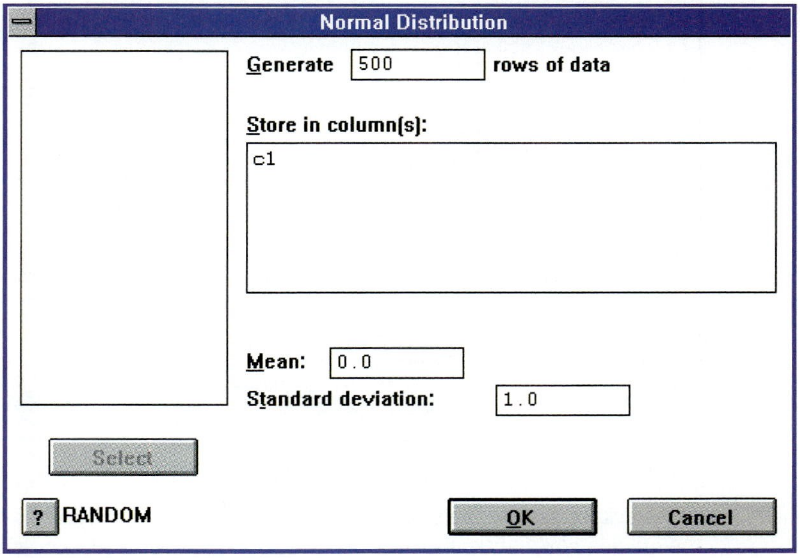

Normal Distribution

Generate `500` rows of data

Store in column(s):
`c1`

Mean: `0.0`
Standard deviation: `1.0`

Select

? RANDOM   OK   Cancel

Figure M3.6

Construct a histogram for these generated values to observe that indeed the histogram resembles that of a normal data set. An example is shown in Fig. M3.7. Column C1 was renamed *GVALUES*.

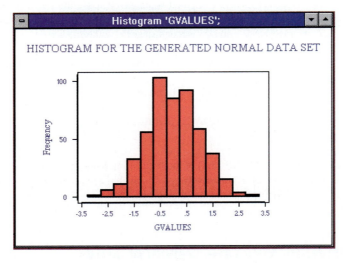

Figure M3.7

Recall that the empirical rule states that approximately 68 percent of the observations lie within $\bar{x} \pm s$. In this example $\bar{x} = 0$ and $s = 1$. Thus, approximately 68 percent of the generated values should lie between -1 and $+1$. To verify this, select Manip→Copy Columns. The Copy dialog box is shown in Fig. M3.8 with the appropriate selections.

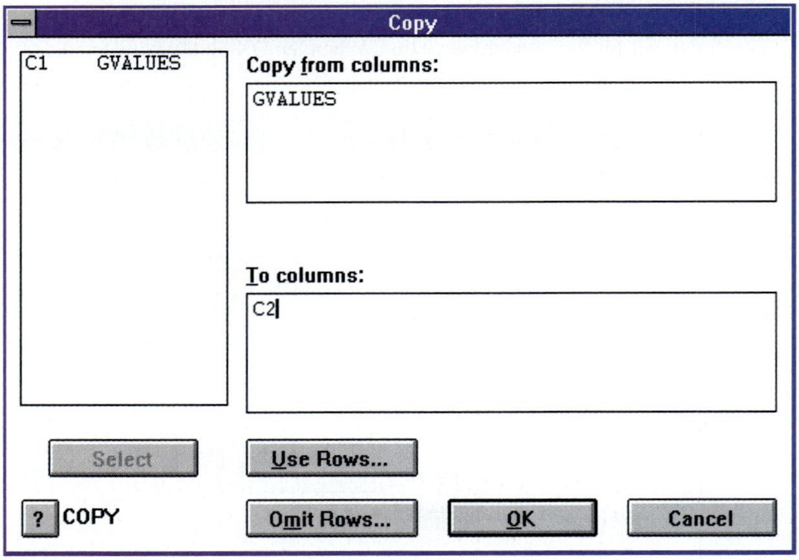

Figure M3.8

Click on the Use rows button to specify the values to be copied into C2. The Copy—Use rows dialog box will be displayed, as in Fig. M3.9. Select Use rows with column equal to, and in the text box type in $-1:1$. This will select only values in C1 between -1 and $+1$ and will place them in C2. Select the OK button, and these values will be displayed in the Session window. Divide the number of values in C2 by the number of values in C1. This ratio should be approximately .68, or 68 percent. That is, approximately 68 percent of the values in a normal data set will be within 1 standard deviation of the mean ($\bar{x} \pm s$).

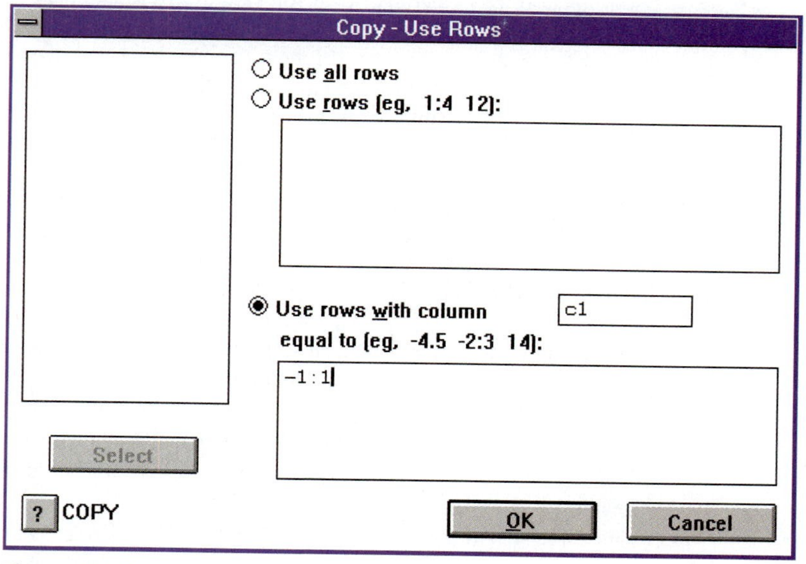

Figure M3.9

Repeat for $-2:2$ and $-3:3$, and establish that approximately 95 percent and 99.7 percent of the data values will lie between 2 and 3 standard deviations of the mean, respectively. Alternatively, you can compute these percentages by using the Session window. If the selected values are stored in K4, in this window type:

```
MTB>LET K4 = N(C2)/N(C1)*100
MTB>PRINT K4
```

where N( ) represents *the number of*. This will compute the percentages for you. Alternatively, you could use the sequence Calc→Mathematical Expressions and define K4 in that dialog box.

Computer Exercises

1. Use Minitab or any other statistical software to help find the *sample* mean, median, mode, standard deviation, variance, first quartile, third quartile, 43rd percentile, 69th percentile, and 90th percentile for the data set in Prob. 10, Sec. 3.3.1.

2. Use Minitab or any other statistical software to construct a frequency table for the information given in Fig. 3.8 (weights of 103 male health club members) of the text. Save this table in the Data window (refer to the lab for Chap. 2).

(a) Use the Session window to help find the sample mean and the median for the data set.

(b) Construct several histograms for this data set.

(c) Discuss what type of data set you are dealing with by observing these histograms. Present these results in a report.

3. Use Minitab or any other statistical software to investigate the empirical rule. Generate several sets of data (at least 10) by using Calc→Random Data→Normal. Vary the mean and standard deviation, and follow the procedure described in this lab. Generate histograms and compute the average values for the proportions of values that lie between 1, 2, and 3 sample standard deviations from the sample mean. Present these results in a report, and discuss your observations as related to the empirical rule.

4. Collect a set of paired data values (at least 30 pairs). Present a scatter diagram for this data set, and fit a line to it. Using the fitted line, estimate values for the Y variable from the X variable. Compute the correlation coefficient and interpret its value. Present your results in a report.

5. Use Minitab or any other statistical software to help find the sample mean, median, mode, standard deviation, variance, first quartile, third quartile, 35th percentile, 57th percentile, and 80th percentile for the data set in the stem-and-leaf plot (weights of 200 members of a health club) of Example 3.23, Sec. 3.6.

CHAPTER 4

Probability

Probability

> *Probability is the very guide of life.*
> Cicero, De Natura

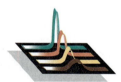

This chapter starts with consideration of an experiment whose outcome cannot be predicted with certainty. We define the events of this experiment. We then introduce the concept of the probability of an event, which is the probability that the outcome of the experiment is contained in the event. An interpretation of the probability of an event as being a long-term relative frequency is given. Properties of probabilities are discussed. The conditional probability of one event, given the occurrence of a second event, is introduced. We see what it means for events to be independent.

4.1   INTRODUCTION

To gain information about the current leader in the next gubernatorial election, a representative sample of 100 voters has been polled. If 62 of those polled are in favor of the Republican candidate, can we conclude that a majority of the state's voters favor this candidate? Or, is it possible that *by chance* the sample contained a much greater proportion of this candidate's supporters than is contained in the general population, and that the Democratic candidate is actually the current choice of a majority of the electorate?

To answer the above questions, it is necessary to know something about the chance that as many as 62 people in a representative sample of size 100 would favor a candidate who, in fact, is not favored by a majority of the entire population. Indeed, as a general rule, to be able to draw valid inferences about a population from a sample, one needs to know how likely it is that certain events will occur under various circumstances. The determination of the likelihood, or chance, that an event will occur is the subject matter of *probability.*

4.2   SAMPLE SPACE AND EVENTS OF AN EXPERIMENT

The word *probability* is a commonly used term which relates to the chance that a particular event will occur when some experiment is performed, where we use the word *experiment* in a very broad sense. Indeed, an *experiment* for us is *any* process that produces an observation, or *outcome.*

We are often concerned with an experiment whose outcome is not predictable, with certainty, in advance. Even though the outcome of the experiment will not be known in advance, we will suppose that the set of all possible outcomes is known. This set of all possible outcomes of the experiment is called the *sample space* and is denoted by *S*.

Definitions

An *experiment* is any process that produces an observation or *outcome.*
The set of all possible outcomes of an experiment is called the *sample space.*

Example 4.1   Some examples of experiments and their sample spaces are as follows.

(a) If the outcome of the experiment is the gender of a child, then

$$S = \{g, b\}$$

where outcome g means that the child is a girl and b that it is a boy.

(b) If the experiment consists of flipping two coins and noting whether they land heads or tails, then

$$S = \{(H, H), (H, T), (T, H), (T, T)\}$$

The outcome is (H, H) if both coins land heads, (H, T) if the first coin lands heads and the second tails, (T, H) if the first is tails and the second is heads, and (T, T) if both coins land tails.

(c) If the outcome of the experiment is the order of finish in a race among 7 horses having positions 1, 2, 3, 4, 5, 6, 7, then

$$S = \{\text{all orderings of } 1, 2, 3, 4, 5, 6, 7\}$$

The outcome (4, 1, 6, 7, 5, 3, 2) means, for instance, that the number 4 horse comes in first, the number 1 horse comes in second, and so on.

(d) Consider an experiment that consists of rolling two six-sided dice and noting the sides facing up. Calling one of the dice die 1 and the other die 2, we can represent the outcome of this experiment by the pair of upturned values on these dice. If we let (i, j) denote the outcome in which die 1 has value i and die 2 has value j, then the sample space of this experiment is

$$S = \{(1, 1), (1, 2), (1, 3), (1, 4), (1, 5), (1, 6), (2, 1), (2, 2), (2, 3),$$
$$(2, 4), (2, 5), (2, 6), (3, 1), (3, 2), (3, 3), (3, 4), (3, 5), (3, 6),$$
$$(4, 1), (4, 2), (4, 3), (4, 4), (4, 5), (4, 6), (5, 1), (5, 2), (5, 3),$$
$$(5, 4), (5, 5), (5, 6), (6, 1), (6, 2), (6, 3), (6, 4), (6, 5), (6, 6)\}$$

Any set of outcomes of the experiment is called an *event*. That is, an event is a subset of the sample space. Events will be denoted by the capital letters A, B, C, and so on.

Example 4.2   In Example 4.1(a), if $A = \{g\}$, then A is the event that the child is a girl. Similarly, if $B = \{b\}$, then B is the event that the child is a boy.
   In Example 4.1(b), if $A = \{(H, H), (H, T)\}$, then A is the event that the first coin lands on heads.
   In Example 4.1(c), if

$$A = \{\text{all outcomes in } S \text{ starting with } 2\}$$

then A is the event that horse number 2 wins the race.
   In Example 4.1(d), if

$$A = \{(1, 6), (2, 5), (3, 4), (4, 3), (5, 2), (6, 1)\}$$

then A is the event that the sum of the dice is 7.

Definition

Any set of outcomes of the experiment is called an *event*. We designate events by the letters A, B, C, and so on. We say that the event A *occurs* whenever the outcome is contained in A.

For any two events A and B, we define the new event $A \cup B$, called the *union* of events A and B, to consist of all outcomes that are in A or in B or in both A and B. That is, the event $A \cup B$ will occur if *either* A or B occurs.

In Example 4.1(a), if $A = \{g\}$ is the event that the child is a girl and $B = \{b\}$ is the event that it is a boy, then $A \cup B = \{g, b\}$. That is, $A \cup B$ is the whole sample space S.

In Example 4.1(c), let

$A = \{$all outcomes starting with 4$\}$

be the event that the number 4 horse wins; and let

$B = \{$all outcomes whose second element is 2$\}$

be the event that the number 2 horse comes in second. Then $A \cup B$ is the event that either the 4 horse wins or the 2 horse comes in second or both.

A graphical representation of events that is very useful is the *Venn diagram*. The sample space S is represented as consisting of all the points in a large rectangle, and events are represented as consisting of all the points in circles within the rectangle. Events of interest are indicated by shading appropriate regions of the diagram. The colored region of Fig. 4.1 represents the union of events A and B.

For any two events A and B, we define the *intersection* of A and B to consist of all outcomes that are both in A and in B. That is, the intersection will occur if *both* A and B occur. We denote the intersection of A and B by $A \cap B$. The colored region of Fig. 4.2 represents the intersection of events A and B.

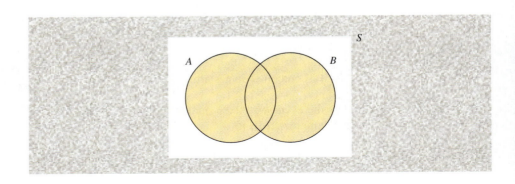

Figure 4.1
A Venn
diagram: shaded
region is $A \cup B$.

In Example 4.1(b), if $A = \{(H, H), (H, T)\}$ is the event that the first coin lands heads and $B = \{(H, T), (T, T)\}$ is the event that the second coin lands tails, then $A \cap B = \{(H, T)\}$ is the event that the first coin lands heads and the second lands tails.

In Example 4.1(c), if A is the event that the number 2 horse wins and B is the event that the number 3 horse wins, then the event $A \cap B$ does not contain any outcomes and so cannot occur. We call the event without any outcomes the *null* event. If the intersection of A and B is the null event, then since A and B cannot simultaneously occur, we say that A and B are *disjoint* or *mutually exclusive*. Two disjoint events are pictured in the Venn diagram of Fig. 4.3.

For any event A we define the event A^c, called the *complement* of A, to consist of all outcomes in the sample space that are not in A. That is, A^c will occur when A does not, and vice versa. For instance, in Example 4.1(a), if $A = \{g\}$ is the event that the child is a girl, then $A^c = \{b\}$ is the event that it is a boy. Also note that the complement of the sample space is the null set, that is, $S^c = \varnothing$. Figure 4.4 indicates A^c, the complement of event A.

We can also define unions and intersections of more than two events. For instance, the union of events A, B, and C, written $A \cup B \cup C$, consists of all the outcomes of the experiment that are in A or in B or in C. Thus, $A \cup B \cup C$ will occur if at least one of these events occurs. Similarly, the intersection $A \cap B \cap C$ consists of the outcomes that are in all the events A, B, and C. Thus, the intersection will occur only if all the events occur.

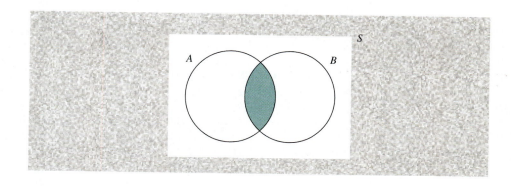

Figure 4.2
Shaded region
is $A \cap B$.

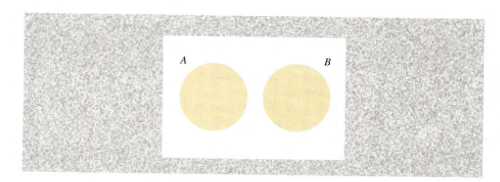

Figure 4.3
A and B are
disjoint events.

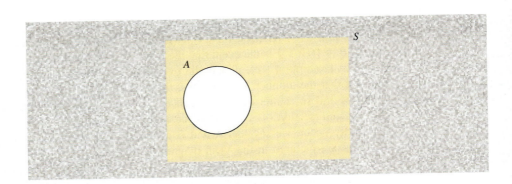

Figure 4.4
Shaded region
is A^c.

We say that events A, B, and C are *disjoint* if no two of them can simultaneously occur.

Problems

1. A box contains three balls—one red, one blue, and one yellow. Consider an experiment that consists of withdrawing a ball from the box, replacing it, and withdrawing a second ball.
   (a) What is the sample space of this experiment?
   (b) What is the event that the first ball drawn is yellow?
   (c) What is the event that the same ball is drawn twice?

2. Repeat Prob. 1 when the second ball is drawn without replacement of the first ball.

3. Audrey and her boyfriend Charles must both choose which colleges they will attend in the coming fall. Audrey was accepted at the University of Michigan (MI), Reed College (OR), San Jose State College (CA), Yale University (CT), and Oregon State University (OR). Charles was accepted at Oregon State University and San Jose State College. Let the outcome of the experiment consist of the colleges that Audrey and Charles choose to attend.
   (a) List all the outcomes in sample space S.
   (b) List all the outcomes in the event that Audrey and Charles attend the same school.
   (c) List all the outcomes in the event that Audrey and Charles attend different schools.
   (d) List all the outcomes in the event that Audrey and Charles attend schools in the same state.

4. An experiment consists of flipping a coin 3 times and each time noting whether it lands heads or tails.

(a) What is the sample space of this experiment?

(b) What is the event that tails occur more often than heads?

5. Family members have decided that their next vacation will be either in France or in Canada. If they go to France, they can either fly or take a boat. If they go to Canada, they can drive, take a train, or fly. Letting the outcome of the experiment be the location of their vacation and their mode of travel, list all the points in sample space S. Also list all the outcomes in A, where A is the event that the family flies to the destination.

6. The New York Yankees and the Chicago White Sox are playing three games this weekend. Assuming that all games are played to a conclusion and that we are interested only in which team wins each game, list all the outcomes in sample space S. Also list all the outcomes in A, where A is the event that the Yankees win more games than the White Sox.

7. Let $S = \{1, 2, 3, 4, 5, 6\}$, $A = \{1, 3, 5\}$, $B = \{4, 6\}$, and $C = \{1, 4\}$. Find

(a) $A \cap B$

(b) $B \cup C$

(c) $A \cup (B \cap C)$

(d) $(A \cup B)^c$

Note: The operations within parentheses are performed first. For instance, in (c) first determine the intersection of B and C, and then take the union of A and that set.

8. A cafeteria offers a three-course meal. One chooses a main course, a starch, and a dessert. The possible choices are given below.

Meal	Choices
Main course	Chicken or roast beef
Starch course	Pasta or rice or potatoes
Dessert	Ice cream or gelatin or apple pie

An individual is to choose one course from each category.

(a) List all the outcomes in the sample space.

(b) Let A be the event that ice cream is chosen. List all the outcomes in A.

(c) Let B be the event that chicken is chosen. List all the outcomes in B.

(d) List all the outcomes in the event $A \cap B$.

(e) Let C be the event that rice is chosen. List all the outcomes in C.

(f) List all the outcomes in the event $A \cap B \cap C$.

9. A hospital administrator codes patients according to whether they have insurance and according to their condition, which is rated as good, fair, serious, or

critical. The administrator records a 0 if a patient has no insurance and a 1 if he or she does, and then records one of the letters *g, f, s,* or *c* depending on the patient's condition. Thus, for instance, the coding 1, *g* is used for a patient with insurance who is in good condition. Consider an experiment that consists of the coding of a new patient.

(a) List the sample space of this experiment.

(b) Specify the event corresponding to the patient's being in serious or critical condition and having no medical insurance.

(c) Specify the event corresponding to the patient's being in either good or fair condition.

(d) Specify the event corresponding to the patient's having insurance.

10. The following pairs of events *E* and *F* relate to the same experiment. Tell in each case whether *E* and *F* are disjoint events.

(a) A die is rolled. Event *E* is that it lands on an even number, and *F* is the event that it lands on an odd number.

(b) A die is rolled. Event *E* is that it lands on 3, and *F* is the event that it lands on an even number.

(c) A person is chosen. Event *E* is that this person was born in the United States, and *F* is the event that this person is a U.S. citizen.

(d) A man is chosen. Event *E* is that he is over 30 years of age, and *F* is the event that he has been married for over 30 years.

(e) A woman waiting in line to register her car at the department of motor vehicles is chosen. Event *E* is that the car is made in the United States, and *F* is the event that it is made in a foreign country.

11. Let *A* be the event that a rolled die lands on an even number.

(a) Describe in words the event A^c.

(b) Describe in words the event $(A^c)^c$.

(c) In general, let *A* be an event. What is the complement of its complement? That is, what is $(A^c)^c$?

12. Two dice are rolled. Let *A* be the event that the sum of the dice is even, let *B* be the event that the first die lands on 1, and let *C* be the event that the sum of the dice is 6. Describe the following events.

(a) $A \cap B$

(b) $A \cup B$

(c) $B \cap C$

(d) B^c

(e) $A^c \cap C$

(f) $A \cap B \cap C$

13. Let *A*, *B*, and *C* be events. Use Venn diagrams to represent the event that of *A*, *B*, and *C*

(a) Only A occurs.
(b) Both A and B occur, but C does not.
(c) At least 1 event occurs.
(d) At least 2 of the events occur.
(e) All 3 events occur.

4.3 PROPERTIES OF PROBABILITY

It is an empirical fact that if an experiment is continually repeated under the same conditions, then, for any event A, the proportion of times that the outcome is contained in A approaches some value as the number of repetitions increases. For example, if a coin is continually flipped, then the proportion of flips landing on tails will approach some value as the number of flips increases. It is this long-run proportion, or *relative frequency,* that we often have in mind when we speak of the probability of an event.

Consider an experiment whose sample space is S. We suppose that for each event A there is a number, denoted $P(A)$ and called the *probability* of event A, that is in accord with the following three properties.

PROPERTY 1:   For any event A, the probability of A is a number between 0 and 1. That is,

$$0 \leq P(A) \leq 1$$

PROPERTY 2:   The probability of sample space S is 1. Symbolically,

$$P(S) = 1$$

PROPERTY 3:   The probability of the union of disjoint events is equal to the sum of the probabilities of these events. For instance, if A and B are disjoint, then

$$P(A \cup B) = P(A) + P(B)$$

The quantity $P(A)$ represents the probability that the outcome of the experiment is contained in event A. Property 1 states that the probability that the outcome of the experiment is contained in A is some value between 0 and 1. Property 2 states that, with probability 1, the outcome of the experiment will be an element of sample space S. Property 3 states that if events A and B cannot simultaneously occur, then the probability that the outcome of the experiment is contained in either A or B is equal to the sum of the probability that it is in A and the probability that it is in B.

If we interpret $P(A)$ as the long-run relative frequency of event A, then the above conditions are satisfied. The proportion of experiments in which the outcome is contained in A would certainly be a number between 0 and 1. The proportion of experiments in which the outcome is contained in S is 1 since all outcomes are contained in sample space S. Finally, if A and B have no outcomes in common, then the proportion of experiments whose outcome is in either A or B is equal to the proportion whose outcome is in A plus the proportion whose outcome is in B. For instance, if the proportion of time that a pair of rolled dice sums to 7 is 1/6 and the proportion of time that they sum to 11 is 1/18, then the proportion of time that they sum to either 7 or 11 is $1/6 + 1/18 = 2/9$.

Properties 1, 2, and 3 can be used to establish some general results concerning probabilities. For instance, since A and A^c are disjoint events whose union is the entire sample space, we can write

$$S = A \cup A^c$$

Using properties 2 and 3 now yields the following.

$$
\begin{aligned}
1 &= P(S) && \text{by property 2} \\
&= P(A \cup A^c) \\
&= P(A) + P(A^c) && \text{by property 3}
\end{aligned}
$$

Therefore, we see that

$$P(A^c) = 1 - P(A)$$

In words, the probability that the outcome of the experiment is not contained in A is 1 minus the probability that it is. For instance, if the probability of obtaining heads on the toss of a coin is .4, then the probability of obtaining tails is .6.

The following formula relates the probability of the union of events A and B, which are not necessarily disjoint, to $P(A)$, $P(B)$, and the probability of the intersection of A and B. It is often called the *addition rule of probability.*

Addition rule

For any events A and B,

$$P(A \cup B) = P(A) + P(B) - P(A \cap B)$$

To see why the addition rule holds, note that $P(A \cup B)$ is the probability of all outcomes that are either in A or in B. On the other hand, $P(A) + P(B)$ is the probability of all the outcomes that are in A plus the probability of all the outcomes that are in B. Since any outcome that is in both A and B is counted twice in $P(A) + P(B)$ and only once in $P(A \cup B)$ (see Fig. 4.5), it follows that

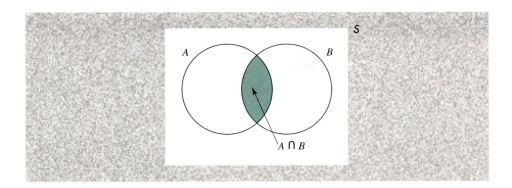

Figure 4.5
$P(A \cup B) =$
$P(A) + P(B) -$
$P(A \cap B).$

$$P(A) + P(B) = P(A \cup B) + P(A \cap B)$$

Subtracting $P(A \cap B)$ from both sides of the preceding equation gives the addition rule.

Example 4.3 illustrates the use of the addition rule.

Example 4.3   A certain retail establishment accepts either the American Express or the VISA credit card. A total of 22 percent of its customers carry an American Express card, 58 percent carry a VISA credit card, and 14 percent carry both. What is the probability that a customer will have at least one of these cards?

Solution Let A denote the event that the customer has an American Express card, and let B be the event that she or he has a VISA card. The information given above yields

$$P(A) = .22 \qquad P(B) = .58 \qquad P(A \cap B) = .14$$

By the additive rule, the desired probability $P(A \cup B)$ is

$$P(A \cup B) = .22 + .58 - .14 = .66$$

That is, 66 percent of the establishment's customers carry at least one of the cards that it will accept.

As an illustration of the interpretation of probability as a long-run relative frequency, we have simulated 10,000 flips of a perfectly symmetric coin. The total numbers of heads and tails that occurred in the first 10, 50, 100, 500, 2000, 6000, 8000, and 10,000 flips, along with the proportion of them that was heads, are presented in Table 4.1. Note how the proportion of the flips that lands heads becomes very close to .5 as the number of flips increases.

The results of 10,000 simulated rolls of a perfectly symmetric die are presented in Table 4.2.

| Table 4.1 | **10,000 Flips of a Symmetric Coin** |

n	Number of heads in first n flips	Number of tails in first n flips	Proportion of first n flips that lands on heads
10	3	7	.3
50	21	29	.42
100	46	54	.46
500	248	252	.496
2,000	1,004	996	.502
6,000	3,011	2,989	.5018
8,000	3,974	4,026	.4968
10,000	5,011	4,989	.5011

| Table 4.2 | **10,000 Rolls of a Symmetric Die** |

	i					
	1	**2**	**3**	**4**	**5**	**6**
Frequency of outcome	1724	1664	1628	1648	1672	1664
Relative frequency	.1724	.1664	.1628	.1648	.1672	.1664

Note: $1/6 = 0.166667$.

Problems

1. Suppose the sample space of an experiment is

$$S = \{1, 2, 3, 4, 5, 6\}$$

Let A_i denote the event consisting of the single outcome i, and suppose that

$$P(A_1) = .1 \qquad P(A_4) = .15$$
$$P(A_2) = .2 \qquad P(A_5) = .1$$
$$P(A_3) = .15 \qquad P(A_6) = .3$$

That is, the outcome of the experiment is 1 with probability .1, it is 2 with probability .2, it is 3 with probability .15, and so on. Let events E, F, and G be as follows:

$$E = \{1, 3, 5\} \qquad F = \{2, 4, 6\} \qquad G = \{1, 4, 6\}$$

Historical Perspective

(Bettmann)

Pierre Fermat

(Bettmann)

Blaise Pascal

The notion that chance, or probability, can be treated numerically is relatively recent. Indeed, for most of recorded history it was felt that what occurred in life was determined by forces that were beyond one's ability to understand. It was only during the first half of the 17th century, near the end of the Renaissance, that people became curious about the world and the laws governing its operation. Among the curious were the gamblers. A group of Italian gamblers, unable to answer certain questions concerning dice, approached the famous scientist Galileo. Galileo, though busy with other work, found their problems to be of interest and not only provided solutions but also wrote a short treatise on games of chance.

A few years later a similar story took place in France, where a gambler known as Chevalier de Mere resided. De Mere, a strong amateur mathematician as well as a gambler, had an ac-

quaintance with the brilliant mathematician Blaise Pascal. It was to Pascal that de Mere turned for help in his more difficult gaming questions. One particular problem, known as the *problem of the points,* concerned the equitable division of stakes when two players are interrupted in the midst of a game of chance. Pascal found this problem particularly intriguing and, in 1654, wrote to the mathematician Pierre Fermat about it. Their resulting exchange of letters not only led to a solution of this problem but also laid the framework for the solution of many other problems connected with games of chance. Their celebrated correspondence, cited by some as the birth date of probability, stimulated interest in probability among some of the foremost European mathematicians of the time. For instance, the young Dutch genius Ludwig Huyghens came to Paris to discuss the new subject, and activity in this new field grew rapidly.

Find
(a) $P(E), P(F), P(G)$
(b) $P(E \cup F)$
(c) $P(E \cup G)$
(d) $P(F \cup G)$
(e) $P(E \cup F \cup G)$
(f) $P(E \cap F)$
(g) $P(F \cap G)$
(h) $P(E \cap G)$
(i) $P(E \cap F \cap G)$

2. If A and B are disjoint events for which $P(A) = .2$ and $P(B) = .5$, find
(a) $P(A^c)$
(b) $P(A \cup B)$
(c) $P(A \cap B)$
(d) $P(A^c \cap B)$

3. Phenylketonuria is a genetic disorder that produces mental retardation. About one child in every 10,000 live births in the United States has phenylketonuria. What is the probability that the next child born in a Houston hospital has phenylketonuria?

4. A certain person encounters three traffic lights when driving to work. Suppose that the following represent the probabilities of the total number of red lights that she has to stop for:

$$P\{0 \text{ red lights}\} = .14$$
$$P\{1 \text{ red light}\} = .36$$
$$P\{2 \text{ red lights}\} = .34$$
$$P\{3 \text{ red lights}\} = .16$$

(a) What is the probability that she stops for at least one red light when driving to work?
(b) What is the probability that she stops for more than two red lights?

5. If A and B are disjoint events, is the following possible?

$$P(A) + P(B) = 1.2$$

What if A and B are not disjoint?

6. If the probability of drawing a king from a deck of pinochle cards is 1/6 and the probability of drawing an ace is 1/6, what is the probability of drawing either an ace or a king?

7. Suppose that the demand for Christmas trees from a certain dealer will be

1100	with probability .2
1400	with probability .3
1600	with probability .4
2000	with probability .1

Find the probability that the dealer will be able to sell his entire stock if he purchases
(a) 1100 trees
(b) 1400 trees
(c) 1600 trees
(d) 2000 trees

8. The Japanese automobile company Lexus has established a reputation for quality control. Recent statistics indicate that a newly purchased Lexus ES 300 will have

0 defects with probability .12
1 defect with probability .18
2 defects with probability .25
3 defects with probability .20

4 defects   with probability .15
5 or more defects   with probability .10

If you purchase a new Lexus ES 300, find the probability that it will have
(a) 2 or fewer defects
(b) 4 or more defects
(c) Between (inclusive) 1 and 3 defects

Let p denote the probability it will have an even number of defects. Whereas the information given above does not enable us to specify the value of p, find the
(d) Largest
(e) Smallest

value of p which is consistent with the preceding.

9. When typing a five-page manuscript, a certain typist makes

0 errors   with probability .20
1 error   with probability .35
2 errors   with probability .25
3 errors   with probability .15
4 or more errors   with probability .05

If you give such a manuscript to this typist, find the probability that it will contain
(a) 3 or fewer errors
(b) 2 or fewer errors
(c) 0 errors

10. The table given below is a modern version of a *life table,* which was first developed by John Graunt in 1662. It gives the probabilities that a newly born member of a certain specified group will die in his or her ith decade of life, for i ranging from 1 to 10. The first decade starts with birth and ends with an individual's 10th birthday. The second decade starts at age 10 and ends at the 20th birthday, and so on.

Life Table

Decade	Probability of death	Decade	Probability of death
1 0-10	.062	6	.124
2 10-20	.012	7	.215
3 20-30	.024	8	.271
4 30-40	.033	9	.168
5	.063	10	.028

For example, the probability that a newborn child dies in her or his fifties is .124. Find the probability that a newborn will
(a) Die between the ages of 30 and 60
(b) Not survive to age 40
(c) Survive to age 80

11. The family picnic scheduled for tomorrow will be postponed if it is either cloudy or rainy. The weather report states that there is a 40 percent chance of rain tomorrow, a 50 percent chance of cloudiness, and a 20 percent chance that it will be both cloudy and rainy. What is the probability that the picnic will be postponed?

12. In Example 4.3, what proportion of customers has neither an American Express nor a VISA card?

13. It is estimated that 30 percent of all adults in the United States are obese and that 3 percent suffer from diabetes. If 2 percent of the population both is obese and suffers from diabetes, what percentage of the population either is obese or suffers from diabetes?

14. Welds of tubular joints can have two types of defects, which we call A and B. Each weld produced has defect A with probability .064, defect B with probability .043, and both defects with probability .025. Find the proportion of welds that has
(a) Either defect A or defect B
(b) Neither defect

15. A customer that goes to the suit department of a certain store will purchase a suit with probability .3. The customer will purchase a tie with probability .2 and will purchase both a suit and a tie with probability .1. What proportion of customers purchases neither a suit nor a tie?

16. Anita has a 40 percent chance of receiving an A grade in statistics, a 60 percent chance of receiving an A in physics, and an 86 percent chance of receiving an A in either statistics or physics. Find the probability that she
(a) Does not receive an A in either statistics or physics
(b) Receives A's in both statistics and physics

17. This problem uses a Venn diagram (facing page) to present a formal proof of the addition rule. Events A and B are represented by circles in the Venn diagram.
    In terms of A and B, describe the region labeled
(a) I
(b) II
(c) III

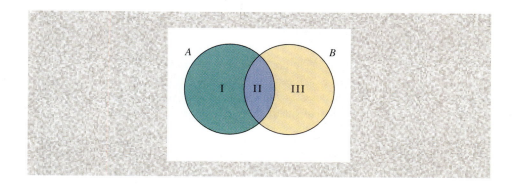

A Venn
diagram:
dividing up
$A \cup B$.

Express, in terms of $P(\text{I})$, $P(\text{II})$, and $P(\text{III})$,
(d) $P(A \cup B)$
(e) $P(A)$
(f) $P(B)$
(g) $P(A \cap B)$
(h) Conclude that

$$P(A \cup B) = P(A) + P(B) - P(A \cap B)$$

4.4 EXPERIMENTS HAVING EQUALLY LIKELY OUTCOMES

For certain experiments it is natural to assume that each outcome in the sample space
S is equally likely to occur. That is, if sample space S consists of N outcomes, say,
$S = \{1, 2, \ldots, N\}$, then it is often reasonable to suppose that

$$P(\{1\}) = P(\{2\}) = \cdots = P(\{N\})$$

In the above, $P(\{i\})$ is the probability of the event consisting of the single outcome
i; that is, it is the probability that the outcome of the experiment is i.

Using the properties of probability, we can show that the above implies that the
probability of any event A is equal to the proportion of the outcomes in the sample
space that is in A. That is,

$$P(A) = \frac{\text{number of outcomes in } S \text{ that are in } A}{N}$$

Example 4.4   In a survey of 420 members of a retirement center, it was found
that 144 are smokers and 276 are not. If a member is selected in such a way that each

of the members is equally likely to be the one selected, what is the probability that person is a smoker?

Solution There are 420 outcomes in the sample space of the experiment of selecting a member of the center. Namely, the outcome is the person selected. Since there are 144 outcomes in the event that the selected person is a smoker, it follows that the probability of this event is

$$P\{\text{smoker}\} = \frac{144}{420} = \frac{12}{35}$$

Example 4.5   Suppose that when two dice are rolled, each of the 36 possible outcomes given in Example 4.1(d) is equally likely. Find the probability that the sum of the dice is 6, and that it is 7.

Solution If we let A denote the event that the sum of the dice is 6 and B that it is 7, then

$$A = \{(1, 5), (2, 4), (3, 3), (4, 2), (5, 1)\}$$

and

$$B = \{(1, 6), (2, 5), (3, 4), (4, 3), (5, 2), (6, 1)\}$$

Therefore, since A contains 5 outcomes and B contains 6, we see that

$$P(A) = P\{\text{sum is } 6\} = 5/36$$
$$P(B) = P\{\text{sum is } 7\} = 6/36 = 1/6$$

Example 4.6   One man and one woman are to be selected from a group that consists of 10 married couples. If all possible selections are equally likely, what is the probability that the woman and man selected are married to each other?

Solution Once the man is selected, there are 10 possible choices of the woman. Since one of these 10 choices is the wife of the man chosen, we see that the desired probability is 1/10.

When each outcome of the sample space is equally likely to be the outcome of the experiment, we say that an element of the sample space is *randomly selected*.

Example 4.7   An elementary school is offering two optional language classes, one in French and the other in Spanish. These classes are open to any of the 120 upper-grade students in the school. Suppose there are 32 students in the French class, 36 in the Spanish class, and a total of 8 who are in both classes. If an upper-grade student is randomly chosen, what is the probability that this student is enrolled in at least one of these classes?

Solution Let A and B denote, respectively, the events that the randomly chosen student is enrolled in the French class and is enrolled in the Spanish class. We will determine $P(A \cup B)$, the probability that the student is enrolled in either French or Spanish, by using the addition rule

$$P(A \cup B) = P(A) + P(B) - P(A \cap B)$$

Since 32 of the 120 students are enrolled in the French class, 36 of the 120 are in the Spanish class, and 8 of the 120 are in both classes, we have

$$P(A) = \frac{32}{120} \qquad P(B) = \frac{36}{120} \qquad \text{and} \qquad P(A \cap B) = \frac{8}{120}$$

Therefore,

$$P(A \cup B) = \frac{32}{120} + \frac{36}{120} - \frac{8}{120} = \frac{1}{2}$$

That is, the probability that a randomly chosen student is taking at least one of the language classes is 1/2.

Example 4.8   Table 4.3 lists the 1989 earnings frequencies of all full-time workers who are at least 15 years old, classified according to their annual salary and gender. Suppose one of these workers is randomly chosen. Find the probability that this person is

(a) A woman   (b) A man
(c) A man earning under $15,000   (d) A woman earning over $25,000

Table 4.3   **Earnings of Workers by Sex, 1989**

Earnings group	Number Women	Number Men	Distribution (percent) Women	Distribution (percent) Men	Likelihood of a woman in each earnings group (percent)
$2,499 or less	427,000	548,000	1.4	1.1	1.3
$2,500 to $4,999	440,000	358,000	1.4	.7	2.0
$5,000 to $7,499	1,274,000	889,000	4.1	1.8	2.3
$7,500 to $9,999	1,982,000	1,454,000	6.3	2.9	2.2
$10,000 to $14,999	6,291,000	5,081,000	20.1	10.2	2.0
$15,000 to $19,999	6,555,000	6,386,000	20.9	12.9	1.6
$20,000 to $24,999	5,169,000	6,648,000	16.5	13.4	1.2
$25,000 to $49,999	8,255,000	20,984,000	26.3	42.1	.6
$50,000 and over	947,000	7,377,000	3.0	14.9	.2
Total	31,340,000	49,678,000	100.0	100.0	—

Source: Department of Commerce, Bureau of the Census.

Solution

(a) Since 31,340,000 of the 31,340,000 + 49,678,000 = 81,018,000 workers are women, it follows that the probability that a randomly chosen worker is a woman is

$$\frac{31,340,000}{81,018,000} \approx .3868$$

That is, there is approximately a 38.7 percent chance that the randomly selected worker is a woman.

(b) Since the event that the randomly selected worker is a man is the complement of the event that the worker is a woman, we see from (a) that the probability is approximately 1 − .3868 = .6132.

(c) Since (in thousands) the number of men earning under $15,000 is

$$548 + 358 + 889 + 1454 + 5081 = 8330$$

we see that the desired probability is 8330/81,018 ≈ .1028. That is, there is approximately a 10.3 percent chance that the person selected is a man with an income under $15,000.

(d) The probability that the person selected is a woman with an income above $25,000 is

$$\frac{8255 + 947}{81,018} \approx .1136$$

That is, there is approximately an 11.4 percent chance that the person selected is a woman with an income above $25,000.

Problems

1. In an experiment involving smoke detectors, an alarm was set off at a college dormitory at 3 a.m. Out of 216 residents of the dormitory, 128 slept through the alarm. If one of the residents is randomly chosen, what is the probability that this person did not sleep through the alarm?

2. Among 32 dieters following a similar routine, 18 lost weight, 5 gained weight, and 9 remained the same weight. If one of these dieters is randomly chosen, find the probability that he or she
   (a) Gained weight
   (b) Lost weight
   (c) Neither lost nor gained weight

3. One card is to be selected at random from an ordinary deck of 52 cards. Find the probability that the selected card is
   (a) An ace   (b) Not an ace
   (c) A spade   (d) The ace of spades

4. The following table lists the 10 countries with the highest production of meat.

Country	Meat production (thousands of metric tons)
China	20,136
United States	17,564
Russia	12,698
Germany	6,395
France	3,853
Brazil	3,003
Argentina	2,951
Britain	2,440
Italy	2,413
Australia	2,373

Source: Statistical Abstract of the United States, 1990.

Suppose a World Health Organization committee is formed to discuss the long-term ramifications of producing such quantities of meat. Suppose further that it consists of one representative from each of these countries. If the chair of this committee is then randomly chosen, find the probability that this person will be from a country whose production of meat (in thousands of metric tons)
   (a) Exceeds 10,000
   (b) Is under 3500
   (c) Is between 4000 and 6000
   (d) Is less than 2000

5. The following data (taken from the U.S. National Center for Health Statistics, *Health Promotion and Disease Prevention,* 1985) relates the proportion of the population 18 years and older that has certain health characteristics.

Characteristic	Percentage of population having characteristic
Sleeps 6 hours or less per day	22.0
Never eats breakfast	24.3
Snacks every day	39.0
Not physically active	16.4
Occasionally a heavy drinker	37.5
Smokes	30.1
Overweight by more than 30 percent	13.0

Suppose a person who is at least 18 years old is randomly chosen. Find the probability that she or he
(a) Sleeps more than 6 hours per night
(b) Does not smoke
(c) Never eats breakfast
(d) Does not snack every day
(e) Is not overweight by more than 30 percent

6. A bag containing pennies and dimes has 4 times as many dimes as pennies. One coin is drawn. Assuming that the drawn coin is equally likely to be any of the coins, what is the probability that it is a dime?

7. A total of 44 out of 100 patients at a rehabilitation center are signed up for a special exercise program which consists of a swimming class and a calisthenics class. Each of these 44 patients takes at least one of these classes. Suppose that there are 26 patients in the swimming class and 28 in the calisthenics class. Find the probability that a randomly chosen patient at the center is
(a) Not in the exercise program
(b) Enrolled in both classes

8. Of the families in a certain community, 20 percent have a cat, 32 percent have a dog, and 12 percent have both a cat and a dog.
(a) If a family is chosen at random, what is the probability it has neither a dog nor a cat?
(b) If the community consists of 1000 families, how many of them have either a cat or a dog?

9. Of the students at a girls' school, 60 percent wear neither a ring nor a necklace, 20 percent wear a ring, and 30 percent wear a necklace. If one of them is randomly chosen, find the probability that she is wearing
(a) A ring or a necklace
(b) A ring and a necklace

10. A sports club has 120 members, of whom 44 play tennis, 30 play squash, and 18 play both tennis and squash. If a member is chosen at random, find the probability that this person
(a) Does not play tennis
(b) Does not play squash
(c) Plays neither tennis nor squash

11. In Prob. 10, how many members play either tennis or squash?

12. If two dice are rolled, find the probability that the sum of the dice is
(a) Either 7 or 11
(b) One of the values 2, 3, or 12
(c) An even number

13. Suppose 2 people are randomly chosen from a set of 20 people that consists of 10 married couples. What is the probability that the 2 people are married to each other? (*Hint:* After the initial person is chosen, the next one is equally likely to be any of the remaining people.)

14. Find the probability that a randomly chosen worker in Example 4.8
    (a) Earns under $7500
    (b) Is a woman who earns between $10,000 and $20,000
    (c) Earns under $50,000

15. A real estate agent has a set of 10 keys, one of which will open the front door of a house he is trying to show to a client. If the keys are tried in a completely random order, find the probability that
    (a) The first key opens the door
    (b) All 10 keys are tried

16. A group of 5 girls and 4 boys is randomly lined up.
    (a) What is the probability that the person in the second position is a boy?
    (b) What is the probability that Charles (one of the boys) is in the second position?

17. The following data are from the U.S. National Oceanic and Atmospheric Administration. They give the average number of days in each month with precipitation of .01 inch or more for Washington, D.C.

Jan.	Feb.	Mar.	Apr.	May	June	July	Aug.	Sept.	Oct.	Nov.	Dec.
10	9	11	10	11	10	10	9	8	7	8	9

Find the probability you will encounter rain if you are planning to visit Washington, D.C., next
    (a) January 5
    (b) August 12
    (c) April 15
    (d) May 15
    (e) October 12

18. Suppose that a foreign student is randomly chosen in a manner that results in each being equally likely to be chosen. Use the graphics entitled "Foreign Students: Who They Are" to answer the following questions.
    (a) What is the probability that he or she is studying for either an undergraduate or graduate degree?
    (b) What is the probability that he or she is primarily supported either by their home government or a private (nonfamily) sponsor?
    (c) What is the probability that he or she is studying one of the sciences (that is, physical/life sciences or social sciences or health sciences or math/computer science)?

Foreign Students: Who They Are

The Institute of International Education, a private organization that administers exchange programs, has released its annual profile of foreign college students in the United States. The report found that while Asian students still make up a majority of the foreign students, their numbers are leveling off. Researchers also found that the number of exchange students who are women continues to increase.

Number of Students

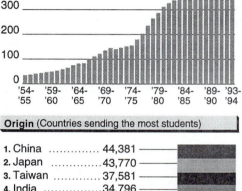

449,749

500 thousand
400
300
200
100
0

'54- '59- '64- '69- '74- '79- '84- '89- '93-
'55   '60   '65   '70   '75   '80   '85   '90   '94

Origin (Countries sending the most students)

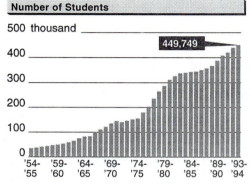

1. China 44,381
2. Japan 43,770
3. Taiwan 37,581
4. India 34,796
5. South Korea 31,076
6. Canada 22,655
7. Hong Kong 13,752
8. Malaysia 13,718
9. Indonesia 11,744
10. Thailand 9,537
All other countries

Demographics

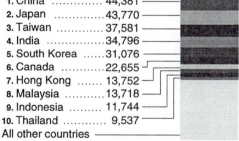

Men   Women
62%   38

At public   At private
colleges   colleges
64%   36

Undergraduate ——   —— Graduate
Intensive   48%   45
English   ⌐Practical training 3
programs 3 ——————   └No degree 2

What They Study

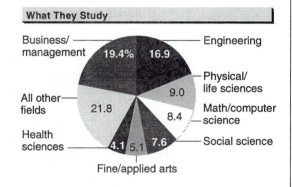

Business/ ———   ——— Engineering
management   19.4%   16.9
                                 ⌐Physical/
                                  life sciences
                            9.0
All other ——   Math/computer
fields   21.8   8.4 ⌐science

Health   ——— Social science
sciences ——   4.1 5.1 7.6
             Fine/applied arts

Who Pays (primary source of financing)

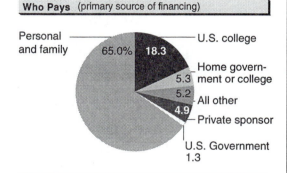

Personal ———   ——— U.S. college
and family   65.0%   18.3
                                 ⌐Home govern-
                            5.3   ment or college
                            5.2 —All other
                            4.9
                                 —Private sponsor

                                 U.S. Government
                                 1.3

Institutions With the Most Students

Institution	Students
Boston University	4,547
University of Southern California	4,189
University of Wisconsin at Madison	4,076
University of Texas at Austin	4,024
New York University	3,636
Ohio State University	3,612
Columbia University	3,585
University of Pennsylvania	3,436
University of Illinois at Champaign-Urbana	3,183
Southern Illinois University	2,832

Source: Institute for International Education

N.Y. Times News Service

If 4.2 percent of the foreign students attend New York University, what is the probability that he or she is
(d) Attending Boston University?
(e) Attending Columbia University?

If 1.2 percent of the foreign students are from Thailand, what is the probability that he or she is
(f) From either China or Taiwan?

4.5   CONDITIONAL PROBABILITY AND INDEPENDENCE

We are often interested in determining probabilities when some partial information concerning the outcome of the experiment is available. In such situations, the probabilities are called *conditional probabilities.*

As an example of a conditional probability, suppose two dice are to be rolled. Then, as noted in Example 4.1(d), the sample space of this experiment is the set of 36 outcomes (i, j), where both i and j range from 1 through 6. The outcome (i, j) results when the first die lands on i and the second on j.

Suppose that each of the 36 possible outcomes is equally likely to occur and thus has probability 1/36. (When this is the case, we say that the dice are *fair.*) Suppose further that the first die lands on 4. Given this information, what is the resulting probability that the sum of the dice is 10? To determine this probability, we reason as follows. Given that the first die lands on 4, there are 6 possible outcomes of the experiment, namely,

$$(4, 1), (4, 2), (4, 3), (4, 4), (4, 5), (4, 6)$$

In addition, since these outcomes initially had the same probabilities of occurrence, they should still have equal probabilities. That is, given that the first die lands on 4, the *conditional* probability of each of the above outcomes should be 1/6. Since in only one of the above outcomes is the sum of the dice equal to 10, namely, the outcome $(4, 6)$, it follows that the conditional probability that the sum is 10, given that the first die lands on 4, is 1/6.

If we let B denote the event that the sum of the dice is 10 and let A denote the event that the first die lands on 4, then the probability obtained above is called the *conditional probability of B given that A has occurred.* It is denoted by

$$P(B|A)$$

A general formula for $P(B|A)$ can be derived by an argument similar to the one used above. Suppose that the outcome of the experiment is contained in A. Now, in order for the outcome to also be in B, it must be in both A and B; that is, it must be in $A \cap B$. However, since we know that the outcome is in A, it follows that A becomes our new (or reduced) sample space and the probability that event $A \cap B$ occurs is the probability of $A \cap B$ relative to the probability of A. That is (see Fig. 4.6),

$$P(B|A) = \frac{P(A \cap B)}{P(A)}$$

The above definition of conditional probability is consistent with the interpretation of probability as being a long-run relative frequency. To show this, suppose that a large number, call it n, of repetitions of the experiment are performed. We will now argue that if we consider only those experiments in which A occurs, then $P(B|A)$ will equal the long-run proportion of them in which B also occurs. To see this, note that since $P(A)$ is the long-run proportion of experiments in which A occurs, it follows that in n repetitions of the experiment, A will occur approximately $nP(A)$ times. Similarly, in approximately $nP(A \cap B)$ of these experiments, both A and B will occur. Hence, out of the approximately $nP(A)$ experiments for which the outcome is contained in A, approximately $nP(A \cap B)$ of them will also have their outcomes in B. Therefore, of those experiments whose outcomes are in A, the proportion whose outcome is also in B is approximately equal to

$$\frac{nP(A \cap B)}{nP(A)} = \frac{P(A \cap B)}{P(A)}$$

Since this approximation becomes exact as n becomes larger and larger, we see that we have given the appropriate definition of the conditional probability of B given that A has occurred.

Example 4.9 As a further check of the preceding formula for the conditional probability, use it to compute the conditional probability that the sum of a pair of rolled dice is 10, given that the first die lands on 4.

Solution Letting B denote the event that the sum of the dice is 10 and A the event that the first die lands on 4, we have

Figure 4.6
$P(B|A)$
$= \dfrac{P(A \cap B)}{P(A)}$

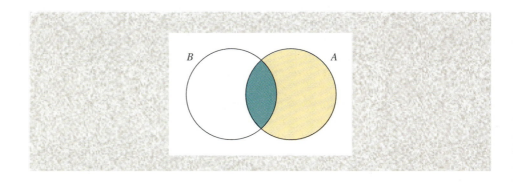

$$P(B|A) = \frac{P(A \cap B)}{P(A)}$$

$$= \frac{P(\{(4,6)\})}{P(\{(4, 1), (4, 2), (4, 3), (4, 4), (4, 5), (4, 6)\})}$$

$$= \frac{1/36}{6/36} = \frac{1}{6}$$

Therefore, we obtain the same result as before.

Example 4.10 The organization that employs Jacobi is organizing a parent-daughter dinner for those employees having at least one daughter. Each of these employees is asked to attend along with one of his or her daughters. If Jacobi is known to have two children, what is the conditional probability that they are both girls given that Jacobi is invited to the dinner? Assume the sample space S is given by

$$S = \{(g, g), (g, b), (b, g), (b, b)\}$$

and that all these outcomes are equally likely, where the outcome (g, b) means, for instance, that Jacobi's oldest child is a girl and youngest is a boy.

Solution Since Jacobi is invited to the dinner, we know that at least one of Jacobi's children is a girl. Letting B denote the event that both of them are girls and A the event that at least one is a girl, we see that the desired probability is $P(B|A)$. This is determined as follows:

$$P(B|A) = \frac{P(A \cap B)}{P(A)}$$

$$= \frac{P(\{g, g\})}{P(\{(g, g), (g, b), (b, g)\})}$$

$$= \frac{1/4}{3/4} = \frac{1}{3}$$

That is, the conditional probability that both of Jacobi's children are girls given that at least one is a girl is 1/3. Many students incorrectly suppose that this conditional probability is 1/2, reasoning that the Jacobi child not attending the dinner is equally likely to be a boy or a girl. Their mistake lies in assuming that these two possibilities are equally likely, for initially there were 4 equally likely outcomes. The information that at least one of the children is a girl is equivalent to knowing that the outcome is not (b, b). Thus we are left with the 3 equally likely outcomes (g, g), (g, b), (b, g) showing that there is only a 1/3 chance that Jacobi has two girls.

Example 4.11 Table 4.4 lists the number (in thousands) of students enrolled in institutions of higher education, categorized by sex and age, in four different years.

Table 4.4   **Enrollment in Institutions of Higher Education**

Sex and age	Number (1000)			
	1970	**1980**	**1985**	**1987**
Total	8,581	12,097	12,247	12,544
Male	5,044	5,874	5,818	5,881
14 to 17 years old	129	99	121	91
18 and 19 years old	1,349	1,375	1,230	1,309
20 and 21 years old	1,095	1,259	1,216	1,089
22 to 24 years old	964	1,064	1,048	1,080
25 to 29 years old	783	993	991	1,016
30 to 34 years old	308	576	574	613
35 years old and over	415	507	639	684
Female	3,537	6,223	6,429	6,663
14 to 17 years old	129	148	113	119
18 and 19 years old	1,250	1,526	1,370	1,455
20 and 21 years old	785	1,165	1,166	1,135
22 to 24 years old	493	925	885	968
25 to 29 years old	292	878	962	931
30 to 34 years old	179	667	687	716
35 years old and over	409	913	1,246	1,339

Source: U.S. Department of Education, National Center for Education Statistics.

(a) Suppose a student is randomly chosen in 1970. What is the probability this student is a woman? Repeat if the student was randomly chosen in 1987.

Find the conditional probability that a randomly chosen student in 1987 is

(b) Over 35, given that this student is a man

(c) Over 35, given that this student is a woman

(d) A woman, given that this student is over 35

(e) A man, given that this student is between 20 and 21

Solution

(a) Since in 1970 there were 3,537,000 women out of a total of 8,581,000 students, it follows that the probability that a randomly chosen student in 1970 was a woman is

$$\frac{3537}{8581} = .4122$$

If the student was randomly chosen in 1987, then the probability that she or he was female is

$$\frac{6663}{12{,}544} = .5312$$

Thus, from 1970 to 1987 the proportion of students who are women increased from 41.22 to 53.12 percent.

(b) Since in 1987 there were a total of 5,881,000 males, of whom 684,000 were over age 35, the desired conditional probability is

$$P(\text{over } 35 \mid \text{man}) = \frac{684}{5881} = .1163$$

(c) By similar reasoning to that used in (b), we see that

$$P(\text{over } 35 \mid \text{woman}) = \frac{1339}{6663} = .2505$$

(d) Since there are a total of $684 + 1339 = 2023$ students who are over age 35, of whom 1339 are women, it follows that

$$P(\text{woman} \mid \text{over } 35) = \frac{1339}{2023} = .6619$$

(e) Since there are a total of $1089 + 1135 = 2224$ students who are between 20 and 21, of whom 1089 are men, it follows that

$$P(\text{man} \mid \text{between } 20 \text{ and } 21) = \frac{1089}{2224} = .4897$$

Since

$$P(B \mid A) = \frac{P(A \cap B)}{P(A)}$$

we obtain, upon multiplying both sides by $P(A)$, the following result, known as the *multiplication rule*.

Multiplication rule

$$P(A \cap B) = P(A)P(B \mid A)$$

This rule states that the probability that both A and B occur is equal to the probability that A occurs multiplied by the conditional probability of B given that A occurs. It is often quite useful for computing the probability of an intersection.

Example 4.12   Suppose that two people are randomly chosen from a group of 4 women and 6 men.

(a) What is the probability that both are women?

(b) What is the probability that one is a woman and the other a man?

Solution

(a) Let A and B denote, respectively, the events that the first person selected is a woman and that the second person selected is a woman. To compute the desired probability $P(A \cap B)$, we start with the identity

$$P(A \cap B) = P(A)P(B|A)$$

Now since the first person chosen is equally likely to be any of the 10 people, of whom 4 are women, it follows that

$$P(A) = \frac{4}{10}$$

Now given that the first person selected is a woman, it follows that the next selection is equally likely to be any of the remaining 9 people, of whom 3 are women. Therefore,

$$P(B|A) = \frac{3}{9}$$

and so

$$P(A \cap B) = \frac{4}{10} \cdot \frac{3}{9} = \frac{2}{15}$$

(b) To determine the probability that the chosen pair consists of 1 woman and 1 man, note first that this can occur in two disjoint ways. Either the first person chosen is a man and the second chosen is a woman, or vice versa. Let us determine the probabilities for each of these cases. Letting A denote the event that the first person chosen is a man and B the event that the second person chosen is a woman, we have

$$P(A \cap B) = P(A)P(B|A)$$

Now, since the first person is equally likely to be any of the 10 people, of whom 6 are men,

$$P(A) = \frac{6}{10}$$

Also, given that the first person is a man, the next selection is equally likely to be any of the remaining 9 people, of whom 4 are women, and so

$$P(B|A) = \frac{4}{9}$$

Therefore,

$$P(\text{man then woman}) = P(A \cap B) = \frac{6}{10} \cdot \frac{4}{9} = \frac{4}{15}$$

By similar reasoning, the probability that the first person chosen is a woman and the second chosen is a man is

$$P(\text{woman then man}) = \frac{4}{10} \cdot \frac{6}{9} = \frac{4}{15}$$

Since the event that the chosen pair consists of a woman and a man is the union of the above two disjoint events, we see that

$$P(1 \text{ woman and } 1 \text{ man}) = \frac{4}{15} + \frac{4}{15} = \frac{8}{15}$$

The conditional probability that B occurs given that A has occurred is not generally equal to the (unconditional) probability of B. That is, knowing that A has occurred generally changes the chances of B's occurrence. In the cases where $P(B|A)$ is equal to $P(B)$, we say that B is *independent* of A.
Since

$$P(A \cap B) = P(A)P(B|A)$$

we see that B is independent of A if

$$P(A \cap B) = P(A)P(B)$$

Since the above equation is symmetric in A and B, it follows that if B is independent of A, then A is also independent of B.
It can also be shown that if A and B are independent, then the probability of B

given that A does not occur is also equal to the (unconditional) probability of B. That is, if A and B are independent, then

$$P(B|A^c) = P(B)$$

Thus, when A and B are independent, any information about the occurrence or non-occurrence of one of these events does not affect the probability of the other.

Events A and B are *independent* if

$$P(A \cap B) = P(A)P(B)$$

If A and B are independent, then the probability that a given one of them occurs is unchanged by information as to whether the other one has occurred.

Example 4.13   Suppose that we roll a pair of fair dice, so each of the 36 possible outcomes is equally likely. Let A denote the event that the first die lands on 3, let B be the event that the sum of the dice is 8, and let C be the event that the sum of the dice is 7.

(a) Are A and B independent?

(b) Are A and C independent?

Solution

(a) Since $A \cap B$ is the event that the first die lands on 3 and the second on 5, we see that

$$P(A \cap B) = P(\{(3, 5)\}) = \frac{1}{36}$$

On the other hand,

$$P(A) = P(\{(3, 1), (3, 2), (3, 3), (3, 4), (3, 5), (3, 6)\}) = \frac{6}{36}$$

and

$$P(B) = P(\{(2, 6), (3, 5), (4, 4), (5, 3), (6, 2)\}) = \frac{5}{36}$$

Therefore, since $1/36 \neq (6/36) \cdot (5/36)$, we see that

$$P(A \cap B) \neq P(A)P(B)$$

and so events A and B are not independent.

   Intuitively, the reason why the events are not independent is that the chance that the sum of the dice is 8 is affected by the outcome of the first die. In particular, the chance that the sum is 8 is enhanced when the first die is 3, since then we still have a chance of obtaining the total of 8 (which we would not have if the first die were 1).

(b) Events A and C are independent. This is seen by noting that

$$P(A \cap C) = P(\{(3, 4)\}) = \frac{1}{36}$$

while

$$P(A) = \frac{1}{6}$$

and

$$P(C) = P(\{(1, 6), (2, 5), (3, 4), (4, 3), (5, 2), (6, 1)\}) = \frac{6}{36}$$

Therefore,

$$P(A \cap C) = P(A)P(C)$$

and so events A and C are independent.

   It is rather intuitive that the event that the sum of the dice is 7 should be independent of the event that the first die lands on 3. For, no matter what the outcome of the first die, there will always be exactly one outcome of the second die that results in the sum being equal to 7. As a result, the conditional probability that the sum is 7 given the value of the first die will always equal 1/6.

Example 4.14   Consider Table 4.4 presented in Example 4.11. Suppose that a 1980 female student is randomly chosen, as is, independently, a 1980 male student. Find the probability that both students are between 22 and 24 years old.

Solution   Since 1064 of the 5874 male students are between 22 and 24 years old, it follows that

$$P(\{\text{male is between 22 and 24}\}) = \frac{1064}{5874} \approx .1811$$

Similarly, since 925 of the 6223 female students are between 22 and 24 years old, we see that

$$P(\{\text{female is between 22 and 24}\}) = \frac{925}{6223} \approx .1486$$

Since the choices of the male and female students are independent, we obtain

$$P(\{\text{both are between ages 22 and 24}\}) = \frac{1064}{5874} \cdot \frac{925}{6223} \approx .027$$

That is, there is approximately a 2.7 percent chance that both students are between 22 and 24 years of age.

While so far we have discussed independence only for pairs of events, this concept can be extended to any number of events. The probability of the intersection of any number of independent events will be equal to the product of their probabilities.

If $A_1, \ldots, A_n$ are independent, then

$$P(A_1 \cap A_2 \cap \cdots \cap A_n) = P(A_1)P(A_2) \cdots P(A_n)$$

Example 4.15   A couple is planning on having three children. Assuming that each child is equally likely to be of either sex and that the sexes of the children are independent, find the probability that

(a) All three children will be girls.

(b) At least one child will be a girl.

Solution

(a) If we let A_i be the event that their ith child is a girl, then

$$\begin{aligned}
P(\text{all girls}) &= P(A_1 \cap A_2 \cap A_3) \\
&= P(A_1)P(A_2)P(A_3) \quad \text{by independence} \\
&= \frac{1}{2} \cdot \frac{1}{2} \cdot \frac{1}{2} = \frac{1}{8}
\end{aligned}$$

(b) The easiest way to compute the probability of at least one girl is by first computing the probability of the complementary event—that all the children are boys. Since, by the same reasoning as used in part (a),

$$P(\text{all boys}) = \frac{1}{8}$$

we see that

$$P(\text{at least one girl}) = 1 - P(\text{all boys}) = \frac{7}{8}$$

Problems

1. It is estimated that 30 percent of all adults in the United States are obese, 3 percent of all adults suffer from diabetes, and 2 percent of all adults both are obese and suffer from diabetes. Determine the conditional probability that a randomly chosen individual
   (a) Suffers from diabetes given that he or she is obese
   (b) Is obese given that she or he suffers from diabetes

2. Suppose a coin is flipped twice. Assume that all four possibilities are equally likely to occur. Find the conditional probability that both coins land heads given that the first one does.

3. Consider Table 4.3 as presented in Example 4.8. Suppose that one of the workers is randomly chosen. Find the conditional probability that this worker
   (a) Is a woman given that he or she earns over $25,000
   (b) Earns over $25,000 given that this worker is a woman

4. Fifty-two percent of the students at a certain college are females. Five percent of the students in this college are majoring in computer science. Two percent of the students are women majoring in computer science. If a student is selected at random, find the conditional probability that
   (a) This student is female, given that the student is majoring in computer science
   (b) This student is majoring in computer science, given that the student is female

Problems 5 and 6 refer to the data in the following table, which describes the age distribution of residents in a northern California county.

Age	Number
0–9	4200
10–19	5100
20–29	6200
30–39	4400
40–49	3600
50–59	2500
60–69	1800
Over 70	1100

5. If a resident is randomly selected from this county, determine the probability that the resident is
   (a) Less than 10 years old
   (b) Between 10 and 20 years old
   (c) Between 20 and 30 years old
   (d) Between 30 and 40 years old

6. Find the conditional probability that a randomly chosen resident is
   (a) Between 10 and 20 years old, given that the resident is less than 30 years old
   (b) Between 30 and 40 years old, given that the resident is older than 30

7. A games club has 120 members, of whom 40 play chess, 56 play bridge, and 26 play both chess and bridge. If a member of the club is randomly chosen, find the conditional probability that she or he
   (a) Plays chess given that he or she plays bridge
   (b) Plays bridge given that she or he plays chess

8. Refer to Table 4.4 which is presented in Example 4.11. Suppose that a student is randomly chosen in 1980. Determine the conditional probability that this student is
   (a) Less than 25 years old, given that the student is a man
   (b) A man, given that this student is less than 25 years old
   (c) Less than 25 years old, given that the student is a woman
   (d) A woman, given that this student is less than 25 years old

9. Repeat Prob. 8, this time supposing that the student is randomly chosen in 1985.

10. Many psychologists believe that birth order and personality are related. To study this hypothesis, 400 elementary school children were randomly selected and then given a test to measure confidence. On the results of this test each of the students was classified as being either confident or not confident. The numbers falling into each of the possible categories are shown:

	Firstborn	**Not firstborn**
Confident	62	60
Not confident	105	173

That is, for instance, out of 167 students who were firstborn children, a total of 62 were rated as being confident. Suppose that a student is randomly chosen from this group.

(a) What is the probability that the student is a firstborn?

(b) What is the probability that the student is rated confident?

(c) What is the conditional probability that the student is rated confident given that the student is a firstborn?

(d) What is the conditional probability that the student is rated confident given that the student is not a firstborn?

(e) What is the conditional probability that the student is a firstborn given that the student is confident?

11. Two cards are randomly selected from a deck of 52 playing cards. What is the conditional probability they are both aces given that they are of different suits?

12. In the U.S. Presidential election of 1984, 68.3 percent of those citizens eligible to vote registered; and of those registering to vote, 59.9 percent actually voted. Suppose a citizen eligible to vote is randomly chosen.

(a) What is the probability that this person voted?

(b) What is the conditional probability that this person registered given that he or she did not vote?

Note: In order to vote, first you must register.

13. There are 30 psychiatrists and 24 psychologists attending a certain conference. Two of these 54 people are randomly chosen to take part in a panel discussion. What is the probability that at least one psychologist is chosen? (*Hint:* You may want to first determine the probability of the complementary event that no psychologists are chosen.)

14. A child has 12 socks in a drawer; 5 are red, 4 are blue, and 3 are green. If 2 socks are chosen at random, find the probability that they are

(a) Both red

(b) Both blue

(c) Both green

(d) The same color

15. Two cards are chosen at random from a deck of 52 playing cards. Find the probability that

(a) Neither one is a spade.   (b) At least one is a spade.

(c) Both are spades.

16. There are n socks in a drawer, of which 3 are red. Suppose that if 2 socks are randomly chosen, then the probability that they are both red is 1/2. Find n.

*17. Suppose the occurrence of A makes it more likely that B will occur. In that case, show that the occurrence of B makes it more likely that A will occur. That is, show that if

$$P(B|A) > P(B)$$

then it is also true that

$$P(A|B) > P(A)$$

18. Two fair dice are rolled.
    (a) What is the probability that at least one of the dice lands on 6?
    (b) What is the conditional probability that at least one of the dice lands on 6 given that their sum is 9?
    (c) What is the conditional probability that at least one of the dice lands on 6 given that their sum is 10?

19. There is a 40 percent chance that a particular company will set up a new branch office in Chicago. If it does, there is a 60 percent chance that Norris will be named the manager. What is the probability that Norris will be named the manager of a new Chicago office?

20. According to a geologist, the probability that a certain plot of land contains oil is .7. Moreover, if oil is present, then the probability of hitting it with the first well is .5. What is the probability that the first well hits oil?

21. At a certain hospital, the probability that a patient dies on the operating table during open heart surgery is .20. A patient who survives the operating table has a 15 percent chance of dying in the hospital from the aftereffects of the opera-tion. What fraction of open-heart surgery patients survives both the operation and its aftereffects?

22. An urn initially contains 4 white and 6 black balls. Each time a ball is drawn, its color is noted and then it is replaced in the urn along with another ball of the same color. What is the probability that the first 2 balls drawn are black?

23. Reconsider Prob. 7.
    (a) If a member is randomly chosen, what is the probability that the chosen person plays either chess or bridge?
    (b) How many members play neither chess nor bridge?

    If two members are randomly chosen, find the probability that

(c) They both play chess.

(d) Neither one plays chess or bridge.

(e) Both play either chess or bridge.

24. Consider Table 4.4 as given in Example 4.11. Suppose that in 1985 a female student and a male student are independently and randomly chosen.
    (a) Find the probability that exactly one of them is over 30 years old.
    (b) Given that exactly one of them is over 30 years old, find the conditional probability that the male is older.

25. José and Jim go duck hunting together. Suppose that José hits the target with probability .3 and Jim, independently, with probability .1. They both fire one shot at a duck.
    (a) Given that exactly one shot hits the duck, what is the conditional probability that it is José's shot? that it is Jim's?
    (b) Given that the duck is hit, what is the conditional probability that José hit it? that Jim hit it?

26. A couple has two children. Let A denote the event that their older child is a girl, and let B denote the event that their younger child is a boy. Assuming that all 4 possible outcomes are equally likely, show that A and B are independent.

27. A simplified model for the movement of the price of a stock supposes that on each day the stock's price either moves up 1 unit with probability p or moves down 1 unit with probability $1 - p$. The changes on different days are assumed to be independent. Suppose that for a certain stock p is equal to 1/2. (Therefore, for instance, if the stock's price at the end of today is 100 units, then its price at the end of tomorrow will equally likely be either 101 or 99.)
    (a) What is the probability that after 2 days the stock will be at its original price?
    (b) What is the probability that after 3 days the stock's price will have increased by 1 unit?
    (c) If after 3 days the stock's price has increased by 1 unit, what is the conditional probability that it went up on the first day?

28. A male New York resident is randomly selected. Which of the following pairs of events A and B can reasonably be assumed to be independent?
    (a) A: He is a journalist.
        B: He has brown eyes.
    (b) A: He had a headache yesterday.
        B: He was in an accident yesterday.
    (c) A: He is wearing a white shirt.
        B: He is late to work.

29. A coin that is equally likely to land on heads or on tails is successively flipped until tails appear. Assuming that the successive flips are independent, what is

the probability that the coin will have to be tossed at least 5 times? (*Hint:* Fill in the missing word in the following sentence. The coin will have to be tossed at least 5 times if the first _____ flips all land on heads.)

30. A die is thrown until a 5 appears. Assuming that the die is equally likely to land on any of its six sides and that the successive throws are independent, what is the probability that it takes more than six throws?

31. Suppose that the probability of getting a busy signal when you call a friend is .1. Would it be reasonable to suppose that the probability of getting successive busy signals when you call two friends, one right after the other, is 0.01? If not, can you think of a condition under which this would be a reasonable supposition?

32. Two fields contain 9 and 12 plots of land, as shown below.

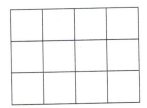

For an agricultural experiment, one plot from each field will be selected at random, independently of each other.
(a) What is the probability that both selected plots are corner plots?
(b) What is the probability that neither plot is a corner plot?
(c) What is the probability at least one of the selected plots is a corner plot?

33. A card is to be randomly selected from a deck of 52 playing cards. Let A be the event that the card selected is an ace, and let B be the event that the card is a spade. Show that A and B are independent.

34. A pair of fair dice is rolled. Let A be the event that the sum of the dice is equal to 7. Is A independent of the event that the first die lands on 1? on 2? on 4? on 5? on 6?

35. What is the probability that two strangers have the same birthday?

36. A U.S. publication reported that 4.78 percent of all deaths in 1988 were caused by accidents. What is the probability that 3 randomly chosen deaths were all due to accidents?

37. Each relay in the circuits shown below will close with probability .8. If all relays function independently, what is the probability that a current flows between A and B for the respective circuits? (The circuit in part (a) of the figure, which needs both of its relays to close, is called a *series circuit*. The circuit in part (b), which needs at least one of its relays to close, is called a *parallel circuit*.)

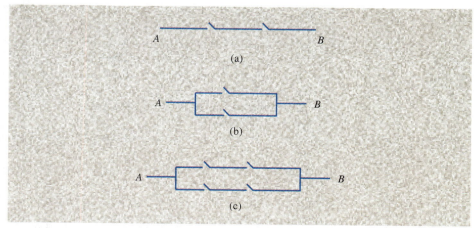

Hint: For parts (b) and (c) use the addition rule.

38. An urn contains 5 white and 5 black balls. Two balls are randomly selected from this urn. Let A be the event that the first ball is white and B be the event that the second ball is black. Are A and B independent events? Explain your reasoning.

39. Suppose in Prob. 38 that the first ball is returned to the urn before the second is selected. Will A and B be independent in this case? Again, explain your answer.

40. Suppose that each person who is asked whether she or he is in favor of a certain proposition will answer yes with probability .7 and no with probability .3. Assume that the answers given by different people are independent. Of the next four people asked, find the probability that
(a) All give the same answer.
(b) The first two answer no and the final two yes.
(c) At least one answers no.
(d) Exactly three answer yes.
(e) At least one answers yes.

41. The following data, obtained from the U.S. National Oceanic and Atmospheric Administration, give the average number of days with precipitation of .01 inch or more in different months for the cities of Mobile, Phoenix, and Los Angeles.

Average Number of Days with Precipitation of .01 Inch or More

City	January	April	July
Mobile	11	7	16
Phoenix	4	2	4
Los Angeles	6	3	1

Suppose that in the coming year you are planning to visit Phoenix on January 4, Los Angeles on April 10, and Mobile on July 15.
(a) What is the probability that it will rain on all three trips?
(b) What is the probability it will be dry on all three trips?
(c) What is the probability that you encounter rain in Phoenix and Mobile but not in Los Angeles?
(d) What is the probability that you encounter rain in Mobile and Los Angeles but not in Phoenix?
(e) What is the probability that you encounter rain in Phoenix and Los Angeles but not in Mobile?
(f) What is the probability that it rains in exactly two of your three trips?

42. Each computer chip produced by machine A is defective with probability .10, whereas each chip produced by machine B is defective with probability .05. If one chip is taken from machine A and one from machine B, find the probability (assuming independence) that
(a) Both chips are defective.
(b) Both are not defective.
(c) Exactly one of them is defective.

If it happens that exactly one of the two chips is defective, find the probability that it was the one from
(d) Machine A
(e) Machine B

43. Genetic testing has enabled parents to determine if their children are at risk for cystic fibrosis (CF), a degenerative neural disease. A child who receives a CF gene from both parents will develop the disease by his or her teenage years and will not live to adulthood. A child who receives either zero or one CF gene will not develop the disease; however, if she or he does receive one CF gene, it may be passed on to subsequent offspring. If an individual has a CF gene, then each of his or her children will receive that gene with probability 1/2.
(a) If both parents possess the CF gene, what is the probability that their child will develop cystic fibrosis?
(b) What is the probability that a 25-year-old person who does not have CF but whose sibling does, carries the gene?

*4.6   BAYES' THEOREM

For any two events A and B, we have the following representation for A:

$$A = (A \cap B) \cup (A \cap B^c)$$

That the above is valid is easily seen by noting that for an outcome to be in A, either it must be in both A and B or it must be in A but not in B (see Fig. 4.7). Since $A \cap B$ and $A \cap B^c$ are mutually exclusive (why?), we have by Property 3 (see Sec. 4.3)

$$P(A) = P(A \cap B) + P(A \cap B^c)$$

Since

$$P(A \cap B) = P(A|B)P(B)$$

and

$$P(A \cap B^c) = P(A|B^c)P(B^c)$$

we have thus shown the following equality:

$$P(A) = P(A|B)P(B) + P(A|B^c)\,P(B^c) \qquad (4.1)$$

This equality states that the probability of event A is a weighted average of the conditional probability of A given that B occurs and the conditional probability of A given that B does not occur; each conditional probability is weighted by the probability of the event on which it is conditioned. It is a very useful formula for it often enables us to compute the probability of an event A by first "conditioning" on whether a second event B occurs.

Before illustrating the use of Eq. (4.1), we first consider the problem of how to reevaluate an initial probability in light of additional evidence. Suppose there is a certain hypothesis under consideration; let H denote the event that the hypothesis is true and $P(H)$ the probability that the hypothesis is true. Now, suppose that additional evidence, call it E, concerning this hypothesis becomes available. We thus want to determine $P(H|E)$, the conditional probability that the hypothesis is true given the new evidence E. Now, by the definition of conditional probability,

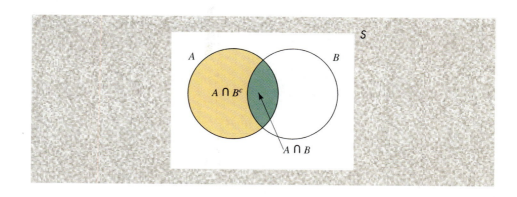

Figure 4.7
$A = (A \cap B)$
$\cup\ (A \cap B^c)$

$$P(H|E) = \frac{P(H \cap E)}{P(E)}$$

$$= \frac{P(E|H)P(H)}{P(E)}$$

By making use of Eq. (4.1), we can compute $P(E)$ by conditioning on whether the hypothesis is true. This yields the following identity, known as *Bayes' theorem.*

Bayes' theorem

$$P(H|E) = \frac{P(E|H)P(H)}{P(E|H)P(H) + P(E|H^c)P(H^c)}$$

Example 4.16   An insurance company believes that people can be divided into two classes—those who are prone to have accidents and those who are not. The data indicate that an accident-prone person will have an accident in a 1-year period with probability .1; the probability for all others is .05. Suppose that the probability is .2 that a new policyholder is accident-prone.

(a) What is the probability that a new policyholder will have an accident in the first year?

(b) If a new policyholder has an accident in the first year, what is the probability that he or she is accident-prone?

Solution   Let H be the event that the new policyholder is accident-prone, and let A denote the event that she or he has an accident in the first year. We can compute $P(A)$ by conditioning on whether the person is accident-prone:

$$P(A) = P(A|H)P(H) + P(A|H^c)P(H^c)$$
$$= (.1)(.2) + (.05)(.8) = .06$$

Therefore, there is a 6 percent chance that a new policyholder will have an accident in the first year.

We compute $P(H|A)$ as follows:

$$P(H|A) = \frac{P(H \cap A)}{P(A)}$$

$$= \frac{P(A|H)P(H)}{P(A)}$$

$$= \frac{(.1)(.2)}{.06} = \frac{1}{3}$$

Therefore, given that a new policyholder has an accident in the first year, the conditional probability that the policyholder is prone to accidents is 1/3.

Example 4.17   A blood test is 99 percent effective in detecting a certain disease when the disease is present. However, the test also yields a *false-positive* result for 2 percent of the healthy patients tested. (That is, if a healthy person is tested, then with probability .02 the test will say that this person has the disease.) Suppose .5 percent of the population has the disease. Find the conditional probability that a randomly tested individual actually has the disease given that his or her test result is positive.

Solution Let D denote the event that the person has the disease, and let E be the event that the test is positive. We want to determine $P(D|E)$, which can be accomplished by using Bayes' theorem as follows.

$$P(D|E) = \frac{P(E|D)P(D)}{P(E|D)P(D) + P(E|D^c)P(D^c)}$$

$$= \frac{(.99)(.005)}{(.99)(.005) + (.02)(.995)} = .199$$

Thus, there is approximately a 20 percent chance that a randomly chosen person from the population who tests positive actually has the disease. (The reason why it is so low is that the chance that a randomly chosen person is free of the disease yet tests positive is greater than the chance that the person has the disease and tests positive.)

Problems

1. There are two coins on a table. When both are flipped, one coin lands on heads with probability .5 while the other lands on heads with probability .6. A coin is randomly selected from the table and flipped.
   (a) What is the probability it lands on heads?
   (b) Given that it lands on tails, what is the conditional probability that it was the fair coin (that is, the one equally likely to land heads or tails)?

2. Suppose that when answering a question on a multiple-choice test, a student either knows the answer or guesses at it. If he guesses at the answer, then he will be correct with probability 1/5. If the probability that a student knows the answer is .6, what is the conditional probability that the student knew the answer given that he answered it correctly?

3. The inspector in charge of a criminal investigation is 60 percent certain of the guilt of a certain suspect. A new piece of evidence proving that the criminal was left-handed has just been discovered. Whereas the inspector knows that 18

percent of the population is left-handed, she is waiting to find out whether the suspect is left-handed.

(a) What is the probability that the suspect is left-handed?

(b) If the suspect turns out to be left-handed, what is the probability that the suspect is guilty?

4. Urn 1 contains 4 red and 3 blue balls, and urn 2 contains 2 red and 2 blue balls. A ball is randomly selected from urn 1 and placed in urn 2. A ball is then drawn from urn 2.

(a) What is the probability that the ball drawn from urn 2 is red?

(b) What is the conditional probability that the ball drawn from urn 1 is red given that a blue ball is drawn from urn 2?

5. Consider a diagnostic test that is 97 percent accurate on both those who have and those who do not have the disease. (That is, if a person has the disease, then with probability .97 the diagnosis will be positive; and if the person does not have the disease, then with probability .97 the diagnosis will be negative.) Suppose 2 percent of the population has the disease. What is the conditional probability that a randomly selected member of the population has the disease if that person's diagnosis was positive?

6. There are three cards in a hat. One is colored red on both sides, one is black on both sides, and one is red on one side and black on the other. The cards are thoroughly mixed in the hat, and one card is drawn and placed on a table. If the side facing up is red, what is the conditional probability that the other side is black?

7. A total of 52 percent of voting-age residents of a certain city are Republicans, and the other 48 percent are Democrats. Of these residents, 64 percent of the Republicans and 42 percent of the Democrats are in favor of discontinuing affirmative action city hiring policies. A voting-age resident is randomly chosen.

(a) What is the probability that the chosen person is in favor of discontinuing affirmative action city hiring policies?

(b) If the person chosen is against discontinuing affirmative action hiring policies, what is the probability she or he is a Republican?

8. A person's eye color is determined by a single pair of genes. If both genes are blue-eyed genes, then the person will have blue eyes; if they are both brown-eyed genes, then the person will have brown eyes; and if one gene is blue-eyed and the other is brown-eyed, then the person will have brown eyes. (Because of this latter fact we say that the brown-eyed gene is *dominant* over the blue-eyed one.) A newborn child independently receives one eye gene from each parent, and the gene that the child receives from a parent is equally likely to be either

of the two eye genes of that parent. Suppose that Susan has blue eyes and both her parents have brown eyes.

(a) What is the eye gene pair of Susan's mother? of her father?

(b) Susan's brown-eyed sister is pregnant. If her sister's husband has blue eyes, what is the probability the baby will have blue eyes? *Hint:* What is the probability that Susan's sister has a blue-eyed gene?

Key Terms

Experiment: Any process that produces an observation.

Outcome: The observation produced by an experiment.

Sample space: The set of all possible outcomes of an experiment.

Event: Any set of outcomes of the experiment. An event is a subset of sample space S. The event is said to occur if the outcome of the experiment is contained in it.

Union of events: The union of events A and B, denoted by $A \cup B$, consists of all outcomes that are in A or in B or in both A and B.

Intersection of events: The intersection of events A and B, denoted by $A \cap B$, consists of all outcomes that are in both A and B.

Complement of an event: The complement of event A, denoted by A^c, consists of all outcomes that are not in A.

Mutually exclusive or disjoint: Events are mutually exclusive or disjoint if they cannot occur simultaneously.

Null event: The event containing no outcomes. It is the complement of sample space S.

Venn diagram: A graphical representation of events.

Probability of an event: The probability of event A, denoted by $P(A)$, is the probability that the outcome of the experiment is contained in A.

Addition rule of probability: The formula

$$P(A \cup B) = P(A) + P(B) - P(A \cap B)$$

Conditional probability: The probability of one event given the information that a second event has occurred. We denote the conditional probability of B given that A has occurred by $P(B|A)$.

Multiplication rule: The formula

$$P(A \cap B) = P(A)P(B|A)$$

Independent: Two events are said to be independent if knowing whether a specific one has occurred does not change the probability that the other occurs.

Summary

Let S denote all possible outcomes of an experiment whose outcome is not predictable in advance. S is called the *sample space* of the experiment.

Any set of outcomes, or equivalently any subset of S, is called an event. If A and B are events, then $A \cup B$, called the *union* of A and B, is the event consisting of all outcomes that are in A or in B or in both A and B. The event $A \cap B$ is called the *intersection* of A and B. It consists of all outcomes that are in both A and B.

For any event A, we define the event A^c, called the *complement* of A, to consist of all outcomes in S that are not in A. The event S^c, which contains no outcomes, is designated by $\varnothing$. If $A \cap B = \varnothing$, meaning that A and B have no outcomes in common, then we say that A and B are *disjoint* (also called *mutually exclusive*).

We suppose that for every event A there is a number $P(A)$, called the *probability of A*. These probabilities satisfy the following three properties.

PROPERTY 1: $0 \leq P(A) \leq 1$
PROPERTY 2: $P(S) = 1$
PROPERTY 3: $P(A \cup B) = P(A) + P(B)$ when $A \cap B = \varnothing$

The quantity $P(A)$ represents the probability that the outcome of the experiment is in A. If so, we say that A occurs.

The identity

$$P(A \cup B) = P(A) + P(B) - P(A \cap B)$$

is called the *addition rule of probability.*

We sometimes assume that all the outcomes of an experiment are equally likely. Under this assumption, it can be shown that

$$P(A) = \frac{\text{number of outcomes in } A}{\text{number of outcomes in } S}$$

The conditional probability of B given that A has occurred is denoted by $P(B|A)$ and is given by the following equation:

$$P(B|A) = \frac{P(A \cap B)}{P(A)}$$

Multiplying both sides of the above by $P(A)$ gives the following identity, known as the multiplication rule:

$$P(A \cap B) = P(A)P(B|A)$$

If

$$P(A \cap B) = P(A)P(B)$$

then we say that events A and B are *independent*. If A and B are independent, then the probability of one of them occurring is unchanged by information as to whether the other has occurred.

Review Problems

1. Of 12 bottles in a case of wine, 3 are bad. Suppose 2 bottles are randomly chosen from the case. Find the probability that
   (a) The first bottle chosen is good.
   (b) The second bottle chosen is good.
   (c) Both bottles are good.
   (d) Both bottles are bad.
   (e) One is good, and one is bad.

2. A basketball player makes each of her foul shots with probability .8. Suppose she is fouled and is awarded two foul shots. Assuming that the results of different foul shots are independent, find the probability that she
   (a) Makes both shots
   (b) Misses both shots
   (c) Makes the second shot given that she missed the first

3. Suppose that a basketball player makes her first foul shot with probability .8. However, suppose that the probability that she makes her second shot depends on whether the first shot is successful. Specifically, suppose that if she is successful on her first shot, then her second will be successful with probability .85, whereas if she misses her first shot, then the second will be successful with probability .7. Find the probability that she
   (a) Makes both shots
   (b) Misses both shots
   (c) Makes the first but misses the second shot

4. Of the registered voters in a certain community, 54 percent are women and 46 percent are men. Sixty-eight percent of the registered women voters and 62 percent of the registered men voters voted in the last local election. If a registered voter from this community is randomly chosen, find the probability that this person is
   (a) A woman who voted in the last election
   (b) A man who did not vote in the last election
   (c) What is the conditional probability that this person is a man given that this person voted in the last election?

5. A kindergarten class consists of 24 students—13 girls and 11 boys. Each day one of these children is chosen as "student of the day." The selection is made as follows. At the beginning of the school year, the names of the children are written on slips of paper which are then put in a large urn. On the first day of school, the urn is shaken and a name is chosen to be student of the day. The next day this process is repeated with the remaining 23 slips of paper, and so

on. When each student has been selected once (which occurs on day 24), the process is repeated.
(a) What is the probability that the first selection is a boy?
(b) If the first selection is a boy, what is the probability that the second is a girl?

6. Two cards are chosen from an ordinary deck of 52 playing cards. Find the probability that
   (a) Both are aces.
   (b) Both are spades.
   (c) They are of different suits.
   (d) They are of different denominations.

7. What is the probability of the following outcomes when a fair coin is independently tossed 6 times?
   (a) H H H H H H
   (b) H T H T H T
   (c) T T H H T H

8. Find the probability of getting a perfect score just by guessing on a true/false test with
   (a) 2 questions   (b) 3 questions
   (c) 10 questions

9. A cafeteria offers a three-course meal. One chooses a main course, a starch, and a dessert. The possible choices are given in the table.

Meal	Choices
Main course	Chicken or roast beef
Starch course	Rice or potatoes
Dessert	Melon or ice cream or gelatin

Let the outcome of an experiment be the dinner selection of a person who makes one selection from each of the courses.
(a) List all the outcomes in sample space S.
(b) Suppose the person is allergic to rice and melon. List all the outcomes in the event corresponding to a choice that is acceptable to this person.
(c) If the person randomly chooses a dessert, what is the probability it is ice cream?
(d) If the person makes a random choice in each of the courses, what is the probability that chicken, rice, and melon are chosen?

10. The following is a breakdown by age and sex of the population of the United States. The numbers in each class are in units of 1 million.

	Sex	
Age	Females	Males
Under 25 years	48.8	50.4
Over 25 years	74.5	66.6

Suppose a person is chosen at random. Let A be the event that the person is male and B be the event that the person is under age 25. Find

(a) $P(A)$ and $P(A^c)$ (b) $P(B)$ and $P(B^c)$

(c) $P(A \cap B)$ (d) $P(A \cap B^c)$

(e) $P(A|B)$ (f) $P(B|A)$

11. A person has three keys of which only one fits a certain lock. If she tries the keys in a random order, find the probability that
    (a) The successful key is the first one tried.
    (b) The successful key is the second one tried.
    (c) The successful key is the third one tried.
    (d) The second key works given that the first one did not.

12. Two cards from an ordinary playing deck constitute a blackjack if one card is an ace and the other is a face card, where a face card is 10, jack, queen, or king. What is the probability that a random selection of two cards yields a blackjack? (*Hint:* You might try to compute the probability that the first card is an ace and the second a face card, and the probability that the first is a face card and the second an ace.)

13. A delivery company has 12 trucks, of which 4 have faulty brakes. If an inspector randomly chooses 2 of the trucks for a brake check, what is the probability that neither one has faulty brakes?

14. Suppose that A and B are independent events, and

$$P(A) = .8 \qquad P(B^c) = .4$$

    Find
    (a) $P(A \cap B)$
    (b) $P(A \cup B)$
    (c) $P(B)$
    (d) $P(A^c \cap B)$

15. A deck of 52 cards is shuffled, and the cards are turned face up, one at a time.
    (a) What is the probability that the first card turned up is the ace of spades?
    (b) Let A denote the event that the first card turned up is not the ace of spades, and let B denote the event that the second card turned up is the ace of

spades. Therefore, $A \cap B$ is the event that the second card turned up is the ace of spades. Compute the probability of this event by using

$$P(A \cap B) = P(A)P(B|A)$$

   (c) Fill in the missing word in the following intuitive argument for the solution obtained in part (b): Since all orderings are equally likely, the second card turned up is _____ likely to be any of the 52 cards.
   (d) What is the probability that the 17th card turned up is the ace of spades?

16. Floppy disks go through a two-stage inspection procedure. Each disk is checked first manually and then electronically. If the disk is defective, then a manual inspection will spot the defect with probability .70. A defective disk that passes the manual inspection will be detected electronically with probability .80. What percentage of defective disks is not detected?

17. Assume that business conditions in any year can be classified as either good or bad. Suppose that if business is good this year, then with probability .7 it will also be good next year. Also suppose that if business is bad this year, then with probability .4 it will be good next year. The probability that business will be good this year is .6. Find the probabilities that the following statements are true.
   (a) Business conditions both this year and next will be good.
   (b) Business conditions will be good this year and bad next year.
   (c) Business conditions will be bad both years.
   (d) Business conditions will be good next year.
   (e) Given that business conditions are good next year, what is the conditional probability that they were good this year?

18. Both John and Maureen have one gene for blue eyes and one for brown eyes. A child of theirs will receive one gene for eye color from Maureen and one from John. The gene received from each parent is equally likely to be either of the parent's two genes. Also, the gene received from John is independent of the one received from Maureen. If a child receives a blue gene from both John and Maureen, then that child will have blue eyes; otherwise, it will have brown eyes. Maureen and John have two children.
   (a) What is the probability that their older child has blue eyes?
   (b) What is the probability that the older child has blue and the younger has brown eyes?
   (c) What is the probability that the older has brown and the younger has blue eyes?
   (d) What is the probability that one child has blue eyes and the other has brown eyes?
   (e) What is the probability they both have blue eyes?
   (f) What is the probability they both have brown eyes?

19. It is estimated that for the U.S. adult population as a whole, 55 percent are above ideal weight, 20 percent have high blood pressure, and 60 percent either are above ideal weight or have high blood pressure. Let A be the event that a randomly chosen member of the population is above his or her ideal weight, and let B be the event that this person has high blood pressure. Are A and B independent events?

20. A card is randomly selected from a deck of playing cards. Let A be the event that the card is an ace, and let B be the event that it is a spade. State whether A and B are independent, if the deck is
    (a) A standard deck of 52 cards
    (b) A standard deck, with all 13 hearts removed
    (c) A standard deck, with the hearts from 2 through 9 removed

21. A total of 500 married working couples were polled about whether their annual salaries exceeded $25,000. The following information was obtained:

	Husband	
Wife	**Less than \$25,000**	**More than \$25,000**
Less than $25,000	212	198
More than $25,000	36	54

Thus, for instance, in 36 couples, the wife earned over $25,000 and the husband earned less than $25,000. One of the couples is randomly chosen.
    (a) What is the probability that the husband earns less than $25,000?
    (b) What is the conditional probability that the wife earns more than $25,000 given that the husband earns more than this amount?
    (c) What is the conditional probability that the wife earns more than $25,000 given that the husband earns less than this amount?
    (d) Are the salaries of the wife and husband independent?

22. The probability that a new car battery functions for over 10,000 miles is .8, the probability it functions for over 20,000 miles is .4, and the probability it functions for over 30,000 miles is .1. If a new car battery is still working after 10,000 miles, find the conditional probability that
    (a) Its total life will exceed 20,000 miles.
    (b) Its additional life will exceed 20,000 miles.

23. Of the drivers who stop at a certain gas station, 90 percent purchase either gasoline or oil. A total of 86 percent purchase gasoline, and 8 percent purchase oil.
    (a) What percentage of drivers purchases gasoline and oil?

    Find the conditional probability that a driver

(b) Purchases oil given that she or he purchases gasoline

(c) Purchases gasoline given that he or she purchases oil

(d) Suppose a driver stops at a gas station. Are the events that the driver purchases oil and that the driver purchases gasoline independent?

The following table gives participation rates at various artistic and leisure activities for individuals in different age categories. The data are for the year 1984, and the numbers represent the proportion of the population being considered who satisfied the stated criteria.

	Attended at least once						Visited at least once— art museum or gallery	Read— novel, short stories, poetry, or plays
Characteristic	Jazz performance	Classical music performance	Opera performance	Musical plays	Plays	Ballet performance		
Average	10	13	3	17	12	4	22	56
18–24 years old	14	11	2	15	11	4	22	57
25–34 years old	15	12	2	16	12	5	26	59
35–44 years old	10	16	4	21	14	6	27	62
45–54 years old	8	15	4	20	13	3	22	57
55–64 years old	5	11	3	18	10	4	19	50
65–74 years old	3	13	3	13	10	4	16	50
75 years old and over	1	10	1	8	7	2	10	48
Male	10	11	2	15	11	3	21	48
Female	9	14	3	19	12	5	23	63

Source: U.S. National Endowment for the Arts, *1985 Survey of Public Participation in the Arts.*

Problems 24 to 26 refer to the preceding table.

24. Suppose an 18- to 24-year-old is randomly chosen, as is a 35- to 44-year-old. Find the probability that
    (a) Both attended a jazz performance in 1984.
    (b) Exactly one attended a jazz performance in 1984.
    (c) Given that exactly one of them attended a jazz performance in 1984, what is the conditional probability that it was the younger person who attended?

25. Suppose that a man and a woman are randomly chosen. Find the probability that
    (a) Exactly one attended a ballet performance in 1984.
    (b) At least one attended an opera.
    (c) Both attended a musical play.

26. Suppose an individual is randomly chosen. Is the above table informative enough for us to determine the probability that this individual attended both a jazz and a classical music performance in 1984? If not, under what assumption would we be able to determine this probability? Compute the probability under this assumption, and then tell whether you think it is a reasonable assumption in this situation.

CHAPTER 5

Discrete Random Variables

> *His sacred majesty, chance, decides everything.*
> Voltaire

We continue our study of probability by introducing random variables—quantities whose values are determined by the outcome of the experiment. The expected value of a random variable is defined, and its properties are explored. The concept of variance is introduced. An important special type of random variable, known as the *binomial,* is studied.

5.1   INTRODUCTION

The National Basketball Association (NBA) draft lottery involves the 11 teams that had the worst won-lost records during the preceding year. Sixty-six Ping-Pong balls are placed in an urn. Each of these balls is inscribed with the name of a team; 11 have the name of the team with the worst record, 10 have the name of the team with the second worst record, 9 have the name of the team with the third worst record, and so on (with 1 ball having the name of the team with the 11th worst record). A ball is then chosen at random, and the team whose name is on the ball is given the first pick in the draft of players about to enter the league. All the other balls belonging to this team are then removed, and another ball is chosen. The team to which this ball "belongs" receives the second draft pick. Finally, another ball is chosen, and the team named on this ball receives the third draft pick. The remaining draft picks 4 through 11 are then awarded to the 8 teams that did not "win the lottery" in inverse order of their won-lost records. For instance, if the team with the worst record did not receive any of the 3 lottery picks, then that team would receive the fourth draft pick.

The outcome of this draft lottery is the order in which the 11 teams get to select players. However, rather than being concerned mainly about the actual outcome, we are sometimes more interested in the values of certain specified quantities. For instance, we may be primarily interested in finding out which team gets the first choice or in learning the draft number of our home team. These quantities of interest are known as *random variables,* and a special type, called *discrete,* will be studied in this chapter.

Random variables are introduced in Sec. 5.2. In Sec. 5.3 we consider the notion of the expected value of a random variable. We see that this represents, in a sense made precise, the average value of the random variable. Properties of the expected value are presented in Sec. 5.3.

Section 5.4 is concerned with the variance of a random variable, which is a measure of the amount by which a random variable tends to differ from its expected value. The concept of independent random variables is introduced in this section.

Section 5.5 deals with a very important type of discrete random variable that is called *binomial.* We see how such random variables arise and study their properties.

The first ball drawn in the 1993 NBA draft lottery belonged to the Orlando Magic, even though the Magic had finished the season with the 11th worst record and so had only 1 of the 66 balls!

5.2   RANDOM VARIABLES

When a probability experiment is performed, often we are not interested in all the details of the experimental result, but rather are interested in the value of some numerical quantity determined by the result. For instance, in tossing dice, often we care about only their sum and are not concerned about the values on the individual dice. Also an investor might not be interested in all the changes in the price of a stock on a given day, but rather might care about only the price at the end of the day. These quantities of interest that are determined by the result of the experiment are known as *random variables.*

Since the value of a random variable is determined by the outcome of the experiment, we may assign probabilities to its possible values.

Example 5.1   The outcome of the NBA draft lottery experiment, which was discussed in Sec. 5.1, is the specification of the teams that are to receive the first, second, and third picks in the draft. For instance, outcome $(3, 1, 4)$ could mean that the team with the third worst record received pick number 1, the team with the worst record received pick number 2, and the team with the fourth worst record received pick number 3. If we let X denote the team that received draft pick 1, then X would equal 3 if the outcome of the experiment were $(3, 1, 4)$.

Clearly, X can take on any integral value between 1 and 11 inclusive. It will equal 1 if the first ball chosen is one of the 11 balls that belong to the team with the worst record, it will equal 2 if the first ball is one of the 10 balls that belong to the team with the second worst record, and so on. Since each of the 66 balls is equally likely to be the first ball chosen, it follows that

$$P\{X = 1\} = \frac{11}{66} \qquad P\{X = 7\} = \frac{5}{66}$$

$$P\{X = 2\} = \frac{10}{66} \qquad P\{X = 8\} = \frac{4}{66}$$

$$P\{X = 3\} = \frac{9}{66} \qquad P\{X = 9\} = \frac{3}{66}$$

$$P\{X = 4\} = \frac{8}{66} \qquad P\{X = 10\} = \frac{2}{66}$$

$$P\{X = 5\} = \frac{7}{66} \qquad P\{X = 11\} = \frac{1}{66}$$

$$P\{X = 6\} = \frac{6}{66}$$

Example 5.2   Suppose we are about to learn the sexes of the three children of a certain family. The sample space of this experiment consists of the following 8 outcomes:

$$\{(b, b, b), (b, b, g), (b, g, b), (b, g, g), (g, b, b), (g, b, g), (g, g, b), (g, g, g)\}$$

The outcome (g, b, b) means, for instance, that the youngest child is a girl, the next youngest is a boy, and the oldest is a boy. Suppose that each of these 8 possible outcomes is equally likely, and so each has probability 1/8.

If we let X denote the number of female children in this family, then the value of X is determined by the outcome of the experiment. That is, X is a random variable whose value will be 0, 1, 2, or 3. We now determine the probabilities that X will equal each of these four values.

Since X will equal 0 if the outcome is (b, b, b), we see that

$$P\{X = 0\} = P(\{(b, b, b)\}) = \frac{1}{8}$$

Since X will equal 1 if the outcome is (b, b, g) or (b, g, b) or (g, b, b), we have

$$P\{X = 1\} = P(\{(b, b, g), (b, g, b), (g, b, b)\}) = \frac{3}{8}$$

Similarly,

$$P\{X = 2\} = P(\{(b, g, g), (g, b, g), (g, g, b)\}) = \frac{3}{8}$$

$$P\{X = 3\} = P(\{(g, g, g)\}) = \frac{1}{8}$$

A random variable is said to be *discrete* if its possible values constitute a sequence of separated points on the number line. Thus, for instance, any random variable that can take on only a finite number of different values is discrete.

In this chapter we will study discrete random variables. Let X be such a quantity, and suppose that it has n possible values, which we will label $x_1, x_2, \ldots, x_n$. As in Examples 5.1 and 5.2, we will use the notation $P\{X = x_i\}$ to represent the probability that X is equal to x_i. The collection of these probabilities is called the *probability distribution* of X. Since X must take on one of these n values, we know that

$$\sum_{i=1}^{n} P\{X = x_i\} = 1$$

Example 5.3   Suppose that X is a random variable that takes on one of the values 1, 2, or 3. If

$$P\{X = 1\} = .4 \quad \text{and} \quad P\{X = 2\} = .1$$

what is $P\{X = 3\}$?

Solution Since the probabilities must sum to 1, we have

$$1 = P\{X = 1\} + P\{X = 2\} + P\{X = 3\}$$

or

$$1 = .4 + .1 + P\{X = 3\}$$

Therefore,

$$P\{X = 3\} = 1 - .5 = .5$$

A graph of $P\{X = i\}$ is shown in Fig. 5.1.

Example 5.4   A saleswoman has scheduled two appointments to sell encyclope-dias. She feels that her first appointment will lead to a sale with probability .3. She also feels that the second will lead to a sale with probability .6 and that the results from the two appointments are independent. What is the probability distribution of X, the number of sales made?

Solution   The random variable X can take on any of the values 0, 1, or 2. It will equal 0 if neither appointment leads to a sale, and so

$$
\begin{aligned}
P\{X = 0\} &= P\{\text{no sale on first, no sale on second}\} \\
&= P\{\text{no sale on first}\}P\{\text{no sale on second}\} \qquad \text{by independence} \\
&= (1 - .3)(1 - .6) = .28
\end{aligned}
$$

The random variable X will equal 1 either if there is a sale on the first and not on the second appointment or if there is no sale on the first and one sale on the second appointment. Since these two events are disjoint, we have

$$
\begin{aligned}
P\{X = 1\} &= P\{\text{sale on first, no sale on second}\} \\
&\quad + P\{\text{no sale on first, sale on second}\} \\
&= P\{\text{sale on first}\}P\{\text{no sale on second}\} \\
&\quad + P\{\text{no sale on first}\}P\{\text{sale on second}\} \\
&= .3(1 - .6) + (1 - .3).6 = .54
\end{aligned}
$$

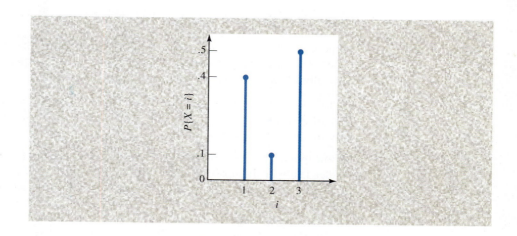

Figure 5.1
A graph of
$P\{X = i\}$.

Finally the random variable X will equal 2 if both appointments result in sales; thus

$$\begin{aligned} P\{X = 2\} &= P\{\text{sale on first, sale on second}\} \\ &= P\{\text{sale on first}\}P\{\text{sale on second}\} \\ &= (.3)(.6) = .18 \end{aligned}$$

As a check on the above, note that

$$P\{X = 0\} + P\{X = 1\} + P\{X = 2\} = .28 + .54 + .18 = 1$$

Problems

1. In Example 5.2, let the random variable Y equal 1 if the family has at least one child of each sex, and let it equal 0 otherwise. Find $P\{Y = 0\}$ and $P\{Y = 1\}$.

2. In Example 5.2, let the random variable W equal the number of girls that came before the first boy. [If the outcome is (g, g, g), take W equal to 3.] Give the possible values of W along with their probabilities. That is, give the probability distribution of W.

3. The following table presents the total number of tornadoes (violent, rotating columns of air with wind speeds over 100 miles per hour) in the United States between 1980 and 1991.

Year	1980	1981	1982	1983	1984	1985	1986	1987	1988	1989	1990	1991
Tornadoes	866	783	1046	931	907	684	764	656	702	856	1133	1132

Source: U.S. National Oceanic and Atmospheric Administration.

Suppose that one of these years is randomly selected, and let X denote the number of tornadoes in that year. Find
(a) $P\{X > 900\}$
(b) $P\{X \leq 800\}$
(c) $P\{X = 852\}$
(d) $P\{700 < X < 850\}$

4. Suppose a pair of dice is rolled. Let X denote their sum. What are the possible values of X? Assuming that each of the 36 possible outcomes of the experiment is equally likely, what is the probability distribution of X?

5. In Prob. 4, let Y denote the smaller of the two numbers appearing on the two dice. (If both dice show the same number, take that as the value of Y.) Determine the probability distribution of Y.

6. Two people are to meet in the park. Each person is equally likely to arrive, independently of the other, at 2:00, 2:30, or 3:00 p.m. Let X equal the time that the first person to arrive has to wait, where X is taken to equal 0 if both people arrive at the same time.
(a) What are the possible values of X?
(b) What are the probabilities that X assumes each of these values?

[handwritten: $P(u, u) = .9$]
[handwritten: $P(H, H) = .49$]
[handwritten: $.9 + .49$]

7. Two volleyball teams are to play a 2-out-of-3 series, in which they continue to play until one has won 2 games. Suppose that the home team wins each game played, independently, with probability .7. Let X denote the number of games played.
   (a) What are the possible values of X?
   (b) What is the probability distribution of X?

[handwritten table:
X	$P(X)$
2	.58
3	.42
]

[handwritten: $H = $ Home team win; $U =$ visiting win; $P(H) = .7$; $P(U) = $]

8. Suppose that 2 batteries are randomly chosen from a bin containing 10 batteries, of which 7 are good and 3 are defective. Let X denote the number of defective batteries chosen. Give the possible values of X along with their probabilities.

9. A shipment of parts contains 120 items of which 10 are defective. Two of these items are randomly chosen and inspected. Let X denote the number that are defective. Find the probability distribution of X.

10. A contractor will bid for two jobs in sequence. She has a .5 probability of winning the first job. If she wins the first job, then she has a .2 chance of winning the second job; if she loses the first job, then she has a .4 chance of winning the second job. (In the latter case, her bid will be higher.) Let X denote the number of jobs that she wins. Find the probability distribution of X.

11. Whenever a certain college basketball player goes to the foul line for two shots, he makes his first shot with probability .75. If he makes the first shot, then he makes the second shot with probability .80; if he misses the first shot, then he makes the second one with probability .70. Let X denote the number of shots he makes when he goes to the foul line for two shots. Find the probability distribution of X.

In Probs. 12, 13, and 14, tell whether the set of numbers $p(i)$, $i = 1, 2, 3, 4, 5$, can represent the probabilities $P\{X = i\}$ of a random variable whose set of values is 1, 2, 3, 4, or 5. If your answer is no, explain why.

12.

i	$p(i)$
1	.4
2	.1
3	.2
4	.1
5	.3

13.

i	$p(i)$
1	.2
2	.3
3	.4
4	−.1
5	.2

14.

i	$p(i)$
1	.3
2	.1
3	.2
4	.4
5	0

15. A sociologist, in a study of 223 households in a small rural town in Iowa, has collected data about the number of children in each household. The data showed that there are 348 children in the town, with the breakdown of the number of children in each household as follows: 38 households have 0 children, 82 have 1 child, 57 have 2 children, 34 have 3 children, 10 have 4 children, and 2 have 5 children. Suppose that one of these households is randomly selected for a more detailed interview. Let X denote the number of children in the household selected. Give the probability distribution of X.

16. Suppose that, in Prob. 15, one of the 348 children of the town is randomly selected. Let Y denote the number of children in the family of the selected child. Find the probability distribution of Y.

17. Suppose that X takes on one of the values 1, 2, 3, 4, or 5. If $P\{X < 3\} = .4$ and $P\{X > 3\} = .5$, find
(a) $P\{X = 3\}$
(b) $P\{X < 4\}$

The following table details some personal health practices based on a 1985 survey. The data are the percentages of the respondents having the stated attributes. Problems 18 to 22 refer to it.

Characteristic	Sleeps 6 hours or less	Never eats break-fast	Snacks every day	Less physically active than contem-poraries	Had 5 or more drinks on any one day	Current smoker	30 percent or more above desirable weight
All persons	22.0	24.3	39.0	16.4	37.5	30.1	13.0
Age							
18–29 years old	19.8	30.4	42.2	17.1	54.4	31.9	7.5
30–44 years old	24.3	30.1	41.4	18.3	39.0	34.5	13.6
45–64 years old	22.7	21.4	37.9	15.3	24.6	31.6	18.1
65 years old and over	20.4	7.5	30.7	13.5	12.2	16.0	13.2
65–74 years old	19.7	9.0	32.4	15.8	(NA)	19.7	14.9
75 years old and over	21.5	5.1	27.8	9.8	(NA)	10.0	10.3
Sex							
Male	22.7	25.2	40.7	16.5	49.3	32.6	12.1
Female	21.4	23.6	37.5	16.3	23.3	27.8	13.7

Source: U.S. National Center for Health Statistics, *Health Promotion and Disease Prevention, United States, 1985.*

18. Suppose that a male and a female are randomly chosen in 1985. Let X denote the number who sleep 6 hours or less, and find the probability distribution of X.

19. Suppose that a female and a male are randomly chosen in 1985. Let X denote the number who never eat breakfast, and find the probability distribution of X.

20. Suppose that a male and a female are randomly chosen in 1985. Let X denote the number who smoke, and find the probability distribution of X.

21. Suppose that one person is randomly chosen from each of the 6 age groups. Let X denote the number who snack every day. Find
    (a) $P\{X = 0\}$
    (b) $P\{X = 6\}$

22. Suppose that one person is randomly chosen from each of the 6 age groups. Let X denote the number who are less physically active than their contemporaries. Find
    (a) $P\{X = 0\}$
    (b) $P\{X = 6\}$

23. An insurance agent has two clients, each of whom has a life insurance policy that pays \$100,000 upon death. Their probabilities of dying this year are .05 and .10. Let X denote the total amount of money that will be paid this year to the client's beneficiaries. Assuming that the event that client 1 dies is independent of the event that client 2 dies, determine the probability distribution of X.

24. A bakery has 3 special cakes at the beginning of the day. The daily demand for this type of cake is

0	with probability .15
1	with probability .20
2	with probability .35
3	with probability .15
4	with probability .10
5 or more	with probability .05

    Let X denote the number of cakes that remain unsold at the end of the day. Determine the probability distribution of X.

5.3 EXPECTED VALUE

A key concept in probability is the expected value of a random variable. If X is a discrete random variable that takes on one of the possible values $x_1, x_2, \ldots, x_n$, then the *expected value* of X, denoted by $E[X]$, is defined by

$$E[X] = \sum_{i=1}^{n} x_i P\{X = x_i\}$$

The expected value of X is a weighted average of the possible values of X, with each value weighted by the probability that X assumes it. For instance, suppose X is equally likely to be either 0 or 1, and so

$$P\{X = 0\} = P\{X = 1\} = \frac{1}{2}$$

Then

$$E[X] = 0\left(\frac{1}{2}\right) + 1\left(\frac{1}{2}\right) = \frac{1}{2}$$

is equal to the ordinary average of the two possible values 0 and 1 that X can assume. On the other hand, if

$$P\{X = 0\} = \frac{2}{3} \qquad P\{X = 1\} = \frac{1}{3}$$

then

$$E[X] = 0\left(\frac{2}{3}\right) + 1\left(\frac{1}{3}\right) = \frac{1}{3}$$

is a weighted average of the two possible values 0 and 1 where the value 0 is given twice as much weight as the value 1, since it is twice as likely that X will equal 0 as it is that X will equal 1.

Definition and terminology

The *expected value* of a discrete random variable X whose possible values are x_1, $x_2, \ldots, x_n$, is denoted by $E[X]$ and is defined by

$$E[X] = \sum_i x_i P\{X = x_i\}$$

Other names used for $E[X]$ are the *expectation* of X and the *mean* of X.

Another motivation for the definition of the expected value relies on the frequency interpretation of probabilities. This interpretation assumes that if a very large number (in theory, an infinite number) of independent replications of an experiment are performed, then the proportion of time that event A occurs will equal $P(A)$. Now consider a random variable X that takes on one of the possible values $x_1, x_2, \ldots, x_n$ with respective probabilities $p(x_1), p(x_2), \ldots, p(x_n)$; and think of X as representing our winnings in a single game of chance. We will now argue that if we play a large number of such games, then our average winning per game will be $E[X]$. To see this, suppose that we play N games, where N is a very large number. Since, by the frequency interpretation of probability, the proportion of games in which we win x_i will approximately equal $p(x_i)$, it follows that we will win x_i in approximately $Np(x_i)$ of the N games. Since this is true for each x_i, it follows that our total winnings in the N games will be approximately equal to

$$\sum_{i=1}^{n} x_i \text{ (number of games we win } x_i) = \sum_{i=1}^{n} x_i \, Np(x_i)$$

Therefore, our average winning per game will be

$$\frac{\sum_{i=1}^{N} x_i \, Np(x_i)}{N} = \sum_{i=1}^{N} x_i p(x_i) = E[X]$$

In other words, if X is a random variable associated with some experiment, then the average value of X over a large number of replications of the experiment is approximately $E[X]$.

Example 5.5   Suppose we roll a die that is equally likely to have any of its 6 sides appear face up. Find $E[X]$, where X is the side facing up.

Solution Since

$$P\{X = i\} = \frac{1}{6} \quad \text{for } i = 1, 2, 3, 4, 5, 6$$

we see that

$$E[X] = 1\left(\frac{1}{6}\right) + 2\left(\frac{1}{6}\right) + 3\left(\frac{1}{6}\right) + 4\left(\frac{1}{6}\right) + 5\left(\frac{1}{6}\right) + 6\left(\frac{1}{6}\right)$$
$$= \frac{21}{6} = 3.5$$

Note that the expected value of X is not one of the possible values of X. Even though we call $E[X]$ the expected value of X, it should be interpreted not as the value that we *expect* X to have, but rather as the average value of X in a large number of repetitions of the experiment. That is, if we continually roll a die, then after a large number of rolls the average of all the outcomes will be approximately 3.5.

Example 5.6   Consider a random variable X that takes on either the value 1 or 0 with respective probabilities p and $1 - p$. That is,

$$P\{X = 1\} = p \qquad P\{X = 0\} = 1 - p$$

Find $E[X]$.

Solution The expected value of this random variable is

$$E[X] = 1(p) + 0(1 - p) = p$$

Example 5.7   An insurance company sets its annual premium on its life insurance policies so that it makes an expected profit of 1 percent of the amount it would have to pay out upon death. Find the annual premium on a $200,000 life insurance policy for an individual who will die during the year with probability .02.

Solution In units of $1000, the bank will set its premium so that its expected profit is 1 percent of 200, or 2. If we let A denote the annual premium, then the profit of the bank will be either

     A if policyholder lives

or

     $A - 200$ if policyholder dies

Therefore, the expected profit is given by

$$\begin{aligned}
E[\text{profit}] &= AP\{\text{policyholder lives}\} + (A - 200)P\{\text{policyholder dies}\} \\
&= A(1 - .02) + (A - 200)(.02) \\
&= A - 200(.02) \\
&= A - 4
\end{aligned}$$

So the company will have an expected profit of $2000 if it charges an annual premium of $6000. ◼ ◼

As seen in Example 5.7, $E[X]$ is always measured in the same units (dollars in that example) as is the random variable X.

The concept of expected value is analogous to the physical concept of the center of gravity of a distribution of mass. Consider a discrete random variable with probabilities given by $p(x_i)$, $i \geq 1$. If we imagine a rod on which weights having masses $p(x_i)$ are placed at points x_i, $i \geq 1$ (see Fig. 5.2), then the point at which the rod would be in balance is known as the *center of gravity*. It can be shown by the laws of mechanics that this point is

$$\sum_i x_i p(x_i) = E[X]$$

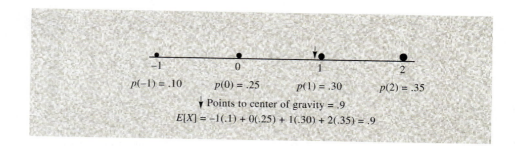

Figure 5.2
Center of
gravity = $E[X]$.

$p(-1) = .10 \qquad p(0) = .25 \qquad p(1) = .30 \qquad p(2) = .35$

Points to center of gravity = .9

$E[X] = -1(.1) + 0(.25) + 1(.30) + 2(.35) = .9$

5.3.1 Properties of Expected Values

Let X be a random variable with expected value $E[X]$. If c is a constant, then the quantities cX and $X + c$ are also random variables and so have expected values. The following useful results can be shown:

$$E[cX] = cE[X]$$
$$E[X + c] = E[X] + c$$

That is, the expected value of a constant times a random variable is equal to the constant times the expected value of the random variable; and the expected value of a constant plus a random variable is equal to the constant plus the expected value of the random variable.

Example 5.8 A married couple works for the same employer. The wife's Christmas bonus is a random variable whose expected value is $1500.

(a) If the husband's bonus is set to equal 80 percent of his wife's, find the expected value of the husband's bonus.

(b) If the husband's bonus is set to equal $1000 more than that of his wife, find its expected value.

Solution Let X denote the bonus (in dollars) to be paid to the wife.

(a) Since the bonus paid to the husband is equal to $.8X$, we have

$$E[\text{bonus to husband}] = E[.8X] = .8E[X] = \$1200$$

(b) In this case the bonus to be paid to the husband is $X + 1000$, and so

$$E[\text{bonus to husband}] = E[X + 1000] = E[X] + 1000 = \$2500$$

A very useful property is that the expected value of the sum of random variables is equal to the sum of the individual expected values.

For any random variables X and Y,

$$E[X + Y] = E[X] + E[Y]$$

Example 5.9   The following are the annual incomes of 7 men and 7 women residents of a certain community.

Annual Income (in $1000)

Men	Women
33.5	24.2
25.0	19.5
28.6	27.4
41.0	28.6
30.5	32.2
29.6	22.4
32.8	21.6

Suppose that a woman and a man are randomly chosen. Find the expected value of the sum of their incomes.

Solution Let X be the man's income and Y the woman's income. Since X is equally likely to be any of the 7 values in the men's column, we see that

$$E[X] = \frac{1}{7}(33.5 + 25 + 28.6 + 41 + 30.5 + 29.6 + 32.8)$$

$$= \frac{221}{7} \approx 31.571$$

Similarly,

$$E[Y] = \frac{1}{7}(24.2 + 19.5 + 27.4 + 28.6 + 32.2 + 22.4 + 21.6)$$

$$= \frac{175.9}{7} \approx 25.129$$

Therefore, the expected value of the sum of their incomes is

$$E[X+Y] = E[X] + E[Y]$$
$$\approx 56.700$$

That is, the expected value of the sum of their incomes is approximately $56,700.

Example 5.10   The following table lists the number of civilian full-time law enforcement employees in eight cities in 1990.

City	Civilian law enforcement employees
Minneapolis, MN	105
Newark, NJ	155
Omaha, NE	149
Portland, OR	195
San Antonio, TX	290
San Jose, CA	357
Tucson, AZ	246
Tulsa, OK	178

Source: Department of Justice, *Uniform Crime Reports for the United States, 1990.*

Suppose that two of the cities are to be randomly chosen and all the civilian law enforcement employees of these cities are to be interviewed. Find the expected number of people who will be interviewed.

Solution Let X be the number of civilian employees in the first city, and let Y be the number in the second city chosen. Since the selection of the cities is random, each of the 8 cities has the same chance to be the first city selected; similarly, each of the 8 cities has the same chance to be the second selection. Therefore, both X and Y are equally likely to be any of the 8 values in the above table, and so

$$E[X] = E[Y] = \frac{1}{8}(105 + 155 + 149 + 195 + 290 + 357 + 246 + 178)$$

$$= \frac{1675}{8}$$

and so

$$E[X + Y] = E[X] + E[Y] = \frac{1675}{4} = 418.75$$

That is, the expected number of interviews that will be needed is 418.75.

By using the frequency interpretation of expected value as being the average value of a random variable over a large number of replications of the experiment, it is easy to intuitively see why the expected value of a sum is equal to the sum of the expected values. For instance, suppose we always make the same two bets on each spin of a roulette wheel, one bet concerning the color of the slot where the ball lands and the other concerning the number on that slot. Let X and Y be the amounts (in dollars) that we lose on the color bet and on the number bet, respectively, in a single spin of the wheel. Then, $X + Y$ is our total loss in a single spin. Now, if in the long run we lose an average of 1 per spin on the color bet (so $E[X] = 1$) and we lose an average of 2 per spin on the number bet (so $E[Y] = 2$), then our average total loss per spin (equal to $E[X + Y]$) will clearly be $1 + 2 = 3$.

The result that the expected value of the sum of random variables is equal to the sum of the expected values holds for not only two but any number of random variables.

Useful result

For any positive integer k and random variables $X_1, \ldots, X_k$,

$$E\left[\sum_{i=1}^{k} X_i\right] = \sum_{i=1}^{k} E[X_i]$$

Example 5.11   A building contractor has sent in bids for three jobs. If the contractor obtains these jobs, they will yield respective profits of 20, 25, and 40 (in units of $1000). On the other hand, for each job the contractor does not win, he will incur a loss (due to time and money already spent in making the bid) of 2. If the probabilities that the contractor will get these jobs are, respectively, .3, .6, and .2, what is the expected total profit?

Solution   Let X_i denote the profit from job i, $i = 1, 2, 3$. Now by interpreting a loss as a negative profit, we have

$$P\{X_1 = 20\} = .3 \quad P\{X_1 = -2\} = 1 - .3 = .7$$

Therefore,

$$E[X_1] = 20(.3) - 2(.7) = 4.6$$

Similarly,

$$E[X_2] = 25(.6) - 2(.4) = 14.2$$

and

$$E[X_3] = 40(.2) - 2(.8) = 6.4$$

The total profit is $X_1 + X_2 + X_3$, and so

$$E[\text{total profit}] = E[X_1 + X_2 + X_3]$$
$$= E[X_1] + E[X_2] + E[X_3]$$
$$= 4.6 + 14.2 + 6.4$$
$$= 25.2$$

Therefore, the expected total profit is $25,200.

Problems

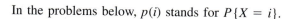

In the problems below, $p(i)$ stands for $P\{X = i\}$.

1. Find the expected value of X when
   (a) $p(1) = 1/3, p(2) = 1/3, p(3) = 1/3$
   (b) $p(1) = 1/2, p(2) = 1/3, p(3) = 1/6$
   (c) $p(1) = 1/6, p(2) = 1/3, p(3) = 1/2$

2. Find $E[X]$ when
   (a) $p(1) = .1, p(2) = .3, p(3) = .3, p(4) = .2, p(5) = .1$
   (b) $p(1) = .3, p(2) = .1, p(3) = .2, p(4) = .1, p(5) = .3$
   (c) $p(1) = .2, p(2) = 0, p(3) = .6, p(4) = 0, p(5) = .2$
   (d) $p(3) = 1$

3. A distributor makes a profit of $30 on each item that is received in perfect condition and suffers a loss of $6 on each item that is received in less-than-perfect condition. If each item received is in perfect condition with probability .4, what is the distributor's expected profit per item?

4. In a certain liability suit, a lawyer has to decide whether to charge a straight fee of $1200 or to take the case on a contingency basis, in which case she will receive a fee of $5000 only if her client wins the case. Determine whether the straight fee or the contingency arrangement will result in a higher expected fee when the probability that the client will win the case is
   (a) 1/2
   (b) 1/3
   (c) 1/4
   (d) 1/5

5. Suppose X can take on any of the values 1, 2, or 3. Find $E[X]$ if

$$p(1) = .3 \quad \text{and} \quad p(2) = .5$$

6. Let X be a random variable that is equally likely to take on any of the values 1, 2, ..., n. That is,

$$P\{X = i\} = \frac{1}{n} \quad i = 1, \ldots, n$$

(a) If $n = 2$, find $E[X]$.

(b) If $n = 3$, find $E[X]$.

(c) If $n = 4$, find $E[X]$.

(d) For general n, what is the value of $E[X]$?

(e) Verify your answer in part (d) by making use of the algebraic identity

$$\sum_{i=1}^{n} i = \frac{n(n + 1)}{2}$$

7. A pair of fair dice is rolled. Find the expected value of the

   (a) Smaller

   (b) Larger

   of the two upturned faces. (If both dice show the same number, then take this to be the value of both the smaller and the larger of the upturned faces.)

8. A computer software firm has been told by its local utility company that there is a 25 percent chance that the electricity will be shut off at some time during the next working day. The company estimates that it will cost $400 in lost revenues if employees do not use their computers tomorrow, and it will cost $1200 if the employees suffer a cutoff in power while using them. If the company wants to minimize the expected value of its loss, should it risk using the computers?

9. An engineering firm must decide whether to prepare a bid for a construction project. It will cost $800 to prepare a bid. If it does prepare a bid, then the firm will make a gross profit (excluding the preparation cost) of $0 if it does not get the contract, $3000 if it gets the contract and the weather is bad, or $6000 if it gets the contract and the weather is not bad. If the probability of getting the contract is .4, and the probability that the weather will be bad is .6, what is the company's expected net profit if it prepares a bid?

10. All blood donated to a blood bank is tested before it is used. To reduce the total number of tests, the bank takes small samples of the blood of four separate donors and pools these samples. The pooled blood is analyzed. If it is deemed acceptable, then the bank stores the blood of these four people for future use. If it is deemed unacceptable, then the blood from each of the four donors is separately tested. Therefore, either one test or five tests are needed to handle the blood of four donors. Find the expected number of tests needed if each donor's blood is independently unacceptable with probability .1.

11. Two people are randomly chosen from a group of 10 men and 20 women. Let X denote the number of men chosen, and let Y denote the number of women.

    (a) Find $E[X]$.

    (b) Find $E[Y]$.

    (c) Find $E[X + Y]$.

12. If the two teams in a World Series have the same chance of winning each game, independently of the results of previously played games, then the probabilities that the series will end in 4, 5, 6, or 7 games are, respectively, 1/8, 1/4, 5/16, and 5/16. What is the expected number of games played in such a series?

13. A company that operates a chain of hardware stores is planning to open a new store in one of two locations. If it chooses the first location, the company thinks it will make a first-year profit of $40,000 if the store is successful and will have a first-year loss of $10,000 if the store is unsuccessful. At the second location, the company thinks it will make a first-year profit of $60,000 if the store is successful and a first-year loss of $25,000 if the store is unsuccessful.
    (a) If the probability of success is 1/2 for both locations, which location will result in a larger expected first-year profit?
    (b) Repeat part (a), this time assuming that the probability that the store is successful is 1/3.

14. If it rains tomorrow, you will earn $200 by doing some tutoring; if it is dry, you will earn $300 by doing construction work. If the probability of rain is 1/4, what is the expected amount that you will earn tomorrow?

15. If you have a 1/10 chance of gaining $400 and a 9/10 chance of gaining −$50 (that is, of losing $50), what is your expected gain?

16. If an investment has a .4 probability of making a $30,000 profit and a .6 probability of losing $15,000, does this investment have a positive expected gain?

17. It costs $40 to test a certain component of a machine. If a defective component is installed, it costs $950 to repair the damage that results to the machine. From the point of view of minimizing the expected cost, determine whether the component should be installed without testing if it is known that its probability of being defective is
    (a) .1
    (b) .05
    (c) .01
    (d) What would the probability of a defective component be if one were indifferent between testing and installing the component untested?

18. A fair bet is one in which the expected gain is equal to 0. If you bet 1 unit on a number in roulette, then you will gain 35 units if the number appears and will lose 1 unit if it does not. If the roulette wheel is perfectly balanced, then the probability that your number will appear is 1/38. What is the expected gain on a 1-unit bet? Is it a fair bet?

19. A school, holding a raffle, will sell each ticket for $1. The school will give out seven prizes—1 for $100, 2 for $50, and 4 for $25. Suppose you purchase one ticket. If a total of 500 tickets are sold, what is your expected gain? [*Hint:* Your gain is −1 (if you do not win a prize), 24 (if you win a $25 prize), 49 (if you win a $50 prize), or 99 (if you win a $100 prize).]

20. A roulette wheel has 18 numbers colored red, 18 colored black, and 2 (zero and double zero) that are uncolored. If you bet 1 unit on the outcome red, then either you win 1 if a red number appears or you lose 1 if a red number does not appear. What is your expected gain?

21. The first player to win 2 sets is the winner of the tennis match. Suppose that whatever happened in the previous sets, each player has probability 1/2 of winning the next set. Determine the expected number of sets played.

22. Suppose in Prob. 21 that the players are not of equal ability and that player 1 wins each set, independently of the results of earlier sets, with probability 1/3.
    (a) Find the expected number of sets played.
    (b) What is the probability that player 1 wins?

23. An insurance company sells a life insurance policy that pays $250,000 if the insured dies, for an annual premium of $1400. If the probability that the policyholder dies in the course of the year is .005, what is the company's expected annual profit from that policyholder?

24. In Example 5.8, find in both (a) and (b) the expected value of the sum of the bonuses earned by the wife and husband.

25. If $E[X] = \mu$, what is $E[X - \mu]$?

26. Four buses carrying 148 students from the same school arrive at a football stadium. The buses carry, respectively, 40, 33, 50, and 25 students. One of the students is randomly selected. Let X be the number of students who were on the bus carrying the selected student. One of 4 bus drivers is also randomly chosen. Let Y be the number of students who were on his or her bus.
    (a) Calculate $E[X]$ and $E[Y]$.
    (b) Can you give an intuitive reason why $E[X]$ is larger than $E[Y]$?

27. A small nursery must decide on the number of Christmas trees to stock. The trees cost $6 each and are to be sold for $20. Unsold trees are worthless. The nursery estimates that the probability distribution for the demand on trees is as follows:

Amount demanded	1200	1500	1800
Probability	.5	.2	.3

    Determine the nursery's expected profit if it purchases
    (a) 1200 trees
    (b) 1500 trees
    (c) 1800 trees

28. Repeat Prob. 27, this time assuming that any unsold tree must be disposed of at a cost of $2 per tree.

29. The daily demand at a bakery for a certain cake is given below:

Daily demand	0	1	2	3	4
Probability	.15	.25	.30	.15	.15

    It costs the bakery $4 to bake each cake, and it sells for $20. Any cakes left unsold at the end of the day are thrown away. Would the bakery have a higher expected profit if it baked 2 or 3 or 4 cakes daily?

30. If $E[X] = 5$ and $E[Y] = 12$, find
    (a) $E[3X + 4Y]$
    (b) $E[2 + 5Y + X]$
    (c) $E[4 + Y]$

31. Determine the expected sum of a pair of fair dice by
    (a) Using the probability distribution of the sum
    (b) Using Example 5.5 along with the fact that the expected value of the sum of random variables is equal to the sum of their expected values

32. A husband's year-end bonus will be

0	with probability .3
$1000	with probability .6
$2000	with probability .1

    His wife's bonus will be

$1000	with probability .7
$2000	with probability .3

    Let S be the sum of their bonuses, and find $E[S]$.

33. The following data gives the number of polluting incidents reported in and around U.S. waters from 1978 to 1985.

Year	Number of incidents
1978	14,495
1979	13,134
1980	11,155
1981	10,564
1982	10,414
1983	11,346
1984	13,026
1985	11,023

    Source: U.S. Coast Guard, *Marine Safety Information Systems,* 1988.

    Suppose that a congressional committee has decided to randomly choose 2 of the above years and then document each of the incidents that occurred in either of these years. Determine the expected number of such incidents.

34. Repeat Prob. 33, this time supposing that the committee is to randomly choose 3 of the years.

35. A small taxi company has 4 taxis. In a month's time, each taxi will get 0 traffic tickets with probability .3, 1 traffic ticket with probability .5, or 2 traffic tickets

with probability .2. What is the expected number of tickets per month amassed by the fleet of 4 taxis?

36. Suppose that 2 batteries are randomly selected from a drawer containing 8 good and 2 defective batteries. Let W denote the number of defective batteries selected.
    (a) Find $E[W]$ by first determining the probability distribution of W.
    Let X equal 1 if the first battery chosen is defective, and let X equal 0 otherwise. Also let Y equal 1 if the second battery is defective and equal 0 otherwise.
    (b) Give an equation relating X, Y, and W.
    (c) Use the equation in (b) to obtain $E[W]$.

5.4   VARIANCE OF RANDOM VARIABLES

It is useful to be able to summarize the properties of a random variable by a few suitably chosen measures. One such measure is the expected value. However, while the expected value gives the weighted average of the possible values of the random variable, it does not tell us anything about the variation, or spread, of these values. For instance, consider random variables U, V, and W whose values and probabilities are as follows:

$$U = 0 \quad \text{with probability 1}$$

$$V = \begin{cases} -1 & \text{with probability } 1/2 \\ 1 & \text{with probability } 1/2 \end{cases}$$

$$W = \begin{cases} -10 & \text{with probability } 1/2 \\ 10 & \text{with probability } 1/2 \end{cases}$$

Whereas all three random variables have expected value 0, there is clearly less spread in the values of U than in V and less spread in the values of V than in W.

Since we expect a random variable X to take on values around its mean $E[X]$, a reasonable way of measuring the variation of X is to consider how far X tends to be from its mean on the average. That is, we could consider $E[|X - \mu|]$, where $\mu = E[X]$ and $|X - \mu|$ is the absolute value of the difference between X and μ. However, it turns out to be more convenient to consider not the absolute value but the square of the difference.

Definition

If X is a random variable with expected value μ, then the *variance* of X, denoted by Var (X), is defined by

$$\text{Var}(X) = E[(X - \mu)^2]$$

Upon expanding $(X - \mu)^2$ to obtain $X^2 - 2\mu X + \mu^2$ and then taking the expected value of each term, we obtain after a little algebra the following useful computational formula for Var (X):

$$\text{Var}(X) = E[X^2] - \mu^2 \tag{5.1}$$

where

$$\mu = E[X]$$

Using Eq. (5.1) is usually the easiest way to compute the variance of X.

Example 5.12   Find Var (X) when the random variable X is such that

$$X = \begin{cases} 1 & \text{with probability } p \\ 0 & \text{with probability } 1 - p \end{cases}$$

Solution In Example 5.6 we showed that $E[X] = p$. Therefore, using the computational formula for the variance, we have

$$\text{Var}(X) = E[X^2] - p^2$$

Now,

$$X^2 = \begin{cases} 1^2 & \text{if } X = 1 \\ 0^2 & \text{if } X = 0 \end{cases}$$

Since $1^2 = 1$ and $0^2 = 0$, we see that

$$\begin{aligned} E[X^2] &= 1 \cdot P\{X = 1\} + 0 \cdot P\{X = 0\} \\ &= 1 \cdot p = p \end{aligned}$$

Hence,

$$\text{Var}(X) = p - p^2 = p(1 - p)$$

Example 5.13   The return from a certain investment (in units of $1000) is a random variable X with probability distribution

$$P\{X = -1\} = .7 \qquad P\{X = 4\} = .2 \qquad P\{X = 8\} = .1$$

Find Var (X), the variance of the return.

Solution Let us first compute the expected return as follows:

$$\begin{aligned} \mu = E[X] &= -1(.7) + 4(.2) + 8(.1) \\ &= .9 \end{aligned}$$

That is, the expected return is \$900. To compute Var (X), we use the formula

$$\text{Var}\,(X) = E[X^2] - \mu^2$$

Now, since X^2 will equal $(-1)^2$, 4^2, or 8^2 with respective probabilities of .7, .2, and .1, we have

$$\begin{aligned} E[X^2] &= 1(.7) + 16(.2) + 64(.1) \\ &= 10.3 \end{aligned}$$

Therefore,

$$\begin{aligned} \text{Var}\,(X) &= 10.3 - (.9)^2 \\ &= 9.49 \end{aligned}$$

5.4.1 Properties of Variances

For any random variable X and constant c, it can be shown that

$$\begin{aligned} \text{Var}\,(cX) &= c^2\,\text{Var}\,(X) \\ \text{Var}\,(X + c) &= \text{Var}\,(X) \end{aligned}$$

That is, the variance of the product of a constant and a random variable is equal to the constant squared times the variance of the random variable; and the variance of the sum of a constant and a random variable is equal to the variance of the random variable.

Whereas the expected value of the sum of random variables is always equal to the sum of the expectations, the corresponding result for variances is generally not true. For instance, consider the following.

$$\begin{aligned} \text{Var}\,(X + X) &= \text{Var}\,(2X) \\ &= 2^2\,\text{Var}\,(X) \\ &\neq \text{Var}\,(X) + \text{Var}\,(X) \end{aligned}$$

However, there is an important case in which the variance of the sum of random variables is equal to the sum of the variances, and this occurs when the random variables are independent. Before presenting this result, we must introduce the concept of independent random variables.

We say that X and Y are independent if knowing the value of one of them does not change the probabilities of the other. That is, if X takes on one of the values x_i, $i \geq 1$, and Y takes on one of the values y_j, $j \geq 1$, then X and Y are independent if the events that X is equal to x_i and that Y is equal to y_j are independent events for all x_i and y_j.

Definition

Random variables X and Y are *independent* if knowing the value of one of them does not change the probabilities of the other.

It turns out that the variance of the sum of independent random variables is equal to the sum of their variances.

Useful result

If X and Y are independent random variables, then

$$\text{Var} \ (X + Y) = \text{Var} \ (X) + \text{Var} \ (Y)$$

More generally, if $X_1, X_2, \ldots, X_k$ are independent random variables, then

$$\text{Var} \left(\sum_{i=1}^{k} X_i \right) = \sum_{i=1}^{k} \text{Var} \ (X_i)$$

Example 5.14   Determine the variance of the sum obtained when a pair of fair dice is rolled.

Solution Number the dice, and let X be the value of the first die and Y the value of the second die. Then the desired quantity is Var $(X + Y)$. Since the outcomes of the two dice are independent, we know that

$$\text{Var} \ (X + Y) = \text{Var} \ (X) + \text{Var} \ (Y)$$

To compute Var (X), the variance of the face of the first die, recall that it was shown in Example 5.5 that

$$E[X] = \frac{7}{2}$$

Since X^2 is equally likely to be any of the values $1^2, 2^2, 3^2, 4^2, 5^2,$ or 6^2, we have

$$E[X^2] = \frac{1}{6}(1 + 4 + 9 + 16 + 25 + 36) = \frac{91}{6}$$

Therefore,

$$\text{Var}(X) = E[X^2] - \left(\frac{7}{2}\right)^2$$

$$= \frac{91}{6} - \frac{49}{4}$$

$$= \frac{35}{12}$$

Since Y has the same probability distribution as X, it also has variance 35/12, and so

$$\text{Var}(X + Y) = \frac{35}{12} + \frac{35}{12} = \frac{35}{6}$$

The positive square root of the variance is called the *standard deviation* (SD).

Definition

The quantity SD (X), defined by

$$\text{SD}(X) = \sqrt{\text{Var}(X)}$$

is called the *standard deviation* of X.

The standard deviation, like the expected value, is measured in the same units as is the random variable. That is, if the value of X is given in terms of miles, then so will the expected value and the standard deviation, too. To compute the standard deviation of a random variable, compute the variance and then take its square root.

Example 5.15 The annual gross earnings of a certain rock singer are a random variable with an expected value of $400,000 and a standard deviation of $80,000. The singer's manager receives 15 percent of this amount. Determine the expected value and standard deviation of the amount received by the manager.

Solution If we let X denote the earnings (in units of $1000) of the singer, then the manager earns .15X. Its expected value is obtained as follows:

$$E[.15X] = .15E[X] = 60$$

To compute the standard deviation, first determine the variance:

$$\text{Var}(.15X) = (.15)^2 \text{Var}(X)$$

Taking the square root of both sides of the preceding gives

$$SD\ (.15X) = .15\ SD\ (X) = 12$$

Therefore, the amount received by the manager is a random variable with an expected value of $60,000 and a standard deviation of $12,000.

Problems

1. Determine the variances of random variables U, V, and W, defined at the beginning of Sec. 5.4.

2. Let $p(i) = P\{X = i\}$. Consider
   (a) $p(0) = .50, p(1) = .50$
   (b) $p(0) = .60, p(1) = .40$
   (c) $p(0) = .90, p(1) = .10$

   In which case do you think Var (X) would be largest? And in which case would it be smallest? Determine the actual variances and check your answers.

3. Suppose that, for some constant c, $P\{X = c\} = 1$. Find Var (X).

4. Find the variances of the random variables specified in Prob. 1 of Sec. 5.3.

5. Find Var (X) for X given in Prob. 5 of Sec. 5.3.

6. If the probability that you earn $300 is 1/3 and the probability that you earn $600 is 2/3, what is the variance of the amount that you earn?

7. Find the variance of the number of sets played in the situation described in Prob. 21 of Sec. 5.3.

8. A small electronics company that started up 4 years ago has 60 employees. The following is a frequency table relating the number of years (rounded up) that these employees have been with the company.

Number of years	Frequency
1	12
2	25
3	16
4	7

Suppose one of these workers is randomly chosen. Let X denote the number of

years he or she has been with the company. Find
(a) $E[X]$
(b) Var (X)

9. The vacation time received by a worker of a certain company depends on the economic performance of the company. Suppose that Fong, an employee of this company, will receive

0 weeks' vacation	with probability .4
1 week's vacation	with probability .2
2 weeks' vacation	with probability .4

Suppose also that Fontanez, another employee, will receive

0 weeks' vacation	with probability .3
1 week's vacation	with probability .4
2 weeks' vacation	with probability .3

Let X denote the number of weeks of vacation for Fong and Y denote the number of weeks for Fontanez.
(a) Which do you think is larger, Var (X) or Var (Y)?
(b) Find Var (X).
(c) Find Var (Y).

10. Find the variance of the profit earned by the nursery in Prob. 27(b) of Sec. 5.3.

11. Two fair coins are tossed. Determine Var (X) when X is the number of heads that appear.
(a) Use the definition of the variance.
(b) Use the fact that the variance of the sum of independent random variables is equal to the sum of the variances.

12. Find the variance of the number of tickets obtained by the fleet of taxis, as described in Prob. 35 of Sec. 5.3. Assume that the numbers of tickets received by each of the taxis are independent.

13. A lawyer must decide whether to charge a fixed fee of $2000 or to take a contingency fee of $8000 if she wins the case (and $0 if she loses). She estimates that her probability of winning is .3. Determine the standard deviation of her fee if
(a) She takes the fixed fee.
(b) She takes the contingency fee.

14. Find the standard deviation of the amount of money you will earn in Prob. 14 of Sec. 5.3.

15. The following is a frequency table giving the number of courses being taken by all 210 first-year students at a certain college.

Number of classes	Frequency
1	2
2	15
3	37
4	90
5	49
6	14
7	3

Let X denote the number of courses taken by a randomly chosen student. Find
(a) $E[X]$
(b) SD (X)

16. The amount of money that Robert earns has expected value $30,000 and standard deviation $3000. The amount of money that his wife Sandra earns has expected value $32,000 and standard deviation $5000. Determine the
    (a) Expected value
    (b) Standard deviation
    of the total earnings of this family. In answering part (b), assume that Robert's earnings and Sandra's earnings are independent. [*Hint:* In answering part (b), first find the variance of the family's total earnings.]

17. If Var $(X) = 4$, what is SD $(3X)$? [*Hint:* First find Var $(3X)$.]

18. If Var $(2X + 3) = 16$, what is SD (X)?

19. If X and Y are independent random variables, both having variance 1, find
    (a) Var $(X + Y)$
    (b) Var $(X - Y)$

5.5   BINOMIAL RANDOM VARIABLES

One of the most important types of random variables is the binomial, which arises as follows. Suppose that n independent subexperiments (or *trials*) are performed, each of which results in either a "success" with probability p or a "failure" with probability $1 - p$. If X is the total number of successes that occur in n trials, then X is said to be a *binomial* random variable with parameters n and p.

Before presenting the general formula for the probability that a binomial random variable X takes on each of its possible values $0, 1, \ldots, n$, we consider a special case. Suppose that $n = 3$ and that we are interested in the probability that X is equal to 2. That is, we are interested in the probability that 3 independent trials, each of which is a success with probability p, will result in a total of 2 successes. To determine this probability, consider all the outcomes that give rise to exactly 2 successes, namely,

$$(s, s, f), (s, f, s), (f, s, s)$$

The outcome (s, f, s) means, for instance, that the first trial is a success, the second a failure, and the third a success. Now, by the assumed independence of the trials, it follows that each of these outcomes has probability $p^2(1 - p)$. For instance, if S_i is the event that trial i is a success and F_i is the event that trial i is a failure, then

$$
\begin{aligned}
P(s, f, s) &= P(S_1 \cap F_2 \cap S_3) \\
&= P(S_1)P(F_2)P(S_3) \qquad \text{by independence} \\
&= p(1 - p)p
\end{aligned}
$$

Since each of the 3 outcomes that result in a total of 2 successes consists of 2 successes and 1 failure, it follows in a similar fashion that each occurs with probability $p^2(1 - p)$. Therefore, the probability of a total of 2 successes in the 3 trials is $3p^2(1 - p)$.

Consider now the general case in which we have n independent trials. Let X denote the number of successes. To determine $P\{X = i\}$, consider any outcome that results in a total of i successes. Since this outcome will have a total of i successes and $n - i$ failures, it follows from the independence of the trials that its probability will be $p^i(1 - p)^{n-i}$. That is, each outcome that results in $X = i$ will have the same probability $p^i(1 - p)^{n-i}$. Therefore, $P\{X = i\}$ is equal to this common probability multiplied by the number of different outcomes that result in i successes. Now, it can be shown that there are $n!/[i!(n - i)!]$ different outcomes that result in a total of i successes and $n - i$ failures, where $n!$ (read "n factorial") is equal to 1 when $n = 0$ and is equal to the product of the natural numbers from 1 to n otherwise. That is,

$$
\begin{aligned}
0! &= 1 \\
n! &= n \cdot (n - 1) \cdots 3 \cdot 2 \cdot 1 \qquad \text{if } n > 0
\end{aligned}
$$

A binomial random variable with parameters n and p represents the number of successes in n independent trials, when each trial is a success with probability p. If X is such a random variable, then for $i = 0, \ldots, n$,

$$
P\{X = i\} = \frac{n!}{i!(n - i)!} p^i(1 - p)^{n-i}
$$

As a check of the preceding equation, note that it states that the probability that there are no successes in n trials is

$$
\begin{aligned}
P\{X = 0\} &= \frac{n!}{0!(n - 0)!} p^0(1 - p)^{n-0} \\
&= (1 - p)^n \qquad \text{since } 0! = p^0 = 1
\end{aligned}
$$

However, the above is clearly correct since the probability that there are 0 successes, and so all the trials are failures, is, by independence, $(1 - p)(1 - p) \cdots (1 - p) = (1 - p)^n$.

The probabilities of three binomial random variables with respective parameters $n = 10$, $p = .5$, $n = 10$, $p = .3$, and $n = 10$, $p = .6$ are presented in Fig. 5.3.

Example 5.16   Three fair coins are flipped. If the outcomes are independent, determine the probability that there are a total of i heads, for $i = 0, 1, 2, 3$.

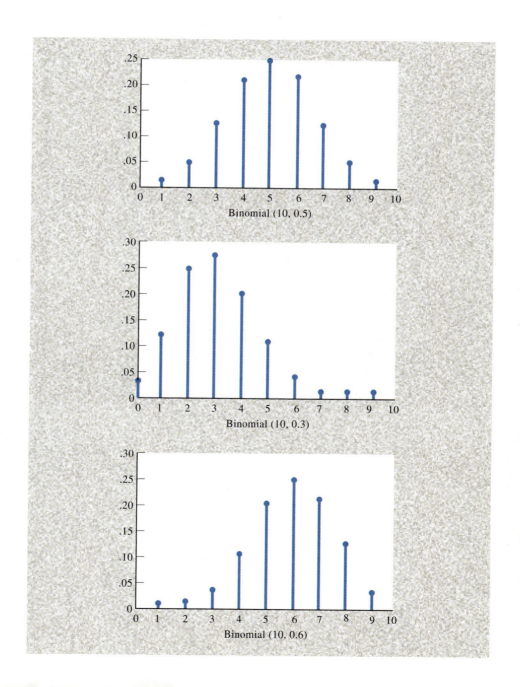

Figure 5.3
Binomial
probabilities.

Solution If we let X denote the number of heads ("successes"), then X is a binomial random variable with parameters $n = 3$, $p = .5$. By the preceding we have

$$P\{X = 0\} = \frac{3!}{0!3!}\left(\frac{1}{2}\right)^0\left(\frac{1}{2}\right)^3 = \left(\frac{1}{2}\right)^3 = \frac{1}{8}$$

$$P\{X = 1\} = \frac{3!}{1!2!}\left(\frac{1}{2}\right)^1\left(\frac{1}{2}\right)^2 = 3\left(\frac{1}{2}\right)^3 = \frac{3}{8}$$

$$P\{X = 2\} = \frac{3!}{2!1!}\left(\frac{1}{2}\right)^2\left(\frac{1}{2}\right)^1 = 3\left(\frac{1}{2}\right)^3 = \frac{3}{8}$$

$$P\{X = 3\} = \frac{3!}{3!0!}\left(\frac{1}{2}\right)^3\left(\frac{1}{2}\right)^0 = \left(\frac{1}{2}\right)^3 = \frac{1}{8}$$

Example 5.17 Suppose that a particular trait (such as eye color or handedness) is determined by a single pair of genes, and suppose that d represents a dominant gene and r a recessive gene. A person with the pair of genes (d, d) is said to be *pure dominant,* one with the pair (r, r) is said to be *pure recessive,* and one with the pair (d, r) is said to be *hybrid.* The pure dominant and the hybrid are alike in appearance. When two individuals mate, the resulting offspring receives one gene from each parent, and this gene is equally likely to be either of the parent's two genes.

(a) What is the probability that the offspring of two hybrid parents has the opposite (recessive) appearance?

(b) Suppose two hybrid parents have 4 offspring. What is the probability 1 of the 4 offspring has the recessive appearance?

Solution

(a) The offspring will have the recessive appearance if it receives a recessive gene from each parent. By independence, the probability of this is $(1/2)(1/2) = 1/4$.

(b) Assuming the genes obtained by the different offspring are independent (which is the common assumption in genetics), it follows from part (a) that the number of offspring having the recessive appearance is a binomial random variable with parameters $n = 4$ and $p = 1/4$. Therefore, if X is the number of offspring that have the recessive appearance, then

$$\begin{aligned}
P\{X = 1\} &= \frac{4!}{1!3!}\left(\frac{1}{4}\right)^1\left(\frac{3}{4}\right)^3 \\
&= 4\left(\frac{1}{4}\right)\left(\frac{3}{4}\right)^3 \\
&= \frac{27}{64}
\end{aligned}$$

Suppose that X is a binomial random variable with parameters n and p, and suppose we want to calculate the probability that X is less than or equal to some value j. In principle, we could compute this as follows:

$$P\{X \le j\} = \sum_{i=0}^{j} P\{X = i\} = \sum_{i=0}^{j} \frac{n!}{i!(n-i)!} p^i(1-p)^{n-i}$$

The amount of computation called for in the preceding equation can be rather large. To relieve this, Table D.5 (found in App. D) gives the values of $P\{X \le j\}$ for $n \le 20$ and for various values of p. In addition, you can use Program 5-1. In this program you enter the binomial parameters and the desired value of j, and you get as output the probability that the binomial is less than or equal to j, the probability that the binomial is equal to j, and the probability that the binomial is greater than or equal to j.

Example 5.18

(a) Determine $P\{X \le 12\}$ when X is a binomial random variable with parameters 20 and .4.

(b) Determine $P\{Y \ge 10\}$ when Y is a binomial random variable with parameters 16 and .5.

Solution
From Table D.5, we see that

(a) $P\{X \le 12\} = .9790$

(b) $P\{Y \ge 10\} = 1 - P\{Y < 10\} = 1 - P\{Y \le 9\} = 1 - .7728 = .2272$

We could also have run Program 5-1 to obtain the following:

The probability that a binomial (20, .4) is less than or equal to 12 is .978969

The probability that a binomial (16, .5) is greater than or equal to 10 is .2272506

5.5.1   Expected Value and Variance of a Binomial Random Variable

A binomial (n, p) random variable X is equal to the number of successes in n independent trials when each trial is a success with probability p. As a result, we can represent X as

$$X = \sum_{i=1}^{n} X_i$$

where X_i is equal to 1 if trial i is a success and is equal to 0 if trial i is a failure. Since

$$P\{X_i = 1\} = p \qquad P\{X_i = 0\} = 1 - p$$

it follows from the results of Examples 5.6 and 5.12 that

$$E[X_i] = p \qquad \text{Var } (X_i) = p(1 - p)$$

Therefore, using the fact that the expectation of the sum of random variables is equal to the sum of their expectations, we see that

$$E[X] = np$$

Also, since the variance of the sum of independent random variables is equal to the sum of their variances, we have

$$\text{Var } (X) = np(1 - p)$$

In summary,

If X is binomial with parameters n and p, then

$$E[X] = np$$
$$\text{Var } (X) = np(1 - p)$$

Example 5.19   Suppose that each screw produced is independently defective with probability .01. Find the expected value and variance of the number of defective screws in a shipment of size 1000.

Solution The number of defective screws in the shipment of size 1000 is a binomial random variable with parameters $n = 1000$, $p = .01$. Hence, the expected number of defective screws is

$$E[\text{number of defectives}] = 1000(.01) = 10$$

and the variance of the number of defective screws is

$$\text{Var (number of defectives)} = 1000(.01)(.99) = 9.9$$

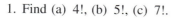

Historical Perspective

(Bettmann)

Jacques Bernoulli

Independent trials having a common success probability p were first studied by the Swiss mathematician Jacques Bernoulli (1654–1705). In his book *Ars Conjectandi* (the Art of Conjecturing), published by his nephew Nicholas eight years after his death in 1713, Bernoulli showed that if the number of such trials were large, then the proportion of them that were successes would be close to p with a probability near 1.

Jacques Bernoulli was from the first generation of the most famous mathematical family of all time. Altogether there were anywhere between 8 and 12 Bernoullis, spread over three generations, who made fundamental contribu-

tions to probability, statistics, and mathematics. One difficulty in knowing their exact number is the fact that several had the same name. (For example, two of the sons of Jacques' brother Jean were named Jacques and Jean.) Another difficulty is that several of the Bernoullis were known by different names in different places. Our Jacques (sometimes written Jaques), for instance, was also known as Jakob (sometimes written Jacob) and as James Bernoulli. But whatever their number, their influence and output were prodigious. Like the Bachs of music, the Bernoullis of mathematics were a family for the ages!

Problems

1. Find (a) 4!, (b) 5!, (c) 7!.

2. Find

   (a) $\dfrac{8!}{3!5!}$ (b) $\dfrac{7!}{3!4!}$ (c) $\dfrac{9!}{4!5!}$

3. Given that $9! = 362{,}880$, find $10!$.

4. Use the probability distribution of a binomial random variable with parameters n and p to show that

$$P\{X = n\} = p^n$$

Then argue directly why the above is valid.

5. If X is a binomial random variable with parameters $n = 8$ and $p = .4$, find
  (a) $P\{X = 3\}$
  (b) $P\{X = 5\}$
  (c) $P\{X = 7\}$

6. Each ball bearing produced is independently defective with probability .05. If a sample of 5 is inspected, find the probability that
   (a) None are defective.
   (b) 2 or more are defective.

7. Suppose you will be attending 6 hockey games. If each game independently will go to overtime with probability .10, find the probability that
   (a) At least 1 of the games will go into overtime.
   (b) At most 1 of the games will go into overtime.

8. A satellite system consists of 4 components and can function if at least 2 of them are working. If each component independently works with probability .8, what is the probability the system will function?

9. A communications channel transmits the digits 0 and 1. Because of static, each digit transmitted is independently incorrectly received with probability .1. Suppose an important single-digit message is to be transmitted. To reduce the chance of error, the string of digits 0 0 0 0 0 is to be transmitted if the message is 0 and the string 1 1 1 1 1 is to be transmitted if the message is 1. The receiver of the message uses "majority rule" to decode; that is, she decodes the message as 0 if there are at least 3 zeros in the message received and as 1 otherwise.
   (a) For the message to be incorrectly decoded, how many of the 5 digits received would have to be incorrect?
   (b) What is the probability that the message is incorrectly decoded?

10. A multiple-choice examination has 3 possible answers for each of 5 questions. What is the probability that a student will get 4 or more correct answers just by guessing?

11. A man claims to have extrasensory perception (ESP). As a test, a fair coin is to be flipped 8 times, and he is asked to predict the outcomes in advance. Suppose he gets 6 correct answers. What is the probability that he would have gotten at least this number of correct answers if he had no ESP but had just guessed?

12. Each diskette produced by a certain company will be defective with probability .05 independently of the others. The company sells the diskettes in packages of 10 and offers a money-back guarantee that all the diskettes in a package will be nondefective. Suppose that this offer is always taken up.
   (a) What is the probability that a package is returned?
   (b) If someone buys 3 packages, what is the probability that exactly 1 of them is returned?

13. Four fair dice are to be rolled. Find the probability that
   (a) 6 appears at least once.
   (b) 6 appears exactly once.
   (c) 6 appears at least twice.

14. Statistics indicate that alcohol is a factor in 55 percent of fatal automobile accidents. Of the next 3 fatal automobile accidents, find the probability that alcohol is a factor in
    (a) All 3   (b) Exactly 2
    (c) At least 1

15. Individuals who have two sickle cell genes will develop the disease called *sickle cell anemia,* while individuals having none or one sickle cell gene will not be harmed. If two people, both of whom have one sickle cell gene, have a child, then that child will receive two sickle cell genes with probability 1/4. Suppose that both members of each of three different couples have exactly one sickle cell gene. If each of these couples has a child, find the probability that
    (a) None of the children receives two sickle cell genes.
    (b) Exactly one of the children receives two sickle cell genes.
    (c) Exactly two of the children receive two sickle cell genes.
    (d) All three children receive two sickle cell genes.

16. Let X be a binomial random variable with parameters $n = 20$ and $p = .6$. Find
    (a) $P\{X \leq 14\}$ (b) $P\{X < 10\}$
    (c) $P\{X \geq 13\}$ (d) $P\{X > 10\}$
    (e) $P\{9 \leq X \leq 16\}$ (f) $P\{7 < X < 15\}$

17. A fair die is to be rolled 20 times. Find the expected value of the number of times
    (a) 6 appears.   (b) 5 or 6 appears.
    (c) An even number appears.   (d) Anything else but 6 appears.

18. Find the variances of the random variables in Prob. 17.

19. The probability that a fluorescent bulb burns for at least 500 hours is .90. Of 8 such bulbs, find the probability that
    (a) All 8 burn for at least 500 hours.
    (b) Exactly 7 burn for at least 500 hours.
    (c) What is the expected value of the number of bulbs that burn for at least 500 hours?
    (d) What is the variance of the number of bulbs that burn for at least 500 hours?

20. If a fair coin is flipped 500 times, what is the standard deviation of the number of times that a head appears?

21. The FBI has reported that 44 percent of murder victims are killed with handguns. If 4 murder victims are randomly selected, find
    (a) The probability that they were all killed by handguns
    (b) The probability that none were killed by handguns
    (c) The probability that at least two were killed by handguns
    (d) The expected number killed by handguns
    (e) The standard deviation of the number killed by handguns

22. The expected number of heads in a series of 10 flips of a fair coin is 6. What is the probability there are 8 heads?

23. If X is a binomial random variable with expected value 4 and variance 2.4, find
    (a) $P\{X = 0\}$
    (b) $P\{X = 12\}$

24. If X is a binomial random variable with expected value 4.5 and variance .45, find
    (a) $P\{X = 3\}$
    (b) $P\{X \geq 4\}$

25. Find the mean and standard deviation of a binomial random variable with parameters
    (a) $n = 100$, $p = .5$ (b) $n = 100$, $p = .4$
    (c) $n = 100$, $p = .6$ (d) $n = 50$, $p = .5$
    (e) $n = 150$, $p = .5$ (f) $n = 200$, $p = .25$

*5.6 HYPERGEOMETRIC RANDOM VARIABLES

Suppose that n batteries are to be randomly selected from a bin of N batteries, of which Np are functional and the other $N(1 - p)$ are defective. The random variable X, equal to the number of functional batteries in the sample, is then said to be a hypergeometric random variable with parameters n, N, p.

We can interpret the preceding experiment as consisting of n trials, where trial i is considered a success if the ith battery withdrawn is a functional battery. Since each of the N batteries is equally likely to be the ith one withdrawn, it follows that trial i is a success with probability $Np/N = p$. Therefore, X can be thought of as representing the number of successes in n trials where each trial is a success with probability p. What distinguishes X from a binomial random variable is that these trials are not independent. For instance, suppose that two batteries are to be withdrawn from a bin of five batteries, of which one is functional and the others defective. (That is, $n = 2$, $N = 5$, $p = 1/5$.) Then the probability that the second battery withdrawn is functional is 1/5. However, if the first one withdrawn is functional, then the conditional probability that the second one is functional is 0 (since when the second battery is chosen all four remaining batteries in the bin are defective). That is, when the selections of the batteries are made without replacing the previously chosen ones, the trials are not independent, so X is not a binomial random variable.

By using the result that each of the n trials is a success with probability p, it can be shown that the expected number of successes is np. That is,

$$E[X] = np$$

In addition, it can be shown that the variance of the hypergeometric random variable is given by

$$\text{Var}(X) = \frac{N - n}{N - 1} np(1 - p)$$

Thus, whereas the expected value of the hypergeometric random variable with parameters n, N, p is the same as that of the binomial random variable with parameters n, p, its variance is smaller than that of the binomial by the factor $(N - n)/(N - 1)$.

Example 5.20 If 6 people are randomly selected from a group consisting of 12 men and 8 women, then the number of women chosen is a hypergeometric random variable with parameters $n = 6$, $N = 20$, $p = 8/20 = .4$. Its mean and variance are

$$E[X] = 6(.4) = 2.4 \qquad \text{Var}(X) = \frac{14}{19} 6(.4)(.6) \approx 1.061$$

Similarly, the number of men chosen is a hypergeometric random variable with parameters $n = 6$, $N = 20$, $p = .6$. ▪▪

Suppose now that N, the number of batteries in the urn, is large in comparison to n, the number to be selected. For instance, suppose that 20 batteries are to be randomly chosen from a bin containing 10,000 batteries of which 90 percent are functional. In this case, no matter which batteries were previously chosen each new selection will be defective with a probability that is approximately equal to .9. For instance, the first battery selected will be functional with probability .9. If the first battery is functional then the next one will also be functional with probability $8999/9999 \approx .89999$, whereas if the first battery is defective then the second one will be functional with probability $9000/9999 \approx .90009$. A similar argument holds for the other selections, and thus we may conclude that when N is large in relation to n, then the n trials are approximately independent, which means that X is approximately a binomial random variable.

When N is large in relation to n, a hypergeometric random variable with parameters n, N, p approximately has a binomial distribution with parameters n and p.

Problems

In the following problems, state whether the random variable X is binomial or hypergeometric. Also give its parameters (n and p if it is binomial or n, N, and p if it is hypergeometric).

1. A lot of 200 items contains 18 defectives. Let X denote the number of defectives in a sample of 20 items.

2. A restaurant knows from past experience that 15 percent of all reservations do not show. Twenty reservations are expected tonight. Let X denote the number that show.

3. In one version of the game of lotto each player selects six of the numbers from 1 to 54. The organizers also randomly select six of these numbers. These latter six are called the winning numbers. Let X denote how many of a given player's six selections are winning numbers.

4. Each new fuse produced is independently defective with probability .05. Let X denote the number of defective fuses in the last 100 produced.

5. Suppose that a collection of 100 fuses contains 5 that are defective. Let X denote the number of defectives discovered when 20 of them are randomly chosen and inspected.

6. A deck of cards is shuffled and the cards are successively turned over. Let X denote the number of aces in the first 10 cards.

7. A deck of cards is shuffled and the top card is turned over. The card is then returned to the deck and the operation repeated. This continues until a total of 10 cards have been turned over. Let X denote the number of aces that have appeared.

Key Terms

Random variable: A quantity whose value is determined by the outcome of a probability experiment.

Discrete random variable: A random variable whose possible values constitute a sequence of disjoint points on the number line.

Expected value of a random variable: A weighted average of the possible values of a random variable; the weight given to a value is the probability that the random variable is equal to that value. Also called the **expectation** or the **mean** of the random variable.

Variance of a random variable: The expected value of the square of the difference between the random variable and its expected value.

Standard deviation of a random variable: The square root of the variance.

Independent random variables: A set of random variables having the property that knowing the values of any subset of them does not affect the probabilities of the remaining ones.

Binomial random variable with parameters n and p: A random variable equal to the number of successes in n independent trials when each trial is a success with probability p.

Summary

A *random variable* is a quantity whose value is determined by the outcome of a probability experiment. If its possible values can be written as a sequence of distinct numbers, then the random variable is called *discrete*.

Let X be a random variable whose possible values are x_i, $i = 1, \ldots, n$; and suppose X takes on the value x_i with probability $P\{X = x_i\}$. The *expected value* of X, also referred to as the *mean* of X or as the *expectation* of X, is denoted by $E[X]$ and is defined as

$$E[X] = \sum_{i=1}^{n} x_i P\{X = x_i\}$$

If X is a random variable and c is a constant, then

$$E[cX] = cE[X]$$
$$E[X + c] = E[X] + c$$

For any random variables $X_1, \ldots, X_k$,

$$E[X_1 + X_2 + \cdots + X_k] = E[X_1] + E[X_2] + \cdots + E[X_k]$$

The random variables X and Y are *independent* if knowing the value of one of them does not change the probabilities for the other.

The *variance* of a random variable measures the average squared distance of the random variable from its mean. Specifically, if X has mean $\mu = E[X]$, then the variance of X, denoted by Var (X), is defined as

$$\text{Var}(X) = E[(X - \mu)^2]$$

A property of the variance is that for any constant c and random variable X,

$$\text{Var}(cX) = c^2 \text{Var}(X)$$
$$\text{Var}(X + c) = \text{Var}(X)$$

Whereas the variance of the sum of random variables in general is not equal to the sum of their variances, it is true in the special case where the random variables are independent. That is,

$$\text{Var}(X + Y) = \text{Var}(X) + \text{Var}(Y)$$

if X and Y are independent.

The square root of the variance is called the *standard deviation* and is denoted by SD (X). That is,

$$SD\ (X) = \sqrt{\text{Var}\ (X)}$$

Consider n independent trials in which each trial results in a success with probability p. If X is the total number of successes, then X is said to be a *binomial* random variable with parameters n and p. Its probabilities are given by

$$P\{X = i\} = \frac{n!}{i!(n - i)!} p^i(1 - p)^{n-i} \qquad i = 0, \ldots, n$$

In the above, $n!$ (called n *factorial*) is defined by

$$0! = 1 \qquad n! = n(n - 1) \cdots 3 \cdot 2 \cdot 1$$

The mean and variance of a binomial random variable with parameters n and p are

$$E[X] = np \qquad \text{and} \qquad \text{Var}\ (X) = np(1 - p)$$

Review Problems

1. If $P\{X \le 4\} = .8$ and $P\{X = 4\} = .2$, find
   (a) $P\{X \ge 4\}$
   (b) $P\{X < 4\}$

2. If $P\{X \le 6\} = .7$ and $P\{X < 6\} = .5$, find
   (a) $P\{X = 6\}$
   (b) $P\{X > 6\}$

3. A graduating law student is not certain whether he actually wants to practice law or go into business with his family. He has decided to base his decision on whether he can pass the bar examination. He has decided to give himself at most 4 attempts at the examination; he will practice law if he passes the examination or go into the family business if he fails on all 4 tries. Suppose that each time he takes the bar examination he is successful, independently of his previous results, with probability .3. Let X denote the number of times he takes the bar examination.
   (a) What are the possible values that X can assume?
   (b) What is the probability distribution of X?
   (c) What is the probability that he passes the bar examination?
   (d) Find $E[X]$.
   (e) Find Var (X).

4. Suppose that X is either 1 or 2. If $E[X] = 1.6$, find $P\{X = 1\}$.

5. A gambling book recommends the following "winning strategy" for the game of roulette. It recommends that a gambler bet $1 on red. If red appears (which has probability 18/38 of occurring), then the gambler should take her $1 profit and quit. If the gambler loses this bet (which has probability 20/38 of occurring) and so is behind $1, then she should make a $2 bet on red and then quit. Let X denote the gambler's final winnings.
   (a) Find $P\{X > 0\}$.
   (b) Are you convinced that the strategy is a winning strategy? Why or why not?
   (c) Find $E[X]$.

6. Two people are to meet in the park. Each person is equally likely to arrive, independently of the other, at 3:00, 4:00, or 5:00 p.m. Let X equal the time that the first person to arrive has to wait, where X is taken to equal 0 if they both arrive at the same time. Find $E[X]$.

7. There is a .3 probability that a used-car salesman will sell a car to his next customer. If he does, then the car that is purchased is equally likely to cost $4000 or $6000. Let X denote the amount of money that the customer spends.
   (a) Find the probability distribution of X.
   (b) Find $E[X]$.
   (c) Find Var (X).
   (d) Find SD (X).

8. Suppose that 2 batteries are randomly chosen from a bin containing 12 batteries, of which 8 are good and 4 are defective. What is the expected number of defective batteries chosen?

9. A company is preparing a bid for a contract to supply a city's schools with notebook supplies. The cost to the company of supplying the material is $140,000. It is considering two alternate bids: to bid high (25 percent above cost) or to bid low (10 percent above cost). From past experience the company knows that if it bids high, then the probability of winning the contract is .15, whereas if it bids low, then the probability of winning the contract is .40. Which bid will maximize the company's expected profit?

10. If $E[3X + 10] = 70$, what is $E[X]$?

11. The probability that a vacuum cleaner saleswoman makes no sales today is 1/3, the probability she makes 1 sale is 1/2, and the probability she makes 2 sales is 1/6. Each sale made is independent and equally likely to be either a standard cleaner which costs $500 or a deluxe cleaner which costs $1000. Let X denote the total dollar value of all sales.
   (a) Find $P\{X = 0\}$.   (b) Find $P\{X = 500\}$.
   (c) Find $P\{X = 1000\}$.   (d) Find $P\{X = 1500\}$.
   (e) Find $P\{X = 2000\}$.   (f) Find $E[X]$.
   (g) Suppose that the saleswoman receives a 20 percent commission on the sales that she makes. Let Y denote the amount of money she earns. Find $E[Y]$.

12. The 5 families living on a certain block have a total of 12 children. One of the families has 4 children, one has 3, two have 2, and one has 1. Let X denote the number of children in a randomly selected family, and let Y denote the number of children in the family of a randomly selected child. That is, X refers to an experiment in which each of the five families is equally likely to be selected, whereas Y refers to one in which each of the 12 children is equally likely to be selected.

    (a) Which do you think has the larger expected value, X or Y?
    (b) Calculate $E[X]$ and $E[Y]$.

13. A financier is evaluating two investment possibilities. Investment A will result in

$20,000 profit	with probability 1/4
$10,000 profit	with probability 1/4
$15,000 loss	with probability 1/2

    Investment B will result in

$30,000 profit	with probability 1/8
$20,000 profit	with probability 1/4
$15,000 loss	with probability 3/8
$40,000 loss	with probability 1/4

    (a) What is the expected profit of investment A?
    (b) What is the expected profit of investment B?
    (c) What is the investor's expected profit if she or he invests in both A and B?

14. If Var $(X) = 4$, find
    (a) Var $(2X + 14)$
    (b) SD $(2X)$
    (c) SD $(2X + 14)$

15. Suppose $E[X] = \mu$ and SD $(X) = \sigma$. Let

$$Y = \frac{X - \mu}{\sigma}$$

    (a) Show that $E[Y] = 0$.
    (b) Show that Var $(Y) = 1$.

    The random variable Y is called the *standardized* version of X. That is, given a random variable, if we subtract its expected value and then divide the result by its standard deviation, then the resulting random variable is said to be standardized. The standardized variable has expected value 0 and variance 1.

16. A manager has two clients. The gross annual earnings of his first client are a random variable with expected value $200,000 and standard deviation $60,000. The gross annual earnings of his second client are a random variable with expected value $140,000 and standard deviation $50,000. If the manager's fee is 15 percent of his first client's gross earnings and 20 percent of his second client's gross earnings, find the
    (a) Expected value
    (b) Standard deviation
    of the manager's total fee. In part (b) assume that the earnings of the two clients are independent.

17. A weighted coin that comes up heads with probability .6 is flipped n times. Find the probability that the total number of heads in these flips exceeds the total number of tails when
    (a) $n = 1$ (b) $n = 3$
    (c) $n = 5$ (d) $n = 7$
    (e) $n = 9$ (f) $n = 19$

18. Each customer who enters a television store will buy a normal-size television with probability .3, buy an extra-large television with probability .1, or not buy any television with probability .6. Find the probability that the next 5 customers
    (a) Purchase a total of 3 normal-size sets
    (b) Do not purchase any extra-large sets
    (c) Purchase a total of 2 sets

19. A saleswoman has a 60 percent chance of making a sale each time she visits a computer store. She visits 3 stores each month. Assume that the outcomes of successive visits are independent.
    (a) What is the probability she makes no sales next month?
    (b) What is the probability she makes 2 sales next month?
    (c) What is the probability that she makes at least 1 sale in each of the next 3 months?

20. Let X be a binomial random variable such that

$$E[X] = 6 \quad \text{and} \quad \text{Var}(X) = 2.4$$

    Find
    (a) $P\{X > 2\}$
    (b) $P\{X \leq 9\}$
    (c) $P\{X = 12\}$

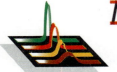

Discrete Random Variables

Purpose

Use Minitab to

1. Investigate discrete random variables.

2. Investigate the binomial distribution.

Procedures

First, load the Minitab (Windows version) software as in the Minitab lab for Chap. 1.

1. DISCRETE RANDOM VARIABLES AND THEIR PROBABILITY DISTRIBUTIONS

Example 1   Use Minitab to generate a discrete probability distribution.

Suppose we want to generate values for the random variable from 0 to 5. Select Calc→Set Patterned Data, and the Set patterned data dialog box will be displayed. Enter the appropriate values in the text boxes, as shown in Fig. M5.1. Note that we are saving the generated values from 0 to 5 in C1; and since we are using an increment of 1, all values from 0 to 5 will be listed. Observe that we are repeating the values and the entire list only once and that the Patterned sequence check box is selected. Select the OK button, and the values 0 to 5 will be listed in C1. Rename this column and call it x by typing x in the first cell in column C1. That is, x represents the values of a random variable X.

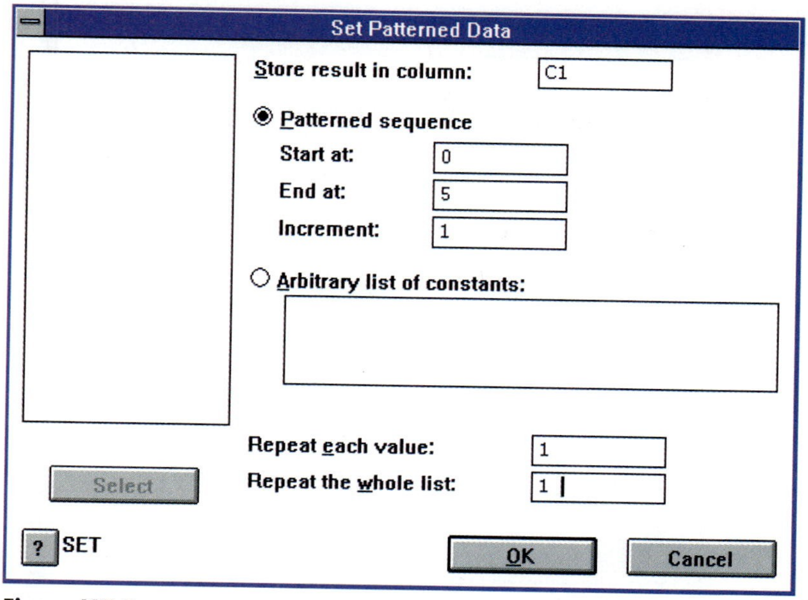

Figure M5.1

Next let us generate a frequency count for these individual values. Since frequency counts are integer values, we need to generate integer values. So select Calc→Random Data→Integer, and the Integer distribution dialog box will appear. Suppose we want frequency counts from 1 to 500 for these six values (0 to 5), and we want to store them in column C2. Type the value of 6 in the Generate rows of data text box. C2 in the Store in column(s) box, the value of 1 in the Minimum value box, and 500 in the Maximum value box. Select the OK button, and the values will be generated and stored in C2. Rename this column and call it f (frequency) by typing f in the first cell in column C2.

To obtain the probability distribution of X, we need to compute the relative frequencies for the values of X. To compute the relative frequencies, select Calc→Mathematical Expressions, and the Mathematical expressions dialog box will be displayed. Save the relative frequencies in C3, and compute them by typing the expression in the Expression text box, as shown in Fig. M5.2. Select the OK button, and the values will be generated and displayed in the Data window. Observe that column C3 was renamed $P(x)$. Thus, this output represents a discrete probability distribution since we have the values x of the random variable X and the corresponding probability $P(x)$ of the occurrence of these x values.

Example 2 Compute the expected value and variance for the random variable in Example 1.
To compute the expected value (mean), select Calc→Mathematical Expressions. In the Mathematical expressions dialog box, as in Fig. M5.2, type in Sum ('x'*'P(x)') in the Expression text box, and store the computed value in column C4. Observe that this is the equation of the expected value of the random variable X. Select the OK button, and the value 2.83967 (expected value of X) will be displayed in C4 for this discrete distribution that was created.

To compute the variance, in the Expression text box, type in the expression Sum(('x'-C4)**2*'P(x)'), and store the value in column C5. Observe that this is the equation for the variance of the random variable X. Select the OK button, and the value 2.38974 will be displayed in C5. To compute the standard deviation of X, just take the square root of 2.38974.

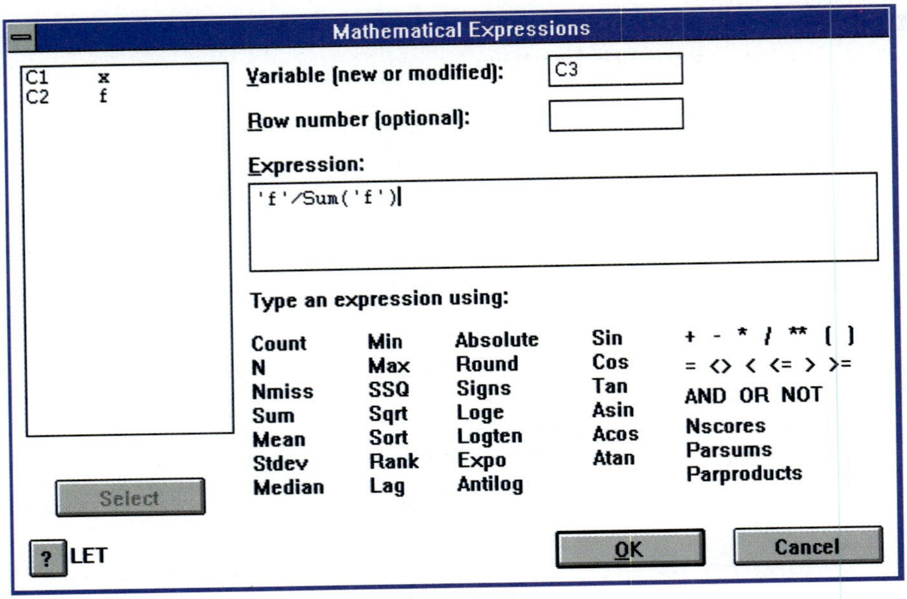

Figure M5.2

2. INVESTIGATING THE BINOMIAL DISTRIBUTION

Example 3 Use Minitab to generate 500 values of a binomial random variable with $n = 25$ and varying values of p. Investigate the shape of these distributions.

Select Calc→Random Data→Binomial. The Binomial distribution dialog box will be displayed. Type in the appropriate values, as shown in Fig. M5.3. Here we are generating 500 values of the random variable from a binomial distribution with $n = 25$ and a selected $p = .1$,

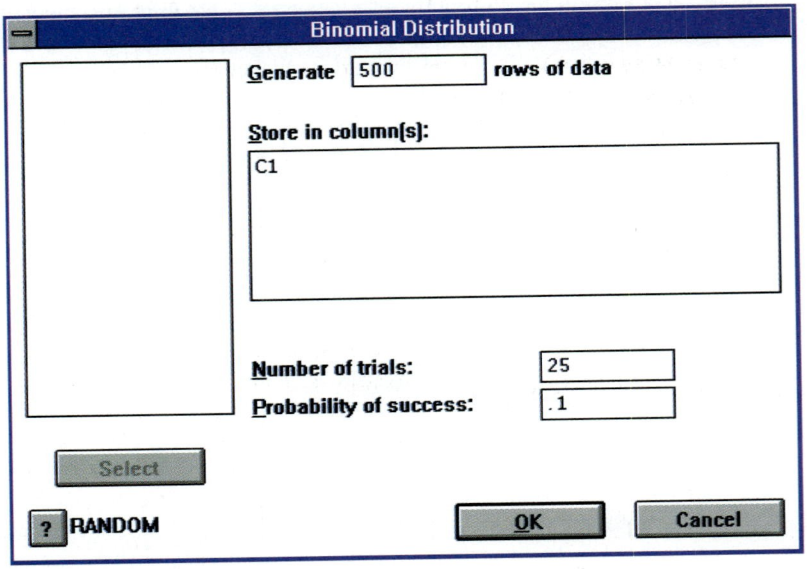

Figure M5.3

and we are storing these values in C1. Select the OK button, and the values will be generated and stored in the Data window in column C1. Use the procedures in the Minitab labs for Chaps. 2 and 3 to construct a line graph and a histogram for the generated values in C1. Repeat for $p = .2, .3, .4, .5, .6, .7, .8, .9$. Observe the shapes of these displays.

Example 4   Repeat Example 3 with $n = 100$.

3. USING MINITAB TO COMPUTE BINOMIAL PROBABILITIES

Example 5   Use Minitab to compute the probability that the binomial random variable X will equal 3 when $n = 10$ and $p = 4$. That is, we need to compute $P(X = 3)$ when $n = 10$ and $p = .4$.

Select Calc→Probability Distributions→Binomial, and the Binomial distribution dialog box will appear. Enter the appropriate numbers in the respective text boxes, as shown in Fig. M5.4. Observe that the Probability and Input constant check boxes are selected.

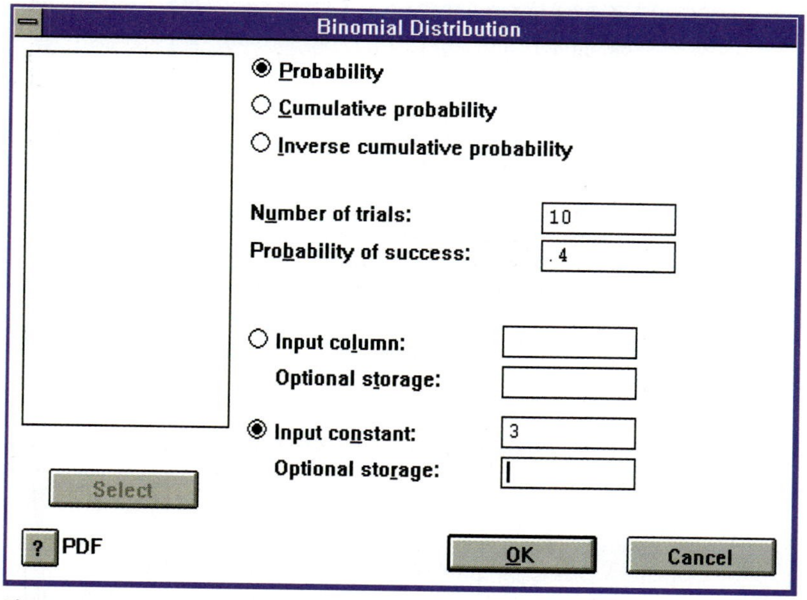

```
┌─────────────────────────────────────────────────────────┐
│ ▬              Binomial Distribution                      │
├─────────────────────────────────────────────────────────┤
│ ┌──────────────┐   ◉ Probability                         │
│ │              │   ○ Cumulative probability              │
│ │              │   ○ Inverse cumulative probability      │
│ │              │                                         │
│ │              │   Number of trials:        ┌─────────┐  │
│ │              │   Probability of success:  │ 10      │  │
│ │              │                            └─────────┘  │
│ │              │                            ┌─────────┐  │
│ │              │                            │ .4      │  │
│ │              │                            └─────────┘  │
│ │              │   ○ Input column:          ┌─────────┐  │
│ │              │                            └─────────┘  │
│ │              │     Optional storage:      ┌─────────┐  │
│ │              │                            └─────────┘  │
│ │              │   ◉ Input constant:        ┌─────────┐  │
│ └──────────────┘                            │ 3       │  │
│ ┌──────────┐        Optional storage:       ┌─────────┐  │
│ │  Select  │                                │ │       │  │
│ └──────────┘                                └─────────┘  │
│ ┌─┐                          ┌─────────┐   ┌──────────┐  │
│ │?│ PDF                      │   OK    │   │  Cancel  │  │
│ └─┘                          └─────────┘   └──────────┘  │
└─────────────────────────────────────────────────────────┘
```

Figure M5.4

Select the OK button, and the probability will be computed and displayed in the Session window since no storage column was specified.

From this Session window, $P(X = 3) = .2150$. This value, of course, may be obtained from tables in the text or by using a scientific calculator. However, like Program 5-1 on the text diskette, Minitab will compute the probabilities for *any value of p*. Try $p = .1134, .5557,$ and .978642.

Example 6   Compute the probabilities for the binomial random variable X, where the possible values of X are $0, 1, \ldots, 15$ with $p = .5743$.

This would be quite tedious to do even with a scientific calculator. Observe that you cannot use the tables since $p = .5743$ is not one of the probability values. To use Minitab, enter the values 0, 1, 2, . . . , 15 in C1; or use Calc→Set Patterned Data as in Example 1. Rename C1 and call it x by typing the symbol x in the first cell of column C1. Select Calc→Probability Distributions→Binomial. The Binomial distribution dialog box will be displayed. Select the Probability option, with the Number of trials being 15 and Probability of success being 0.5743. Also select the Input column text box and type in x.

Select the OK button, and the probability table shown in Fig. M5.5 will be generated. This will be displayed in the Session window since the generated values were not saved in any specified column. Figure M5.5 displays the values of the binomial random variable and the associated probabilities of occurrence.

```
┌──────────────────────── Session ────────────────────────┐
│                                                          │
│  Probability Density Function                            │
│                                                          │
│  Binomial with n = 15 and p = .574300                    │
│                                                          │
│            x            P(X = x)                         │
│          .00              .0000                          │
│         1.00              .0001                          │
│         2.00              .0005                          │
│         3.00              .0031                          │
│         4.00              .0124                          │
│         5.00              .0367                          │
│         6.00              .0824                          │
│         7.00              .1430                          │
│         8.00              .1929                          │
│         9.00              .2024                          │
│        10.00              .1639                          │
│        11.00              .1005                          │
│        12.00              .0452                          │
│        13.00              .0141                          │
│        14.00              .0027                          │
│        15.00              .0002                          │
│                                                          │
└──────────────────────────────────────────────────────────┘
```

Figure M5.5

4. USING MINITAB TO COMPUTE CUMULATIVE BINOMIAL PROBABILITIES

Example 7 Suppose you want to compute the cumulative probabilities for the binomial random variable X, where the possible values of X are 0, 1, . . . , 15 with $p = .5743$.

Select Calc→Probability Distributions→Binomial, and the Binomial distribution dialog box will be displayed. Select the Cumulative probability option with the number of trials being 15 and the Probability of success being 0.5743. Select the Input column option and type x in the text box. Also, type C2 in the Optional storage box. Select the OK button, and the probabilities will be computed and displayed in the Data window in column C2 since C2 was used in the Optional storage text box.

Observe that the probabilities accumulate to 1, as shown in Fig. M5.6. The probabilities in C2 correspond to $P(X \leq x)$ for $x = 1, 2, . . . , 15$. For example, the probability that $X \leq 5$ is $P(X \leq 5) = .05266$.

Data		
C1	**C2**	**C3**
x		
1	1	.00006
2	2	.00058
3	3	.00363
4	4	.01599
5	5	.05266
6	6	.13510
7	7	.27811
8	8	.47103
9	9	.67346
10	10	.83732
11	11	.93779
12	12	.98298
13	13	.99705
14	14	.99976
15	15	1.00000

Figure M5.6

Note

1. If you only want to compute $P(X \leq 5)$, just select Input constant and type in 5 in the corresponding text box. Select OK, and the probability will be displayed in the Session window.

2. Suppose we want to compute $P(X > 5)$. We know that $P(X > 5) = 1 - P(X \leq 5)$, so we can use the cumulative probabilities to help compute this complement.

3. You can use the procedures in Sec. 1 to compute the expected value, variance, and standard deviation for a binomial random variable.

Computer Exercises

1. Use Minitab or any other statistical software to generate a discrete probability distribution for a random variable X with values 0, 1, 2, 3, ..., 50. Compute the expected value for this distribution (random variable). Repeat this experiment 10, 20, 30, 40, and 50 times, and record the expected values for these distributions. Generate histograms and line graphs for these distributions to observe the shapes of the distributions. Generate histograms for the computed means. Present a report with your observations. Include *at most* five histograms for the distributions to illustrate your observations; however, include all five histograms for the expected values.

2. Present the results in Example 3 in a report. Include the line graphs and histograms, and discuss any observations from these graphs.

3. Present the results in Example 4 in a report. Include the line graphs and histograms, and discuss any observations from these graphs.

4. Use Minitab or any other statistical software to generate binomial probabilities for $n = 10$ and $p = .1, .3, .5, .7,$ and $.9$. Discuss any observations from these generated tables. Type in C2, etc., in the Optional storage text box to save the generated probabilities. Construct line graphs and histograms to help present your observations in a report.

5. In Example 7 plot the cumulative probabilities along the Y axis and the x values along the X axis, using connect, symbols, etc. Present any other graphs that you think may be appropriate. Present these graphs in a report, and discuss any observations from these graphs.

6. Use Minitab or any other statistical software to generate a discrete probability distribution for a random variable X with values $0, 1, 2, 3, \ldots, 50$. Compute the variance for this distribution (random variable). Repeat this experiment 10, 20, 30, 40, and 50 times, and record the variance for these distributions. Generate histograms and line graphs for these distributions to observe the shapes of the distributions. Generate histograms for the computed variances. Present a report with your observations. Include *at most* five histograms for the distributions to illustrate your observations; however, include all five histograms for the variances.

7. Let X and Y be two independent binomial random variables. For X, let $n = 10$ and $p = .4$; and for Y, let $n = 15$ and $p = .7$. Use Minitab or any other statistical software to generate 500 values of X, and store in C1 if you use Minitab. Find the column mean. Compare this value to the theoretical mean np for the random variable X. Repeat the procedure for the random variable Y, and store these generated values in C2. Next add the values in C1 and C2 and save in C3. Find the mean for C3, and compare to the sum of the means for C1 and C2. Repeat this experiment 20, 30, and 40 times. Present three separate tables with the computed means for C1, C2, and C3. Find the average of these values in each table and compare. What general conclusions can you draw? Are your conclusions restricted to only binomial random variables?

8. Repeat Exercise 7 for the variance.

CHAPTER 6

Normal Random
Variables

Normal Random Variables

> *Among other peculiarities of the nineteenth century is this one, by initiating the systematic collection of statistics it has made the quantitative study of social forces possible.*
> Alfred North Whitehead

W e introduce continuous random variables, which are random variables that can take on any value in an interval. We show how their probabilities are determined from an associated curve known as a *probability density function.* A special type of continuous random variable, known as a *normal random variable,* is studied. The standard normal random variable is introduced, and a table is presented which enables us to compute the probabilities of that variable. We show how any normal random variable can be transformed to a standard one, enabling us to determine its probabilities. We present the additive property of normal random variables. The percentiles of normal random variables are studied.

6.1   INTRODUCTION

In this chapter we introduce and study the normal distribution. Both from a theoretical and from an applied point of view, this distribution is unquestionably the most important in all statistics.

The normal distribution is one of a class of distributions that are called *continuous*. Continuous distributions are introduced in Sec. 6.2. In Sec. 6.3 we define what is meant by a normal distribution and present an approximation rule concerning its probabilities. In Sec. 6.4, we consider the standard normal distribution, which is a normal distribution having mean 0 and variance 1, and we show how to determine its probabilities by use of a table. In Sec. 6.5 we show how any normal random variable can be transformed to a standard normal, and we use this transformation to determine the probabilities of that variable. The additive property of normal random variables is discussed in Sec. 6.6, and in Sec. 6.7 we consider their percentiles.

The normal distribution was introduced by the French mathematician Abraham De Moivre in 1733.

6.2   CONTINUOUS RANDOM VARIABLES

Whereas the possible values of a discrete random variable can be written as a sequence of isolated values, a *continuous random variable* is one whose set of possible values is an interval. That is, a continuous random variable is able to take on any value within some interval. For example, such variables as the time it takes to complete a scientific experiment or the weight of an individual are usually considered to be continuous random variables.

Every continuous random variable X has a curve associated with it. This curve, formally known as a *probability density function,* can be used to obtain probabilities associated with the random variable. This is accomplished as follows. Consider any two points a and b, where a is less than b. The probability that X assumes a value that lies between a and b is equal to the area under the curve between a and b. That is,

$$P\{a \leq X \leq b\} = \text{area under curve between } a \text{ and } b$$

Figure 6.1 presents a probability density function.

Since X must assume some value, it follows that the total area under the density curve must equal 1. Also, since the area under the graph of the probability density function between points a and b is the same regardless of whether the endpoints a and b are themselves included, we see that

$$P\{a \leq X \leq b\} = P\{a < X < b\}$$

Historical Perspective

Abraham De Moivre (1667–1754)

Abraham De Moivre

Today there is no shortage of statistical consultants, many of whom ply their trade in the most elegant of settings. However, the first of their breed worked, in the early years of the 18th century, out of a dark, grubby betting shop in Long Acres, London, known as Slaughter's Coffee House. He was Abraham De Moivre, a Protestant refugee from Catholic France, and for a price, he would compute the probability of gambling bets in all types of games of chance.

Although De Moivre, the discoverer of the normal curve, made his living at the coffee shop, he was a mathematician of recognized abilities. Indeed, he was a member of the Royal Society and was reported to be an intimate of Isaac Newton.

This is Karl Pearson imagining De Moivre at work at Slaughter's Coffee House:

> I picture De Moivre working at a dirty table in the coffee house with a broken-down gambler beside him and Isaac Newton walking through the crowd to his corner to fetch out his friend. It would make a great picture for an inspired artist.

That is, the probability that a continuous random variable lies in some interval is the same whether you include the endpoints of the interval or not.

The probability density curve of a random variable X is a curve that never goes below the x axis and has the property that the total area between it and the x axis is equal to 1. It determines the probabilities of X in that the area under the curve between points a and b is equal to the probability that X is between a and b.

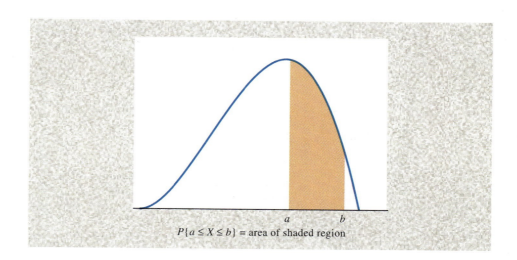

$P\{a \leq X \leq b\}$ = area of shaded region

Figure 6.1
Probability density function of X.

Problems

1. Figure 6.2 is a probability density function for the random variable that represents the time (in minutes) it takes a repairer to service a television. The numbers in the regions represent the areas of those regions. What is the probability that the repairer takes
   (a) Less than 20
   (b) Less than 40
   (c) More than 50
   (d) Between 40 and 70
   minutes to complete a repair?

2. A random variable is said to be a *uniform* random variable in the interval (a, b) if its set of possible values is this interval and if its density curve is a horizontal line. That is, its density curve is as given in Fig. 6.3.
   (a) Explain why the height of the density curve is $1/(b - a)$. (*Hint:* Remember that the total area under the density curve must equal 1, and recall the formula for the area of a rectangle.)
   (b) What is $P\{X \leq (a + b)/2\}$?

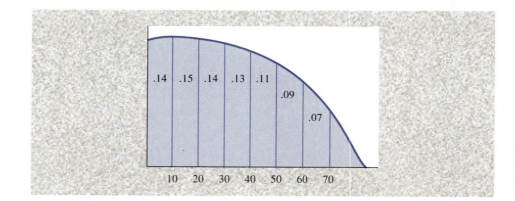

Figure 6.2
Probability density function of X.

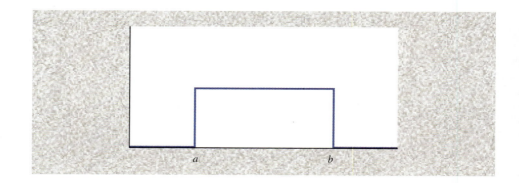

Figure 6.3
Density curve of the uniform (a, b) random variable.

3. Suppose that X is a uniform random variable over the interval $(0, 1)$. That is, $a = 0$ and $b = 1$ for the random variable in Prob. 2. Find

(a) $P\left\{X > \dfrac{1}{3}\right\}$

(b) $P\{X \le .7\}$

(c) $P\{.3 < X \le .9\}$

(d) $P\{.2 \le X < .8\}$

4. You are to meet a friend at 2 p.m. However, while you are always exactly on time, your friend is always late and indeed will arrive at the meeting place at a time uniformly distributed between 2 and 3 p.m. Find the probability that you will have to wait

(a) At least 30 minutes

(b) Less than 15 minutes

(c) Between 10 and 35 minutes

(d) Less than 45 minutes

5. Suppose in Prob. 4 that your friend will arrive at a time that is uniformly distributed between 1:30 and 3 p.m. Find the probability that

(a) You are the first to arrive.

(b) Your friend will have to wait more than 15 minutes.

(c) You will have to wait over 30 minutes.

6. Suppose that the number of minutes of playing time of a certain college basketball player in a randomly chosen game has the density curve given below.

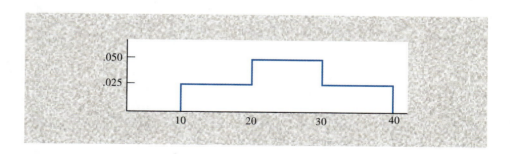

Find the probability that the player plays

(a) Over 20 minutes

(b) Less than 25 minutes

(c) Between 15 and 35 minutes

(d) More than 35 minutes

7. Let X denote the number of minutes played by the basketballer of Prob. 6. Find

(a) $P\{20 < X < 30\}$

(b) $P\{X > 50\}$

(c) $P\{20 < X < 40\}$

(d) $P\{15 < X < 25\}$

8. It is now 2 p.m., and Joan is planning on studying for her statistics test until 6 p.m., when she will have to go out to dinner. However, she knows that she will probably have interruptions and thinks that the amount of time she will actually spend studying in the next 4 hours is a random variable whose probability density curve is as follows:

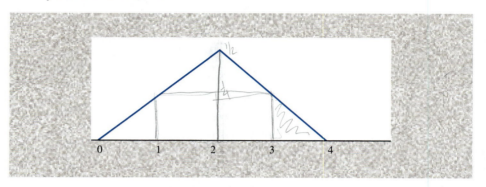

(a) What is the height of the curve at the value 2? (*Hint:* You will have to recall the formula for the area of a triangle.)
(b) What is the probability she will study more than 3 hours?
(c) What is the probability she will study between 1 and 3 hours?

6.3 NORMAL RANDOM VARIABLES

The most important type of random variable is the normal random variable. The probability density function of a normal random variable X is determined by two parameters, namely, the expected value and the standard deviation of X. We designate these values as μ and σ, respectively. That is, we will let

$$\mu = E[X] \quad \text{and} \quad \sigma = \text{SD}(X)$$

The normal probability density function is a bell-shaped density curve that is symmetric about the value μ. Its variability is measured by σ. The larger σ is, the more variability there is in this curve. Figure 6.4 presents three different normal probability density functions. Note how the curves flatten out as σ increases.

Because the probability density function of a normal random variable X is symmetric about its expected value μ, it follows that X is equally likely to be on either side of μ. That is,

$$P\{X < \mu\} = P\{X > \mu\} = \frac{1}{2}$$

Not all bell-shaped symmetric density curves are normal. The normal density curves are specified by a particular formula, namely, the height of the curve above point x on the abscissa is

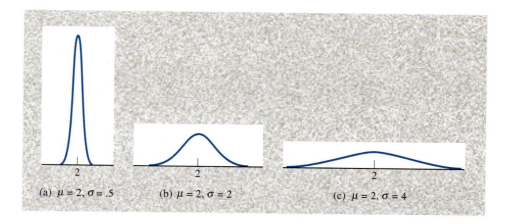

Figure 6.4
Three normal
probability
density
functions.

(a) $\mu = 2, \sigma = .5$ (b) $\mu = 2, \sigma = 2$ (c) $\mu = 2, \sigma = 4$

$$\frac{1}{\sqrt{2\pi}\sigma} e^{-(x-\mu)^2/\alpha\sigma^2}$$

Although we will not make direct use of the above formula, it is interesting to note that it involves two of the famous constants of mathematics, namely, π (the area of a circle of radius 1) and e (which is the base of the natural logarithms). Also note that this formula is completely specified by the mean value μ and the standard deviation σ.

A normal random variable having mean value 0 and standard deviation 1 is called a *standard normal* random variable, and its density curve is called the *standard normal curve*. Figure 6.5 presents the standard normal curve. In this text we will use (and reserve) the letter Z to represent a standard normal random variable.

In Sec. 6.5 we will show how to determine probabilities concerning an arbitrary normal random variable by relating them to probabilities about the standard normal

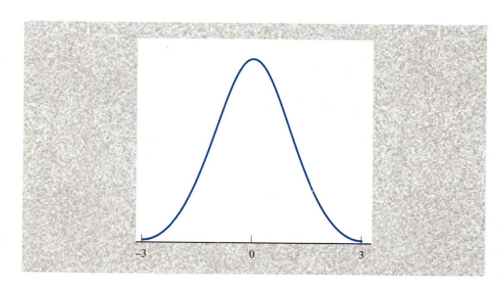

Figure 6.5
The standard
normal curve.

random variable. In doing so, we will show the following useful approximation rule for normal probabilities.

Approximation rule
A normal random variable with mean μ and standard deviation σ will be
Between $\mu - \sigma$ and $\mu + \sigma$ with approximate probability .68
Between $\mu - 2\sigma$ and $\mu + 2\sigma$ with approximate probability .95
Between $\mu - 3\sigma$ and $\mu + 3\sigma$ with approximate probability .997

This approximation rule is illustrated in Fig. 6.6. It often enables us to obtain a quick feel for a data set.

Example 6.1   Test scores on the Scholastic Aptitude Test (SAT) verbal portion are normally distributed with a mean score of 504. If the standard deviation of a score is 84, then we can conclude that approximately 68 percent of all scores are between 504 $-$ 84 and 504 $+$ 84. That is, approximately 68 percent of the scores are between 420 and 588. Also, approximately 95 percent of them are between 504 $-$ 168 $=$ 336 and 504 $+$ 168 $=$ 672; and approximately 99.7 percent are between 252 and 756.

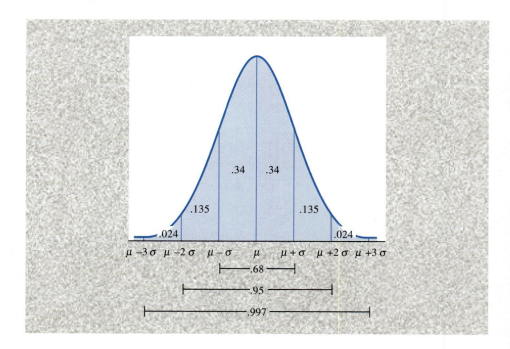

Figure 6.6
Approximate areas under a normal curve.

The approximation rule is the theoretical basis of the empirical rule of Sec. 3.6. The connection between these rules will become apparent in Chap. 8, when we show how a sample mean and sample standard deviation can be used to estimate the quantities μ and σ.

By using the symmetry of the normal curve about the value μ, we can obtain other facts from the approximation rule. For instance, since the area between μ and $\mu + \sigma$ is the same as that between $\mu - \sigma$ and μ, it follows from this rule that a normal random variable will be between μ and $\mu + \sigma$ with approximate probability $.68/2 = .34$.

Problems

1. The systolic blood pressures of adults, in the appropriate units, are normally distributed with a mean of 128.4 and a standard deviation of 19.6.
   (a) Give an interval in which the blood pressures of approximately 68 percent of the population fall.
   (b) Give an interval in which the blood pressures of approximately 95 percent of the population fall.
   (c) Give an interval in which the blood pressures of approximately 99.7 percent of the population fall.

2. The heights of a certain population of males are normally distributed with mean 69 inches and standard deviation 6.5 inches. Approximate the proportion of this population whose height is less than 82 inches.

Problems 3 through 16 are multiple-choice problems. Give the answer which you think is closest to the true answer. Remember, Z always refers to a standard normal random variable. Draw a picture in each case to justify your answer.

3. $P\{-2 < Z < 2\}$ is approximately
   (a) .68   (b) .95   (c) .975   (d) .50

4. $P\{Z > -1\}$ is approximately
   (a) .50   (b) .95   (c) .84   (d) .16

5. $P\{Z > 1\}$ is approximately
   (a) .50   (b) .95   (c) .84   (d) .16

6. $P\{Z > 3\}$ is approximately
   (a) .30   (b) .05   (c) 0   (d) .99

7. $P\{Z < 2\}$ is approximately
   (a) .95   (b) .05   (c) .975   (d) .025

In Probs. 8 to 11, X is a normal random variable with expected value 15 and standard deviation 4.

8. The probability that X is between 11 and 19 is approximately
   (a) .50   (b) .95   (c) .68   (d) .34

9. The probability that X is less than 23 is approximately
   (a) .975   (b) .95   (c) .68   (d) .05

10. The probability that X is less than 11 is approximately
    (a) .34   (b) .05   (c) .16   (d) .50

11. The probability that X is greater than 27 is approximately
    (a) .05   (b) 0   (c) .01   (d) .32

12. Variable X is a normal random variable with standard deviation 3. If the probability that X is between 7 and 19 is .95, then the expected value of X is approximately
    (a) 16   (b) 15   (c) 14   (d) 13

13. Variable X is a normal random variable with standard deviation 3. If the probability that X is less than 16 is .84, then the expected value of X is approximately
    (a) 16   (b) 15   (c) 14   (d) 13

14. Variable X is a normal random variable with standard deviation 3. If the probability that X is greater than 16 is .975, then the expected value of X is approximately
    (a) 20   (b) 22   (c) 23   (d) 25

15. Variable X is a normal random variable with expected value 100. If the probability that X is greater than 90 is .84, then the standard deviation of X is approximately
    (a) 5   (b) 10   (c) 15   (d) 20

16. Variable X is a normal random variable with expected value 100. If the probability that X is greater than 130 is .025, then the standard deviation of X is approximately
    (a) 5   (b) 10   (c) 15   (d) 20

17. If X is normal with expected value 100 and standard deviation 2, and Y is normal with expected value 100 and standard deviation 4, is X or is Y more likely to
    (a) Exceed 104   (b) Exceed 96
    (c) Exceed 100

18. If X is normal with expected value 100 and standard deviation 2, and Y is normal with expected value 105 and standard deviation 10, is X or is Y more likely to
    (a) Exceed 105
    (b) Be less than 95

19. The scores on a particular job aptitude test are normal with expected value 400 and standard deviation 100. If a company will consider only those applicants scoring in the top 5 percent, determine whether it should consider one whose score is
    (a) 400   (b) 450
    (c) 500   (d) 600

6.4 PROBABILITIES ASSOCIATED WITH A STANDARD NORMAL RANDOM VARIABLE

Let Z be a standard normal random variable. That is, Z is a normal random variable with mean 0 and standard deviation 1. The probability that Z is between two numbers a and b is equal to the area under the standard normal curve between a and b. Areas under this curve have been computed, and tables have been prepared that enable us to find these probabilities. One such table is Table 6.1, page 288.

For each nonnegative value of x, Table 6.1 specifies the probability that Z is less than x. For instance, suppose we want to determine $P\{Z < 1.22\}$. To do this, first we must find the entry in the table corresponding to $x = 1.22$. This is done by first searching the left-hand column to find the row labeled 1.2 and then searching the top row to find the column labeled .02. The value that is in both the row labeled 1.2 and the column labeled .02 is the desired probability. Since this value is .8888, we see that

$$P\{Z < 1.22\} = .8888$$

A portion of Table 6.1 illustrating the preceding is presented here:

			↓			
x	.00	.01	.02	.03	$\cdots$	.09
.0	.5000	.5040				
⋮						
1.1	.8413					
→1.2	.8849	.8869	.8888			
1.3	.9032					

We can also use Table 6.1 to determine the probability that Z is greater than x. For instance, suppose we want to determine the probability that Z is greater than 2. To accomplish this, we note that either Z is less than or equal to 2 or Z is greater than 2, and so

$$P\{Z \leq 2\} + P\{Z > 2\} = 1$$

or

$$P\{Z > 2\} = 1 - P\{Z \leq 2\}$$
$$= 1 - .9772$$
$$= .0228$$

In other words, the probability that Z is larger than x can be obtained by subtracting from 1 the probability that Z is smaller than x. That is, for any x,

$$P\{Z > x\} = 1 - P\{Z \leq x\}$$

Table 6.1   Standard Normal Probabilities

x	.00	.01	.02	.03	.04	.05	.06	.07	.08	.09
.0	.5000	.5040	.5080	.5120	.5160	.5199	.5239	.5279	.5319	.5359
.1	.5398	.5438	.5478	.5517	.5557	.5596	.5636	.5675	.5714	.5753
.2	.5793	.5832	.5871	.5910	.5948	.5987	.6026	.6064	.6103	.6141
.3	.6179	.6217	.6255	.6293	.6331	.6368	.6406	.6443	.6480	.6517
.4	.6554	.6591	.6628	.6664	.6700	.6736	.6772	.6808	.6844	.6879
.5	.6915	.6950	.6985	.7019	.7054	.7088	.7123	.7157	.7190	.7224
.6	.7257	.7291	.7324	.7357	.7389	.7422	.7454	.7486	.7517	.7549
.7	.7580	.7611	.7642	.7673	.7704	.7734	.7764	.7794	.7823	.7852
.8	.7881	.7910	.7939	.7967	.7995	.8023	.8051	.8078	.8106	.8133
.9	.8159	.8186	.8212	.8238	.8264	.8289	.8315	.8340	.8365	.8389
1.0	.8413	.8438	.8461	.8485	.8508	.8531	.8554	.8577	.8599	.8621
1.1	.8643	.8665	.8686	.8708	.8729	.8749	.8770	.8790	.8810	.8830
1.2	.8849	.8869	.8888	.8907	.8925	.8944	.8962	.8980	.8997	.9015
1.3	.9032	.9049	.9066	.9082	.9099	.9115	.9131	.9147	.9162	.9177
1.4	.9192	.9207	.9222	.9236	.9251	.9265	.9279	.9292	.9306	.9319
1.5	.9332	.9345	.9357	.9370	.9382	.9394	.9406	.9418	.9429	.9441
1.6	.9452	.9463	.9474	.9484	.9495	.9505	.9515	.9525	.9535	.9545
1.7	.9554	.9564	.9573	.9582	.9591	.9599	.9608	.9616	.9625	.9633
1.8	.9641	.9649	.9656	.9664	.9671	.9678	.9686	.9693	.9699	.9706
1.9	.9713	.9719	.9726	.9732	.9738	.9744	.9750	.9756	.9761	.9767
2.0	.9772	.9778	.9783	.9788	.9793	.9798	.9803	.9808	.9812	.9817
2.1	.9821	.9826	.9830	.9834	.9838	.9842	.9846	.9850	.9854	.9857
2.2	.9861	.9864	.9868	.9871	.9875	.9878	.9881	.9884	.9887	.9890
2.3	.9893	.9896	.9898	.9901	.9904	.9906	.9909	.9911	.9913	.9916
2.4	.9918	.9920	.9922	.9925	.9927	.9929	.9931	.9932	.9934	.9936
2.5	.9938	.9940	.9941	.9943	.9945	.9946	.9948	.9949	.9951	.9952
2.6	.9953	.9955	.9956	.9957	.9959	.9960	.9961	.9962	.9963	.9964
2.7	.9965	.9966	.9967	.9968	.9969	.9970	.9971	.9972	.9973	.9974
2.8	.9974	.9975	.9976	.9977	.9977	.9978	.9979	.9979	.9980	.9981
2.9	.9981	.9982	.9982	.9983	.9984	.9984	.9985	.9985	.9986	.9986
3.0	.9987	.9987	.9987	.9988	.9988	.9989	.9989	.9989	.9990	.9990
3.1	.9990	.9991	.9991	.9991	.9992	.9992	.9992	.9992	.9993	.9993
3.2	.9993	.9993	.9994	.9994	.9994	.9994	.9994	.9995	.9995	.9995
3.3	.9995	.9995	.9995	.9996	.9996	.9996	.9996	.9996	.9996	.9997
3.4	.9997	.9997	.9997	.9997	.9997	.9997	.9997	.9997	.9997	.9998

Data value in table is $P\{Z < x\}$.

Example 6.2   Find

(a) $P\{Z < 1.5\}$

(b) $P\{Z \geq .8\}$

Solution

(a) From Table 6.1,

$$P\{Z < 1.5\} = .9332$$

(b) From Table 6.1, $P\{Z < .8\} = .7881$ and so

$$P\{Z \geq .8\} = 1 - .7881 = .2119$$

While Table 6.1 specifies $P\{Z < x\}$ for only nonnegative values of x, it can be used even when x is negative. Probabilities for negative x are obtained from the table by making use of the symmetry about zero of the standard normal curve. For instance, suppose we want to calculate the probability that Z is less than -2. By symmetry (see Fig. 6.7) that is the same as the probability that Z is greater than 2; and so

$$\begin{aligned} P\{Z < -2\} &= P\{Z > 2\} \\ &= 1 - P\{Z < 2\} \\ &= 1 - .9772 = .0028 \end{aligned}$$

In general, for any value of x,

$$P\{Z < -x\} = P\{Z > x\} = 1 - P\{Z < x\}$$

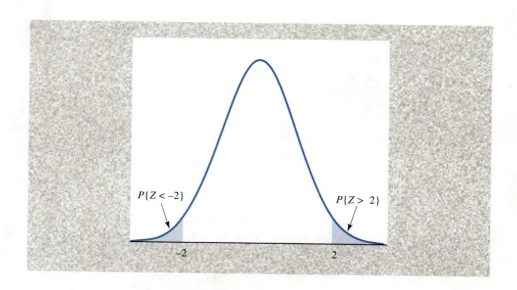

Figure 6.7
$P\{Z < -2\} =$
$P\{Z > 2\}$

We can determine the probability that Z lies between a and b, for $a < b$, by determining the probability that Z is less than b and then subtracting from this the probability that Z is less than a. (See Fig. 6.8.)

Example 6.3   Find

(a) $P\{1 < Z < 2\}$

(b) $P\{-1.5 < Z < 2.5\}$

Solution

(a) $P\{1 < Z < 2\} = P\{Z < 2\} - P\{Z < 1\}$
$$= .9772 - .8413$$
$$= .1359$$

(b) $P\{-1.5 < Z < 2.5\} = P\{Z < 2.5\} - P\{Z < -1.5\}$
$$= P\{Z < 2.5\} - P\{Z > 1.5\}$$
$$= .9938 - (1 - .9332)$$
$$= .9270$$

Let a be positive and consider $P\{|Z| > a\}$, the probability that a standard normal is, in absolute value, larger than a. Since $|Z|$ will exceed a if either $Z > a$ or $Z < -a$, we see that

$$P\{|Z| > a\} = P\{Z > a\} + P\{Z < -a\}$$
$$= 2P\{Z > a\}$$

where the last equality uses the symmetry of the standard normal density curve (see Fig. 6.9).

Example 6.4   Find $P\{|Z| > 1.8\}$.

Solution

$$P\{|Z| > 1.8\} = 2P\{Z > 1.8\}$$
$$= 2(1 - .9641)$$
$$= .0718$$

Another easily established result is that for any positive value of a

$$P\{-a < Z < a\} = 2P\{Z < a\} - 1$$

The verification of the above is left as an exercise.

Table 6.1 is also listed as Table D.1 in App. D. In addition, Program 6-1 can be used to obtain normal probabilities. You enter the value x, and the program outputs $P\{Z < x\}$.

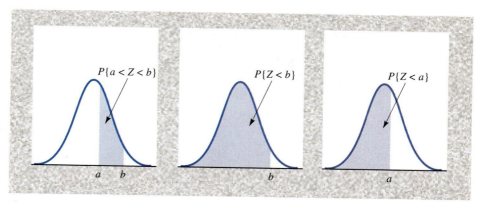

Figure 6.8
$P\{a < Z < b\} = P\{Z < b\} - P\{Z < a\}$

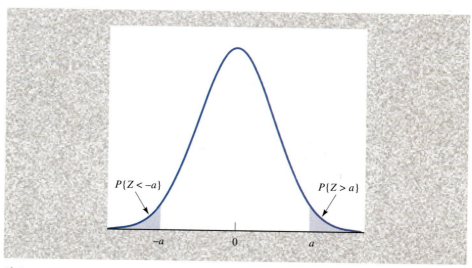

Figure 6.9
$P\{Z > a\} = P\{Z < -a\}$

Example 6.5   Determine $P\{Z > .84\}$.

Solution We can either use Table 6.1 or run Program 6-1 as follows:

We can either use Table 6.1 or run Program 6-1, which computes the probability that a standard normal random variable is less than x. Running Program 6-1, we learn that if the desired value of x is .84, the probability is .7995459.

The desired probability is $1 - .80 = .20$. That is, there is a 20 percent chance that a standard normal random variable will exceed .84.

Problems

1. For a standard normal random variable Z find
   (a) $P\{Z < 2.2\}$ (b) $P\{Z > 1.1\}$
   (c) $P\{0 < Z < 2\}$ (d) $P\{-.9 < Z < 1.2\}$
   (e) $P\{Z > -1.96\}$ (f) $P\{Z < -.72\}$
   (g) $P\{|Z| < 1.64\}$ (h) $P\{|Z| > 1.20\}$
   (i) $P\{-2.2 < Z < 1.2\}$

2. Show that $-Z$ is also a standard normal random variable. *Hint:* It suffices to show that, for all x,

$$P\{-Z < x\} = P\{Z < x\}$$

3. Find the value of the question mark:

$$P\{-3 < Z < -2\} = P\{2 < Z < ?\}$$

Use a picture to show that your answer is correct.

4. Use a picture of the standard normal curve to show that

$$P\{Z > -2\} = P\{Z < 2\}$$

5. Argue, using either pictures or equations, that for any positive value of a,

$$P\{-a < Z < a\} = 2P\{Z < a\} - 1$$

6. Find
   (a) $P\{-1 < Z < 1\}$
   (b) $P\{|Z| < 1.4\}$

7. Find the value of x, to two decimal places, for which
   (a) $P\{Z > x\} = .05$ (b) $P\{Z > x\} = .025$
   (c) $P\{Z > x\} = .005$ (d) $P\{Z < x\} = .50$
   (e) $P\{Z < x\} = .66$ (f) $P\{|Z| < x\} = .99$
   (g) $P\{|Z| < x\} = .75$ (h) $P\{|Z| > x\} = .90$
   (i) $P\{|Z| > x\} = .50$

6.5   FINDING NORMAL PROBABILITIES: CONVERSION TO THE STANDARD NORMAL

Let X be a normal random variable with mean μ and standard deviation σ. We can determine probabilities concerning X by using the fact that the variable Z defined by

$$Z = \frac{X - \mu}{\sigma}$$

has a standard normal distribution. That is, if we *standardize* a normal random variable by subtracting its mean and then dividing by its standard deviation, the resulting variable has a standard normal distribution.

The value of the standardized variable tells us how many standard deviations the original variable is from its mean. For instance, if the standardized variable Z has value 2, then

$$Z = \frac{X - \mu}{\sigma} = 2$$

or

$$X - \mu = 2\sigma$$

That is, X is larger than its mean by 2 standard deviations.

We can compute any probability statement about X by writing an equivalent statement in terms of $Z = (X - \mu)/\sigma$ and then making use of Table 6.1 or Program 6-1. For instance, suppose we want to compute $P\{X < a\}$. Since $X < a$ is equivalent to the statement

$$\frac{X - \mu}{\sigma} < \frac{a - \mu}{\sigma}$$

we see that

$$P\{X < a\} = P\left\{\frac{X - \mu}{\sigma} < \frac{a - \mu}{\sigma}\right\}$$

$$= P\left\{Z < \frac{a - \mu}{\sigma}\right\}$$

where Z is a standard normal random variable.

Example 6.6   IQ examination scores for sixth-graders are normally distributed with mean value 100 and standard deviation 14.2.

(a) What is the probability a randomly chosen sixth-grader has a score greater than 130?

(b) What is the probability a randomly chosen sixth-grader has a score between 90 and 115?

Solution Let X denote the score of a randomly chosen student. We compute probabilities concerning X by making use of the fact that the standardized variable

$$Z = \frac{X - 100}{14.2}$$

has a standard normal distribution.

(a) $P\{X > 130\} = P\left\{\dfrac{X - 100}{14.2} > \dfrac{130 - 100}{14.2}\right\}$

$\qquad\qquad\qquad = P\{Z > 2.1127\}$

$\qquad\qquad\qquad = .017$

(b) The inequality $90 < X < 115$ is equivalent to

$$\dfrac{90 - 100}{14.2} < \dfrac{X - 100}{14.2} < \dfrac{115 - 100}{14.2}$$

or, equivalently,

$$-.7042 < Z < 1.056$$

Therefore,

$$\begin{aligned}
P\{90 < X < 115\} &= P\{-.7042 < Z < 1.056\}\\
&= P\{Z < 1.056\} - P\{Z < -.7042\}\\
&= .854 - .242\\
&= .612
\end{aligned}$$

Example 6.7   Let X be normal with mean μ and standard deviation σ. Find

(a) $P\{|X - \mu| > \sigma\}$

(b) $P\{|X - \mu| > 2\sigma\}$

(c) $P\{|X - \mu| > 3\sigma\}$

Solution   The statement $|X - \mu| > a\sigma$ is, in terms of the standardized variable $Z = (X - \mu)/\sigma$, equivalent to the statement $|Z| > a$. Using this fact, we obtain the following results.

(a) $\begin{aligned}[t]
P\{|X - \mu| > \sigma\} &= P\{|Z| > 1\}\\
&= 2P\{Z > 1\}\\
&= 2(1 - .8413)\\
&= .3174
\end{aligned}$

(b) $\begin{aligned}[t]
P\{|X - \mu| > 2\sigma\} &= P\{|Z| > 2\}\\
&= 2P\{Z > 2\}\\
&= .0456
\end{aligned}$

(c) $\begin{aligned}[t]
P\{|X - \mu| > 3\sigma\} &= P\{|Z| > 3\}\\
&= 2P\{Z > 3\}\\
&= .0026
\end{aligned}$

Thus, we see that the probability that a normal random variable differs from its mean by more than 1 standard deviation is (to two decimal places) .32; or equivalently, the complementary probability that it is within 1 standard deviation of its mean is .68. Similarly, parts (b) and (c) imply, respectively, that the probability the random variable is within 2 standard deviations of its mean is .95 and the probability that it is within 3 standard deviations of its mean is .997. Thus we have verified the approximation rule presented in Sec. 6.3.

6.6   ADDITIVE PROPERTY OF NORMAL RANDOM VARIABLES

The fact that $Z = (X - \mu)/\sigma$ is a standard normal random variable follows from the fact that if one either adds or multiplies a normal random variable by a constant, then the resulting random variable remains normal. As a result, if X is normal with mean μ and standard deviation σ, then $Z = (X - \mu)/\sigma$ also will be normal. It is now easy to verify that Z has expected value 0 and variance 1.

An important fact about normal random variables is that the sum of independent normal random variables is also a normal random variable. That is, if X and Y are independent normal random variables with respective parameters μ_x, σ_x and μ_y, σ_y, then $X + Y$ also will be normal. Its mean value will be

$$E[X + Y] = E[X] + E[Y] = \mu_x + \mu_y$$

Its variance is

$$\text{Var}(X + Y) = \text{Var}(X) + \text{Var}(Y) = \sigma_x^2 + \sigma_y^2$$

That is, we have the following result.

Suppose X and Y are independent normal random variables with means μ_x and μ_y and with standard deviations σ_x and σ_y, respectively. Then $X + Y$ is normal with mean

$$E[X + Y] = \mu_x + \mu_y$$

and standard deviation

$$\text{SD}(X + Y) = \sqrt{\sigma_x^2 + \sigma_y^2}$$

Example 6.8   Suppose the amount of time a light bulb works before burning out is a normal random variable with mean 400 hours and standard deviation 40 hours. If an individual purchases two such bulbs, one of which will be used as a spare to replace the other when it burns out, what is the probability that the total life of the bulbs will exceed 750 hours?

Solution   We need to compute the probability that $X + Y > 750$, where X is the life of the bulb used first and Y is the life of the other bulb. Variables X and Y are both normal with mean 400 and standard deviation 40. In addition, we will suppose they are independent, and so $X + Y$ is also normal with mean 800 and standard deviation $\sqrt{40^2 + 40^2} = \sqrt{3200}$. Therefore, $Z = (X + Y - 800)/\sqrt{3200}$ has a standard normal distribution. Thus, we have

$$P\{X + Y > 750\} = P\left\{ \frac{X + Y - 800}{\sqrt{3200}} > \frac{750 - 800}{\sqrt{3200}} \right\}$$
$$= P\{Z > -.884\}$$
$$= P\{Z < .884\}$$
$$= .81$$

Therefore, there is an 81 percent chance that the total life of the two bulbs exceeds 750 hours.

Example 6.9   Data from the U.S. Department of Agriculture indicate that the annual amount of apples eaten by a randomly chosen woman in 1987 is normally distributed with a mean of 19.9 pounds and a standard deviation of 3.2 pounds, whereas the amount eaten by a randomly chosen man is normally distributed with a mean of 20.7 pounds and a standard deviation of 3.4 pounds. Suppose a man and a woman are randomly chosen. What is the probability that the woman ate a greater amount of apples in 1987 than the man?

Solution   Let X denote the amount eaten by the woman and Y the amount eaten by the man. We want to determine $P\{X > Y\}$, or equivalently $P\{X - Y > 0\}$. Now X is a normal random variable with mean 19.9 and standard deviation 3.2. Also $-Y$ is a normal random variable (since it is equal to the normal random variable Y multiplied by the constant -1) with mean -20.7 and standard deviation $|-1|(3.4) = 3.4$. Therefore, their sum $X + (-Y) (= X - Y)$ is normal with mean

$$E[X - Y] = 19.9 + (-20.7) = -.8$$

and standard deviation

$$SD (X - Y) = \sqrt{(3.2)^2 + (3.4)^2} = 4.669$$

Thus, if we let $W = X - Y$, then

$$P\{W > 0\} = P\left\{\frac{W + .8}{4.669} > \frac{.8}{4.669}\right\}$$
$$= P\{Z > 0.17\}$$
$$= 1 - .5675 = .4325$$

That is, with probability .4325 the randomly chosen woman would have eaten a greater amount of apples than the randomly chosen man.

Problems

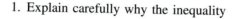

1. Explain carefully why the inequality

   $$x > a$$

   is equivalent to the inequality

   $$\frac{x - \mu}{\sigma} > \frac{a - \mu}{\sigma}$$

   What fact are we using about the quantity σ? (*Hint:* Would the above inequalities be equivalent if σ were negative?)

2. If X is normal with mean 10 and standard deviation 3, find
   (a) $P\{X > 12\}$
   (b) $P\{X < 13\}$
   (c) $P\{8 < X < 11\}$
   (d) $P\{X > 7\}$
   (e) $P\{|X - 10| > 5\}$
   (f) $P\{X > 10\}$
   (g) $P\{X > 20\}$

3. The length of time that a new hair dryer functions before breaking down is normally distributed with mean 40 months and standard deviation 8 months. The manufacturer is thinking of guaranteeing each dryer for 3 years. What proportion of dryers will not meet this guarantee?

4. The scores on a scholastic achievement test were normally distributed with mean 520 and standard deviation 94.
   (a) If your score was 700, by how many standard deviations did it exceed the average score?
   (b) What percentage of examination takers received a higher score than you did?

5. The number of bottles of shampoo sold monthly by a certain discount drugstore is a normal random variable with mean 212 and standard deviation 40. Find the probability that next month's shampoo sales will be

    (a) Greater than 200

    (b) Less than 250

    (c) Greater than 200 but less than 250

6. The life of a certain automobile tire is normally distributed with mean 35,000 miles and standard deviation 5000 miles.

    (a) What proportion of such tires last between 30,000 and 40,000 miles?

    (b) What proportion of such tires last over 40,000 miles?

    (c) What proportion last over 50,000 miles?

7. Suppose you purchased such a tire as described in Prob. 6. If the tire is in working condition after 40,000 miles, what is the conditional probability that it will still be working after an additional 10,000 miles?

8. The pulse rate of young adults is normally distributed with a mean of 72 beats per minute and a standard deviation of 9.5 beats per minute. If the requirements for the military rule out anyone whose rate is over 95 beats per minute, what percentage of the population of young adults does not meet this standard?

9. The time required to complete a certain loan application form is a normal random variable with mean 90 minutes and standard deviation 15 minutes. Find the probability that an application form is filled out in

    (a) Less than 75 minutes

    (b) More than 100 minutes

    (c) Between 90 and 120 minutes

10. The bolts produced by a manufacturer are specified to be between 1.09 and 1.11 inches in diameter. If the production process results in the diameter of bolts being a normal random variable with mean 1.10 inches and standard deviation .005 inch, what percentage of bolts does not meet the specifications?

11. The activation pressure of a valve produced by a certain company is a normal random variable with expected value 26 pounds per square inch and standard deviation 4 pounds per square inch. What percentage of the valves produced by this company has activation pressures between 20 and 32 pounds per square inch?

12. You are planning on junking your old car after it runs an additional 20,000 miles. The battery on this car has just failed, and you must decide which of two types of batteries, costing the same amount, to purchase. After some research you have discovered that the lifetime of the first battery is normally distributed with mean life 24,000 and standard deviation 6000 miles, and the lifetime of the second battery is normally distributed with mean life 22,000 and standard deviation 2000 miles.

    (a) If all you care about is that the battery purchased lasts at least 20,000 miles, which one should you buy?

    (b) What if you wanted the battery to last for 21,000 miles?

13. The lifetime of a color television picture tube is a normal random variable with mean 8.2 years and standard deviation 1.4 years. Find the percentage of such tubes that last
    (a) More than 10 years
    (b) Less than 5 years
    (c) Between 5 and 10 years

14. The annual rainfall in Cincinnati, Ohio, is normally distributed with mean 40.14 inches and standard deviation 8.7 inches.
    (a) What is the probability that this year's rainfall exceeds 42 inches?
    (b) What is the probability that the sum of the next 2 years' rainfall exceeds 84 inches?
    (c) What is the probability that the sum of the next 3 years' rainfall exceeds 126 inches?
    (d) For parts (b) and (c), what independence assumptions are you making?

15. The height of adult women in the United States is normally distributed with mean 64.5 inches and standard deviation 2.4 inches. Find the probability that a randomly chosen woman is
    (a) Less than 63 inches tall
    (b) Less than 70 inches tall
    (c) Between 63 and 70 inches tall
    (d) Alice is 72 inches tall. What percentage of women are shorter than Alice?
    (e) Find the probability that the average of the heights of two randomly chosen women is greater than 67.5 inches.

16. The weight of an introductory chemistry textbook is a normal random variable with mean 3.5 pounds and standard deviation 2.2 pounds, whereas the weight of an introductory economics textbook is a normal random variable with mean 4.6 pounds and standard deviation 1.3 pounds. If Alice is planning on taking introductory courses in both chemistry and economics, find the probability that
    (a) The total weight of her two books will exceed 9 pounds.
    (b) Her economics book will be heavier than her chemistry book.
    (c) What assumption have you made?

6.7 PERCENTILES OF NORMAL RANDOM VARIABLES

For any α between 0 and 1, we define z_α to be that value for which

$$P\{Z > z_\alpha\} = \alpha$$

In words, the probability that a standard normal random variable is greater than z_α is equal to α (see Fig. 6.10).

We can determine the value of z_α by using Table 6.1. For instance, suppose we want to find $z_{.025}$. Since

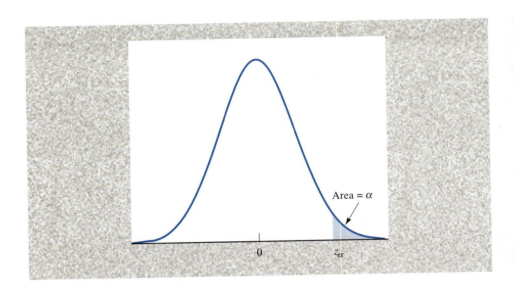

Figure 6.10
$P\{Z > z_\alpha\} = \alpha$

$$P\{Z < z_{.025}\} = 1 - P\{Z > z_{.025}\} = .975$$

we search in Table 6.1 for the entry .975, and then we find the value x that corresponds to this entry. Since the value .975 is found in the row labeled 1.9 and the column labeled .06, we see that

$$z_{.025} = 1.96$$

That is, 2.5 percent of the time a standard normal random variable will exceed 1.96.

Since 97.5 percent of the time a standard normal random variable will be less than 1.96, we say that 1.96 is the 97.5 percentile of the standard normal distribution. In general, since $100(1 - \alpha)$ percent of the time a standard normal random variable will be less than z_α, we call z_α the $100(1 - \alpha)$ percentile of the standard normal distribution.

The quantity z_α is called the $100(1 - \alpha)$ *percentile* of the standard normal distribution.

Suppose now that we want to find $z_{.05}$. If we search Table 6.1 for the value .95, we do not find this exact value. Rather, we see that

$$P\{Z < 1.64\} = .9495$$

and

$$P\{Z < 1.65\} = .9505$$

Therefore, it would seem that $z_{.05}$ lies roughly halfway between 1.64 and 1.65, and so we could approximate it by 1.645. In fact, it turns out that, to three decimal places, this is the correct answer, and so

$$z_{.05} = 1.645$$

The values of $z_{.10}$, $z_{.05}$, $z_{.025}$, $z_{.01}$, and $z_{.005}$, are, as we will see in later chapters, of particular importance in statistics. Their values are as follows:

$$z_{.10} = 1.280 \qquad z_{.025} = 1.960 \qquad z_{.005} = 2.576$$
$$z_{.05} = 1.645 \qquad z_{.01} \ = 2.330$$

For all other values of α, we can use Table 6.1 to find z_α by searching for the entry that is closest to $1 - \alpha$. In addition, Program 6-2 can be used to obtain z_α.

Example 6.10   Find

(a) $z_{.25}$

(b) $z_{.80}$

Solution

(a) The 75th percentile $z_{.25}$ is the value for which

$$P\{Z > z_{.25}\} = .25$$

or, equivalently,

$$P\{Z < z_{.25}\} = .75$$

The closest entry to .75 in Table 6.1 is the entry .7486 which corresponds to the value .67. Thus, we see that

$$z_{.25} \approx .67$$

A more precise value could be obtained by running Program 6-2. This gives the following.

If a is equal to .25, we learn that the value of z (.25) is .6741893.

(b) We are asked to find the value $z_{.80}$ such that

$$P\{Z > z_{.80}\} = .80$$

Now the value of $z_{.80}$ will be negative (why is this?), and so it is best to write the equivalent equation (see Fig. 6.11)

$$P\{Z < -z_{.80}\} = .80$$

From Table 6.1 we see that

$$-z_{.80} \approx .84$$

and so

$$z_{.80} \approx -.84$$

We can easily obtain the percentiles of any normal random variable by converting to the standard normal. For instance, suppose we want to find the value x for which

$$P\{X < x\} = .95$$

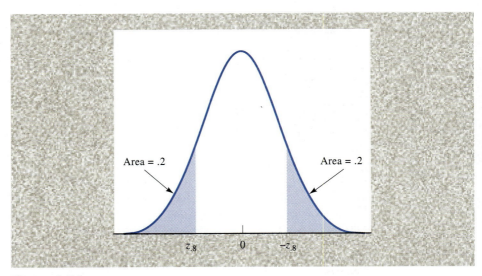

Figure 6.11
$P\{Z < -z_{.8}\} = .80$

when X is normal with mean 40 and standard deviation 5. By writing the inequality $X < x$ in terms of the standardized variable $Z = (X - 40)/5$, we see that

$$.95 = P\{X < x\}$$

$$= P\left\{\frac{X - 40}{5} < \frac{x - 40}{5}\right\}$$

$$= P\left\{Z < \frac{x - 40}{5}\right\}$$

But $P\{Z < z_{.05}\} = .95$, and so we obtain

$$\frac{x - 40}{5} = z_{.05} = 1.645$$

and so the desired value of x is

$$x = 5(1.645) + 40 = 48.225$$

Example 6.11 An IQ test produces scores that are normally distributed with mean value 100 and standard deviation 14.2. The top 1 percent of all scores is in what range?

Solution We want to find the value of x for which

$$P\{X > x\} = .01$$

when X is normal with mean 100 and standard deviation 14.2. Now

$$P\{X > x\} = P\left\{\frac{X - 100}{14.2} > \frac{x - 100}{14.2}\right\}$$

$$= P\left\{Z > \frac{x - 100}{14.2}\right\}$$

Since $P\{Z > z_{.01}\} = .01$, it follows that the above probability will equal .01 if

$$\frac{x - 100}{14.2} = z_{.01} = 2.33$$

and so

$$x = 14.2(2.33) + 100 = 133.086$$

That is, the top 1 percent consists of all those having scores above 134.

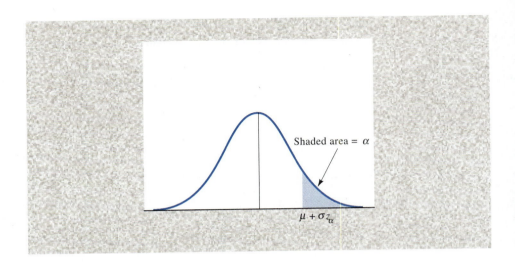

Figure 6.12
$P\{X > \mu + \sigma z_\alpha\} = \alpha$

Figure 6.12 illustrates the result

$$P\{X > \mu + \sigma z_\alpha\} = \alpha$$

when X is a normal random variable with mean μ and standard deviation σ.

Problems

1. Find to two decimal places:
   (a) $z_{.07}$ (b) $z_{.12}$
   (c) $z_{.30}$ (d) $z_{.03}$
   (e) $z_{.65}$ (f) $z_{.50}$
   (g) $z_{.95}$ (h) $z_{.008}$

2. Find the value of x for which
   (a) $P\{|Z| > x\} = .05$
   (b) $P\{|Z| > x\} = .025$
   (c) $P\{|Z| > x\} = .005$

3. If X is a normal random variable with mean 50 and standard deviation 6, find the approximate value of x for which
   (a) $P\{X > x\} = .5$ (b) $P\{X > x\} = .10$
   (c) $P\{X > x\} = .025$ (d) $P\{X < x\} = .05$
   (e) $P\{X < x\} = .88$

4. Scores on an examination for real estate brokers are normally distributed with mean 420 and standard deviation 66. If the real estate board wants to designate the highest 10 percent of all scores as *excellent*, at what score should excellence begin?

5. Suppose in Prob. 4 that the board wants only the highest 25 percent to pass. What should be set as the passing score?

6. The time it takes for high school boys to run 1 mile is normally distributed with mean 460 seconds and standard deviation 40 seconds. All those falling in the slowest 20 percent are deemed to need additional training. What is the critical time above which one is deemed to need additional training?

7. In Prob. 6, the fastest 5 percent all ran the mile in less than x seconds. What is the smallest value of x for which the preceding is a true statement?

8. Repeat Prob. 7, replacing *fastest 5 percent* by *fastest 1 percent.*

9. The amount of radiation that can be absorbed by an individual before death ensues varies from individual to individual. However, over the entire population this amount is normally distributed with mean 500 roentgens and standard deviation 150 roentgens. Above what dosage level will only 5 percent of those exposed survive?

10. Transmissions on a new car will last for a normally distributed amount of time with mean 70,000 miles and standard deviation 10,000 miles. A warranty on this part is to be provided by the manufacturer. If the company wants to limit warranty work to no more than 20 percent of the cars sold, what should be the length (in miles) of the warranty period?

11. The attendance at home football games of a certain college is a normal random variable with mean 52,000 and standard deviation 4000. Which of the following statements are true?
    (a) Over 80 percent of the games have an attendance of over 46,000.
    (b) The attendance exceeds 58,000 less than 10 percent of the time.

12. Scores on the quantitative part of the Graduate Record Examination were normally distributed with a mean score of 510 and a standard deviation of 92. How high a score was necessary to be in the top
    (a) 10
    (b) 5
    (c) 1
    percent of all scores?

13. The fasting blood glucose level (per 100 milliliters of blood) of diabetics is normally distributed with mean 106 milligrams and standard deviation 8 milligrams. In order for the blood glucose level of a diabetic to be in the lower 20 percent of all diabetics, that person's blood level must be less than what value?

Key Terms

Continuous random variable: A random variable that can take on any value in some interval.

Probability density function: A curve associated with a continuous random variable. The probability that the random variable is between two points is equal to the area under the curve between these points.

Normal random variable: A type of continuous random variable whose probability density function is a bell-shaped symmetric curve.

Standard normal random variable: A normal random variable having mean 0 and variance 1.

100p percentile of a continuous random variable: The probability that the random variable is less than this value is p.

Summary

A *continuous* random variable is one that can assume any value within some interval. Its probabilities can be obtained from its *probability density curve.* Specifically, the probability that the random variable will lie between points a and b will equal the area under the density curve between a and b.

A *normal* random variable X has a probability density curve that is specified by two parameters, the mean μ and the standard deviation σ of X. The density curve is a bell-shaped curve that is symmetric about μ and that spreads out more as the value of σ gets larger.

A normal random variable will take on a value that is within 1 standard deviation of its mean approximately 68 percent of the time; it will take on a value that is within 2 standard deviations of its mean approximately 95 percent of the time; and it will take on a value that is within 3 standard deviations of its mean approximately 99.7 percent of the time.

A normal random variable having mean 0 and standard deviation 1 is called a *standard normal* random variable. We let Z designate such a random variable. Probabilities of a standard normal random variable can be obtained from Table 6.1 (reprinted as Table D.1). For any nonnegative value x, specified up to two decimal places, this table gives the probability that a standard normal random variable is less than x. For negative x, this probability can be obtained by making use of the symmetry of the normal curve about 0. This results in the equality

$$P\{Z < x\} = P\{Z > -x\}$$

The value of $P\{Z > -x\} = 1 - P\{Z < -x\}$ can now be obtained from Table 6.1.

Program 6-1 can also be used to obtain probabilities of standard normal random variables.

If X is normal with mean μ and standard deviation σ, then Z, defined by

$$Z = \frac{X - \mu}{\sigma}$$

Historical Perspective

Karl F. Gauss

(Bettmann)

The Normal Curve

The normal distribution was introduced by the French mathematician Abraham De Moivre in 1733. De Moivre, who used this distribution to approximate probabilities connected with coin tossing, called it the *exponential bell-shaped curve*. Its usefulness, however, only became truly apparent in 1809 when the famous German mathematician K. F. Gauss used it as an integral part of his approach to predicting the location of astronomical entities. As a result, it became common after this time to call it the *gaussian distribution.*

During the middle to late 19th century, however, most statisticians started to believe that the majority of data sets would have histograms conforming to the gaussian bell-shaped form. Indeed, it came to be accepted that it was "normal" for any well-behaved data set to follow this curve. As a result, following the lead of Karl Pearson, people began referring to the gaussian curve by simply calling it the *normal* curve. (For an explanation as to why so many data sets

conform to the normal curve, the interested student will have to wait to read Secs. 7.3 and 12.6.)

Karl Friedrich Gauss (1777–1855), one of the earliest users of the normal curve, was one of the greatest mathematicians of all time. Listen to the words of the well-known mathematical historian E. T. Bell, as expressed in his 1954 book *Men of Mathematics.* In a chapter entitled "The Prince of Mathematicians," he writes:

> Archimedes, Newton, and Gauss; these three are in a class by themselves among the great mathematicians, and it is not for ordinary mortals to attempt to rank them in order of merit. All three started tidal waves in both pure and applied mathematics. Archimedes esteemed his pure mathematics more highly than its applications; Newton appears to have found the chief justification for his mathematical inventions in the scientific uses to which he put them; while Gauss declared it was all one to him whether he worked on the pure or on the applied side.

has a standard normal distribution. This fact enables us to compute probabilities of X by transforming them to probabilities concerning Z. For instance,

$$P\{X < a\} = P\left\{\frac{X - \mu}{\sigma} < \frac{a - \mu}{\sigma}\right\}$$

$$= P\left\{Z < \frac{a - \mu}{\sigma}\right\}$$

For any value of α between 0 and 1 the quantity z_α is defined as that value for which

$$P\{Z > z_\alpha\} = \alpha$$

Thus, a standard normal will be less than z_α with probability $1 - \alpha$. That is, $100(1 - \alpha)$ percent of the time Z will be less than z_α. The quantity z_α is called the $100(1 - \alpha)$ percentile of the standard normal distribution.

The values z_α for specified α can be obtained either from Table 6.1 or by running Program 6-2. The percentiles of an arbitrary normal random variable X with mean μ and standard deviation σ can be obtained from the standard normal percentiles by using the fact that $Z = (X - \mu)/\sigma$ is a standard normal distribution. For instance, suppose we want to find the value x for which

$$P\{X > x\} = \alpha$$

Thus, we want to find x for which

$$\alpha = P\left\{\frac{X - \mu}{\sigma} > \frac{x - \mu}{\sigma}\right\}$$

$$= P\left\{Z > \frac{x - \mu}{\sigma}\right\}$$

Therefore, since $P\{Z > z_\alpha\} = \alpha$, we can conclude that

$$\frac{x - \mu}{\sigma} = z_\alpha$$

or

$$x = \mu + \sigma z_\alpha$$

Review Problems

1. The heights of adult males are normally distributed with a mean of 69 inches and a standard deviation of 2.8 inches. Let X denote the height of a randomly chosen male adult. Find
   (a) $P\{X > 65\}$
   (b) $P\{62 < X < 72\}$
   (c) $P\{|X - 69| > 6\}$
   (d) $P\{63 < X < 75\}$
   (e) $P\{X > 72\}$
   (f) $P\{X < 60\}$
   (g) x if $P\{X > x\} = .01$
   (h) x if $P\{X < x\} = .95$
   (i) x if $P\{X < x\} = .40$

2. Find
   (a) $z_{.04}$
   (b) $z_{.22}$
   (c) $P\{Z > 2.2\}$
   (d) $P\{Z < 1.6\}$
   (e) $z_{.78}$

3. In tests conducted on jet pilots, it was found that their blackout thresholds are normally distributed with a mean of 4.5*g* and a standard deviation of .7*g*. If only those pilots whose thresholds are in the top 25 percent are to be allowed to apply to become astronauts, what is the cutoff point?

4. In Prob. 3, find the proportion of jet pilots who have blackout thresholds
   (a) Above 5*g*
   (b) Below 4*g*
   (c) Between 3.7*g* and 5.2*g*

5. The working life of a certain type of light bulb is normally distributed with a mean of 500 hours and a standard deviation of 60 hours.
   (a) What proportion of such bulbs lasts more than 560 hours?
   (b) What proportion lasts less than 440 hours?
   (c) If a light bulb is still working after 440 hours of operation, what is the conditional probability that its lifetime exceeds 560 hours?
   (d) Fill in the missing number in the following sentence. Ten percent of these bulbs will have a lifetime of at least _____ hours.

6. The American Cancer Society has stated that a 25-year-old man who smokes a pack of cigarettes a day gives up, on average, 5.5 years of life. Assuming that the number of years lost is normally distributed with mean 5.5 and standard deviation 1.5, find the probability that the decrease in life of such a man is
   (a) Less than 2 years
   (b) More than 8 years
   (c) Between 4 and 7 years

7. Suppose that the yearly cost of upkeep for condominium owners at a certain complex is normal with mean $3000 and standard deviation $600. Find the probability that an owner's total cost over the next 2 years will
   (a) Exceed $5000
   (b) Be less than $7000
   (c) Be between $5000 and $7000
   Assume that costs incurred in different years are independent random variables.

8. The speeds of cars traveling on New Jersey highways are normally distributed with mean 60 miles per hour and standard deviation 5 miles per hour. If New Jersey police follow a policy of ticketing only the fastest 5 percent, at what speed do the police start to issue tickets?

9. The gross weekly sales at a certain used-car lot are normal with mean $18,800 and standard deviation $9000.
   (a) What is the probability that next week's sales exceed $20,000?
   (b) What is the probability that weekly sales will exceed $20,000 in each of the next 2 weeks?
   (c) What is the probability that the total sales in the next 2 weeks exceed $40,000?
   In parts (b) and (c) assume that the sales totals in different weeks are independent.

10. The yearly number of miles accumulated by an automobile in a large car rental company's fleet is normal with mean 18,000 miles and standard deviation 1700 miles. At the end of the year the company sells 80 percent of its year-old cars, keeping the 20 percent with the lowest mileage. Do you think a car whose year-end mileage is 17,400 is likely to be kept?

11. An analysis of the scores of professional football games has led some researchers to conclude that a team that is favored by x points will outscore its opponent by a random number of points that is approximately normally distributed with mean x and standard deviation 14. Thus, for instance, the difference in the points scored by a team that is favored by 5 points and its opponent will be a normal random variable with mean 5 and standard deviation 14. Assuming that this theory is correct, determine the probability that
    (a) A team that is favored by 7 points wins the game.
    (b) A team that is a 4-point underdog wins the game.
    (c) A team that is a 14-point favorite loses the game.

12. U.S. Department of Agriculture data for 1987 indicate that the amount of tomatoes consumed per year by a randomly chosen woman is a normal random variable with a mean of 14.0 pounds and a standard deviation of 2.7 pounds, while the amount eaten yearly by a randomly chosen man is a normal random variable with a mean of 14.6 pounds and a standard deviation of 3.0 pounds. Suppose a man and a woman are randomly chosen. Find the probability that
    (a) The woman ate more than 14.6 pounds of tomatoes in 1987.
    (b) The man ate less than 14 pounds of tomatoes in 1987.
    (c) The woman ate more and the man ate less than 15 pounds of tomatoes in 1987.
    (d) The woman ate more tomatoes in 1987 than did the man.

13. Suppose in Prob. 12 that a person, equally likely to be either a man or a woman, is chosen. Find the probability that this person is
    (a) A woman who ate less than 14 pounds of tomatoes in 1987
    (b) A man who ate more than 14 pounds of tomatoes in 1987

Normal Random Variables

Purpose

Use Minitab to

1. Display graphs for normal random variables.

2. Compute probabilities associated with a normal random variable.

3. Investigate the additive property of normal random variables.

4. Compute the percentiles of normal random variables.

Procedures

First, load the Minitab (windows version) software as in the Minitab lab for Chap. 1.

1. NORMAL RANDOM VARIABLES

Example 1 Use Minitab to generate a set of 500 standard normal values (Z scores), and investigate the shape of the distribution for these values.

Select Calc→Random Data→Normal, and the Normal distribution dialog box will be displayed. In the Generate text box type 500, in the Store in column(s) text box type C1, in the Mean text box type 0, and in the Standard deviation text box type 1.0. Column C1 was renamed ZVALUES. Select the OK button, and the values will be listed in column C1 in the Data window.

Next we investigate the shape of the distribution for these generated values. We know that these values were generated from a standard normal distribution, so we expect the histogram and curve to be normal. However, these procedures could apply to any observed set of data when you want to fit a curve to it.

The first step is to sort ZVALUES. Select Manip→Sort. Save the sorted values in C2 (SORTED). Next select Calc→Probability Distributions→Normal. Type 0 in the Mean text box and 1 in the Standard deviation text box. Select C2 (SORTED) for the Input column text box, and in Optional storage type in C3 (PDFZV). Select the OK button. Next select Calc→Mathematical Expressions. In the Variable text box, type C3 (PDFZV); and in the Expression text box, type 'PDFZV'*COUNT('ZVALUES')*.51 and select OK. *Note:* The value .51 was chosen in order for the normal curve to be drawn properly. You will have to adjust this value in your work when you are employing these precedures. Select Graph→Histogram, and type C1 in the X text box. Select Options→CutPoint→Midpoint/Cutpoint Positions and type −3:3/.5 in the text box. That is, we are telling Minitab that the Z values lie between −3 and +3, and we want the width of the intervals to be .5. Click the OK button and select Annotation→Line. In the Points text box, type C2 (SORTED) − C3 (PDFZV), and click on the OK button twice. The resulting output is shown in Fig. M6.1.

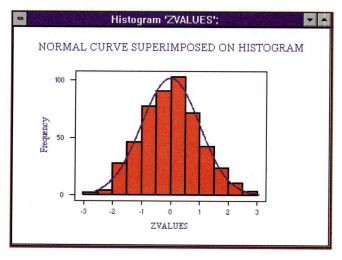

Figure M6.1

Observe that since these are *simulated* Z scores, you should not expect the histogram to be perfectly symmetric, and hence the normal curve will not be a perfect fit.

2. COMPUTING PROBABILITIES FOR A NORMAL RANDOM VARIABLE

Example 2 Use Minitab to find (a) $P(Z \le 1.37)$, (b) $P(Z > -1.1)$, and (c) $P(-1.25 < Z \le 1.63)$.

(a) Select Calc→Probability Distributions→Normal. The Normal distribution dialog box will be displayed. Since we are dealing with the standard normal random variable z, type 0 in the Mean text box and 1 in the Standard deviation text box. Figure M6.2 shows the dialog box with both the Cumulative probability and Input constant check boxes highlighted. We use Cumulative probability since we are computing $P(Z \le 1.37)$. The value of 1.37 is the input constant. Click on the OK button, and the probability will be computed and displayed in the Session window since we did not store the computed value with the Optional storage feature.

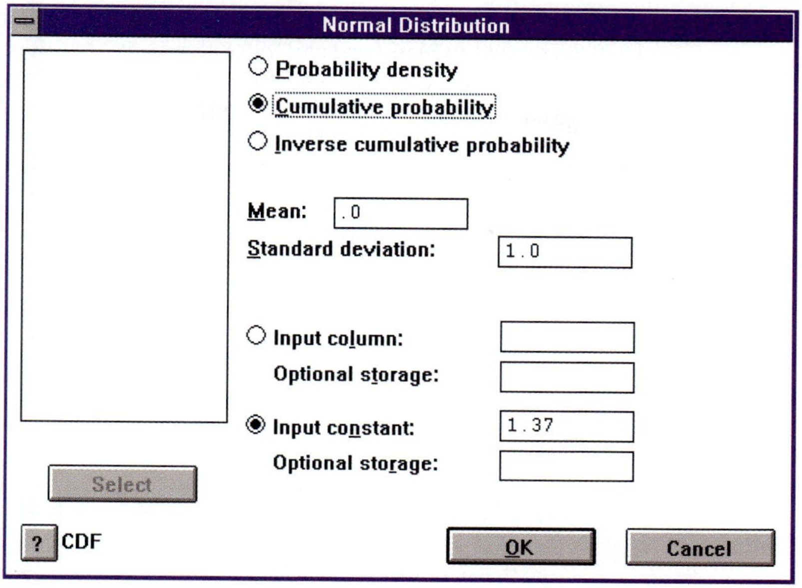

Figure M6.2

Observe that x represents z in the Session window since the mean is 0 and the standard deviation is 1. Thus, $P(Z \le 1.37) = .9147$.

(b) To find $P(Z > -1.1)$, observe that $P(Z > -1.1) = 1 - P(Z \le -1.1)$. Thus apply the procedures as in (a) to $P(Z \le -1.1)$, and subtract the value from 1. Your answer should be $1 - .1357 = .8643$.

(c) To find $P(-1.25 < Z \le 1.63)$, observe that the area under the standard normal curve between -1.25 and 1.63 equals the area to the left of 1.63 *minus* the area to the left of -1.25. That is, $P(-1.25 < Z \le 1.63) = P(Z \le 1.63) - P(Z < -1.25)$. Using the above procedures, we get $P(Z \le 1.63) - P(Z < -1.25) = .9484 - .1056 = .8428$.

Example 3 Use Minitab to find (a) $P(X \le 3.793)$, (b) $P(X > -2.106)$, and (c) $P(-2.253 < X \le 6.637)$ if the random variable X is normally distributed with mean $\mu = 2$ and a standard deviation of $\sigma = 4$.

Use the above procedures with $\mu = 2$ and $\sigma = 4$ in Fig. M6.3. Show that $P(X \le 3.793) = .6730$, $P(X > -2.106) = 1 - .1523 = .8477$, and $P(-2.253 < X \le 6.637) = .8768 - .1438 = .7360$.

Note You could have used the Optional storage in Fig. M6.2 to help find these differences.

3. ADDITIVE PROPERTY OF NORMAL RANDOM VARIABLES

Example 4 Suppose that the random variable X has a normal distribution with mean 3 and standard deviation 5, and Y has a normal distribution with mean 4 and standard deviation 2. Use Minitab to generate data from these distributions, and establish that $X + Y$ has a normal distribution with mean $3 + 4 = 7$ and variance $5 + 2 = 7$. Assume that X and Y are independent.

Generate 1,000 values for the random variable X and 1,000 values for the random variable Y by using the sequence Calc→Random Data→Normal. Enter the appropriate values in Fig. M6.1. Save the values in C1 and C2, and rename X and Y, respectively. Draw histograms for X and Y, and observe that the distributions are indeed normal (or approximately). Next add the values in C1 and C2 by using Calc→Mathematical Expressions. In the Mathematical expressions dialog box, type C3 in the Variable (new or modified) text box and C1 + C2 in the Expression text box. When you select the OK button, C3 will display the sum of the values in these columns. Rename C3 by calling it $X + Y$. Draw a histogram for $X + Y$, and observe that the distribution is normal (or approximately). Thus visually you can observe that the sum of two independent normal random variables is also normal. Next, select Stat→Basic Statistics→Descriptive Statistics. In the Descriptive statistics dialog box, enter C1 to C3 in the Variables text box. Select OK and observe that the mean and variance for C3 will be approximately the sum for that of C1 and C2. Thus the sum $X + Y$ is normal (approximately) with a mean being the sum of the means and a variance equal to the sum of the variances. Note these values will be *approximate* since we are using simulation techniques to illustrate these concepts.

4. COMPUTING PERCENTILES FOR NORMAL RANDOM VARIABLES

Example 5   Use Minitab to find $z_{.05}$, the 95th percentile for the standard normal distribution.

Recall that z_α is defined to be $P\{Z > z_\alpha\} = \alpha$, where z_α is called the $100(1 - \alpha)$ *percentile* of the standard normal distribution. Now, to use Minitab to compute values of z_α, we use the complement $1 - \alpha$. Select Calc→Probability Distributions→Normal, and the Normal distribution dialog box will be displayed. Select the Inverse cumulative probability check box. Since we are dealing with the standard normal distribution, type 0 in the Mean text box and 1 in the Standard deviation text box. Select the Input constant check box, and type .95 (= $1 - .05$) in the text box.

Select the OK button, and the Session window, as shown in Fig. M6.3, will display the value of $z_{.05}$ as 1.6449.

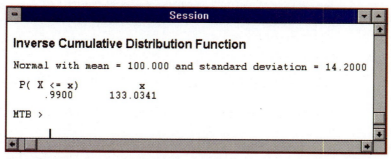

Figure M6.3

Example 6   Repeat Example 6.11, in Sec. 6.7 (normal IQ test scores).

To use Minitab to help solve this type of problem, we can use the procedures of Example 5. Since we need $P\{X > x\} = .01$, in Fig. M6.7 select Inverse cumulative probability, and type 100 in the text box for the Mean and 14.2 in the Standard deviation text box. Select the Input constant check box, and type in .99, since the area to the right of x is .01 and thus the area to the left will be .99. Select the OK button, and the Session window shown in Fig. M6.4 will display a value of 133.0341 for x.

```
┌─────────────────────────────────────────────────────────────┐
│ ▬                        Session                    ▼ ▲      │
├─────────────────────────────────────────────────────────────┤
│                                                        ↑     │
│ Inverse Cumulative Distribution Function                     │
│                                                              │
│ Normal with mean = 0 and standard deviation = 1.00000        │
│                                                              │
│   P( X <= x )           x                                    │
│       .9500          1.6449                                  │
│                                                              │
│ MTB > |                                                ▼     │
└─────────────────────────────────────────────────────────────┘
 ← ┃                                                    →
```

Figure M6.4

Computer Exercises

1. Collect a large set of data (price of a particular stock over a long time, heights of baseball players, etc.). Use Minitab or any other statistical software to construct a histogram for the data set. Use the histogram to estimate the mean and standard deviation for the population from which your data set was sampled. Use the procedures in Sec. 1 to help you superimpose a normal curve, using the estimated mean and standard deviation. Present your observations and results in a report. *Note:* You should try to observe values of a random variable that are normal or approximately normal.

2. Use Minitab or any other statistical software to help you work Chap. 6 Review Probs. 1 to 10.

3. Repeat Example 4 first 10, then 20, 30, 40, and 50 times. Sum these values and save in C11, C21, etc. In each set of simulations, select your own population means and variances. Be sure to keep a record of them. Compute and record the column means for C1–C11, C1–C21, etc. Find the sum of the column means for each set of simulations C1–C10, C1–C20, etc. Compare to the means for C11, C21, ... Also compare the means for C11, C21, ... to the sum of the theoretical means (means you used for the simulations). Present histograms and normal probability plots for C11, C21, ... Also present tables with the computed means for the five sets of simulations, the average for these computed means, and the means for C11, C21, ..., along with the theoretical means. Discuss your observations in a report. Can you make a general statement from these observations? Is this statement restricted to only normal random variables?

4. Use Minitab or any other statistical software to help find the percentiles in Prob. 1, Sec. 6.7.

5. Use Minitab or any other statistical software to help find the *x* values in Probs. 2 and 3 in Sec. 6.7.

6. Repeat Exercise 3 for the variance.

7. Use Minitab or any other statistical software to help solve Probs. 1 and 7 in Sec. 6.4.

CHAPTER 7

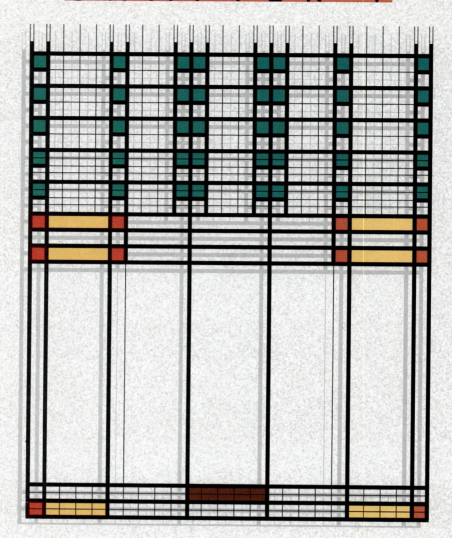

Distributions of
Sampling Statistics

He uses statistics as a drunken man uses lampposts—for support rather than illumination.
Andrew Lang
(Scottish author)

I could prove God statistically.
George Gallup, U.S. pollster

We introduce the concept of sampling from a population distribution. The sample mean and sample variance are studied, and their expectations and variances are given. The central limit theorem is presented and applied to show that the distribution of the sample mean is approximately normal.

We consider samples taken from a finite population in which certain members have a particular characteristic of interest. We show that when the population size is large, the number of members of the sample who have the characteristic is approximately a binomial random variable. The central limit theorem is used to show that the probabilities of such a random variable can be approximated by the probabilities of a normal random variable.

We present the distribution of the sample variance in the case where the underlying population distribution is normal.

317

7.1   A PREVIEW

If you bet $1 on a number at a roulette table in a U.S. casino, then either you will win $35 if your number appears on the roulette wheel or you will lose $1 if it does not. Since the wheel has 38 slots—numbered 0, 00, and each of the integers from 1 to 36—it follows that the probability that your number appears is 1/38. As a result, your expected gain on the bet is

$$E[\text{gain}] = 35\left(\frac{1}{38}\right) - 1\left(\frac{37}{38}\right) = -\frac{2}{38} = -.0526$$

That is, your expected loss on each spin of the wheel is approximately 5.3 cents.

Suppose you continually place bets at the roulette table. How lucky do you have to be in order to be winning money at the end of your play? Well, it depends on how long you continue to play. Indeed, after 100 plays you will be ahead with probability .4916. On the other hand, after 1000 plays your chance of being ahead drops to .39. After 100,000 plays not only will you almost certainly be losing (your probability of being ahead is approximately .002), but also you can be 95 percent certain that your average loss per play will be 5.26 ± 1.13 cents (read as 5.26 plus or minus 1.13 cents). In other words, even if you did not know it to begin with, if you play long enough, you will learn that the average loss per game is around 5.26 cents.

7.2   INTRODUCTION

One of the key concerns of statistics is the drawing of conclusions from a set of observed data. These data will usually consist of a sample of certain elements of a population, and the objective will be to use the sample to draw conclusions about the entire population.

Suppose that each member of a population has a numerical value associated with it. To use sample data to make inferences about the values of the entire population, it is necessary to make some assumptions about the population values and about the relationship between the sample and the population. One such assumption is that there is an underlying probability distribution for the population values. That is, the values of different members of the population are assumed to be independent random variables having a common distribution. In addition, the sample data are assumed to be independent values from this distribution. Thus, by observing the sample data we are able to learn about this underlying population distribution.

Definition

If $X_1, \ldots, X_n$ are independent random variables having a common probability distribution, we say they constitute a *sample* from that distribution.

In most applications, the population distribution will not be completely known, and one will attempt to use the sample data to make inferences about it. For instance, a manufacturer may be producing a new type of battery to be used in a particular electric powered automobile. These batteries will each last for a random number of miles having some unknown probability distribution. To learn about this underlying probability distribution, the manufacturer could build and road-test a set of batteries. The resulting data, consisting of the number of miles of use obtained from each battery, would then constitute a sample from this distribution.

In this chapter we are concerned with the probability distributions of certain statistics that arise from a sample, where a statistic is a numerical quantity whose value is determined by the sample. Two important statistics that we will consider are the sample mean and the sample variance. In Sec. 7.3, we consider the sample mean and present formulas for the expectation and variance of this statistic. We also note that when the sample size is at least moderately large, the probability distribution of the sample mean can be approximated by a normal distribution. This result, which follows from one of the most important theoretical results in probability theory, known as the *central limit theorem,* will be discussed in Sec. 7.4. In Sec. 7.5 we concern ourselves with situations in which we are sampling from a finite population of objects, and we explain what it means for the sample to be a *random* sample. When the population size is large in relation to the sample size, then we often treat the population as if it were infinite. We illustrate and explain exactly when this can be done and what the consequences are. In Sec. 7.6, we consider the distribution of the sample variance from a sample chosen from a normal population.

7.3   SAMPLE MEAN

Consider a population of elements, each of which has a numerical value attached to it. For instance, the population might consist of the adults of a specified community, and the value attached to each adult might be her or his annual income, or height, or age, and so on. We often suppose that the value associated with any member of the population can be regarded as being the value of a random variable having expectation μ and variance σ^2. The quantities μ and σ^2 are called the *population mean* and the *population variance,* respectively. Let $X_1, X_2, \ldots, X_n$ be a sample of values from this population. The sample mean is defined by

$$\overline{X} = \frac{X_1 + \cdots + X_n}{n}$$

Since the value of the sample mean $\overline{X}$ is determined by the values of the random variables in the sample, it follows that $\overline{X}$ is also a random variable. Its expectation can be shown to be

$$E[\overline{X}] = \mu$$

That is, the expected value of the sample mean $\bar{X}$ is equal to the population mean μ. In addition, it can be shown that the variance of the sample mean is

$$\text{Var } (\bar{X}) = \frac{\sigma^2}{n}$$

Thus we see that the sample mean $\bar{X}$ has the same expected value as an individual data value, but its variance is smaller than that of an individual data value by the factor $1/n$, where n is the size of the sample. Therefore, we can conclude that $\bar{X}$ is also centered on the population mean μ, but its spread becomes more and more reduced as the sample size increases. Figure 7.1 plots the probability density function of the sample mean from a standard normal population for a variety of sample sizes.

Example 7.1   Let us check the above formulas for the expected value and variance of the sample mean by considering a sample of size 2 from a population whose values are equally likely to be either 1 or 2. That is, if X is the value of a member of the population, then

$$P\{X = 1\} = \frac{1}{2}$$

$$P\{X = 2\} = \frac{1}{2}$$

The population mean and variance are obtained as follows:

$$\mu = E[X] = 1\left(\frac{1}{2}\right) + 2\left(\frac{1}{2}\right) = 1.5$$

and

$$\sigma^2 = \text{Var } (X) = E[(X - \mu)^2]$$

$$= (1 - 1.5)^2\left(\frac{1}{2}\right) + (2 - 1.5)^2\left(\frac{1}{2}\right)$$

$$= \frac{1}{4}$$

To obtain the probability distribution of the sample mean $(X_1 + X_2)/2$, note that the pair of values X_1, X_2 can assume any of four possible pairs of values

$$(1, 1), (1, 2), (2, 1), (2, 2)$$

where the pair $(2, 1)$ means, for instance, that $X_1 = 2$, $X_2 = 1$. By the independence of X_1 and X_2 it follows that the probability of any given pair is 1/4. Therefore, we see

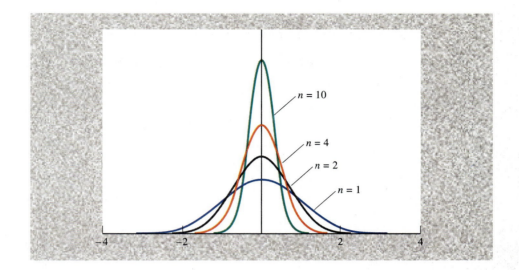

Figure 7.1
Densities of sample means from a standard normal population.

that the possible values of $\bar{X} = (X_1 + X_2)/2$ along with their respective probabilities are as follows:

$$P\{\bar{X} = 1\} = P\{(1, 1)\} = \frac{1}{4}$$

$$P\{\bar{X} = 1.5\} = P\{(1, 2) \text{ or } (2, 1)\} = \frac{2}{4} = \frac{1}{2}$$

$$P\{\bar{X} = 2\} = P\{(2, 2)\} = \frac{1}{4}$$

Therefore,

$$E[\bar{X}] = 1\left(\frac{1}{4}\right) + 1.5\left(\frac{1}{2}\right) + 2\left(\frac{1}{4}\right) = \frac{6}{4} = 1.5$$

Also

$$\text{Var}\,(\bar{X}) = E[(\bar{X} - 1.5)^2]$$

$$= (1 - 1.5)^2\left(\frac{1}{4}\right) + (1.5 - 1.5)^2\left(\frac{1}{2}\right) + (2 - 1.5)^2\left(\frac{1}{4}\right)$$

$$= \frac{1}{16} + 0 + \frac{1}{16} = \frac{1}{8}$$

which, since $\mu = 1.5$ and $\sigma^2 = 1/4$, verifies that $E[\bar{X}] = \mu$ and $\text{Var}\,(\bar{X}) = \sigma^2/2$.

Figure 7.2 plots the population probability distribution alongside the probability distribution of the sample mean of a sample of size 2.

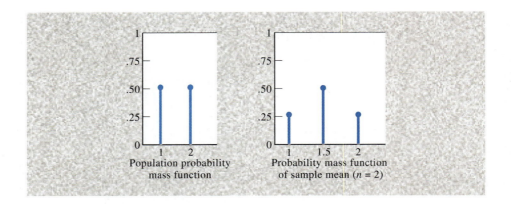

Figure 7.2
Probability
mass functions.

The standard deviation of a random variable, which is equal to the square root of its variance, is a direct indicator of the spread in the distribution. It follows from the identity

$$\text{Var } (\bar{X}) = \frac{\sigma^2}{n}$$

that SD $(\bar{X})$, the standard deviation of the sample mean $\bar{X}$, is given by

$$\text{SD } (\bar{X}) = \sqrt{\frac{\sigma^2}{n}} = \frac{\sigma}{\sqrt{n}}$$

In the preceding formula, σ is the population standard deviation, and n is the sample size.

The *standard deviation* of the sample mean is equal to the population standard deviation divided by the square root of the sample size.

Summing up, we have seen in this section that the expectation of the sample mean from a sample of size n will equal the population mean, and the variance of the sample mean will equal the population variance reduced by the factor $1/n$. Now, whereas knowledge of the mean and variance of a statistic tells us quite a bit about its probability distribution, it still leaves much unanswered. We will, however, show in Sec. 7.4 that the probability distribution of a sample mean is approximately normal, and as we know a normal distribution is completely specified by its mean and variance.

Problems

1. Consider the population described in Example 7.1. Plot the possible values along with their probabilities of the sample mean of a sample of size
   (a) $n = 3$
   (b) $n = 4$
   In both cases also derive the standard deviation of the sample mean.

2. Suppose that X_1 and X_2 constitute a sample of size 2 from a population in which a typical value X is equal to either 1 or 2 with respective probabilities

$$P\{X = 1\} = .7 \qquad P\{X = 2\} = .3$$

   (a) Compute $E[X]$.
   (b) Compute Var (X).
   (c) What are the possible values of $\bar{X} = (X_1 + X_2)/2$?
   (d) Determine the probabilities that $\bar{X}$ assumes the values in (c).
   (e) Using (d), directly compute $E[\bar{X}]$ and Var $(\bar{X})$.
   (f) Are your answers to (a), (b), and (e) consistent with the formulas presented in this section?

3. Consider a population whose probabilities are given by

$$p(1) = p(2) = p(3) = \frac{1}{3}$$

   (a) Determine $E[X]$.
   (b) Determine SD (X).
   (c) Let $\bar{X}$ denote the sample mean of a sample of size 2 from this population. Determine the possible values of $\bar{X}$ along with their probabilities.
   (d) Use the result of part (c) to compute $E[\bar{X}]$ and SD $(\bar{X})$.
   (e) Are your answers consistent?

4. The amount of money withdrawn in each transaction at an automatic teller of a branch of the Bank of America has mean $80 and standard deviation $40. What are the mean and standard deviation of the average amount withdrawn in the next 20 transactions?

5. A producer of cigarettes claims that the mean nicotine content in its cigarettes is 2.4 milligrams with a standard deviation of .2 milligram. Assuming these figures are correct, find the expected value and variance of the sample mean nicotine content of
   (a) 36   (b) 64   (c) 100   (d) 900
   randomly chosen cigarettes.

6. The lifetime of a certain type of electric bulb has expected value 475 hours and standard deviation 60 hours. Determine the expected value and standard deviation of the sample mean of
   (a) 100   (b) 200   (c) 400
   such lightbulbs.

7. The weight of a randomly chosen person riding a ferry has expected value 155 pounds and standard deviation 28 pounds. The ferry has the capacity to carry 100 riders. Find the expected value and standard deviation of the total passenger weight load of a ferry at capacity.

7.4   CENTRAL LIMIT THEOREM

In the previous section we showed that if we take a sample of size n from a population whose elements have mean μ and standard deviation σ, then the sample mean $\bar{X}$ will have mean μ and standard deviation $\sigma/\sqrt{n}$. In this section, we consider one of the most important results in probability theory, known as the *central limit theorem,* which states that the sum (and thus also the average) of a large number of independent random variables is approximately normally distributed.

Central limit theorem

Let $X_1, X_2, \ldots, X_n$ be a sample from a population having mean μ and standard deviation σ. For n large, the sum

$$X_1 + X_2 + \cdots + X_n$$

will approximately have a normal distribution with mean $n\mu$ and standard deviation $\sigma\sqrt{n}$.

Example 7.2   An insurance company has 10,000 ($= 10^4$) automobile policyholders. If the expected yearly claim per policyholder is $260 with a standard deviation of $800, approximate the probability that the total yearly claim exceeds $2.8 million ($= \$2.8 \times 10^6$).

Solution   Number the policyholders, and let X_i denote the yearly claim of policyholder $i, i = 1, \ldots, 10^4$. By the central limit theorem, $X = \sum_{i=1}^{10^4} X_i$ will have an approximately normal distribution with mean $10^4 \times 260 = 2.6 \times 10^6$ and standard deviation $800\sqrt{10^4} = 800 \times 10^2 = 8 \times 10^4$. Hence,

$$P\{X > 2.8 \times 10^6\} = P\left\{\frac{X - 2.6 \times 10^6}{8 \times 10^4} > \frac{2.8 \times 10^6 - 2.6 \times 10^6}{8 \times 10^4}\right\}$$

$$\approx P\left\{Z > \frac{.2 \times 10^6}{8 \times 10^4}\right\}$$

$$= P\left\{Z > \frac{20}{8}\right\}$$

$$= P\{Z > 2.5\} = .0062$$

That is, there are only 6 chances out of 1000 that the total yearly claim will exceed $2.8 million.

The preceding version of the central limit theorem is by no means the most general, for it can be shown that $\sum_{i=1}^{n} X_i$ will approximately have a normal distribution even in cases where the random variables X_i have different distributions. Indeed, provided that all the random variables tend to be of roughly the same magnitude, so that none of them tends to dominate the value of the sum, it can be shown that the sum of a large number of independent random variables will approximately have a normal distribution.

Not only does the central limit theorem give us a method for approximating the distribution of the sum of random variables, but also it helps explain the remarkable fact that the empirical frequencies of so many naturally occurring populations exhibit a bell-shaped (that is, a normal) curve. Indeed, one of the first uses of the central limit theorem was to provide a theoretical justification of the empirical fact that measurement errors tend to be normally distributed. That is, by regarding an error in measurement as being composed of the sum of a large number of small independent errors, the central limit theorem implies that it should be approximately normal. For instance, the error in a measurement in astronomy can be regarded as being equal to the sum of small errors caused by such things as

1. Temperature effects on the measuring device

2. Bending of the device caused by the rays of the sun

3. Elastic effects

4. Air currents

5. Air vibrations

6. Human errors

and so on. Therefore, by the central limit theorem, the total measurement error will approximately follow a normal distribution. From this it follows that a histogram of errors resulting from a series of measurements of the *same* object will tend to follow a bell-shaped normal curve.

The central limit theorem also partially explains why many data sets related to biological characteristics tend to be approximately normal. For instance, consider a particular couple, call them Maria and Peter Fontanez, and consider the heights of their daughters (say, when they are 20 years old). Now the height of a given daughter can be thought of as being composed of the sum of a large number of roughly independent random variables—relating, among other things, to the random set of genes that the daughter received from her parents as well as environmental factors. Since each of these variables plays only a small role in determining the total height, it seems reasonable, based on the central limit theorem, that the height of a Fontanez daughter will be normally distributed. If the Fontanez family has many daughters, then a histogram of their heights should roughly follow a normal curve. (The same thing is true for the sons of Peter and Maria, but the normal curve of the sons would have different parameters from the one of the daughters. The central limit theorem cannot be used to conclude that a plot of the heights of all the Fontanez children would follow a normal curve, since the gender factor does not play a "small" role in determining height.)

Thus, the central limit theorem can be used to explain why the heights of the many daughters of a particular pair of parents will follow a normal curve. However, by itself the theorem does not explain why a histogram of the heights of a collection of daughters from different parents will follow a normal curve. To see why not, suppose that this collection includes both a daughter of Maria and Peter Fontanez and a daughter of Henry and Catherine Silva. By the same argument given before, the height of the Silva daughter will be normally distributed as will the height of the Fontanez daughter. However, the parameters of these two normal distributions—one for each family—will be different. (For instance, if Catherine and Henry are both around 6 feet tall while Maria and Peter are both about 5 feet tall, then it is clear that the heights of their daughters will have different normal distributions.) By the same reasoning, we can conclude that the heights of a collection of many women, from different families, will all come from different normal distributions. It is, therefore, by no means apparent that a plot of those heights would itself follow a normal curve. (A more complete explanation of why biological data sets often follow a normal curve will be given in Chap. 12.)

7.4.1   Distribution of the Sample Mean

The central limit theorem can be used to approximate the probability distribution of the sample mean. That is, let $X_1, \ldots, X_n$ be a sample from a population having mean μ and variance σ^2, and let

$$\overline{X} = \frac{\sum\limits_{i=1}^{n} X_i}{n}$$

be the sample mean. Since a constant multiple of a normal random variable is also normal, it follows from the central limit theorem that $\overline{X}$ (which equals $\sum\limits_{i=1}^{n} X_i$ multiplied

H i s t o r i c a l   P e r s p e c t i v e

The application of the central limit theorem to show that measurement errors are approximately normally distributed is regarded as an important contribution to science. Indeed, in the 17th and 18th centuries, the central limit theorem was often called the *law of frequency of errors*.

The *law of frequency of errors* was considered a major advance by scientists. Listen to the words of Francis Galton (taken from his book *Natural Inheritance*, published in 1889).

I know of scarcely anything so apt to impress the imagination as the wonderful form of cosmic order expressed by the "Law of Frequency of Error." The Law would have been personified by the Greeks and deified, if they had known of it. It reigns with serenity and in complete self-effacement amidst the wildest confusion. The huger the mob and the greater the apparent anarchy, the more perfect is its sway. It is the supreme law of unreason.

by the constant $1/n$) also will be approximately normal when the sample size n is large. Since $\bar{X}$ has expectation μ and standard deviation $\sigma/\sqrt{n}$, the standardized variable

$$\frac{\bar{X} - \mu}{\sigma/\sqrt{n}}$$

approximately has a standard normal distribution.

Let $\bar{X}$ be the sample mean of a sample of size n from a population having mean μ and variance σ^2. By the central limit theorem,

$$P\{\bar{X} \le a\} = P\left\{\frac{\bar{X} - \mu}{\sigma/\sqrt{n}} \le \frac{a - \mu}{\sigma/\sqrt{n}}\right\}$$

$$\approx P\left\{Z \le \frac{a - \mu}{\sigma/\sqrt{n}}\right\}$$

where Z is a standard normal.

Example 7.3   The blood cholesterol levels of a population of workers have mean 202 and standard deviation 14.

(a) If a sample of 36 workers is selected, approximate the probability that the sample mean of their blood cholesterol levels will lie between 198 and 206.

(b) Repeat (a) when the sample size is 64.

Solution

(a) It follows from the central limit theorem that $\bar{X}$ is approximately normal with mean $\mu = 202$ and standard deviation $\sigma/\sqrt{n} = 14/\sqrt{36} = 7/3$. Thus the standardized variable

$$W = \frac{\bar{X} - 202}{7/3}$$

approximately has a standard normal distribution. To compute $P\{198 \leq \bar{X} \leq 206\}$, first we must write the inequality in terms of the standardized variable W. This results in the equality

$$P\{198 \leq \bar{X} \leq 206\} = P\left\{\frac{198 - 202}{7/3} \leq \frac{\bar{X} - 202}{7/3} \leq \frac{206 - 202}{7/3}\right\}$$

$$= P\{-1.714 \leq W \leq 1.714\}$$
$$\approx P\{-1.714 \leq Z \leq 1.714\} \quad \text{where } Z \text{ is a standard}$$
$$\text{normal random variable}$$
$$= 2P\{Z \leq 1.714\} - 1$$
$$= .913$$

where $\approx$ means "is approximately equal to" and the final equality follows from Table D.1 in App. D (or from Program 6-1).

(b) For a sample size of 64, the sample mean $\bar{X}$ will have mean 202 and standard deviation $14/\sqrt{64} = 7/4$. Hence, writing the desired probability in terms of the standardized variable

$$\frac{\bar{X} - 202}{7/4}$$

yields

$$P\{198 \leq \bar{X} \leq 206\} = P\left\{\frac{198 - 202}{7/4} \leq \frac{\bar{X} - 202}{7/4} \leq \frac{206 - 202}{7/4}\right\}$$

$$\approx P\{-2.286 \leq Z \leq 2.286\}$$
$$= 2P\{Z \leq 2.286\} - 1$$
$$= .978$$

Thus, we see that increasing the sample size from 36 to 64 increases the probability that the sample mean will be within 4 of the population mean from .913 to .978.

Example 7.4 An astronomer is interested in measuring, in units of light years, the distance from her observatory to a distant star. However, the astronomer knows that due to differing atmospheric conditions and normal errors, each time a measurement is made, it will yield not the exact distance, but an estimate of it. As a result, she is planning on making a series of 10 measurements and using the average of these measurements as her estimated value for the actual distance. If the values of the measurements constitute a sample from a population having mean d (the actual distance) and a standard deviation of 3 light years, approximate the probability that the astronomer's estimated value of the distance will be within .5 light year of the actual distance.

Solution The probability of interest is

$$P\{-.5 < \bar{X} - d < .5\}$$

where $\bar{X}$ is the sample mean of the 10 measurements. Since $\bar{X}$ has mean d and standard deviation $3/\sqrt{10}$, the above probability should be written in terms of the standardized variable

$$\frac{\bar{X} - d}{3/\sqrt{10}}$$

This gives

$$P\{-.5 < \bar{X} - d < .5\} = P\left\{\frac{-.5}{3/\sqrt{10}} < \frac{\bar{X} - d}{3/\sqrt{10}} < \frac{.5}{3/\sqrt{10}}\right\}$$

$$\approx P\left\{\frac{-.5}{3/\sqrt{10}} < Z < \frac{.5}{3/\sqrt{10}}\right\}$$

$$= P\{-.527 < Z < .527\}$$

$$= 2P\{Z < .527\} - 1 = .402$$

Therefore, we see that with 10 measurements there is a 40.2 percent chance that the estimated distance will be within plus or minus .5 light year of the actual distance.

7.4.2 How Large a Sample Is Needed?

The central limit theorem leaves open the question of how large the sample size n needs to be for the normal approximation to be valid, and indeed the answer depends on the population distribution of the sample data. For instance, if the underlying population distribution is normal, then the sample mean $\bar{X}$ will also be normal, no matter what the sample size is. A general rule of thumb is that you can be confident of the normal approximation whenever the sample size n is at least 30. That is, practically speaking, no matter how nonnormal the underlying population distribution is, the sam-

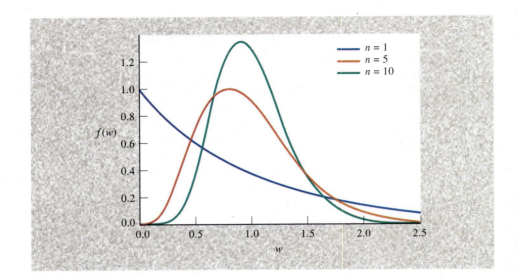

Figure 7.3
Density of the
average of *n*
exponential
random
variables.

ple mean of a sample size of at least 30 will be approximately normal. In most cases the normal approximation is valid for much smaller sample sizes. Indeed, usually a sample size of 5 will suffice for the approximation to be valid. Figure 7.3 presents the distribution of the sample means from a certain underlying population distribution (known as the *exponential distribution*) for samples sizes *n* = 1, 5, and 10.

Problems

1. Consider a sample from a population having mean 128 and standard deviation 16. Compute the approximate probability that the sample mean will lie between 124 and 132 when the sample size is
   (a) *n* = 9
   (b) *n* = 25
   (c) *n* = 100

2. Frequent fliers of a particular airline fly a random number of miles each year, having mean and standard deviation (in thousands of miles) of 23 and 11, respectively. As a promotional gimmick, the airline has decided to randomly select 20 of these fliers and give them, as a bonus, a check of $10 for each 1000 miles flown. Approximate the probability that the total amount paid out is
   (a) Between $4500 and $5000
   (b) More than $5200

3. In Example 7.2, approximate the probability that the yearly payout of the insurance company is between $2.5 and $2.7 million.

4. If you place a $1 bet on a number of a roulette wheel, then either you win $35 with probability 1/38 or you lose $1 with probability 37/38. Let X denote your gain on a bet of this type.

Historical Perspective

*Pierre Simon,
Marquis de Laplace*

The central limit theorem was originally stated and proved by the French mathematician Pierre Simon, the Marquis de Laplace, who came to this theorem from his observations that errors of measurement (which usually can be regarded as being the sum of a large number of tiny forces) tend to be normally distributed. Laplace, who was also a famous astronomer (and indeed was called "the Newton of France"), was one of the great early contributors to both probability and statistics. Laplace was a popularizer of the uses of probability in everyday life. He strongly believed in its importance, as is indicated by the following quotation, taken from his published book *Analytical Theory of Probability.*

> We see that the theory of probability is at bottom only common sense reduced to calculation; it makes us appreciate with exactitude what reason-

able minds feel by a sort of instinct, often without being able to account for it. . . . It is remarkable that this science, which originated in the consideration of games of chance, should become the most important object of human knowledge. . . . The most important questions of life are, for the most part, really only problems of probability.

An interesting footnote to the central limit theorem is that, because of it, most scientists in the late 19th and early 20th centuries believed that almost all data sets were normal. In the words of the famous French mathematician Henri Poincaré,

> Everyone believes it: experimentalists believe that it is a mathematical theorem, and mathematicians believe that it is an empirical fact.

(a) Find $E[X]$ and SD (X).

Suppose you continually place bets of the preceding type. Show that

(b) The probability that you will be winning after 1000 bets is approximately .39.

(c) The probability that you will be winning after 100,000 bets is approximately .002.

5. The time it takes to develop a photographic print is a random variable with mean 17 seconds and standard deviation .8 second. Approximate the probability that the total amount of time that it takes to process 100 prints is

(a) More than 1720 seconds

(b) Between 1690 and 1710 seconds

6. A zircon semiconductor is critical to the operation of a superconductor and must be immediately replaced upon failure. Its expected lifetime is 100 hours, and its standard deviation is 34 hours. If 22 of these semiconductors are available, approximate the probability that the superconductor can operate for the next 2000 hours. (That is, approximate the probability that the sum of the 22 lifetimes exceeds 2000.)

7. The amount of paper a print shop uses per job has mean 200 pages and standard deviation 50 pages. There are 2300 sheets of paper on hand and 10 jobs that need to be filled. What is the approximate probability that 10 jobs can be filled with the paper on hand?

8. A highway department has enough salt to handle a total of 80 inches of snowfall. Suppose the daily amount of snow has a mean of 1.5 inches and a standard deviation of .3 inch.
   (a) Approximate the probability that the salt on hand will suffice for the next 50 days.
   (b) What assumption did you make in solving part (a)?
   (c) Do you think this assumption is justified? Explain briefly!

9. Fifty numbers are rounded off to the nearest integer and then summed. If the individual roundoff errors are uniformly distributed between $-.5$ and $.5$, what is the approximate probability that the resultant sum differs from the exact sum by more than 3? (Use the fact that the mean and variance of a random variable that is uniformly distributed between $-.5$ and $.5$ are 0 and $1/12$, respectively.)

10. A six-sided die, in which each side is equally likely to appear, is repeatedly rolled until the total of all rolls exceeds 400. What is the approximate probability that this will require more than 140 rolls? (*Hint:* Relate the above to the probability that the sum of the first 140 rolls is less than 400.)

11. In Example 7.4, approximate the probability that the astronomer's estimate will be within .5 light year of the true distance if
    (a) She makes a total of 100 observations.
    (b) She makes 10 observations, but has found a way of improving the measurement technique so that the standard deviation of each observation is reduced from 3 to 2 light years.

12. Suppose that the number of miles that an electric car battery functions has mean μ and standard deviation 100. Using the central limit theorem, approximate the probability that the average number of miles per battery obtained from a set of n batteries will differ from μ by more than 20 if
    (a) $n = 10$ (b) $n = 20$ (c) $n = 40$ (d) $n = 100$

13. A producer of cigarettes claims that the mean nicotine content in its cigarettes is 2.4 milligrams with a standard deviation of .2 milligram. Assuming these figures are correct, approximate the probability that the sample mean of 100 randomly chosen cigarettes is
    (a) Greater than 2.5 milligrams
    (b) Less than 2.25 milligrams

14. The lifetime of a certain type of electric bulb has expected value 500 hours and standard deviation 60 hours. Approximate the probability that the sample mean of 20 such lightbulbs is less than 480 hours.

15. Consider a sample of size 16 from a population having mean 100 and standard deviation σ. Approximate the probability that the sample mean lies between 96 and 104 when

(a) $\sigma = 16$ (b) $\sigma = 8$ (c) $\sigma = 4$
(d) $\sigma = 2$ (e) $\sigma = 1$

16. An instructor knows from past experience that student examination scores have mean 77 and standard deviation 15. At present, the instructor is teaching two separate classes—one of size 25 and the other of size 64.
    (a) Approximate the probability that the average test score in the class of size 25 lies between 72 and 82.
    (b) Repeat (a) for the class of size 64.
    (c) What is the approximate probability that the average test score in the class of size 25 is higher than that in the class of size 64?
    (d) Suppose the average scores in the two classes are 76 and 83. Which class—the one of size 25 or the one of size 64—do you think was more likely to have averaged 83? Explain your intuition.

7.5   SAMPLING PROPORTIONS FROM A FINITE POPULATION

Consider a population of size N in which certain elements have a particular characteristic of interest. Let p denote the proportion of the population having this characteristic. So Np elements of the population have it, and $N(1 - p)$ do not.

Example 7.5   Suppose that 60 out of a total of 900 students of a particular school are left-handed. If left-handedness is the characteristic of interest, then $N = 900$ and $p = 60/900 = 1/15$.   ▪ ▪

A sample of size n is said to be a *random sample* if it is chosen in a manner so that each of the possible population subsets of size n is equally likely to be in the sample. For instance, if the population consists of the three elements a, b, c, then a random sample of size 2 is one that is chosen so that it is equally likely to be any of the subsets $\{a, b\}$, $\{a, c\}$, or $\{b, c\}$. A random subset can be chosen sequentially by letting its first element be equally likely to be any of the N elements of the population, then letting its second element be equally likely to be any of the remaining $N - 1$ elements of the population, and so on.

Definition

A sample of size n selected from a population of N elements is said to be a *random sample* if it is selected in such a manner that the sample chosen is equally likely to be any of the subsets of size n.

The mechanics of using a computer to choose a random sample are explained in App. C. (In addition, Program A-1 on the enclosed disk can be used to accomplish this task.)

Suppose now that a random sample of size n has been chosen from a population of size N. For $i = 1, \ldots, n$, let

$$X_i = \begin{cases} 1 & \text{if the } i\text{th member of the sample has the characteristic} \\ 0 & \text{otherwise} \end{cases}$$

Consider now the sum of the X_i, that is, consider

$$X = X_1 + X_2 + \cdots + X_n$$

Since the term X_i contributes 1 to the sum if the ith member of the sample has the characteristic and contributes 0 otherwise, it follows that the sum is equal to the number of members of the sample that possess the characteristic. (For instance, suppose $n = 3$ and $X_1 = 1$, $X_2 = 0$, and $X_3 = 1$. Then members 1 and 3 of the sample possess the characteristic, and member 2 does not. Hence, exactly 2 of the sample members possess it, as indicated by $X_1 + X_2 + X_3 = 2$.) Similarly, the sample mean

$$\bar{X} = \frac{X}{n} = \frac{\sum\limits_{i=1}^{n} X_i}{n}$$

will equal the *proportion* of members of the sample who possess the characteristic. Let us now consider the probabilities associated with the statistic $\bar{X}$.

Since the ith member of the sample is equally likely to be any of the N members of the population, of which Np have the characteristic, it follows that

$$P\{X_i = 1\} = \frac{Np}{N} = p$$

Also

$$P\{X_i = 0\} = 1 - P\{X_i = 1\} = 1 - p$$

That is, each X_i is equal to either 1 or 0 with respective probabilities p and $1 - p$.

Note that the random variables $X_1, X_2, \ldots, X_n$ are not independent. For instance, since the second selection is equally likely to be any of the N members of the population, of which Np have the characteristic, it follows that the probability that the second selection has the characteristic is $Np/N = p$. That is, without any knowledge of the outcome of the first selection,

$$P\{X_2 = 1\} = p$$

However, the conditional probability that $X_2 = 1$, given that the first selection has the characteristic, is

$$P\{X_2 = 1 \mid X_1 = 1\} = \frac{Np - 1}{N - 1}$$

which is seen by noting that if the first selection has the characteristic, then the second selection is equally likely to be any of the remaining $N - 1$ elements of which $Np - 1$ have the characteristic. Similarly, the probability that the second selection has the characteristic, given that the first one does not, is

$$P\{X_2 = 1 \mid X_1 = 0\} = \frac{Np}{N - 1}$$

Thus, knowing whether the first element of the random sample has the characteristic changes the probability for the next element. However, when the population size N is large in relation to the sample size n, this change will be very slight. For instance, if $N = 1000$ and $p = .4$, then

$$P\{X_2 = 1 \mid X_1 = 1\} = \frac{399}{999} = .3994$$

which is very close to the unconditional probability that $X_2 = 1$; namely,

$$P\{X_2 = 1\} = .4$$

Similarly, the probability that the second element of the sample has the characteristic, given that the first does not, will be given by

$$P\{X_2 = 1 \mid X_1 = 0\} = \frac{400}{999} = .4004$$

which is again very close to .4.

Indeed, it can be shown that when the population size N is large with respect to the sample size n, then $X_1, X_2, \ldots, X_n$ are approximately independent. Now if we think of each X_i is representing the result of a trial that is a success if X_i equals 1 and is a failure otherwise, it follows that $X = \sum_{i=1}^{n} X_i$ can be thought of as representing the total number of successes in n trials. Hence, if the X_i's are independent, then X represents the number of successes in n independent trials where each trial is a success with probability p. In other words, X is a binomial random variable with parameters n and p.

If we let X denote the number of members of the population who have the characteristic, then it follows from the preceding that if the population size N is large in relation to the sample size n, then the distribution of X is approximately a binomial distribution with parameters n and p.

For the remainder of this text we will suppose that the underlying population is large in relation to the sample size, and we will take the distribution of X to be binomial.

By using the formulas given in Sec. 5.5.1 for the mean and standard deviation of a binomial random variable, we see that

$$E[X] = np \quad \text{and} \quad SD(X) = \sqrt{np(1-p)}$$

Since $\bar{X}$, the proportion of the sample that has the characteristic, is equal to X/n, we see from the above that

$$E[\bar{X}] = \frac{E[X]}{n} = p$$

and

$$SD(\bar{X}) = \frac{SD(X)}{n} = \sqrt{\frac{p(1-p)}{n}}$$

Example 7.6   Suppose that 50 percent of the population is planning on voting for candidate A in an upcoming election. If a random sample of size 100 is chosen, then the proportion of those in the sample who favor candidate A has expected value

$$E[\bar{X}] = .50$$

and standard deviation

$$SD(\bar{X}) = \sqrt{\frac{.50(1-.50)}{100}} = \sqrt{\frac{1}{400}} = .05$$

7.5.1   Probabilities Associated with Sample Proportions: The Normal Approximation to the Binomial Distribution

Again, let $\bar{X}$ denote the proportion of members of a random sample of size n who have a certain characteristic. To determine the probabilities connected with the random variable $\bar{X}$, we make use of the fact that $X = n\bar{X}$ is binomial with parameters n and p. Now, binomial probabilities can be approximated by making use of the central limit theorem. Indeed, from an historical point of view, one of the most important applications of the central limit theorem was in computing binomial probabilities.

To see how this is accomplished, let X denote a binomial random variable having parameters n and p. Since X can be thought of as being equal to the number of successes in n independent trials when each trial is a success with probability p, it follows that it can be represented as

$$X = X_1 + X_2 + \cdots + X_n$$

where

$$X_i = \begin{cases} 1 & \text{if trial } i \text{ is a success} \\ 0 & \text{if trial } i \text{ is a failure} \end{cases}$$

Now, in Examples 5.6 and 5.12, we showed that

$$E[X_i] = p \quad \text{and} \quad \text{Var}(X_i) = p(1 - p)$$

Hence, it follows that X/n can be regarded as the sample mean of a sample of size n from a population having mean p and standard deviation $\sqrt{p(1 - p)}$. Thus, from the central limit theorem, we see that for n large

$$\frac{X/n - p}{\sqrt{p(1 - p)/n}} = \frac{X - np}{\sqrt{np(1 - p)}}$$

will approximately have a standard normal distribution. (Figure 7.4 graphically illustrates how the probability distribution of a binomial random variable with parameters n and p becomes more and more "normal" as n becomes larger and larger.)

From a practical point of view, the normal approximation to the binomial is quite good provided n is large enough that the quantities np and $n(1 - p)$ are both greater than 5.

Example 7.7 Suppose that exactly 46 percent of the population favors a particular candidate. If a random sample of size 200 is chosen, what is the probability that at least 100 favor this candidate?

Solution If X is the number who favor the candidate, then X is a binomial random variable with parameters $n = 200$ and $p = .46$. The desired probability is $P\{X \geq 100\}$. To employ the normal approximation, first we note that since the binomial is a discrete and the normal is a continuous random variable, it is best to compute $P\{X = i\}$ as $P\{i - .5 \leq X \leq i + .5\}$ when applying the normal approximation (this is called the *continuity correction*). Therefore, to compute $P\{X \geq 100\}$, we should use the normal approximation on the equivalent probability $P\{X \geq 99.5\}$. Considering the standardized variable

$$\frac{X - 200(.46)}{\sqrt{200(.46)(.54)}} = \frac{X - 92}{7.0484}$$

we obtain the following normal approximation to the desired probability.

$$\begin{aligned} P\{X \geq 100\} &= P\{X \geq 99.5\} \\ &= P\left\{ \frac{X - 92}{7.0484} \geq \frac{99.5 - 92}{7.0484} \right\} \\ &\approx P\{Z > 1.0641\} \\ &= .144 \quad \text{(from Table D.1 or Program 6-1)} \end{aligned}$$

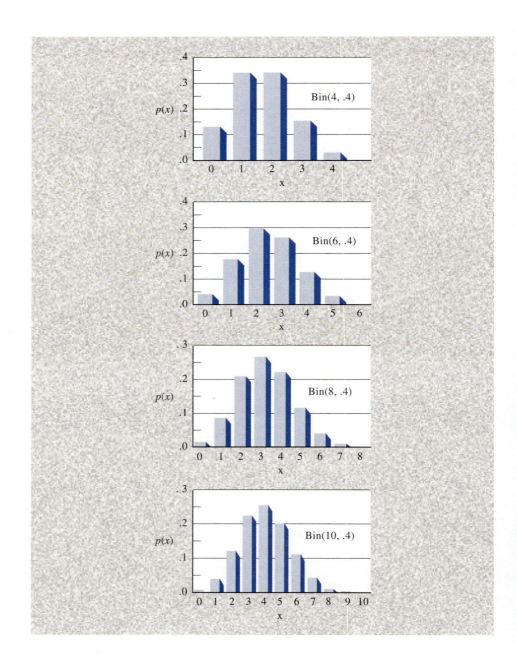

Figure 7.4
Probability
mass functions
of binomial
random
variables
become more
normal with
increasing *n*.

The exact value of the desired probability could, of course, have been obtained from Program 5-1. Running this program shows that the exact probability that a binomial random variable with parameters $n = 200$ and $p = .46$ is greater than or equal to 100 is .1437. Thus, in this problem, the normal approximation gives an answer that is correct to three decimal places.

*A Cautionary Tale: Be Sure You Are Sampling From the Right Population

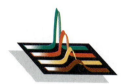

Company X, which is not located near any public transportation, and all of whose employees drive to work, is concerned that not enough people are utilizing car pools. The company has decided that if the average number of workers per car is less than 3, then it will organize its own car pool service and, at the same time, begin charging those employees who drive non-car-pool automobiles a stiff parking fee. To determine if such a change is justified, 100 workers were chosen at random and were queried as to the number of workers in the car in which they drove to work that day. The average answer was 3.4; that is, the sum of the 100 answers divided by 100 was 3.4. On the basis of this, the company chose not to change its policy. Did the company make the correct decision?

The above question is very tricky because the company, when selecting its random sample of 100 workers, has chosen a random sample from the wrong population. Since it wanted to learn about the average number of workers per car, the company should have chosen a random sample from the population of cars arriving in the parking lot—not from the population of workers. To see why, consider an extreme case where there are only 2 cars and 5 workers, with one of the cars containing 4 workers and the

other containing 1 worker. Now, if we average over the 2 cars, then the average number of workers per car is clearly $(4 + 1)/2 = 2.5$. However, if we average over all the workers, then since 4 of the 5 workers ride in a car containing 4 workers, it follows that the average is $(4 + 4 + 4 + 4 + 1)/5 = 3.4$.

In general, by randomly choosing workers (as opposed to cars) it follows that cars containing more riders will tend to be more heavily represented (by their riders) in the sample than will those cars having fewer riders. As a result, the average number of riders in the cars of the randomly chosen workers will tend to be larger than the average number of workers per car.

To obtain a correct estimate of the average number of workers per car, the random sample should have been created by randomly choosing among the cars in the parking lot and then ascertaining how many workers were in each car.

Because the wrong random sample was chosen, the company cannot conclude that the average number of workers per car is at least 3. Indeed, a new sample chosen in the manner noted above will have to be taken before the company can decide whether any changes are needed.

Problems

1. Suppose that 60 percent of the residents of a city are in favor of teaching evolution in the high school. Determine the mean and the standard deviation of the proportion of a random sample of size n that is in favor when
   (a) $n = 10$ (b) $n = 100$
   (c) $n = 1000$ (d) $n = 10,000$

2. Ten percent of all electrical batteries are defective. In a random selection of 8 of these batteries, find the probability that
   (a) There are no defective batteries.
   (b) More than 15 percent of the batteries are defective.
   (c) Between 8 and 12 percent of the batteries are defective.

3. Suppose there was a random selection of $n = 50$ batteries in Prob. 2. Determine approximate probabilities for parts (a), (b), and (c) of that problem.

4. Consider Prob. 1. Find the probability that over 55 percent of the members of the sample are in favor of the proposal if the sample size is
   (a) $n = 10$
   (b) $n = 100$
   (c) $n = 1000$
   (d) $n = 10,000$

The following table gives the 1986 unemployment rates in various U.S. industries. Problems 5, 6, and 7 are based on it.

Industry	1986 Unemployment rate
Agriculture	12.5
Mining	13.5
Construction	13.1
Manufacturing	7.1
Transportation	5.1
Insurance and real estate	3.5

Source: U.S. Bureau of Labor Statistics, *Employment and Earnings.*

5. Suppose that a random sample of 400 agricultural workers was selected in 1986. Approximate the probability that
   (a) Forty or fewer were unemployed.
   (b) More than 50 were unemployed.

6. Suppose that a random sample of 400 manufacturing workers was selected in 1986. Approximate the probability that
   (a) Thirty or fewer were unemployed.
   (b) More than 40 were unemployed.

7. Suppose that a random sample of 400 insurance and real estate workers was selected in 1986. Approximate the probability that
   (a) Ten or fewer were unemployed.
   (b) More than 25 were unemployed.

8. If 65 percent of the population of a certain community is in favor of a proposed increase in school taxes, find the approximate probability that a random sample of 100 people will contain

(a) At least 45 who are in favor of the proposition
(b) Fewer than 60 who are in favor
(c) Between 55 and 75 who are in favor

9. The ideal size of a first-year class at a particular college is 160 students. The college, from past experience, knows that on average only 40 percent of those accepted for admission will actually attend. Based on this, the college employs a policy of initially accepting 350 applicants. Find the normal approximation to the probability that this will result in
(a) More than 160 accepted students attending
(b) Fewer than 150 accepted students attending

10. An airline company experiences a 6 percent rate of no-shows among passengers holding reservations. If 260 people hold reservations on a flight in which the airplane can hold a maximum of 250 people, approximate the probability that the company will be able to accommodate everyone having a reservation who shows up.

The following table lists the likely fields of study as given by the entering college class of 1987.

Field of study	Percentage
Arts and humanities	9
Biological sciences	4
Business	27
Education	9
Engineering	10
Physical sciences	2
Social sciences	9
Professional	11
Technical	3
Other	16

Source: Higher Educational Institute, University of California, Los Angeles, C, *The American Freshman National Norms,* annual.

Problems 11 through 14 are based on the preceding table. In each of these problems suppose that a random sample of 200 entering students is chosen.

11. What is the probability that 22 or more students are planning to major in arts and humanities?

12. What is the probability that more than 60 students are planning to major in business?

13. What is the probability that 30 or more are planning to major in one of the sciences (biological, physical, or social)?

14. What is the probability that fewer than 15 students are planning to major in engineering?

15. Let X be a binomial random variable with parameters $n = 100$ and $p = .2$. Approximate the following probabilities.
(a) $P\{X \le 25\}$
(b) $P\{X > 30\}$
(c) $P\{15 < X < 22\}$

16. Let X be a binomial random variable with parameters $n = 150$ and $p = .6$. Approximate the following probabilities.
(a) $P\{X \le 100\}$
(b) $P\{X > 75\}$
(c) $P\{80 < X < 100\}$

17. A recent study has shown that 54 percent of all incoming first-year students at major universities do not graduate within 4 years of their entrance. Suppose a random sample of 500 entering first-year students is to be surveyed after 4 years.
(a) What is the approximate probability that fewer than half graduate within 4 years?
(b) What is the approximate probability that more than 175 but fewer than 225 students graduate within 4 years?

The following table gives the percentages of individuals, categorized by gender, who follow certain negative health practices.

	Sleep 6 hours or less per night	Smoker	Never eat breakfast	Are 30% or more overweight
Males	22.7	32.6	25.2	12.1
Females	21.4	27.8	23.6	13.7

Source: U.S. National Center for Health Statistics, *Health Promotion and Disease Prevention,* 1985.

Problems 18, 19, and 20 are based on the preceding table.

18. Suppose a random sample of 300 males is chosen. Approximate the probability that
(a) At least 75 never eat breakfast.
(b) Fewer than 100 smoke.

19. Suppose a random sample of 300 females is chosen. Approximate the probability that
(a) At least 25 are overweight by 30 percent or more.
(b) Fewer than 50 sleep 6 hours or less nightly.

*20. Suppose random samples of 300 females and 300 males are chosen. Approximate the probability that there are more smokers in the sample of men than in the sample of woman. (*Hint:* Let X and Y denote, respectively, the numbers of men and women in the samples who are smokers. Write the desired probability as $P\{X - Y > 0\}$, and recall that the difference of two independent normal random variables is also a normal random variable.)

7.6 DISTRIBUTION OF THE SAMPLE VARIANCE OF A NORMAL POPULATION

Before discussing the distribution of the sample variance of a normal population, we need to introduce the concept of the chi-squared distribution, which is the distribution of the sum of the squares of independent standard normal random variables.

Definition

If $Z_1, \ldots, Z_n$ are independent standard normal random variables, then the random variable

$$\sum_{i=1}^{n} Z_i^2$$

is said to be a *chi-squared* random variable with n *degrees of freedom.*

Figure 7.5 presents the chi-squared density functions for three different values of the degree of freedom parameter n.

To determine the expected value of a chi-squared random variable, note first that for a standard normal random variable Z

$$\begin{aligned}
1 &= \text{Var}\,(Z) \\
&= E[Z^2] - (E[Z])^2 \\
&= E[Z^2] \quad \text{since } E[Z] = 0
\end{aligned}$$

Hence, $E[Z^2] = 1$ and so

$$E\left[\sum_{i=1}^{n} Z_i^2\right] = \sum_{i=1}^{n} E[Z_i^2] = n$$

The expected value of a chi-squared random variable is equal to its number of degrees of freedom.

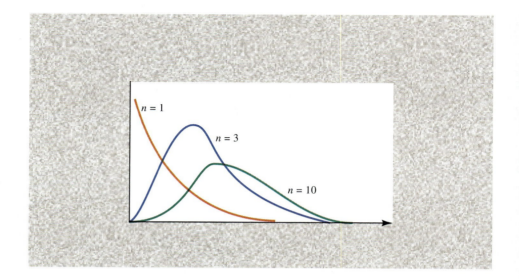

Figure 7.5
Chi-squared
density function
with n degrees
of freedom,
$n = 1, 3, 10$.

Suppose now that we have a sample $X_1, \ldots, X_n$ from a normal population having mean μ and variance σ^2. Consider the sample variance S^2 defined by

$$S^2 = \frac{\sum\limits_{i=1}^{n}(X_i - \bar{X})^2}{n - 1}$$

The following result can be proved:

Theorem

$$\frac{(n-1)S^2}{\sigma^2} = \frac{\sum\limits_{i=1}^{n}(X_i - \bar{X})^2}{\sigma^2}$$

has a chi-squared distribution with $n - 1$ degrees of freedom.

Although a mathematical proof of the above theorem is beyond the scope of this text, we can obtain some understanding of why it is true. This understanding will also be useful in guiding our intuition as we continue our studies in later chapters. To begin, let us consider the standardized variables $(X_i - \mu)/\sigma$, $i = 1, \ldots, n$, where μ is the population mean. Since these variables are independent standard normals, it follows that the sum of their squares

$$\frac{\sum_{i=1}^{n}(X_i - \mu)^2}{\sigma^2}$$

has a chi-squared distribution with n degrees of freedom. Now, if we substitute the sample mean $\bar{X}$ for the population mean μ above, then the new quantity

$$\frac{\sum_{i=1}^{n}(X_i - \bar{X})^2}{\sigma^2}$$

will remain a chi-squared random variable, but it will lose 1 degree of freedom because the population mean (μ) is replaced by its estimator (the sample mean $\bar{X}$).

Problems

1. The following data sets come from normal populations whose standard deviation σ is specified. In each case, determine the value of a statistic whose distribution is chi squared, and tell how many degrees of freedom this distribution has.
   (a) 104, 110, 100, 98, 106; $\sigma = 4$
   (b) 1.2, 1.6, 2.0, 1.5, 1.3, 1.8; $\sigma = .5$
   (c) 12.4, 14.0, 16.0; $\sigma = 2.4$

2. Explain why a chi-squared random variable having n degrees of freedom will approximately have the distribution of a normal random variable when n is large. (*Hint:* Use the central limit theorem.)

Key Terms

A sample from a population distribution: If $X_1, \ldots, X_n$ are independent random variables having a common distribution F, we say that they constitute a sample from the population distribution F.

Statistic: A numerical quantity whose value is determined by the sample.

Sample mean: If $X_1, \ldots, X_n$ are a sample, then the sample mean is

$$\bar{X} = \frac{\sum_{i=1}^{n} X_i}{n}$$

Sample variance: If $X_1, \ldots, X_n$ are a sample, then the sample variance is

$$S^2 = \frac{\sum_{i=1}^{n}(X_i - \bar{X})^2}{n-1}$$

Central limit theorem: A theorem stating that the sum of a sample of size n from a population will approximately have a normal distribution when n is large.

Random sample: A sample of n members of a population is a random sample if it is obtained in such a manner that each of the possible subsets of n members is equally likely to be the chosen sample.

Chi-squared distribution with n degrees of freedom: The distribution of the sum of the squares of n independent standard normals.

Summary

If $\bar{X}$ is the sample mean of a sample of size n from a population having mean μ and standard deviation σ, then its mean and standard deviation are

$$E[\bar{X}] = \mu \quad \text{and} \quad \text{SD}(\bar{X}) = \frac{\sigma}{\sqrt{n}}$$

The central limit theorem states that the sample mean of a sample of size n from a population having mean μ and standard deviation σ will, for large n, have an approximately normal distribution with mean μ and standard deviation $\sigma/\sqrt{n}$.

Consider a random sample of size n from a population of N individuals in which Np of them have a certain characteristic. Let X denote the number of members of the sample who have the characteristic. When N is large in relation to n, X will approximately be a binomial random variable with parameters n and p. In this text, we will always suppose that this is the case.

The proportion of the sample having the characteristic, namely, $\bar{X} = X/n$, has a mean and a standard deviation given by

$$E[\bar{X}] = p \quad \text{and} \quad \text{SD}(\bar{X}) = \sqrt{\frac{p(1-p)}{n}}$$

It follows from the central limit theorem that a binomial random variable with parameters n and p can, for reasonably large n, be approximated by a normal random variable with mean np and standard deviation $\sqrt{np(1-p)}$. The approximation should be quite accurate provided that n is large enough that both np and $n(1-p)$ are larger than 5.

If S^2 is the sample variance from a sample of size n from a normal population having variance σ^2, then $(n-1)S^2/\sigma^2$ has a chi-squared distribution with $n-1$ degrees of freedom.

The expected value of a chi-squared random variable is equal to its number of degrees of freedom.

Review Problems

1. The sample mean and sample standard deviation of all student scores on the last Scholastic Aptitude Test (SAT) examination were, respectively, 517 and 120. Find the approximate probability that a random sample of 144 students would have an average score exceeding
   (a) 507
   (b) 517
   (c) 537
   (d) 550

2. Let $\bar{X}$ denote the sample mean of a sample of size 10 from a population whose probability distribution is given by

$$P\{X = i\} = \begin{cases} .1 & \text{if } i = 1 \\ .2 & \text{if } i = 2 \\ .3 & \text{if } i = 3 \\ .4 & \text{if } i = 4 \end{cases}$$

   Compute
   (a) The population mean μ
   (b) The population standard deviation σ
   (c) $E[\bar{X}]$
   (d) Var $(\bar{X})$
   (e) SD $(\bar{X})$

3. In Prob. 2, suppose the sample size was 2. Find the probability distribution of $\bar{X}$, and use it to compute $E[\bar{X}]$ and SD $(\bar{X})$. Check your answers by using the values of μ and σ.

4. The mean and standard deviation of the lifetime of a type of battery used in electric cars are, respectively, 225 and 24 minutes. Approximate the probability that a set of 10 batteries, used one after the other, will last for more than
   (a) 2200 minutes
   (b) 2350 minutes
   (c) 2500 minutes
   (d) What is the probability they will last between 2200 and 2350 minutes?

5. Suppose that 12 percent of the members of a population is left-handed. In a random sample of 100 individuals from this population,
   (a) Find the mean and standard deviation of the number of left-handed people.
   (b) Find the probability that this number is between 10 and 14 inclusive.

6. The weight of a randomly chosen person riding a ferry has expected value 155 and standard deviation 28 pounds. The ferry's capacity is 100 riders. Find the probability that, at capacity, the total passenger load exceeds 16,000 pounds.

7. The monthly telephone bill of a student residing in a dormitory has an expected value of $15 with a standard deviation of $7. Let X denote the sum of the monthly telephone bills of a sample of 20 such students.
   (a) What is $E[X]$?
   (b) What is SD (X)?
   (c) Approximate the probability that X exceeds $300.

8. A recent newspaper article claimed that the average salary of newly graduated seniors majoring in chemical engineering is $34,000, with a standard deviation of $3000. Suppose a random sample of 12 such graduates revealed an average salary of $30,500. How likely is it that an average salary as low as or lower than $30,500 would have been observed from this sample if the newspaper article were correct?

9. An advertising agency ran a campaign to introduce a product. At the end of its campaign, it claimed that at least 25 percent of all consumers were now familiar with the product. To verify this claim, the producer randomly sampled 1000 consumers and found that 232 knew of the product. If 25 percent of all consumers actually knew of the product, what is the probability that as few as 232 (that is, 232 or less) in a random sample of 1000 consumers were familiar?

10. A club basketball team will play a 60-game season. Of these games 32 are against class A teams and 28 are against class B teams. The outcomes of all the games are independent. The team will win each game against a class A opponent with probability .5, and it will win each game against a class B opponent with probability .7. Let X denote the total number of victories in the season.
   (a) Is X a binomial random variable?
   (b) Let X_A and X_B denote, respectively, the number of victories against class A and class B teams. What are the distributions of X_A and X_B?
   (c) What is the relationship among X_A, X_B, and X?
   (d) Approximate the probability that the team wins 40 or more games. (*Hint:* Recall that the sum of independent normal random variables is also a normal random variable.)

11. If X is binomial with parameters $n = 80$ and $p = .4$, approximate the following probabilities.
   (a) $P\{X > 34\}$
   (b) $P\{X \leq 42\}$
   (c) $P\{25 \leq X \leq 39\}$

12. Consider the following simple model for daily changes in price of a stock. Suppose that on each day the price either goes up 1 with probability .52 or goes down 1 with probability .48. Suppose the price at the beginning of day 1 is 200. Let X denote the price at the end of day 100.
   (a) Define random variables $X_1, X_2, \ldots, X_{100}$ such that

$$X = \sum_{i=1}^{100} X_i$$

(b) Determine $E[X_i]$.

(c) Determine Var (X_i).

(d) Use the central limit theorem to approximate $P\{X \geq 210\}$.

The following table uses 1989 data concerning the percentages of women and men full-time workers whose annual salaries fall in different salary groupings.

Earnings range ($)	Percentage of women	Percentage of men
4,999 or less	2.8	1.8
5,000–9,999	10.4	4.7
10,000–19,999	41.0	23.1
20,000–25,000	16.5	13.4
25,000–49,999	26.3	42.1
50,000 and over	3.0	14.9

Source: U.S. Department of Commerce, *Bureau of the Census.*

13. Suppose random samples of 1000 women and 1000 men are chosen. Use the preceding table to approximate the probability that
    (a) At least half of the women earn less than $20,000.
    (b) Over half of the men earn $20,000 or more.
    (c) Over half of the women *and* over half of the men earn $20,000 or more.
    (d) Of the women, 250 or fewer earn at least $25,000.
    (e) At least 200 men earn $50,000 or more.
    *(f) More women than men earn between $20,000 and $24,999.

*14. A university administrator wants a quick estimate of the average number of students enrolled per class. Because he does not want the faculty to be aware of his interest, he has decided to enlist the aid of students. He has decided to randomly choose 100 names from the roster of students and have them determine and then report to him the number of students in each of their classes. His estimate of the average number of students per class will be the average number reported per class.
    (a) Will this method achieve the desired goal?
    (b) If the answer to part (a) is yes, explain why. If it is no, give a method that will work.

Distributions of Sampling Statistics

Purpose

Use Minitab to

1. Investigate the sampling distribution of the sample mean.

2. Apply the central limit theorem (CLT).

3. Investigate the normal approximation to the binomial distribution.

4. Investigate the sampling distribution of the sample variance.

Procedures

First, load the Minitab (Windows version) software as in the Minitab lab for Chap. 1.

1. GENERATING RANDOM SAMPLES, COMPUTING SAMPLE MEANS, AND APPLYING THE CLT

Example 1 Use Minitab to simulate the rolling of a regular six-sided die. Roll the die 50 times, and repeat the experiment 200 times. Compute the row means, and use them to discuss the central limit theorem.

Use the mouse to select Calc→Random Data→Integer, and the Integer distribution dialog box will be displayed. At the Generate prompt, type 200 in the text box. Use the mouse to select the Store in column(s) box; type in C1-C50 in the text box. Use the mouse to select the Minimum value box, and type in the value 1 in the text box. Use the mouse to select the Maximum value box; type in the value 6 in the text box. The dialog box with the appropriate entries is shown in Fig. M7.1.

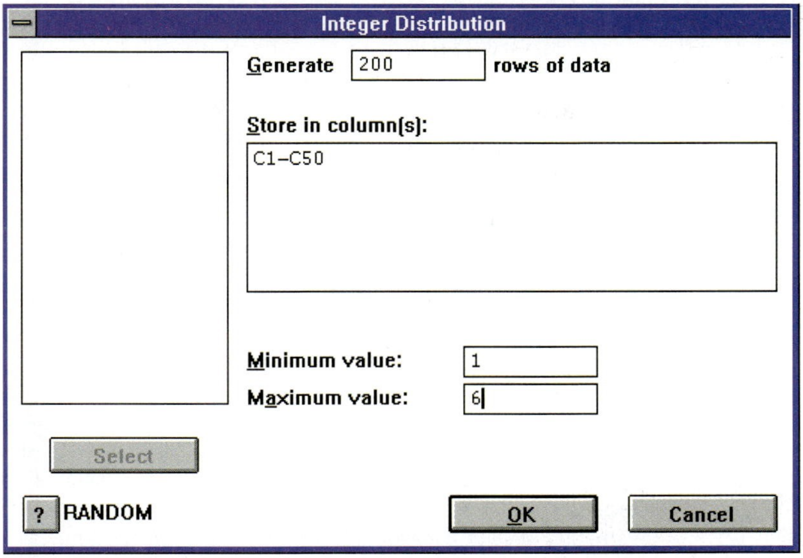

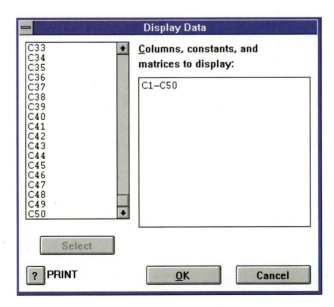

Figure M7.1

Next use the mouse to select the OK button, and the data set will be displayed in the Data window. The preceding instructions will enable Minitab to generate 50 columns (C1 to C50) of 200 rolls of a fair six-sided die. You can use the mouse to click on the Data screen window to observe the generated data. You will have to scroll the screen down as well as to the right to observe all the generated data. Also, you can use the mouse to select File→Display Data. In the Display data dialog box, use the mouse to select C1 through C50 by dragging the arrow from C1 to C50 in the left-hand side box and then clicking on the Select button. C1 to C50 will be displayed in the right-hand side box. The Display data dialog box is shown in Fig. M7.2.

Figure M7.2

Select the OK button, the data will be scrolled on the screen.

Note

1. These rows represent random samples of size 50 from the sampling population.

2. The sampling distribution is a discrete uniform distribution where $P(X = i) =$ $1/6$ for $i = 1, 2, 3, 4, 5, 6$.

   Next select Calc→Row Statistics, and the Row statistics dialog box will be displayed. Select the Mean check box, and suppose we want to consider samples of size 25. In the Input variables text box, select C1-C25 (say) with the mouse, or type in C1-C25. In the Store result in text box, type in C51. Thus, C51 stores the means of these 200 sets of random samples of size 25 generated from the *discrete uniform* distribution.

   To display these computed means in a histogram, select Graph→Histogram. In the Graph variables text box for graph 1, select C51 and OK. The histogram for these average values (C51) should be displayed in a histogram window. Observe the general shape of the histogram, and try to apply the central limit theorem.

   Before you attempt Example 2, erase the data in the occupied columns for Example 1.

Note To delete a list of columns, choose Manip → Erase Variables. This command allows you to delete any combination of columns, stored constants, and matrices. Also, you can use the mouse to highlight the columns and rows you want to delete, and then you press the Delete key.

Example 2 Generate 500 values from the following skewed distribution. Repeat the experiment 50 times and store in columns C3 to C52. Compute the row means and use them to discuss the central limit theorem. The following discrete distribution is given:

x	$P(X = x)$
1	.0249
2	.0741
3	.2132
4	.3621
5	.3257

   Enter the above values in C1 and C2. Rename C1 and C2 as x and $P(X = x)$, respectively, as explained in earlier labs. Note that this is a *discrete distribution*. Next select Calc→Random Data→Discrete, and the Discrete distribution dialog box will be displayed. Generate 500 rows of data, and store in columns C3 to C52 with values in C1 and probability values in C2. The Discrete distribution dialog box is shown in Fig. M7.3 with the appropriate entries. Each of these 50 columns (C3 to C52) represents a random sample of size 500 from the given population. Select a few of the generated columns (C3 to C52) and use the Graph option to display histograms for these columns. They should be skewed. Next, compute row means for C3-C10, C3-C20, C3-C30, C3-C40, and C3-C50 by using the Calc option; and place answers in columns C53, C54, C55, C56, and C57, respectively. Graph these column means and observe their distribution. *These graphs will give you insight into the sampling distribution of the sample mean for different sample sizes (8, 18, 28, 38, 48) from this distribution.* Try to apply the central limit theorem.

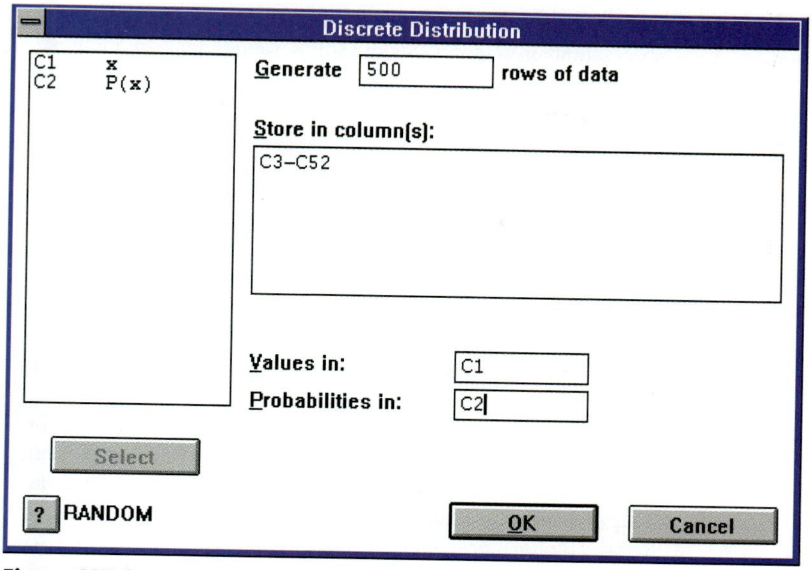

Figure M7.3

2. INVESTIGATING THE NORMAL APPROXIMATION TO THE BINOMIAL DISTRIBUTION

From the discussion in Sec. 7.5.1 of the text, the normal approximation to the binomial is quite good, provided n is large enough that $np \geq 5$ and $n(1 - p) \geq 5$. Here we will investigate this approximation graphically.

Example 3   For $p = .3$ and $n = 4, 8, 12, 16, 20,$ and 24, generate 500 values from the binomial distribution and save in columns C1 to C6. Construct histograms for these columns, and observe the shapes of the histograms. Also construct normal probability plots (NPPs) for C1 to C6 to help with your observations. Employ the procedures in the Minitab lab for Chap. 5 and the current lab to do this example.

Computer Exercises

1. (a) Fill out a distribution table for the discrete uniform distribution in Example 1.
   (b) Compute the population mean μ for the distribution in Example 1.
   (c) Compute the population standard deviation σ for the distribution in Example 1.

2. Use Minitab or any other statistical software to repeat the above process in Example 1. If you use Minitab, compute the means for C1–C2, C1–C10, C1–C20, C1–C30, C1–C40, and C1–C50 and save, respectively, in columns C51, C52, C53, C54, C55, and C56. Draw histograms for these computed column means, and observe their general shapes. These graphs will give you an insight into the sampling distribution of the sample mean for different sample sizes (2, 10, 20, 30, 40, and 50). In addition, generate descriptive statistics values for the sample means in C51 to C56 by selecting Stat→ Descriptive Stats. Observe the sample means, and compare to the population mean [this

should be the value in Exercise 1(b)]. Generate histograms for C51 to C56. Present your observations in a report. Make sure you relate your discussions to the central limit theorem.

3. Compute the mean μ and standard deviation σ for the population distribution given in Example 2.

4. Do the same for Example 2 as you did for Example 1 in Exercise 2. Be careful with your column numbers if you use Minitab.

5. The two examples in this Minitab lab dealt with *discrete distributions*. There are several continuous distributions with normal being the most commonly used. When you are using Minitab, from the Calc→Random Data→Normal selection sequence, generate 500 samples of size 10 from a normal distribution with mean $\mu = 21$ and standard deviation $\sigma = 4$. Select sample sizes of 2 to 10, and compute their means. Draw histograms for these sample means. Discuss your results and observations in a report. Are there any differences between your observations here and the observations for the discrete distributions of Exercises 2 and 4? Explain.

6. Suppose that the manufacturer of a certain brand of 60-watt bulb advertises that the *mean* life of the bulb is 1,200 hours. Assume that the lifetimes of the bulbs are *exponentially distributed* (so the standard deviation is also 1,200 hours). Use Minitab or any other statistical software to generate 500 random samples, each having size 50, from this population. If you use Minitab, save in columns C1 to C50. (Use Calc→Random Data→Exponential and follow the procedures in the Minitab lab for Chap. 6.)
   (a) Produce a histogram for column C1. Describe the shape of the distribution of C1 values. Does this shape represent the distribution of the lifetimes of the bulbs? (It should be a decreasing histogram.)
   (b) Generate the 500 sample means, using columns C1–C10, C1–C20, C1–C30, C1–C40, and C1–C50, and store in C51 to C55, respectively, as in previous exercises. Produce histograms for columns C51 to C55, and find the mean and standard deviation of these columns. Compare with what you would expect from the CLT.
   (c) Sort the values of C31 by selecting Manip→Sort. In the Sort column text box, type in C31. In the Store sorted column(s) in text box, type in C56. In the Sort by column text box, type in C31; and select the OK button. Using the values in C56, estimate the probability that the mean lifetime for these bulbs lies between 950 and 1,100 hours.
   (d) Work the probability exercise in part (c) by using the methods of the text (normal distribution table). Apply the central limit theorem with $n = 500$, $\mu = 1,200$, and $\sigma = 1,200$. Compare your answers in parts (c) and (d).

7. Write up a report for Example 3.

Estimation

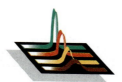

 e learn how to use sample data to estimate a population mean, a population variance, and a population proportion. We discuss point estimates, which are single-value estimates of the parameter. The standard error of these estimates is considered. We also consider interval estimates which contain the parameter with specified degrees of confidence.

8.1   INTRODUCTION

It would not be unusual to see in a daily newspaper that "a recent poll of 1500 randomly chosen Americans indicates that 22 percent of the entire U.S. population is presently dieting, with a margin of error of ± 2 percent." Perhaps you have wondered about such claims. For instance, what exactly does *with a margin of error of ± 2 percent* mean? Also how is it possible, in a nation of over 150 million adults, that the proportion of them presently on diets can be ascertained by a sampling of only 1500 people?

In this chapter we will find the answers to the above questions. In general, we will consider how one can learn about the numerical characteristics of a population by analyzing results from a sample of this population.

Whereas the numerical values of the members of the population can be summarized by a population probability distribution, this distribution is often not completely known. For instance, certain of its parameters, such as its mean or its standard deviation, may be unknown. A fundamental concern in statistics relates to how one can use the results from a sample of the population to estimate these unknown parameters.

For instance, if the items of the population consist of newly manufactured computer chips, then we may be interested in learning about the average functional lifetime of these chips. That is, we would be interested in estimating the population mean of the distribution of the lifetimes of these chips.

In this chapter we will consider ways of estimating certain parameters of the population distribution. To accomplish this, we will show how to use estimators and the estimates they give rise to.

Definition

An *estimator* is a statistic whose value depends on the particular sample drawn. The value of the estimator, called the *estimate,* is used to predict the value of a population parameter.

For instance, if we want to estimate the mean lifetime of a chip, then we could employ the sample mean as an *estimator* of the population mean. If the value of the sample mean were 122 hours, then the *estimate* of the population mean would be 122 hours.

In Sec. 8.2 we consider the problem of estimating a population mean; and in Sec. 8.3, a population proportion. Section 8.4 deals with estimating the population variance. The estimators considered in these sections are called *point estimators* because they are single values which we hope will be close to the parameters they are estimating. In the remaining sections, we consider the problem of obtaining *interval estimators*. In this case, rather than specifying a particular value as our estimate, we specify an interval in which we predict that the parameter lies. We also consider the question of how much confidence to attach to a given interval estimate, that is, how certain we can be that the parameter indeed lies within this interval.

8.2 POINT ESTIMATOR OF A POPULATION MEAN

Let $X_1, \ldots, X_n$ denote a sample from a population whose mean μ is unknown. The sample mean $\bar{X}$ can be used as an estimator of μ. Since, as was noted in Sec. 7.3,

$$E[\bar{X}] = \mu$$

we see that the expected value of this estimator is the parameter we want to estimate. Such an estimator is called *unbiased.*

Definition

An estimator whose expected value is equal to the parameter it is estimating is said to be an *unbiased* estimator of that parameter.

Example 8.1   To estimate the average amount of damages claimed in fires at medium-size apartment complexes, a consumer organization sampled the files of a large insurance company to come up with the following amounts (in thousands of dollars) for 10 claims:

   121, 55, 63, 12, 8, 141, 42, 51, 66, 103

The estimate of the mean amount of damages claimed in all fires of the type being considered is thus

$$\bar{X} = \frac{121 + 55 + 63 + 12 + 8 + 141 + 42 + 51 + 66 + 103}{10}$$

$$= \frac{662}{10} = 66.2$$

That is, we estimate that the mean fire damage claim is \$66,200.

   As we have shown, the sample mean $\bar{X}$ has expected value μ. Since a random variable is not likely to be too many standard deviations away from its expected value, it is important to determine the standard deviation of $\bar{X}$. However, as we have already noted in Sec. 7.3,

$$\text{SD}\,(\bar{X}) = \frac{\sigma}{\sqrt{n}}$$

where σ is the population standard deviation. The quantity $\text{SD}\,(\bar{X})$ is sometimes called the *standard error* of $\bar{X}$ as an estimator of the mean. Since a random variable is unlikely to be more than 2 standard deviations away from its mean (especially when that random variable is approximately normal, as $\overline{X}$ will be when the sample size n is large), we are usually fairly confident that the estimate of the population mean will be correct to within ± 2 standard errors. Note that the standard error decreases by the square root of the sample size; as a result, to cut the standard error in half, we must increase the sample size by a factor of 4.

Example 8.2   Successive tests for the level of potassium in an individual's blood vary because of the basic imprecision of the test and because the actual level itself varies depending on such things as the amount of food recently eaten and the amount of exertion recently undergone. Suppose it is known that, for a given individual, the successive readings of potassium level vary around a mean value μ with a standard deviation of .3. If a set of four readings on a particular individual yields the data

3.6, 3.9, 3.4, 3.5

then the estimate of the mean potassium level of that person is

$$\frac{3.6 + 3.9 + 3.4 + 3.5}{4} = 3.6$$

with the standard error of the estimate being equal to

$$SD\,(\overline{X}) = \frac{\sigma}{\sqrt{n}} = \frac{.3}{2} = .15$$

Therefore, we can be quite confident that the actual mean will not differ from 3.6 by more than .30.

Suppose we wanted the estimator to have a standard error of .05. Then, since this would be a reduction in standard error by a factor of 3, it follows that we would have had to choose a sample 9 times as large. That is, we would have had to take 36 blood potassium readings.

Problems

1. The weights of a random sample of eight participants in the 1990 Boston Marathon were as follows:

   121, 163, 144, 152, 186, 130, 128, 140

   Use these data to estimate the average weight of all the participants in this race.

2. Suppose, in Prob. 1, that the data represented the weights of the top eight finishers in the marathon. Would you still be able to use these data to estimate the average weight of all the runners? Explain!

3. To determine the average amount of money spent by university students on textbooks, a random sample of 13 students was chosen, and the students were questioned. If the amounts (to the nearest dollar) spent were

     122, 46, 168, 52, 212, 154, 166, 103, 97, 111, 44, 51, 73

   what is your estimate of the average amount spent by all students at the university?

4. A random sample of nine preschoolers from a given neighborhood yielded the following data concerning the number of hours per day each one spent watching television.

     3, 0, 5, 3.5, 1.5, 2, 3, 2.5, 2

   Estimate the average number of hours per day spent watching television by preschoolers in that neighborhood.

5. A manufacturer of compact disk players wants to estimate the average lifetime of the lasers in its product. A random sample of 40 is chosen. If the sum of the lifetimes of these lasers is 6624 hours, what is the estimate of the average lifetime of a laser?

6. A proposed study for estimating the average cholesterol level of working adults calls for a sample size of 1000. If we want to reduce the resulting standard error by a factor of 4, what sample size is necessary?

7. It is known that the standard deviation of the weight of a newborn child is 10 ounces. If we want to estimate the average weight of a newborn, how large a sample will be needed for the standard error of the estimate to be less than 3 ounces?

8. The following data represent the number of minutes each of a random sample of 12 recent patients at a medical clinic spent waiting to see a physician.

     46, 38, 22, 54, 60, 36, 44, 50, 35, 66, 48, 30

   Use these data to estimate the average waiting time of all patients at this clinic.

9. The following frequency table gives the household sizes of a random selection of 100 single-family households in a given city.

Household size	Frequency
1	11
2	19
3	28
4	26
5	11
6	4
7	1

Estimate the average size of all single-family households in the city.

10. Does (a) or (b) yield a more precise estimator of μ?
   (a) The sample mean of a sample of size n from a population with mean μ and variance σ^2
   (b) The sample mean of a sample of size $3n$ from a population with mean μ and variance $2\sigma^2$
   (c) How large would the sample in (b) have to be in order to match the precision of the estimator in (a)?

11. Repeat Prob. 10 when (a) and (b) are as follows:
   (a) The sample mean of a sample of size n from a population with mean μ and standard deviation σ
   (b) The sample mean of a sample of size $3n$ from a population with mean μ and standard deviation 3σ

8.3   POINT ESTIMATOR OF A POPULATION PROPORTION

Suppose that we are trying to estimate the proportion of a large population that is in favor of a given proposition. Let p denote the unknown proportion. To estimate p, a random sample should be chosen, and then p should be estimated by the proportion of the sample that is in favor. Calling this estimator $\hat{p}$, we can express it by

$$\hat{p} = \frac{X}{n}$$

where X is the number of members of the sample who are in favor of the proposition and n is the size of the sample.

From the results of Sec. 7.5, we know that

$$E[\hat{p}] = p$$

That is, $\hat{p}$, the proportion of the sample in favor of the proposition, is an unbiased estimator of p, the proportion of the entire population that is in favor. The spread of the estimator $\hat{p}$ about its mean p is measured by its standard deviation, which (again from Sec. 7.5) is equal to

$$\mathrm{SD}\,(\hat{p}) = \sqrt{\frac{p(1 - p)}{n}}$$

The standard deviation of $\hat{p}$ is also called the *standard error* of $\hat{p}$ as an estimator of the population proportion p. By the above formula this standard error will be small whenever the sample size n is large. In fact, since it can be shown that for every value of p

$$p(1 - p) \leq \frac{1}{4}$$

it follows that

$$SD\ (\hat{p}) \leq \sqrt{\frac{1}{4n}} = \frac{1}{2\sqrt{n}}$$

For instance, suppose a random sample of size 900 is chosen. Then no matter what proportion of the population is actually in favor of the proposition, it follows that the standard error of the estimator of this proportion is less than or equal to $1/(2\sqrt{900}) = 1/60$.

The above formula and bound on the standard error assume that we are drawing a random sample of size n from an infinitely large population. When the population size is smaller (as, of course, it will be in practice), then so is the standard error, thus making the estimator even more precise than indicated above.

Example 8.3 A school district is trying to determine its students' reaction to a proposed dress code. To do so, the school selected a random sample of 50 students and questioned them. If 20 were in favor of the proposal, then

(a) Estimate the proportion of all students who are in favor.

(b) Estimate the standard error of the estimate.

Solution

(a) The estimate of the proportion of all students who are in favor of the dress code is $20/50 = .40$.

(b) The standard error of the estimate is $\sqrt{p(1 - p)/50}$, where p is the actual proportion of the entire population that is in favor. Using the estimate for p of .4, we can estimate this standard error by $\sqrt{.4(1 - .4)/50} = .0693$.

Problems

1. In 1985, out of a random sample of 1325 North Americans questioned, 510 said that the Communist party would win a free election if one were held in the Soviet Union. Estimate the proportion of all North Americans who felt the same way at that time.

2. Estimate the standard error of the estimate in Prob. 1.

3. To learn the percentage of members who are in favor of increasing annual dues, a large social organization questioned a randomly chosen sample of 20 members. If 13 members were in favor, what is the estimate of the proportion of all members who are in favor? What is the estimate of the standard error?

4. The following are the results of 20 games of solitaire, a card game that results in either a win (*w*) or a loss (*l*).

   w, l, l, l, w, l, l, w, l, w, w, l, l, l, l, w, l, l, w, l

   (a) Estimate the probability of winning a game of solitaire.
   (b) Estimate the standard error of the estimate in part (a).

5. A random sample of 85 students at a large public university revealed that 35 students owned a car that was less than 5 years old. Estimate the proportion of all students at the university who own a car less than 5 years old. What is the estimate of the standard error of this estimate?

6. A random sample of 100 parents found that 64 are in favor of raising the driving age to 18.
   (a) Estimate the proportion of the entire population of parents who are in favor of raising the driving age to 18.
   (b) Estimate the standard error of the estimate.

7. A random sample of 1000 construction workers revealed that 122 are presently unemployed.
   (a) Estimate the proportion of all construction workers who are unemployed.
   (b) Estimate the standard error of the estimate in part (a).

8. Out of a random sample of 500 architects, 104 were women.
   (a) Estimate the proportion of all architects who are women.
   (b) Estimate the standard error of the estimate in part (a).

9. A random sample of 1200 engineers included 28 Hispanic Americans, 45 African Americans, and 104 females. Estimate the proportion of all engineers who are
   (a) Hispanic American
   (b) African American
   (c) Female

10. In parts (a), (b), and (c) of Prob. 9, estimate the standard error of the estimate.

11. A random sample of 400 death certificates related to teenagers yielded that 98 had died due to a motor vehicle accident.
    (a) Estimate the proportion of all teenage deaths due to motor vehicle accidents.
    (b) Estimate the standard error of the estimate in part (a).

12. A survey is being planned to discover the proportion of the population that is in favor of a new school bond. How large a sample is needed in order to be certain that the standard error of the resulting estimator is less than or equal to .1?

13. Los Angeles has roughly 3 times the voters of San Diego. Each city will be voting on a local education bond initiative. To determine the sentiments of the voters, a random sample of 3000 Los Angeles voters and a random sample of 1000 San Diego voters will be queried. Of the following statements, which is most accurate?
    (a) The resulting estimates of the proportions of people who will vote for the bonds in the two cities were equally accurate.
    (b) The Los Angeles estimate is 3 times as accurate.
    (c) The Los Angeles estimate is roughly 1.7 times as accurate.
    Explain how you are interpreting the word *accurate* is statements (a), (b), and (c).

*14. The city of Chicago had 12,048 full-time law enforcement officers in 1990. To determine the number of African Americans in this group, a random sample of 600 officers was chosen, and it was discovered that 87 were African Americans.
    (a) Estimate the number of African American law enforcement officers who were employed full-time in Chicago in 1990.
    (b) Estimate the standard error of the estimate of part (a).

*8.3.1 Estimating the Probability of a Sensitive Event

Suppose that a company is interested in learning about the extent of illegal drug use among its employees. However, the company recognizes that employees might be reluctant to truthfully answer questions on this subject even if they have been assured that their answers will be kept in confidence. Indeed, even if the company assures workers that responses will not be traced to particular individuals, the employees might still remain suspicious and not answer truthfully. Given this background, how can the company elicit the desired information?

We now present a method which will enable the company to gather the desired information while at the same time protecting the privacy of those questioned. The method is to employ a randomization technique, and it works as follows: To begin, suppose that the sensitive question is stated in such a way that *yes* is the sensitive answer. For instance, the question could be, Have you used any illegal drugs in the past month? Presumably if the true answer is no, then the worker will not hesitate to give that answer. However, if the real answer is yes, then some workers may still answer no. To relieve any pressure to lie, the following rule for answering should be explained to each worker before the questioning begins: After the question has been posed, the worker is to flip a fair coin, not allowing the questioner to see the result of the flip. If the coin lands on heads, then the worker should answer yes to the question; and if it lands on tails, then the worker should answer the question honestly. It should be explained to the worker that an answer of yes does not mean that he or she is admitting to having used illegal drugs, since that answer may have resulted solely from the coin flip's landing on heads (which will occur 50 percent of the time). In this manner the workers sampled should feel assured that they can play the game truthfully and, at the same time, preserve their privacy.

Let us now analyze the above to see how it can be used to estimate p, the proportion of the workforce that has actually used an illegal drug in the past month. Let $q = 1 - p$ denote the proportion that has not. Let us start by computing the probability that a sampled worker will answer no to the question. Since this will occur only if both (1) the coin toss lands on tails and (2) the worker has not used any illegal drugs in the past month, we see that

$$P\{no\} = \frac{1}{2} \times q = \frac{q}{2}$$

Hence, we can take the fraction of workers sampled who answered no as our estimate of $q/2$; or, equivalently, we can estimate q to be twice the proportion who answered no. Since $p = 1 - q$, this will also result in an estimate of p, the proportion of all workers who have used an illegal drug in the past month.

For instance, if 70 percent of the workers sampled answered the question in the affirmative, and so 30 percent answered no, then we would estimate that q was equal to $2(.3) = .6$. That is, we would estimate that 60 percent of the population has not, and so 40 percent of the population has, used an illegal drug in the past month. If 35 percent of the workers answered no, then we would estimate that q was equal to $2(.35) = .7$ and thus that $p = .3$. Similarly, if 48 percent of the workers answered no, then our estimate of p would be $1 - 2(.48) = .04$.

Thus, by this trick of having each respondent flip a coin, we are able to obtain an estimate of p. However, the "price" we pay is an increased value of the standard error. Indeed, it can be shown that the standard error of the estimator of p is now $\sqrt{(1 + p)(1 - p)/n}$, which is larger than the standard error of the estimator when there is no need to use a coin flip (because all answers will be honestly given).

Problems

1. Suppose the randomization scheme described in this section is employed. If a sample of 50 people results in 32 yes answers, what is the estimate of p?

2. In Prob. 1, what would your estimate of p be if 40 of the 50 people answered yes?

3. When the randomization technique is used, the standard error of the estimator of p is $\sqrt{(1 + p)(1 - p)/n}$. Now, if there was no need to use the randomization technique, because everyone always answered honestly, then the standard error of the estimator of p would be $\sqrt{p(1 - p)/n}$. The ratio of these standard errors is thus

$$\frac{\text{Standard error with randomization}}{\text{Usual standard error}} = \sqrt{\frac{1 + p}{p}}$$

The above ratio is thus an indicator of the price one must pay because of the sensitivity of the question.

(a) Do you think this price would be higher for large or small values of p?

(b) Determine the value of this ratio for $p = .1, .5,$ and $.9$.

8.4   ESTIMATING A POPULATION VARIANCE

Suppose that we have a sample of size n, $X_1, \ldots, X_n$, from a population whose variance σ^2 is unknown, and we are interested in using the sample data to estimate σ^2. The sample variance S^2, defined by

$$S^2 = \frac{\sum_{i=1}^{n} (X_i - \bar{X})^2}{n - 1}$$

is an estimator of the population variance σ^2. To understand why, recall that the population variance is the expected squared difference between an observation and the population mean μ. That is, for $i = 1, \ldots, n$,

$$\sigma^2 = E[(X_i - \mu)^2]$$

Thus, it seems that the natural estimator of σ^2 would be the average of the squared differences between the data and the population mean μ. That is, it seems that the appropriate estimator of σ^2 would be

$$\frac{\sum_{i=1}^{n} (X_i - \mu)^2}{n}$$

The above is indeed the appropriate estimator of σ^2 when the population mean μ is known. However, if the population mean μ is also unknown, then it is reasonable to use the above expression with μ replaced by its estimator, namely, $\bar{X}$. To keep the estimator unbiased, this also leads us to change the denominator from n to $n - 1$; and thus we obtain the estimator S^2.

If the population mean μ is known, then the appropriate estimator of the population variance σ^2 is

$$\frac{\sum\limits_{i=1}^{n} (X_i - \mu)^2}{n}$$

If the population mean μ is unknown, then the appropriate estimator of the population variance σ^2 is

$$S^2 = \frac{\sum\limits_{i=1}^{n} (X_i - \bar{X})^2}{n - 1}$$

S^2 is an unbiased estimator of σ^2, that is,

$$E[S^2] = \sigma^2$$

Since the sample variance S^2 will be used to estimate the population variance σ^2, it is natural to use $\sqrt{S^2}$ to estimate the population standard deviation σ.

The population standard deviation σ is estimated by S, the sample standard deviation.

Example 8.4   A random sample of nine electronic components produced by a certain company yields the following sizes (in suitable units):

1211, 1224, 1197, 1208, 1220, 1216, 1213, 1198, 1197

What are the estimates of the population standard deviation and the population variance?

Solution To answer this, we need to compute the sample variance S^2. Since subtracting a constant value from each data point will not affect the value of this statistic, start by subtracting 1200 from each datum to obtain the following transformed data set:

11, 24, −3, 8, 20, 16, 13, −2, −3

Statistics in Perspective

According to Japanese quality control experts, the key to a successful manufacturing process—whether one is producing automobile parts, electronic equipment, computer chips, screws, or anything else—is to ensure that the production process consistently produces, at a reasonable cost, items that have values close to their *target* values. By this they mean that for any item being produced there is always a certain target value that the manufacturer is shooting at. For instance, when car doors are produced, there is a target value for the door's width. To be competitive, the widths of the doors produced must be consistently close to this value. These experts say that the key to producing items close to the target value is to ensure that the variance of the items produced is minimal. That is, once

Variance Reduction Is the Key to Success in Manufacturing

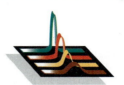

a production process has been established that produces items whose values have a small variance, then the difficult part of reaching the goal of consistently producing items whose values are near the target value has been accomplished.

Experience has led these experts to conclude that it is then a relatively simple matter to fine-tune the process so that the mean value of the item is close to the target value. (For an analogy, these experts are saying that if you want to build a rifle that will enable a shooter to consistently hit a particular target, then you should first concentrate on building a rifle that is extremely stable and will always give the same result when it is pointed in the same direction, and then you should train the shooter to shoot straight.)

Using a calculator on these transformed data shows that the values of the sample variance and sample standard deviation are

$$S^2 = 103 \qquad S = 10.149$$

Therefore, the respective estimates of the population standard deviation and the population variance are 10.149 and 103.

Problems

1. A survey was undertaken to learn about the variation in the weekly number of hours worked by university professors. A sample of 10 professors yielded the following data:

   48, 22, 19, 65, 72, 37, 55, 60, 49, 28

   Use these data to estimate the population standard deviation of the number of hours that college professors work in a week.

2. The following data refer to the widths (in inches) of slots on nine successively produced duralumin forgings, which will be used as a terminal block at the end of an airplane wing span.

   8.751, 8.744, 8.749, 8.750, 8.752, 8.749, 8.764, 8.746, 8.753

   Estimate the mean and the standard deviation of the width of a slot.

3. The following data refer to the amounts (in tons) of chemicals produced daily at a chemical plant. Use them to estimate the mean and the variance of the daily production.

   776, 810, 790, 788, 822, 806, 795, 807, 812, 791

4. Consistency is of great importance in manufacturing baseballs, for one does not want the balls to be either too lively or too dead. The balls are tested by dropping them from a standard height and then measuring how high they bounce. If a sample of 30 balls resulted in the following summary statistics

$$\sum_{i=1}^{30} X_i = 52.1 \qquad \sum_{i=1}^{30} X_i^2 = 136.2$$

   estimate the standard deviation of the size of the bounce. *Hint:* Recall the identity

$$\sum_{i=1}^{n} (x_i - \bar{x})^2 = \sum_{i=1}^{n} x_i^2 - n\bar{x}^2$$

5. Use the data of Prob. 1 of Sec. 8.2 to estimate the standard deviation of the weights of the runners in the 1990 Boston Marathon.

Problems 6, 7, and 8 refer to the following sample data:

   104, 110, 114, 97, 105, 113, 106, 101, 100, 107

6. Estimate the population mean μ and the population variance σ^2.

7. Suppose it is known that the population mean is 104. Estimate the population variance.

8. Suppose it is known that the population mean is 106. Estimate the population standard deviation.

9. Use the data of Prob. 8 of Sec. 8.2 to estimate the standard deviation of the waiting times of patients at the medical clinic.

10. A manufacturer of furniture wants to test a sample of newly developed fire-resistant chairs to learn about the distribution of heat that these chairs can sustain before starting to burn. A sample of seven chairs is chosen, and each is

put, one at a time, in a closed burn room. Once a chair is placed in this room, its temperature is increased, one degree at a time, until the chair bursts into flames. Suppose the burn temperatures for the seven chairs are (in degrees Fahrenheit) as follows:

458, 440, 482, 455, 491, 477, 446

(a) Estimate the mean burn temperature of this type of chair.
(b) Estimate the standard deviation of the burn temperature of this type of chair.

11. Use the data of Prob. 9 of Sec. 8.2 to estimate the standard deviation of the size of a single-family household in the city considered.

12. Suppose that the systolic blood pressure of a worker in the mining industry is normally distributed. Suppose also that a random sample of 13 such workers yielded the following blood pressures:

129, 134, 142, 114, 120, 116, 133, 142, 138, 148, 129, 133, 141

(a) Estimate the mean systolic blood pressure of all miners.
(b) Estimate the standard deviation of the systolic blood pressure.
(c) Use the estimates in parts (a) and (b) along with the fact that the blood pressures are normally distributed to obtain an estimate of the proportion of all miners whose blood pressure exceeds 150.

13. The linear random walk model for the successive daily prices of a stock or commodity supposes that the successive differences of the end-of-day prices of a given stock constitute a random sample from a normal population. The following 20 data values represent the closing prices of crude oil on the New York Mercantile Exchange on 20 consecutive trading days in 1994. Assuming the linear random walk model, use these data to estimate the mean and standard deviation of the population distribution. (Note that the data give rise to 19 values from this distribution, the first being $17.60 - 17.50 = .10$, the second being $17.81 - 17.60 = .21$, and so on.)

17.50, 17.60, 17.81, 17.67, 17.53, 17.39, 17.12, 16.71, 16.70, 16.83,
17.21, 17.24, 17.22, 17.67, 17.83, 17.67, 17.55, 17.68, 17.98, 18.39

*14. Due to a lack of precision in the scale used, the value obtained when a fish is weighed is normal with mean equal to the actual weight of the fish and with standard deviation equal to .1 gram. A sample of 12 *different* fish was chosen, and the fish were weighed, with the following results:

5.5, 6.2, 5.8, 5.7, 6.0, 6.2, 5.9, 5.8, 6.1, 6.0, 5.7, 5.6

Estimate the population standard deviation of the actual weight of a fish.

Hint: First note that, due to the error involved in weighing a fish, each data value is not the true weight of a fish, but rather is the true weight plus an error term. This error term is an independent random variable that has mean 0 and standard deviation .1. Therefore,

Data = true weight + error

and so

Var (data) = Var (true weight) + Var (error)

To determine the variance of the true weight, first estimate the variance of the data.

8.5   INTERVAL ESTIMATORS OF THE MEAN OF A NORMAL POPULATION WITH KNOWN POPULATION VARIANCE

When we estimate a parameter by a point estimator, we do not expect the resulting estimator to exactly equal the parameter, but we expect that it will be "close" to it. To be more specific, we sometimes try to find an interval about the point estimator in which we can be highly confident that the parameter lies. Such an interval is called an *interval estimator.*

Definition

An *interval estimator* of a population parameter is an interval which is predicted to contain the parameter. The *confidence* we ascribe to the interval is the probability that it will contain the parameter.

To determine an interval estimator of a population parameter, we use the probability distribution of the point estimator of that parameter. Let us see how this works in the case of the interval estimator of a normal mean when the population standard deviation is assumed known.

Let $X_1, \ldots, X_n$ be a sample of size n from a normal population having known standard deviation σ, and suppose we want to utilize this sample to obtain a 95 percent confidence interval estimator for the population mean μ. To obtain such an interval, we start with the sample mean $\bar{X}$, which is the point estimator of μ. We now make use of the fact that $\bar{X}$ is normal with mean μ and standard deviation $\sigma/\sqrt{n}$, which implies that the standardized variable

$$Z = \frac{\bar{X} - \mu}{\sigma/\sqrt{n}} = \sqrt{n}\frac{\bar{X} - \mu}{\sigma}$$

has a standard normal distribution. Now, since $z_{.025} = 1.96$, it follows that 95 percent of the time the absolute value of Z is less than or equal to 1.96 (see Fig. 8.1).

Thus, we can write

$$P\left\{\frac{\sqrt{n}}{\sigma}\,|\bar{X} - \mu| \leq 1.96\right\} = .95$$

Upon multiplying both sides of the inequality by $\sigma/\sqrt{n}$, we see that the preceding equation is equivalent to

$$P\left\{|\bar{X} - \mu| \leq 1.96\,\frac{\sigma}{\sqrt{n}}\right\} = .95$$

From the preceding statement we see that, with 95 percent probability, μ and $\bar{X}$ will be within $1.96\,\sigma/\sqrt{n}$ of each other. But this is equivalent to stating that

$$P\left\{\bar{X} - 1.96\,\frac{\sigma}{\sqrt{n}} \leq \mu \leq \bar{X} + 1.96\,\frac{\sigma}{\sqrt{n}}\right\} = .95$$

That is, with 95 percent probability, the interval $\bar{X} \pm 1.96\,\sigma/\sqrt{n}$ will contain the population mean.

The interval from $\bar{X} - 1.96\,\sigma/\sqrt{n}$ to $\bar{X} + 1.96\,\sigma/\sqrt{n}$ is said to be a *95 percent confidence interval estimator* of the population mean μ. If the observed value of $\bar{X}$ is $\bar{x}$, then we call the interval $\bar{x} \pm 1.96\,\sigma/\sqrt{n}$ a *95 percent confidence interval estimate* of μ.

In the long run, 95 percent of the interval estimates so constructed will contain the mean of the population from which the sample is drawn.

Example 8.5   Suppose that if a signal having intensity μ originates at location A, then the intensity recorded at location B is normally distributed with mean μ and standard deviation 3. That is, due to "noise," the intensity recorded differs from the actual intensity of the signal by an amount that is normal with mean 0 and standard deviation 3. To reduce the error, the same signal is independently recorded 10 times. If the successive recorded values are

17, 21, 20, 18, 19, 22, 20, 21, 16, 19

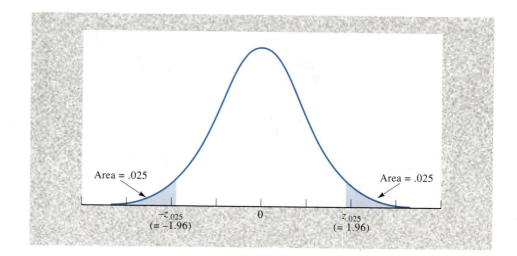

Figure 8.1
$P\{|Z| \leq 1.96\}$
$= P\{-1.96$
$\leq Z \leq 1.96\}$
$= .95.$

construct a 95 percent confidence interval for μ, the actual intensity.

Solution The value of the sample mean is

$$\frac{17 + 21 + 20 + 18 + 19 + 22 + 20 + 21 + 16 + 19}{10} = 19.3$$

Since $\sigma = 3$, it follows that a 95 percent confidence interval estimate of μ is given by

$$19.3 \pm 1.96 \, \frac{3}{\sqrt{10}} = 19.3 \pm 1.86$$

That is, we can assert with 95 percent confidence that the actual intensity of the signal lies between 17.44 and 21.16. A picture of this confidence interval estimate is given in Fig. 8.2. ▪ ▪

We can also consider confidence interval estimators having confidence levels different from .95. Recall that for any value of α between 0 and 1, the probability that a standard normal lies in the interval between $-z_{\alpha/2}$ and $z_{\alpha/2}$ is equal to $1 - \alpha$ (see Fig. 8.3). From this it follows that

$$P\left\{\frac{\sqrt{n}}{\sigma} \, |\bar{X} - \mu| \leq z_{\alpha/2}\right\} = 1 - \alpha$$

By the same logic used previously when $\alpha = .05$ ($z_{.025} = 1.96$), we can show that, with probability $1 - \alpha$, μ will lie in the interval $\bar{X} \pm z_{\alpha/2}\sigma/\sqrt{n}$.

Figure 8.2
Confidence
interval
estimate of μ
for Example
8.5.

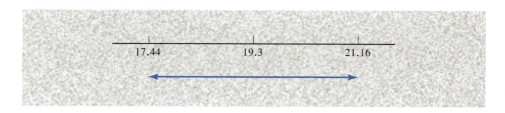

Figure 8.3
$P\{|Z| \leq z_{\alpha/2}\}$
$= P\{-z_{\alpha/2}$
$\leq Z \leq z_{\alpha/2}\}$
$= 1 - \alpha$

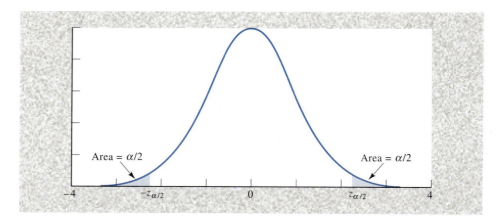

The interval $\bar{X} \pm z_{\alpha/2}\sigma/\sqrt{n}$ is called a $100(1 - \alpha)$ percent confidence interval estimator of the population mean.

Table 8.1 lists the values of $z_{\alpha/2}$ needed to construct 90, 95, and 99 percent confidence interval estimates of μ.

Example 8.6   Determine, for the data of Example 8.5,

(a) A 90 percent confidence interval estimate of μ

(b) A 99 percent confidence interval estimate of μ

Solution We are being asked to construct a $100(1 - \alpha)$ confidence interval estimate, with $\alpha = .10$ in part (a) and $\alpha = .01$ in part (b). Now

$$z_{.05} = 1.645 \qquad \text{and} \qquad z_{.005} = 2.576$$

and so the 90 percent confidence interval estimator is

$$\bar{X} \pm 1.645 \frac{\sigma}{\sqrt{n}}$$

Table 8.1	Confidence level $100(1 - \alpha)$	Corresponding value of α	Value of $z_{\alpha/2}$
	90	.10	$z_{.05} = 1.645$
	95	.05	$z_{.025} = 1.960$
	99	.01	$z_{.005} = 2.576$

and the 99 percent confidence interval estimator is

$$\bar{X} \pm 2.576 \frac{\sigma}{\sqrt{n}}$$

For the data of Example 8.5, $n = 10$, $\bar{X} = 19.3$, and $\sigma = 3$. Therefore, the 90 and 99 percent confidence interval estimates for μ are, respectively,

$$19.3 \pm 1.645 \frac{3}{\sqrt{10}} = 19.3 \pm 1.56$$

and

$$19.3 \pm 2.576 \frac{3}{\sqrt{10}} = 19.3 \pm 2.44$$

Figure 8.4 indicates the 90, 95, and 99 percent confidence interval estimates of μ. Note that the larger the confidence coefficient $100(1 - \alpha)$, the larger the length of this interval. This makes sense because if you want to increase your certainty that the parameter lies in a specified interval, then you will clearly have to enlarge that interval. ∎

Figure 8.4
The 90, 95, and 99 percent confidence interval estimates.

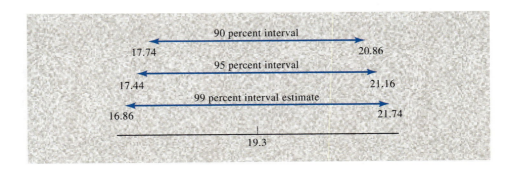

Sometimes we are interested in obtaining a $100(1 - \alpha)$ percent confidence interval whose length is less than or equal to some specified value, and the problem is to choose the appropriate sample size. For instance, suppose we want to determine an interval of length at most b which, with 95 percent certainty, contains the population mean. How large a sample is needed? To answer this, note that since $z_{.025} = 1.96$, a 95 percent confidence interval for μ based on a sample of size n is (see Fig. 8.5)

$$\bar{X} \pm 1.96 \, \frac{\sigma}{\sqrt{n}}$$

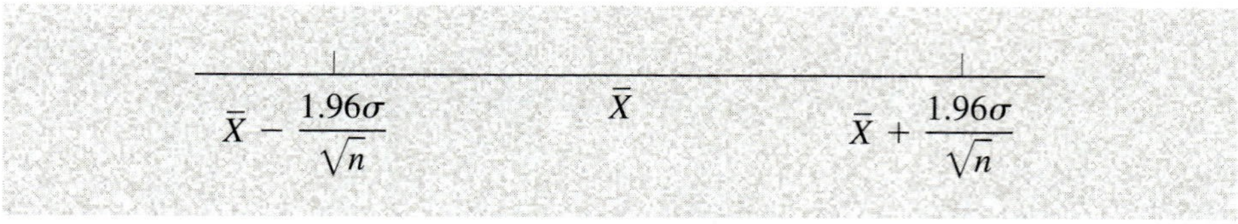

Figure 8.5
The 95 percent confidence interval for μ.

Since the length of this interval is

$$\text{Length of interval} = 2(1.96) \, \frac{\sigma}{\sqrt{n}} = 3.92 \, \frac{\sigma}{\sqrt{n}}$$

we must choose n so that

$$\frac{3.92\sigma}{\sqrt{n}} \leq b$$

or, equivalently,

$$\sqrt{n} \geq \frac{3.92\sigma}{b}$$

Upon squaring both sides of the above, we see that the sample size n must be chosen so that

$$n \geq \left(\frac{3.92\sigma}{b} \right)^2$$

Example 8.7   If the population standard deviation is $\sigma = 2$ and we want a 95 percent confidence interval estimate of the mean μ that is of size less than or equal to $b = .01$, how large a sample is needed?

Solution   We have to select a sample of size n where

$$n \geq \left(\frac{3.92 \times 2}{.1} \right)^2 = (78.4)^2 = 6146.6$$

That is, a sample of size 6147 or larger is needed.

The analysis for determining the required sample size so that the length of a $100(1 - \alpha)$ percent confidence interval is less than or equal to b is exactly the same as given when $\alpha = .05$. The result is as follows:

Determining the necessary sample size

The length of the $100(1 - \alpha)$ percent confidence interval estimator of the population mean will be less than or equal to b when the sample size n satisfies

$$n \geq \left(\frac{2z_{\alpha/2}\sigma}{b}\right)^2$$

The confidence interval estimator is

$$\bar{X} \pm z_{\alpha/2}\frac{\sigma}{\sqrt{n}}$$

Example 8.8   From past experience it is known that the weights of salmon grown at a commercial hatchery are normal with a mean that varies from season to season but with a standard deviation that remains fixed at .3 pound. If we want to be 90 percent certain that our estimate of the mean weight of a salmon is correct to within $\pm.1$ pound, how large a sample is needed? What if we want to be 99 percent certain?

Solution   Since the 90 percent confidence interval estimator from a sample of size n will be $\bar{X} \pm 1.645\sigma/\sqrt{n}$, it follows that we can be 90 percent confident that the point estimator $\bar{X}$ will be within $\pm.1$ of μ whenever the length of this confidence interval is less than or equal to .2 (see Fig. 8.6). Hence, from the preceding we see that n must be chosen so that

$$n \geq \left(\frac{2 \times 1.645 \times .3}{.2}\right)^2 = 24.35$$

That is, a sample size of at least 25 is required.

On the other hand, if we wanted to be 99 percent certain that $\bar{X}$ will be within .1 pound of the true mean, then since $z_{.005} = 2.576$, the sample size n would need to satisfy

$$n \geq \left(\frac{2 \times 2.576 \times .3}{.2}\right)^2 = 59.72$$

That is, a sample of size 60 or more is needed.

Figure 8.6
Confidence interval centered at $\bar{X}$. If length of interval is $2L$, then $\bar{X}$ is within L of any point in the interval.

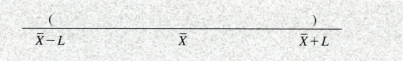

In deriving confidence interval estimators of a normal mean whose variance σ^2 is known, we used the fact that $\bar{X}$ is normally distributed with mean μ and standard deviation $\sigma/\sqrt{n}$. However, by the central limit theorem, this will remain approximately true for the sample mean of any population distribution provided the sample size n is relatively large ($n \geq 30$ is almost always sufficiently large). As a result, we can use the interval $\bar{X} \pm z_{\alpha/2}\, \sigma/\sqrt{n}$ as a $100(1 - \alpha)$ percent confidence interval estimator of the population mean for any population provided the sample size is large enough for the central limit theorem to apply.

Example 8.9   To estimate μ, the average nicotine content of a newly marketed cigarette, 44 of these cigarettes are randomly chosen, and their nicotine contents are determined.

(a) If the average nicotine finding is 1.74 milligrams, what is a 95 percent confidence interval estimator of μ?

(b) How large a sample is necessary for the length of the 95 percent confidence interval to be less than or equal to .3 milligram?

Assume that it is known from past experience that the standard deviation of the nicotine content of a cigarette is equal to .7 milligram.

Solution

(a) Since 44 is a large sample size, we do not have to suppose that the population distribution is normal to assert that a 95 percent confidence interval estimator of the population mean is

$$\bar{X} \pm z_{.025}\, \frac{\sigma}{\sqrt{n}}$$

In this case, the above reduces to

$$1.74 \pm \frac{1.96(.7)}{\sqrt{44}} = 1.74 \pm .207$$

That is, we can assert with 95 percent confidence that the average amount of nicotine per cigarette lies between 1.533 and 1.947 milligrams.

(b) The length of the 95 percent confidence interval estimate will be less than or equal to .3 if the sample size n is large enough that

$$n \geq \left(\frac{2 \times 1.96 \times .7}{.3} \right)^2 = 83.7$$

That is, a sample size of at least 84 is needed.

*8.5.1   Lower and Upper Confidence Bounds

Sometimes we are interested in making a statement to the effect that a population mean is, with a given degree of confidence, greater than some stated value. To obtain such a *lower confidence bound* for the population mean, we again use the fact that

$$Z = \sqrt{n} \frac{\bar{X} - \mu}{\sigma}$$

has a standard normal distribution. As a result, it follows that (see Fig. 8.7)

$$P\left\{ \sqrt{n} \frac{\bar{X} - \mu}{\sigma} < z_\alpha \right\} = 1 - \alpha$$

which can be rewritten as

$$P\left\{ \mu > \bar{X} - z_\alpha \frac{\sigma}{\sqrt{n}} \right\} = 1 - \alpha$$

From this equation, we can conclude the following:

A $100(1 - \alpha)$ percent lower confidence bound for the population mean μ is given by

$$\bar{X} - z_\alpha \frac{\sigma}{\sqrt{n}}$$

That is, with $100(1 - \alpha)$ percent confidence, we can assert that

$$\mu > \bar{X} - z_\alpha \frac{\sigma}{\sqrt{n}}$$

Example 8.10   Suppose in Example 8.8 that we want to specify a value which, with 95 percent confidence, is less than the average weight of a salmon. If a sample of 50 salmon yields an average weight of 5.6 pounds, determine this value.

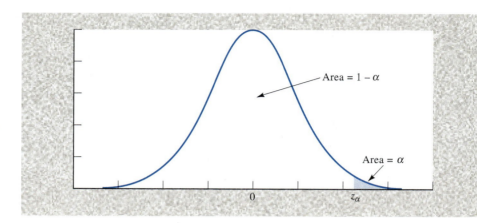

Figure 8.7
$P\{Z \le z_\alpha\} = 1 - \alpha$

Solution We are asked to find a 95 percent lower confidence bound for μ. By the preceding analysis this will be given by

$$\overline{X} - z_{.05} \frac{\sigma}{\sqrt{n}}$$

Since $z_{.05} = 1.645$, $\sigma = .3$, $n = 50$, and $\overline{X} = 5.6$, the lower confidence bound will equal

$$5.6 - \frac{1.645 \; .3}{\sqrt{50}} = 5.530$$

That is, we can assert, with 95 percent confidence, that the mean weight of a salmon is greater than 5.530 pounds.

We can also derive a $100(1 - \alpha)$ percent upper confidence bound for μ. The result is the following.

A $100(1 - \alpha)$ percent upper confidence bound for the population mean μ is given by

$$\overline{X} + z_\alpha \frac{\sigma}{\sqrt{n}}$$

That is, with $100(1 - \alpha)$ percent confidence, we can assert that

$$\mu < \overline{X} + z_\alpha \frac{\sigma}{\sqrt{n}}$$

Example 8.11   In Example 8.9, find a 95 percent upper confidence bound for μ.

Solution A 95 percent upper confidence bound is given by

$$\bar{X} + z_{.05} \frac{\sigma}{\sqrt{n}} = 1.74 + 1.645 \frac{.7}{\sqrt{44}} = 1.914$$

That is, we can assert with 95 percent confidence that the average nicotine content is less than 1.914 milligrams.

Problems

1. An electric scale gives a reading equal to the true weight plus a random error that is normally distributed with mean 0 and standard deviation $\sigma = .1$ ounce. Suppose that the results of five successive weighings of the same object are as follows: 3.142, 3.163, 3.155, 3.150, 3.141.
   (a) Determine a 95 percent confidence interval estimate of the true weight.
   (b) Determine a 99 percent confidence interval estimate of the true weight.

2. Suppose that a hospital administrator states that a statistical experiment has indicated that "With 90 percent certainty, the average weight at birth of all boys born at the certain hospital is between 6.6 and 7.2 pounds." How would you interpret this statement?

3. The polychlorinated biphenyl (PCB) concentration of a fish caught in Lake Michigan was measured by a technique that is known to result in an error of measurement that is normally distributed with standard deviation .08 part per million. If the results of 10 independent measurements of this fish are

   11.2, 12.4, 10.8, 11.6, 12.5, 10.1, 11.0, 12.2, 12.4, 10.6

   give a 95 percent confidence interval estimate of the PCB level of this fish.

4. Suppose in Prob. 3 that 40 measurements are taken, with the same average value resulting as in Prob. 3. Again determine a 95 percent confidence interval estimate of the PCB level of the fish tested.

5. The life of a particular brand of television picture tube is known to be normally distributed with a standard deviation of 400 hours. Suppose that a random sample of 20 tubes resulted in an average lifetime of 9000 hours. Obtain a
   (a) 90 percent
   (b) 95 percent
   confidence interval estimate of the mean lifetime of such a tube.

6. An engineering firm manufactures a space rocket component that will function for a length of time that is normally distributed with a standard deviation of 3.4 hours. If a random sample of nine such components has an average life of 10.8 hours, find a

(a) 95 percent
(b) 99 percent
confidence interval estimate of the mean length of time that these components function.

7. The standard deviation of test scores on a certain achievement test is 11.3. A random sample of 81 students had a sample mean score of 74.6. Find a 90 percent confidence interval estimate for the average score of all students.

8. In Prob. 7, suppose the sample mean score was 74.6 but the sample was of size 324. Again find a 90 percent confidence interval estimate.

9. The standard deviation of the lifetime of a certain type of lightbulb is known to equal 100 hours. A sample of 169 such bulbs had an average life of 1350 hours. Find a
(a) 90 percent
(b) 95 percent
(c) 99 percent
confidence interval estimate of the mean life of this type of bulb.

10. The average life of a sample of 10 tires of a certain brand was 28,400 miles. If it is known that the lifetimes of such tires are normally distributed with a standard deviation of 3,300 miles, determine a 95 percent confidence interval estimate of the mean life.

11. For Prob. 10, how large a sample would be needed to obtain a 99 percent confidence interval estimator of smaller size than the interval obtained in the problem?

12. A pilot study has revealed that the standard deviation of workers' monthly earnings in the chemical industry is $180. How large a sample must be chosen to obtain an estimator of the mean salary that, with 90 percent confidence, will be correct to within ±$20?

13. Repeat Prob. 12 when you require 95 percent confidence.

14. A college admissions officer wanted to know the average Scholastic Aptitude Test (SAT) score of this year's class of entering students. Rather than checking all student folders, she decided to use a randomly chosen sample. If it is known that student scores are normally distributed with a standard deviation of 70, how large a random sample is needed if the admissions officer wants to obtain a 95 percent confidence interval estimate that is of length 4 or less?

*15. In Prob. 7, find a
(a) 90 percent lower confidence bound
(b) 95 percent lower confidence bound
(c) 95 percent upper confidence bound
(d) 99 percent upper confidence bound
for the average test score.

*16. The following are data from a normal population with standard deviation 3:

5, 4, 8, 12, 11, 7, 14, 12, 15, 10

(a) Find a value which, with 95 percent confidence, is larger than the population mean.

(b) Find a value which, with 99 percent confidence, is smaller than the population mean.

*17. Suppose, on the basis of the sample data noted in Prob. 10, that the tire manufacturer advertises, "With 95 percent certainty, the average tire life is over 26,000 miles." Is this false advertising?

8.6 INTERVAL ESTIMATORS OF THE MEAN OF A NORMAL POPULATION WITH UNKNOWN POPULATION VARIANCE

Suppose now that we have a sample $X_1, \ldots, X_n$ from a normal population having an unknown mean μ and an unknown standard deviation σ; and we want to use the sample data to obtain an interval estimator of the population mean μ.

To start, let us recall how we obtained the interval estimator of μ when σ was assumed to be known. This was accomplished by using the fact that Z, the standardized version of the point estimator $\bar{X}$, which is given by

$$Z = \sqrt{n}\,\frac{\bar{X} - \mu}{\sigma}$$

has a standard normal distribution. Since σ is no longer known, it is natural to replace it by its estimator S, the sample standard deviation, and thus to base our confidence interval on the variable T_{n-1} given by

$$T_{n-1} = \sqrt{n}\,\frac{\bar{X} - \mu}{S}$$

The random variable T_{n-1} defined above is said to be a t random variable having $n - 1$ degrees of freedom.

The random variable

$$T_{n-1} = \sqrt{n}\,\frac{\bar{X} - \mu}{S}$$

is said to be a *t random variable having n − 1 degrees of freedom.*

The reason that T_{n-1} has $n-1$ degrees of freedom is that the sample variance S^2, which is being used to estimate σ^2, has a chi-squared distribution with $n-1$ degrees of freedom (see Sec. 7.6).

The density function of a t random variable, like a standard normal random variable, is symmetric about zero. It looks similar to a standard normal density, although it is somewhat more spread out, resulting in its having "larger tails." As the degree of freedom parameter increases, the density becomes more and more similar to the standard normal density. Figure 8.8 depicts the probability density functions of t random variables for a variety of different degrees of freedom.

The quantity $t_{n,\alpha}$ is defined to be such that

$$P\{T_n > t_{n,\alpha}\} = \alpha$$

where T_n is a t random variable with n degrees of freedom (see Fig. 8.9).

Since $P\{T_n < t_{n,\alpha}\} = 1 - \alpha$, it follows that $t_{n,\alpha}$ is the $100(1-\alpha)$ percentile of the t distribution with n degrees of freedom. For instance, $P\{T_n < t_{n,.05}\} = .95$, showing that 95 percent of the time a t random variable having n degrees of freedom will be less than $t_{n,.05}$. The quantity $t_{n,\alpha}$ is analogous to the quantity z_α of the standard normal distribution.

Values of $t_{n,\alpha}$ for various values of n and α are presented in Appendix D, Table D.2. In addition, Program 8-1 will compute the value of these percentiles. Also Program 8-2 can be used to compute the probabilities of a t random variable.

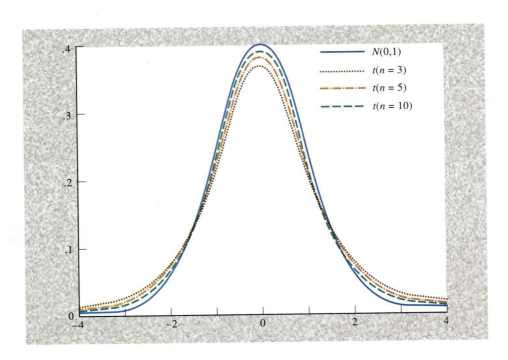

Figure 8.8
Standard normal and t distributions.

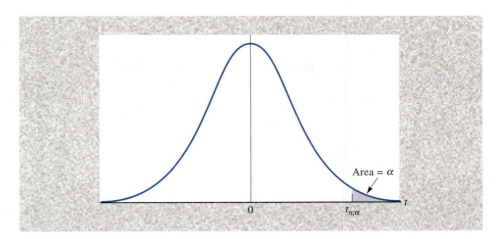

Figure 8.9
The t density
percentile:
$P\{T_n > t_{n,\alpha}\} = \alpha.$

Example 8.12 Find $t_{8,\,.05}$.

Solution The value of $t_{8,\,.05}$ can be obtained from Table D.2. The following is taken from that table:

Values of $t_{n,\alpha}$

n	$\alpha = .10$	$\alpha = .05$	$\alpha = .025$
6	1.440	1.943	2.447
7	1.415	1.895	2.365
→ 8	1.397	1.860	2.306
9	1.383	1.833	2.262

Reading down the $\alpha = .05$ column for the row $n = 8$ shows that $t_{8,\,.05} = 1.860$.

By the symmetry of the t distribution about zero, it follows (see Fig. 8.10) that

$$P\{|T_n| \leq t_{n,\alpha/2}\} = P\{-t_{n,\alpha/2} \leq T_n \leq t_{n,\alpha/2}\} = 1 - \alpha$$

Hence, upon using the result that $\sqrt{n}(\bar{X} - \mu)/S$ has a t distribution with $n - 1$ degrees of freedom, we see that

$$P\left\{\sqrt{n}\,\frac{|\bar{X} - \mu|}{S} \leq t_{n-1,\alpha/2}\right\} = 1 - \alpha$$

In exactly the same manner as we did when σ was known, we can show that the preceding equation is equivalent to

$$P\left\{\bar{X} - t_{n-1,\alpha/2}\,\frac{S}{\sqrt{n}} \leq \mu \leq \bar{X} + t_{n-1,\alpha/2}\,\frac{S}{\sqrt{n}}\right\} = 1 - \alpha$$

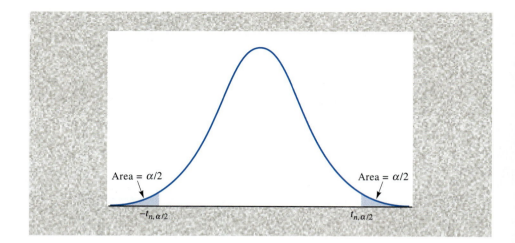

Figure 8.10
$P\{|T_n| \leq t_{n,\alpha/2}\}$
$= P\{-t_{n,\alpha/2}$
$\leq T_n \leq t_{n,\alpha/2}\}$
$= 1 - \alpha.$

Therefore, we showed the following:

A $100(1 - \alpha)$ percent confidence interval estimator for the population mean μ is given by the interval

$$\overline{X} \pm t_{n-1,\alpha/2} \frac{S}{\sqrt{n}}$$

Program 8-3 will compute the desired confidence interval estimate for a given data set.

Example 8.13   The Environmental Protection Agency (EPA) is concerned about the amounts of PCB, a toxic chemical, in the milk of nursing mothers. In a sample of 20 women, the amounts (in parts per million) of PCB were as follows:

16, 0, 0, 2, 3, 6, 8, 2, 5, 0, 12, 10, 5, 7, 2, 3, 8, 17, 9, 1

Use these data to obtain a

(a) 95 percent confidence interval

(b) 99 percent confidence interval

of the average amount of PCB in the milk of nursing mothers.

Solution A simple calculation yields that the sample mean and sample standard deviation are

$$\bar{X} = 5.8 \qquad S = 5.085$$

Since $100(1 - \alpha)$ equals .95 when $\alpha = .05$ and equals .99 when $\alpha = .01$, we need the values of $t_{19, .025}$ and $t_{19, .005}$. From Table D.2 we see that

$$t_{19, .025} = 2.093 \qquad t_{19, .005} = 2.861$$

Hence, the 95 percent confidence interval estimate of μ is

$$5.8 \pm 2.093 \, \frac{5.085}{\sqrt{20}} = 5.8 \pm 2.38$$

and the 99 percent confidence interval estimate of μ is

$$5.8 \pm 2.861 \, \frac{5.085}{\sqrt{20}} = 5.8 \pm 3.25$$

That is, we can be 95 percent confident that the average amount of PCB in the milk of nursing mothers is between 3.42 and 8.18 parts per million; and we can be 99 percent confident that it is between 2.55 and 9.05 parts per million.

The above could also have been solved by running Program 8-3. This yields the following:

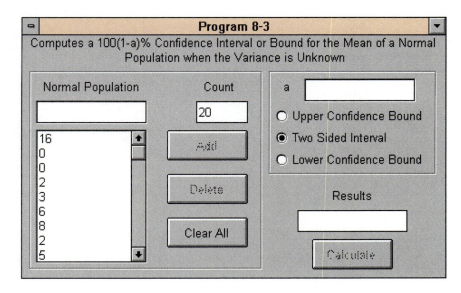

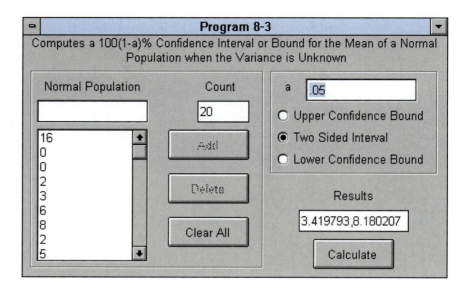

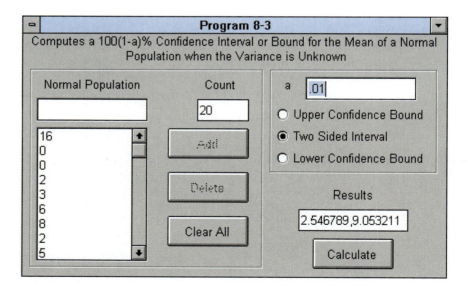

*8.6.1 Lower and Upper Confidence Bounds

Lower and upper confidence bounds for μ are also easily derived, with the following results.

A $100(1 - \alpha)$ percent lower confidence bound for μ is given by

$$\bar{X} - t_{n-1,\alpha} \frac{S}{\sqrt{n}}$$

That is, with $100(1 - \alpha)$ percent confidence the population mean is greater than

$$\bar{X} - t_{n-1,\alpha} \frac{S}{\sqrt{n}}$$

A $100(1 - \alpha)$ percent upper confidence bound for μ is given by

$$\bar{X} + t_{n-1,\alpha} \frac{S}{\sqrt{n}}$$

That is, with $100(1 - \alpha)$ percent confidence the population mean is less than

$$\bar{X} + t_{n-1,\alpha} \frac{S}{\sqrt{n}}$$

Example 8.14   In Example 8.13, find a

(a) 95 percent upper confidence bound

(b) 99 percent lower confidence bound

for the average amount of PCB in nursing mothers.

Solution The sample size in Example 8.13 was equal to 20, and the values of the sample mean and the sample standard deviation were

$$\bar{X} = 5.8 \qquad S = 5.085$$

(a) From Table D.2 we see that

$$t_{19, .05} = 1.729$$

Therefore, the 95 percent upper confidence bound is

$$5.8 + 1.729 \frac{5.085}{\sqrt{20}} = 7.77$$

That is, we can be 95 percent confident that the average PCB level in the milk of nursing mothers is less than 7.77 parts per million.

(b) From Table D.2,

$$t_{19, .01} = 2.539$$

Therefore, the 99 percent lower confidence bound is

$$5.8 - 2.539 \frac{5.085}{\sqrt{20}} = 2.91$$

and so we can be 99 percent confident that the average PCB level in the milk of nursing mothers is greater than 2.91 parts per million.

Program 8-3 will also compute upper and lower confidence bounds having any desired confidence level.

Problems

1. The National Center for Educational Statistics recently chose a random sample of 2000 newly graduated college students and queried each one about the time it took to complete his or her degree. If the sample mean was 5.2 years with a sample standard deviation of 1.2 years, construct
   (a) A 95 percent confidence interval estimate of the mean completion time of all newly graduated students
   (b) A 99 percent confidence interval estimate

2. The manager of a shipping department of a mail-order operation located in New York has been receiving complaints about the length of time it takes for customers in California to receive their orders. To learn more about this potential problem, the manager chose a random sample of 12 orders and then checked to see how many days it took to receive each of these orders. The resulting data were

   15, 20, 10, 11, 7, 12, 9, 14, 12, 8, 13, 16

   (a) Find a 90 percent confidence interval estimate for the mean time it takes for California customers to receive their orders.
   (b) Find a 95 percent confidence interval estimate.

3. A survey was instituted to estimate μ, the mean salary of middle-level bank executives. A random sample of 15 executives yielded the following yearly salaries (in units of $1000):

   88, 121, 75, 39, 52, 102, 95, 78, 69, 82, 80, 84, 72, 115, 106

Find a
(a) 90 percent
(b) 95 percent
(c) 99 percent
confidence interval estimate of μ.

4. The numbers of riders on an intercity bus on 12 randomly chosen days are

   47, 66, 55, 53, 49, 65, 48, 44, 50, 61, 60, 55

   (a) Estimate the mean number of daily riders.
   (b) Estimate the standard deviation of the daily number of riders.
   (c) Give a 95 percent confidence interval estimate for the mean number of daily riders.

5. Use the data of Prob. 1 of Sec. 8.2 to obtain a
   (a) 95 percent
   (b) 99 percent
   confidence interval estimate of the average weight of all participants of the 1990 Boston Marathon.

6. A random sample of 30 General Electric transistors resulted in an average lifetime of 1210 hours with a sample standard deviation of 92 hours. Compute a
   (a) 90 percent
   (b) 95 percent
   (c) 99 percent
   confidence interval estimate of the mean life of all General Electric transistors.

7. In Prob. 10 of Sec. 8.4 determine a 95 percent confidence interval estimate of the population mean burn temperature.

8. The following are the losing scores in seven randomly chosen Super Bowl football games:

   10, 16, 20, 17, 31, 19, 14

   Construct a 95 percent confidence interval estimate of the average losing score in a Super Bowl game.

9. The following are the winning scores in eight randomly chosen Masters Golf Tournaments:

   285, 279, 280, 288, 279, 286, 284, 279

   Use these data to construct a 90 percent confidence interval estimate of the average winning score in the Masters.

10. All the students at a certain school are to be given a psychological task. To determine the average time it will take a student to perform this task, a random

sample of 20 students was chosen and each was given the task. If it took these students an average of 12.4 minutes with a sample standard deviation of 3.3 minutes, find a 95 percent confidence interval estimate for the average time it will take all students in the school to perform this task.

11. A large company self-insures its large fleet of cars against collisions. To determine its mean repair cost per collision, it has randomly chosen a sample of 16 accidents. If the average repair cost in these accidents is $2200 with a sample standard deviation of $800, find a 90 percent confidence interval estimate of the mean cost per collision.

12. An anthropologist measured the heights (in inches) of a random sample of 64 men of a certain tribe, and she found that the sample mean was 72.4 and the sample standard deviation was 2.2. Find a
    (a) 95 percent
    (b) 99 percent
    confidence interval estimate of the average height of all men of the tribe.

13. To determine the average time span of a phone call made during midday, the telephone company has randomly selected a sample of 1200 such calls. The sample mean of these calls is 4.7 minutes, and the sample standard deviation is 2.2 minutes. Find a
    (a) 90 percent   (b) 95 percent
    confidence interval estimate of the mean length of all such calls.

14. Each of 20 science students independently measured the melting point of lead. The sample mean and sample standard deviation of these measurements were (in degrees Celsius) 330.2 and 15.4, respectively. Construct a
    (a) 95 percent   (b) 99 percent
    confidence interval estimate of the true melting point of lead.

15. A random sample of 300 CitiBank VISA cardholder accounts indicated a sample mean debt of $1220 with a sample standard deviation of $840. Construct a 95 percent confidence interval estimate of the average debt of all cardholders.

16. To obtain information about the number of years that Chicago police officers have been on the job, a sample of 46 officers was chosen. Their average time on the job was 14.8 years with a sample standard deviation of 8.2 years. Determine a
    (a) 90 percent
    (b) 95 percent
    (c) 99 percent
    confidence interval estimate for the average time on the job of all Chicago police officers.

17. The following statement was made by an "expert" in statistics. "If a sample of size 9 is chosen from a normal distribution having mean μ, then we can be 95 percent certain that μ will lie within $\bar{X} \pm 1.96S/3$, where $\bar{X}$ is the sample mean and S is the sample standard deviation." Is this statement correct?

18. The geometric random walk model for the price of a stock supposes that the successive differences in the logarithms of the closing prices of the stock constitute a sample from a normal population. This implies that the percentage changes in the successive closing prices constitute a random sample from a population (as opposed to the linear random walk model given in Problem 13 of Sec. 8.4, which supposes that the magnitudes of the changes constitute a random sample). Thus, for instance, under the geometric random walk model, the chance that a stock whose price is 100 will increase to 102 is the same as the chance when its price is 50 that it will increase to 51.

    The following data give the logarithms and the successive differences of the logarithms of the closing crude oil prices of 20 consecutive trading days in 1994. Assuming the applicability of the geometric random walk model, use them to construct a 95 percent confidence interval for the population mean.

Price	log (price)	log (price) difference
17.50	2.862201	
17.60	2.867899	5.697966E−03
17.81	2.87976	1.186109E−02
17.67	2.871868	−7.891655E−03
17.53	2.863914	−7.954597E−03
17.39	2.855895	−8.018494E−03
17.12	2.840247	−1.564789E−02
16.71	2.816007	−2.424002E−02
16.70	2.815409	−5.986691E−04
16.83	2.823163	7.754326E−03
17.21	2.84549	2.232742E−02
17.24	2.847232	1.741886E−03
17.22	2.846071	−1.16086E−03
17.67	2.871868	2.579689E−02
17.83	2.880883	9.01413E−03
17.67	2.871868	−9.01413E−03
17.55	2.865054	−6.81448E−03
17.68	2.872434	7.380247E−03
17.98	2.88926	1.682591E−02
18.39	2.911807	2.254701E−02

19. Twelve successively tested lightbulbs functioned for the following lengths of time (measured in hours):

    35.6, 39.2, 18.4, 42.0, 45.3, 34.5, 27.9, 24.4, 19.9, 40.1, 37.2, 32.9

    (a) Give a 95 percent confidence interval estimate of the mean life of a lightbulb.
    *(b) A claim has been made that the results of this experiment indicate that "One can be 99 percent certain that the mean life exceeds 30 hours." Do you agree with this statement?

20. A school principal was instructed by his board to determine the average number of school days missed by students in the past year. Rather than making a complete survey of all students, the principal drew a random sample of 50 names. He then discovered that the average number of days missed by these 50 students was 8.4 with a sample standard deviation of 5.1.
    (a) What is a 95 percent confidence interval estimate of the average number of days missed by all students?
    *(b) At a subsequent board meeting the principal stated, "With 95 percent confidence I can state that the average number of days missed is less than _____." Fill in the missing number.

*21. In Prob. 3, suppose we want to assert, with 99 percent confidence, that the average salary is greater than v_1. What is the appropriate value of v_1? What would the value of v_2 be if we wanted to assert, with 99 percent confidence, that the average salary was less than v_2?

*22. In Prob. 2, find a number that is, with 95 percent confidence, greater than the average time it takes California customers to receive their orders.

*23. To convince a potential buyer of the worth of her company, an executive has ordered a survey of the daily cash receipts of the business. A sample of 14 days revealed the following values (in units of $100):

    33, 12, 48, 40, 26, 17, 29, 38, 34, 41, 25, 51, 49, 34

    If the executive wants to present these data in the most favorable way, should she present a confidence interval estimate or a one-sided confidence bound? If one-sided, should it be an upper or a lower bound? If you were the executive, how would you complete the following? "I am 95 percent confident that . . ."

*24. To calm the concerns of a group of citizens worried about air pollution in their neighborhood, a government inspector has obtained sample data relating to carbon monoxide concentrations. The data, in parts per million, are as follows:

    101.4, 103.3, 101.6, 111.6, 98.4, 95.0, 93.6

    If these numbers appear reasonably low to the inspector, how should he, when speaking "with 99 percent confidence," present the results to the group?

8.7 INTERVAL ESTIMATORS OF A POPULATION PROPORTION

Suppose that we desire an interval estimator of p, the proportion of individuals in a large population who have a certain characteristic. Suppose further that a random sample of size n is chosen, and it is determined that X of the individuals in the sample have the characteristic. If we let $\hat{p} = X/n$ denote the proportion of the sample having the characteristic, then as previously noted in Sec. 8.3, the expected value and standard deviation of $\hat{p}$ are

$$E\,[\hat{p}] = p$$

$$4\mathrm{SD}\,(\hat{p}) = \sqrt{\frac{p(1 - p)}{n}}$$

When n is large enough that both np and $n(1 - p)$ are greater than 5, we can use the normal approximation to the binomial distribution to assert that an approximate $100(1 - \alpha)$ percent confidence interval estimator of p is given by

$$\hat{p} \pm z_{\alpha/2}\,\mathrm{SD}\,(\hat{p})$$

Although the standard deviation of $\hat{p}$ is not known, since it involves the unknown proportion p, we can estimate it by replacing p by its estimator $\hat{p}$ in the expression for SD $(\hat{p})$. That is, we can estimate SD $(\hat{p})$ by $\sqrt{\hat{p}(1 - \hat{p})/n}$. This gives rise to the following:

An approximate $100(1 - \alpha)$ percent confidence interval estimator of p is given by

$$\hat{p} \pm z_{\alpha/2}\,\sqrt{\frac{\hat{p}(1 - \hat{p})}{n}}$$

where $\hat{p}$ is the proportion of members of the sample of size n who have the characteristic of interest.

Example 8.15   Out of a random sample of 100 students at a university, 82 stated that they were nonsmokers. Based on this, construct a 99 percent confidence interval estimate of p, the proportion of all the students at the university who are nonsmokers.

Solution Since $100(1 - \alpha) = .99$ when $\alpha = .01$, we need the value of $z_{\alpha/2} = z_{.005}$, which from Table D.2 is equal to 2.576. The 99 percent confidence interval estimate of p is thus

$$.82 \pm 2.576\,\sqrt{\frac{.82(1 - .82)}{100}}$$

or

$$.82 \pm .099$$

That is, we can assert with 99 percent confidence that the true percentage of non-smokers is between 72.1 and 91.9 percent.

Example 8.16   On December 24, 1991, *The New York Times* reported that a poll indicated that 46 percent of the population was in favor of the way that President Bush was handling the economy, with a margin of error of ± 3 percent. What does this mean? Can we infer how many people were questioned?

Solution It has become common practice for the news media to present 95 percent confidence intervals. That is, unless it is specifically mentioned otherwise, it is almost always the case that the interval quoted represents a 95 percent confidence interval. Since $z_{.025} = 1.96$, a 95 percent confidence interval for p is given by

$$\hat{p} \pm 1.96 \sqrt{\frac{\hat{p}(1 - \hat{p})}{n}}$$

where n is the sample size. Since $\hat{p}$, the proportion of those in the random sample who are in favor of the President's handling of the economy, is equal to .46, it follows that the 95 percent confidence interval estimate of p, the proportion of the population in favor, is

$$.46 \pm 1.96 \sqrt{\frac{(.46)(.54)}{n}}$$

Since the margin of error is ± 3 percent, it follows that

$$1.96 \sqrt{\frac{(.46)(.54)}{n}} = .03$$

Squaring both sides of the above equation shows that

$$(1.96)^2 \frac{(.46)(.54)}{n} = (.03)^2$$

or

$$n = \frac{(1.96)^2(.46)(.54)}{(.03)^2} = 1060.3$$

That is, approximately 1060 people were sampled, and 46 percent were in favor of President Bush's handling of the economy.

Case Study

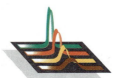

The Aid to Families with Dependent Children (AFDC) program recognizes that errors are inevitable, and so not every family that it funds actually meets the eligibility requirements. However, California holds its counties responsible for overseeing the eligibility requirements and has set a maximum error rate of 4 percent. That is, if over 4 percent of the funded cases in a county are found to be ineligible, then a financial penalty is placed upon the county, with the amount of the penalty determined by the error percentage. Since the state does not have the resources to check every case for eligibility, it uses random sampling to estimate the error percentages.

In 1981, a random sample of 152 cases was chosen in Alameda County, California, and 9 were found to be ineligible. Based on this estimated percentage of $100 \times 9/152 = 5.9$ percent, a penalty of \$949,597 was imposed by the state on the county. The county appealed to the courts, arguing that 9 errors in 152 trials were not sufficient evidence to prove that its error percentage exceeded 4 percent. With help from statistical experts, the court decided that it was unfair to take the point estimate of 5.9 percent as the true error percentage of the county. The court decided it would be fairer to use the lower end of the 95 percent confidence interval estimate. Since the 95 percent confidence interval estimate of the proportion of all funded cases that are ineligible is

$$.059 \pm 1.96 \sqrt{\frac{.059(1 - .059)}{152}} = .059 \pm .037$$

it follows that the lower end of this interval is $.059 - .037 = .022$. Since this lower confidence limit is less than the acceptable value of .04, the court overturned the state's decision and ruled that no penalties were owed.

8.7.1   Length of the Confidence Interval

Since the $100(1 - \alpha)$ percent confidence interval for p goes from

$$\hat{p} - z_{\alpha/2} \sqrt{\frac{\hat{p}(1 - \hat{p})}{n}} \qquad \text{to} \qquad \hat{p} + z_{\alpha/2} \sqrt{\frac{\hat{p}(1 - \hat{p})}{n}}$$

it follows that the length of the interval is as given below.

The length of a $100(1 - \alpha)$ percent confidence interval is

$$2z_{\alpha/2} \sqrt{\frac{\hat{p}(1 - \hat{p})}{n}}$$

where $\hat{p}$ is the proportion of the sample having the characteristic.

Since it can be shown that the product $\hat{p}(1 - \hat{p})$ is always less than or equal to 1/4, it folows from the above equation that an upper bound on the length of the confidence interval is given by $2z_{\alpha/2} \sqrt{1/(4n)}$, which is equivalent to the statement

$$\text{Length of } 100(1 - \alpha) \text{ percent confidence interval} \leq \frac{z_{\alpha/2}}{\sqrt{n}}$$

The preceding bound can be used to determine the appropriate sample size needed to obtain a confidence interval whose length is less than a specified value. For instance, suppose that we want to determine a sufficient sample size so that the length of the resulting $100(1 - \alpha)$ percent confidence interval is less than some fixed value b. In this case, upon using the above inequality, we can conclude that any sample size n for which

$$\frac{z_{\alpha/2}}{\sqrt{n}} < b$$

will suffice. That is, n must be chosen so that

$$\sqrt{n} > \frac{z_{\alpha/2}}{b}$$

Upon squaring both sides, we see that n must be such that

$$n > \left(\frac{z_{\alpha/2}}{b}\right)^2$$

Example 8.17   How large a sample is needed to ensure that the length of the 90 percent confidence interval estimate of p is less than .01?

Solution   To guarantee that the length of the 90 percent confidence interval estimator is less than .01, we need to choose n so that

$$n > \left(\frac{z_{.05}}{.01}\right)^2$$

Since $z_{.05} = 1.645$, this gives

$$n > (164.5)^2 = 27{,}062.25$$

That is, the sample size needs to be at least 27,063 to ensure that the length of the 90 percent confidence interval will be less than .01.

If we let L denote the length of the confidence interval for p,

$$\xleftarrow{\hspace{1.5cm}} L \xrightarrow{\hspace{1.5cm}}$$
$$\hat{p} - \frac{L}{2} \qquad \hat{p} \qquad \hat{p} + \frac{L}{2}$$

then since this interval is centered at $\hat{p}$, it follows that $\hat{p}$ is within $L/2$ of any point in the interval. Therefore, if p lies in the interval, then the distance from $\hat{p}$ to p is at most $L/2$. In Example 8.17 we can thus assert, with 90 percent confidence, that for a sample size as large as 27,063 the observed sample proportion will be within .005 of the true population proportion.

*8.7.2 Lower and Upper Confidence Bounds

Lower and upper confidence bounds for p are easily derived and are given below.

A $100(1 - \alpha)$ percent lower confidence bound for p is given by

$$\hat{p} - z_\alpha \sqrt{\frac{\hat{p}(1 - \hat{p})}{n}}$$

That is, with $100(1 - \alpha)$ percent confidence, the proportion of the population that has the characteristic is greater than this value.

A $100(1 - \alpha)$ percent upper confidence bound for p is given by

$$\hat{p} + z_\alpha \sqrt{\frac{\hat{p}(1 - \hat{p})}{n}}$$

That is, with $100(1 - \alpha)$ percent confidence, the proportion of the population that has the characteristic is less than this value.

Example 8.18   A random sample of 125 individuals working in a large city indicated that 42 are dissatisfied with their working conditions. Construct a 95 percent lower confidence bound on the percentage of all workers in that city who are dissatisfied with their working conditions.

Solution Since $z_{.05} = 1.645$ and $42/125 = .336$, the 95 percent lower bound is given by

$$.336 - 1.645 \sqrt{\frac{.336(.664)}{125}} = .2665$$

That is, we can be 95 percent certain that over 26.6 percent of all workers are dissatisfied with their working conditions.

Problems

1. A random sample of 500 California voters indicated that 302 are in favor of the death penalty. Construct a 99 percent confidence interval estimate of the proportion of all California voters in favor of the death penalty.

2. It is felt that first-time heart attack victims are particularly vulnerable to additional heart attacks during the year following the first attack. To estimate the

proportion of victims who suffer an additional attack within 1 year, a random sample of 300 recent heart attack patients was tracked for 1 year.
(a) If 46 of them suffered an attack within this year, give a 95 percent confidence interval estimate of the desired proportion.
(b) Repeat part (a) if 92 suffered an attack within the year.

3. To estimate p, the proportion of all newborn babies who are male, the gender of 10,000 newborn babies was noted. If 5106 were male, determine a
(a) 90 percent
(b) 99 percent
confidence interval estimate of p.

4. A poll of 1200 voters in 1980 gave Ronald Reagan 57 percent of the vote. Construct a 99 percent confidence interval estimate of the proportion of the population who favored Reagan at the time of the poll.

5. A random sample of 100 Los Angeles residents indicated that 64 were in favor of strict gun control legislation. Determine a 95 percent confidence interval estimate of the proportion of all Los Angeles residents who favor gun control.

6. A random sample of 100 recent recipients of Ph.D.s in science indicated that 42 were optimistic about their future possibilities in science. Find a
(a) 90 percent
(b) 99 percent
confidence interval estimate of the proportion of all recent recipients of Ph.D.s in science who are optimistic.

7. In Prob. 1 of Sec. 8.3, find a 95 percent confidence interval estimate of the proportion of North Americans who believed the Communist party would have won a free election in the Soviet Union in 1985.

8. Using the data of Prob. 4 of Sec. 8.3, find a 90 percent confidence interval estimate of the probability of winning at solitaire.

9. A wine importer has the opportunity to purchase a large consignment of 1947 Chateau Lafite Rothschild wine. Because of the wine's age, some of the bottles may have turned to vinegar. However, the only way to determine whether a bottle is still good is to open it and drink some. As a result, the importer has arranged with the seller to randomly select and open 20 bottles. Suppose 3 of these bottles are spoiled. Construct a 95 percent confidence interval estimate of the proportion of the entire consignment that is spoiled.

10. A sample of 100 cups of coffee from a coffee machine is collected, and the amount of coffee in each cup is measured. Suppose that 9 cups contain less than the amount of coffee specified on the machine. Construct a 90 percent confidence interval estimate of the proportion of all cups dispensed that give less than the specified amount of coffee.

11. A random sample of 400 librarians included 335 women. Give a 95 percent confidence interval estimate of the proportion of all librarians who are women.

12. A random sample of 300 authors included 117 men. Give a 95 percent confidence interval estimate of the proportion of all authors who are men.

13. A random sample of 9 states (West Virginia, New York, Idaho, Texas, New Mexico, Indiana, Utah, Maryland, and Maine) indicated that in 2 of these states the 1990 per capita income exceeded $20,000. Construct a 90 percent confidence interval estimate of the proportion of all states that had a 1990 per capita income in excess of $20,000.

14. A random sample of 1000 psychologists included 457 men. Give a 95 percent confidence interval estimate of the proportion of all psychologists who are men.

15. A random sample of 500 accountants included 42 African Americans, 18 Hispanic Americans, and 246 women. Construct a 95 percent confidence interval estimate of the proportion of all accountants who are
    (a) African American
    (b) Hispanic American
    (c) Female

16. In a poll conducted on January 22, 1991, out of a random sample of 600 people, 450 stated they were in favor of the war against Iraq. Construct a
    (a) 90 percent
    (b) 95 percent
    (c) 99 percent
    confidence interval estimate of p, the proportion of the population in favor of the war at the time.

17. The poll mentioned in Prob. 16 was quoted in the January 28, 1991, San Francisco *Chronicle*, where it was stated that "75 percent of the population are in favor, with a margin of error of plus or minus 4 percentage points."
    (a) Explain why the *Chronicle* should have stated that the margin of error is plus or minus 3.46 percentage points.
    (b) Explain how the *Chronicle* erred to come up with the value of ± 4 percent.

18. A recent newspaper poll indicated that candidate A is favored over candidate B by a 53-to-47 percentage, with a margin of error of ± 4 percent. The newspaper then stated that since the 6-point gap is larger than the margin of error, its readers can be certain that candidate A is the current choice. Is this reasoning correct?

19. A market research firm is interested in determining the proportion of households that are watching a particular sporting event. To accomplish this task, it plans on using a telephone poll of randomly chosen households.
    (a) How large a sample is needed if the company wants to be 90 percent certain that its estimate is correct to within $\pm .02$?
    (b) Suppose there is a sample whose size is the answer in part (a). If 23 percent of the sample were watching the sporting event, do you expect that the 90 percent confidence interval will be exactly of length .02, larger than .02, or smaller than .02?
    (c) Construct the 90 percent confidence interval for part (b).

20. What is the smallest number of death certificates we must randomly sample to estimate the proportion of the U.S. population that dies of cancer, if we want the estimate to be correct to within .01 with 95 percent confidence?

21. Suppose in Prob. 20 that it is known that roughly 20 percent of all deaths are due to cancer. Using this information, determine approximately how many death certificates will have to be sampled to meet the requirements of Prob. 20.

*22. Use the data of Prob. 14 to obtain a 95 percent lower confidence bound for the proportion of all psychologists who are men.

*23. Use the data of Prob. 11 to obtain a 95 percent upper confidence bound for the proportion of all librarians who are women.

*24. A manufacturer is planning on putting out an advertisement claiming that over x percent of the users of his product are satisfied with it. To determine x, a random sample of 500 users was questioned. If 92 percent of these people indicated satisfaction and the manufacturer wants to be 95 percent confident about the validity of the advertisement, what value of x should be used in the advertisement? What value should be used if the manufacturer was willing to be only 90 percent confident about the accuracy of the advertisement?

*25. Use the data of Prob. 15 to obtain a
    (a) 90 percent lower confidence bound
    (b) 90 percent upper confidence bound
    for the proportion of all accountants who are either African American or Hispanic American.

*26. In Prob. 16 construct a
    (a) 95 percent upper confidence bound
    (b) 95 percent lower confidence bound
    for p, the proportion of the population in favor of the war at the time of the poll.

*27. Suppose in Prob. 9 that the importer has decided that purchase of the consignment will be profitable if less than 20 percent of the bottles is spoiled. From the data of this problem, should the importer be
    (a) 95 percent certain
    (b) 99 percent certain
    that the purchase will be profitable?

*28. Refer to the data in Prob. 5. Fill in the missing numbers for these statements:
    (a) With 95 percent confidence, more than _____ percent of all Los Angeles residents favor gun control.
    (b) With 95 percent confidence, less than _____ percent of all Los Angeles residents favor gun control.

Key Terms

Estimator: A statistic used to approximate a population parameter. Sometimes called a *point estimator.*

Estimate: The observed value of the estimator.

Unbiased estimator: An estimator whose expected value is equal to the parameter that it is trying to estimate.

Standard error of an (unbiased) estimator: The standard deviation of the estimator. It is an indication of how close we can expect the estimator to be to the parameter.

Confidence interval estimator: An interval whose endpoints are determined by the data. The parameter will lie within this interval with a certain degree of confidence. This interval is usually centered at the point estimator of the parameter.

$100(1 - \alpha)$ Percent level of confidence: The long-term proportion of time that the parameter will lie within the interval. Equivalently, before the data are observed, the interval estimator will contain the parameter with probability $1 - \alpha$; after the data are observed, the resultant interval estimate contains the parameter with $100(1 - \alpha)$ percent *confidence.*

Lower confidence bound: A number, whose value is determined by the data, which is less than a certain parameter with a given degree of confidence.

Upper confidence bound: A number, whose value is determined by the data, which is greater than a certain parameter with a given degree of confidence.

t Random variable: If $X_1, \ldots, X_n$ are a sample from a normal population having mean μ, then the random variable

$$\sqrt{n}\,\frac{\bar{X} - \mu}{S}$$

is said to be a t random variable with $n - 1$ degrees of freedom, where $\bar{X}$ and S are, respectively, the sample mean and sample standard deviation.

Summary

The sample mean $\bar{X}$ is an unbiased estimator of the population mean μ. Its standard deviation, sometimes referred to as the *standard error* of $\bar{X}$ as an estimator of μ, is given by

$$SD\,(\bar{X}) = \frac{\sigma}{\sqrt{n}}$$

where σ is the population standard deviation.

The statistic $\hat{p}$, equal to the proportion of a random sample having a given characteristic, is the estimate of p, the proportion of the entire population with the characteristic. The standard error of the estimate is

$$\text{SD}(\hat{p}) = \sqrt{\frac{p(1-p)}{n}}$$

where n is the sample size. The standard error can be estimated by

$$\sqrt{\frac{\hat{p}(1-\hat{p})}{n}}$$

The sample variance S^2 is the estimator of the population variance σ^2. Correspondingly, the sample standard deviation S is used to estimate the population standard deviation σ.

If $X_1, \ldots, X_n$ are a sample from a normal population having a known standard deviation σ,

$$\bar{X} \pm z_{\alpha/2} \frac{\sigma}{\sqrt{n}}$$

is a $100(1-\alpha)$ percent confidence interval estimator of the population mean μ. The length of this interval, namely,

$$2z_{\alpha/2} \frac{\sigma}{\sqrt{n}}$$

will be less than or equal to b when the sample size n is such that

$$n \geq \left(\frac{2z_{\alpha/2}\sigma}{b}\right)^2$$

A $100(1-\alpha)$ lower confidence bound for μ is given by

$$\bar{X} - z_\alpha \frac{\sigma}{\sqrt{n}}$$

That is, we can assert with $100(1-\alpha)$ percent confidence that

$$\mu > \bar{X} - z_\alpha \frac{\sigma}{\sqrt{n}}$$

A $100(1 - \alpha)$ upper confidence bound for μ is

$$\bar{X} + z_\alpha \frac{\sigma}{\sqrt{n}}$$

That is, we can assert with $100(1 - \alpha)$ percent confidence that

$$\mu < \bar{X} + z_\alpha \frac{\sigma}{\sqrt{n}}$$

If $X_1, \ldots, n$ are a sample from a normal population whose standard deviation is unknown, a $100(1 - \alpha)$ percent confidence interval estimator of μ is

$$\bar{X} \pm t_{n-1,\alpha/2} \frac{S}{\sqrt{n}}$$

In the above $t_{n-1,\alpha/2}$ is such that

$$P\{T_{n-1} > t_{n-1,\alpha/2}\} = \frac{\alpha}{2}$$

when T_{n-1} is a t random variable with $n - 1$ degrees of freedom.

The $100(1 - \alpha)$ percent lower and upper confidence bounds for μ are, respectively, given by

$$\bar{X} - t_{n-1,\alpha} \frac{S}{\sqrt{n}}$$

and

$$\bar{X} + t_{n-1,\alpha} \frac{S}{\sqrt{n}}$$

To obtain a confidence interval estimate of p, the proportion of a large population with a specific characteristic, take a random sample of size n. If $\hat{p}$ is the proportion of the random sample that has the characteristic, then an approximate $100(1 - \alpha)$ percent confidence interval estimator of p is

$$\hat{p} \pm z_{\alpha/2} \sqrt{\frac{\hat{p}(1 - \hat{p})}{n}}$$

The length of this interval always satisfies

$$\text{Length of confidence interval} \leq \frac{z_{\alpha/2}}{\sqrt{n}}$$

The distance from the center to the endpoints of the 95 percent confidence interval estimator, that is, $1.96\sqrt{\hat{p}(1-\hat{p})/n}$, is commonly referred to as the *margin of error.* For instance, suppose a newspaper states that a new poll indicates that 64 percent of the population consider themselves to be conservationists, with a margin of error of ± 3 percent. By this, the newspaper means that the results of the poll yield that the 95 percent confidence interval estimate of the proportion of the population who consider themselves to be conservationists is .64 $\pm$ .03.

Review Problems

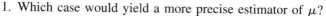

1. Which case would yield a more precise estimator of μ?
   (a) A sample of size n from a population having mean 2μ and variance σ^2
   (b) A sample of size $2n$ from a population having mean μ and standard deviation σ

2. The weights of ball bearings are normally distributed with standard deviation .5 millimeter.
   (a) How large a sample is needed if you want to be 95 percent certain that your estimate of the mean weight of a ball bearing is correct to within $\pm$.1 millimeter?
   (b) Repeat (a) if you want the estimate to be correct to within $\pm$.01 millimeter.
   (c) If a sample of size 8 yields the values

   $$4.1, \ 4.6, \ 3.9, \ 3.3, \ 4.0, \ 3.5, \ 3.9, \ 4.2$$

   give a 95 percent confidence interval estimate for the mean weight.

3. A random sample of 50 people from a certain population was asked to keep a record of the amount of time spent watching television in a specified week. If the sample mean of the resulting data was 24.4 hours and the sample standard deviation was 7.4 hours, give a 95 percent confidence interval estimate for the average time spent watching television by all members of the population that week.

4. Use the first 30 data values in App. A to give a 90 percent confidence interval estimate of the average blood cholesterol level of all the students on the list. Now break up the 30 data values into two groups—one for the females and one for the males. Use the data for each gender separately to obtain 90 percent confidence interval estimates for the mean cholesterol level of the women and of the men. How much confidence would you put in the assertion that the average levels for both the men and the women lie within their respective 90 percent confidence intervals?

5. A standardized test is given annually to all sixth-grade students in the state of Washington. To find out the average score of students in her district, a school supervisor selects a random sample of 100 students. If the sample mean of these students' scores is 320 and the sample standard deviation is 16, give a 95 percent confidence interval estimate of the average score of students in that supervisor's district.

6. An airline is interested in determining the proportion of its customers who are flying for reasons of business. If the airline wants to be 90 percent certain that its estimate will be correct to within 2 percent, how large a random sample should it select?

7. The following data represent the number of drinks sold from a vending machine on a sample of 20 days:

56, 44, 53, 40, 65, 39, 36, 41, 47, 55, 51, 50, 72, 45, 69, 38, 40, 51, 47, 53

(a) Determine a 95 percent confidence interval estimate of the mean number of drinks sold daily.
(b) Repeat part (a) for a 90 percent confidence interval.

8. It is thought that the deepest part of sleep, which is also thought to be the time during which dreams most frequently occur, is characterized by rapid eye movement (REM) of the sleeper. The successive lengths of seven REM intervals of a sleep volunteer were determined at a sleep clinic. The following times in minutes resulted:

37, 42, 51, 39, 44, 48, 29

Give a 99 percent confidence interval estimate for the mean length of a REM interval of the volunteer.

9. A large corporation is analyzing its present health care policy. It is particularly interested in its average cost for delivering a baby. Suppose that a random sample of 24 claims yields that the sample mean of the delivery costs is $1840 and the sample standard deviation is $740. Construct a 95 percent confidence interval estimate of the corporation's present mean cost per delivery.

10. A court-ordered survey yielded the result that out of a randomly chosen sample of 300 farm workers, 144 were in favor of unionizing. Construct a 90 percent confidence interval estimate of the proportion of all farm workers who wanted to be unionized.

11. A sample of nine fastballs thrown by a certain pitcher were measured at speeds of

94, 87, 80, 91, 85, 102, 85, 80, 93

miles per hour.

(a) What is the point estimate of the mean speed of this pitcher's fastball?

(b) Construct a 95 percent confidence interval estimate of the mean speed.

12. A sample of size 9 yields a sample mean of 35. Construct a 95 percent confidence interval estimate of the population mean if the population standard deviation is known to equal

(a) 3

(b) 6

(c) 12

13. Repeat Prob. 12 if the sample size is 36.

14. The following are scores on IQ tests of a random sample of 18 students at a large eastern university.

130, 122, 119, 142, 136, 127, 120, 152, 141, 132, 127, 118, 150, 141, 133, 137, 129, 142

(a) Construct a 90 percent confidence interval estimate of the average IQ score of all students at the university.

(b) Construct a 95 percent confidence interval estimate.

(c) Construct a 99 percent confidence interval estimate.

15. To comply with federal regulations, the state director of education needs to estimate the proportion of all secondary school teachers who are female. If there are 518 females in a random sample of 1000 teachers, construct a 95 percent confidence interval estimate.

16. In Prob. 15, suppose the director had wanted a 99 percent confidence interval estimate whose length was guaranteed to be at most .03. How large a sample would have been necessary?

17. The Census Bureau, to determine the national unemployment rate, uses a random sample of size 50,000. What is the largest possible margin of error?

18. A researcher wants to learn the proportion of the public that favors a certain candidate for office. If a random sample of size 1600 is chosen, what is the largest possible margin of error?

19. A problem of interest in baseball is whether a sacrifice bunt is a good strategy when there is a player on first base and there are no outs. Assuming that the bunter will be out but will be successful in advancing the runner on base, we could compare the probability of scoring a run with a player on first base and no outs to the probability of scoring a run with a player on second base and one out. The following data resulted from a study of randomly chosen major league baseball games played in 1959 and 1960.

Base occupied	Number of outs	Proportion of cases in which no runs are scored	Total number of cases
First	0	.604	1728
Second	1	.610	657

(a) Give a 95 percent confidence interval estimate for the probability of scoring at least one run when there is a player on first base and there are no outs.

(b) Give a 95 percent confidence interval estimate for the probability of scoring at least one run when there is a player on second base and one out.

*20. Use the data of Prob. 15 to construct a

(a) 90 percent
(b) 95 percent
(c) 99 percent

upper confidence bound for the proportion of all secondary school teachers who are female.

*21. Repeat Prob. 20, this time constructing lower confidence bounds. If you were an advocate of greater hiring of female teachers, would you tend to quote an upper or a lower confidence bound?

*22. Suppose that a random sample of nine recently sold houses in a certain city resulted in a sample mean price of $122,000, with a sample standard deviation of $12,000. Give a 95 percent upper confidence bound for the mean price of all recently sold houses in this city.

C H A P T E R   9

Testing Statistical Hypotheses

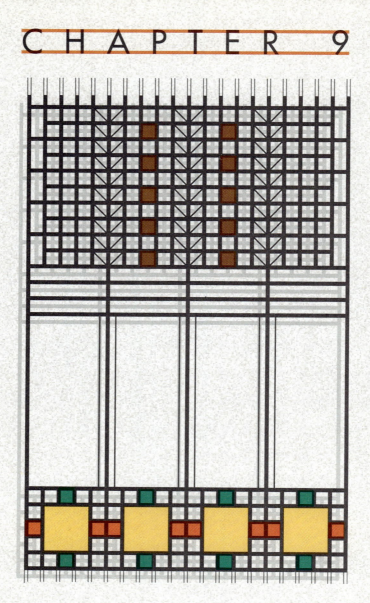

Testing Statistical
Hypotheses

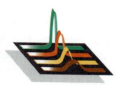

The great tragedy of science—the slaying of a beautiful hypothesis by an ugly set of data.
Thomas H. Huxley, English biologist (Biogenesis and Abiogenesis)

We all learn by experience, and the lesson this time is that you should never lose sight of the alternative.
Sherlock Holmes, in The Adventures of Black Peter by Sir Arthur Conan Doyle

We explain what a statistical hypothesis is and show how sample data can be used to test it. We distinguish between the null and the alternative hypotheses. We explain the significance of rejecting a null hypothesis and of not rejecting it. We introduce the concept of the p value that results from a test.

Tests concerning the mean of a normal population are studied, when the population variance is both known and unknown. One-sided and two-sided tests are considered. Tests concerning a population proportion are presented.

9.1   INTRODUCTION

There has been a great deal of controversy in recent years over the possible dangers of living near a high-level electromagnetic field (EMF). One researcher, after hearing many anecdotal tales concerning the large increases in cancers, especially among children, in communities living near an EMF, decided to study the possible dangers. To do so, she first studied maps giving the locations of electric power lines and then used them to select a fairly large community that was located in a high-level EMF area. She spent time interviewing people in the local schools, hospitals, and public health facilities in order to discover the number of children who had been afflicted with (any type of) cancer in the last 3 years; and she found that there had been 32 such cases.

She then visited a government public health library to learn about the number of cases of childhood cancer that could be expected in a community the size of the one she was considering. She learned that the average number of cases of childhood cancer over a 3-year period in such a community was 16.2, with a standard deviation of 4.7.

Is the discovery of 32 cases of childhood cancers significantly large enough, in comparison with the average number of 16.2, for the researcher to conclude that there is some special factor in the community being studied that increases the chance for children to contract cancer? Or is it possible that there is nothing special about the community and that the greater number of cancers is due solely to chance? In this chapter we will show how such questions can be answered.

9.2   HYPOTHESIS TESTS AND SIGNIFICANCE LEVELS

Statistical inference is the science of drawing conclusions about a population based on information contained in a sample. A particular type of inference is involved with the testing of hypotheses concerning some of the parameters of the population distribution. These hypotheses will usually specify that a population parameter, such as the population mean or variance, has a value that lies in a particular region. We must then decide whether this hypothesis is consistent with data obtained in a sample.

Definition
A *statistical hypothesis* is a statement about the nature of a population. It is often stated in terms of a population parameter.

To test a statistical hypothesis, we must decide whether that hypothesis appears to be consistent with the data of the sample. For instance, suppose that a tobacco firm claims that it has discovered a new way of curing tobacco leaves that will result in the mean nicotine content of a cigarette being less than or equal to 1.5 milligrams. Suppose that a researcher is skeptical of this claim and indeed believes that the mean will exceed 1.5 milligrams. To disprove the claim of the tobacco firm, the researcher has decided to test its hypothesis that the mean is less than or equal to 1.5 milligrams. The statistical hypothesis to be tested, which is called the *null hypothesis* and is denoted by H_0, is thus that the mean nicotine content is less than or equal to 1.5 milligrams. Symbolically, if we let μ denote this mean nicotine content per cigarette, then we can express the null hypothesis as

H_0: $\mu \leq 1.5$

The alternative to the null hypothesis, which the tester is actually trying to establish, is called the *alternative hypothesis* and is designated by H_1. For the above, H_1 is the hypothesis that the mean nicotine content exceeds 1.5 milligrams, and it can be written symbolically as

H_1: $\mu > 1.5$

The *null hypothesis,* denoted by H_0, is a statement about a population parameter. The alternative hypothesis is denoted by H_1. The null hypothesis will be rejected if it appears to be inconsistent with the sample data and will not be rejected otherwise.

To test the null hypothesis that the mean nicotine content per cigarette is less than or equal to 1.5 milligrams, a random sample of cigarettes cured by the new method should be chosen and their nicotine content measured. If the resulting sample data are not "consistent" with the null hypothesis, then we say that the null hypothesis is rejected; if they are "consistent" with the null hypothesis, then the null hypothesis is not rejected.

The decision of whether to reject the null hypothesis is based on the value of a test statistic.

Definition

A *test statistic* is a statistic whose value is determined from the sample data. Depending on the value of this test statistic, the null hypothesis will be rejected or not.

In the cigarette example being considered, the test statistic might be the average nicotine content of the sample of cigarettes. The statistical test would then reject the null hypothesis when this test statistic was sufficiently larger than 1.5. In general, if we let TS denote the test statistic, then to complete our specifications of the test, we must designate the set of values of TS for which the null hypothesis will be rejected.

Definition

The *critical region*, also called the *rejection region,* is that set of values of the test statistic for which the null hypothesis is rejected.

The statistical test of the null hypothesis H_0 is completely specified once the test statistic and the critical region are specified. If TS denotes the test statistic and C denotes the critical region, then the statistical test of the null hypothesis H_0 is as follows:

Reject H_0 if TS is in C

Do not reject H_0 if TS is not in C

For instance, in the nicotine example we have been considering, if it were known that the standard deviation of a cigarette's nicotine content was .8 milligram, then one possible test of the null hypothesis is to use the test statistic $\overline{X}$, equal to the sample mean nicotine level, along with the critical region

$$C = \left\{ \overline{X} \geq 1.5 + \frac{1.312}{\sqrt{n}} \right\}$$

That is, the null hypothesis is to be

Rejected   if $\overline{X} \geq 1.5 + \dfrac{1.312}{\sqrt{n}}$

Not rejected   otherwise

where n is the sample size. (The rationale behind the choice of this particular critical region will become apparent in the next section.)

For instance, if the above test is employed and if the sample size is 36, then the null hypothesis that the population mean is less than or equal to 1.5 will be rejected if $\overline{X} \geq 1.719$ and will not be rejected if $\overline{X} < 1.719$. It is important to note that even when the estimate of μ—namely, the value of the sample mean $\overline{X}$—exceeds 1.5, the null hypothesis may still not be rejected. Indeed, when $n = 36$, a sample mean value of 1.7 will not result in rejection of the null hypothesis. This is true even though such a large value of the sample mean is certainly not evidence in support of the null hypothesis. Nevertheless, it is consistent with the null hypothesis in that if the population mean is 1.5, then there is a reasonable probability that the average of a sample of size 36 will be as large as 1.7. On the other hand, a value of the sample mean as large as 1.9 is so unlikely if the population mean is less than or equal to 1.5 that it will lead to rejection of this hypothesis.

The rejection of the null hypothesis H_0 is a strong statement that H_0 does not appear to be consistent with the observed data. The result that H_0 is not rejected is a weak statement that should be interpreted to mean that H_0 is consistent with the data.

Thus, in any procedure for testing a given null hypothesis, two different types of errors can result. The first, called a *type I error,* is said to result if the test rejects H_0 when H_0 is true. The second, called a *type II error,* is said to occur if the test does not reject H_0 when H_0 is false. Now, it must be understood that the objective of a statistical test of the null hypothesis H_0 is not to determine whether H_0 is true, but rather to determine if its truth is consistent with the resultant data. Therefore, given this objective, it is reasonable that H_0 should be rejected only if the sample data are very unlikely when H_0 is true. The classical way of accomplishing this is to specify a small value α and then require that the test have the property that whenever H_0 is true, its probability of being rejected is less than or equal to α. The value α, called the *level of significance* of the test, is usually set in advance, with commonly chosen values being $\alpha = .10, .05,$ or $.01.$

The classical procedure for testing a null hypothesis is to fix a small significance level α and then require that the probability of rejecting H_0 when H_0 is true is less than or equal to α.

Because of the asymmetry in the test regarding the null and alternative hypotheses, it follows that the only time in which an hypothesis can be regarded as having been "proved" by the data is when the null hypothesis is rejected (thus "proving" that the alternative is true). For this reason the following rule should be noted.

If you are trying to establish a certain hypothesis, then that hypothesis should be designated as the alternative hypothesis. Similarly, if you are trying to discredit a hypothesis, that hypothesis should be designated the null hypothesis.

Thus, for instance, if the tobacco company is running the experiment to prove that the mean nicotine level of its cigarettes is less than 1.5, then it should choose for the null hypothesis

$$H_0: \mu \geq 1.5$$

and for the alternative hypothesis

$$H_1: \mu < 1.5$$

Then the company could use a rejection of the null hypothesis as "proof" of its claim that the mean nicotine content was less than 1.5 milligrams.

Suppose now that we are interested in developing a test of a certain hypothesis regarding θ, a parameter of the population distribution. Specifically, suppose that for a given region R we are trying to test the null hypothesis that θ lies in the region R. That is, we want to test

$$H_0: \theta \text{ lies in } R$$

against the alternative

$$H_1: \theta \text{ does not lie in } R$$

An approach to developing a test of H_0, at level of significance α, is to start by determining a point estimator of θ. The test will reject H_0 when this point estimator is "far away" from the region R. However, to determine how "far away" it needs to be to justify rejection of H_0, first we need to determine the probability distribution of the point estimator when H_0 is true. This will enable us to specify the appropriate critical region so that the probability that the estimator will fall in that region when H_0 is true is less than or equal to α. In the following section we will illustrate this approach by considering tests concerning the mean of a normal population.

Problems

1. Consider a trial in which a jury must decide between hypothesis A that the defendant is guilty and hypothesis B that he or she is innocent.
   (a) In the framework of hypothesis testing and the U.S. legal system, which of the hypotheses should be the null hypothesis?
   (b) What do you think would be the appropriate significance level in this situation?

2. A British pharmaceutical company, Glaxo Holdings, has recently developed a new drug for migraine headaches. Among the claims Glaxo made for its drug, called *somatriptan,* was that the mean time needed for it to enter the bloodstream is less than 10 minutes. To convince the Food and Drug Administration

Historical Perspective

Jerzy Neyman
(Public Information Office,
University of California,
Berkeley)

The concept of significance level was originated by the English statistician Ronald A. Fisher. Fisher also formulated the concept of the null hypothesis as the hypothesis that one is trying to disprove. In Fisher's words, "Every experiment may be said to exist only in order to give the facts a chance of disproving the null hypothesis." The idea of an alternative hypothesis was due to the joint efforts of the Polish-born statistician Jerzy Neyman and his longtime collaborator Egon (son of Karl) Pearson. Fisher, however, did not accept the idea of an alternative hypothesis, arguing that in most scientific applications it was not possible to specify such alternatives, and a great feud ensued between Fisher on one side and Neyman and Pearson on the other. Due to both Fisher's temperament, which was contentious to say the least, and the fact that he was already involved in a controversy with Neyman over the relative benefits of confidence interval estimates, which were originated by Neyman, and Fish-

er's own fiducial interval estimates (which are not much used today), the argument became extremely personal and vitriolic. At one point Fisher called Neyman's position "horrifying for intellectual freedom in the west."

Fisher is famous for his scientific feuds. Aside from the one mentioned above, he carried on a most heated debate with Karl Pearson over the relative merits of two different general approaches for obtaining point estimators, called the *method of moments* and the *method of maximum likelihood*. Fisher, who was a founder of the field of population genetics, also carried out a long-term feud with Sewell Wright, another influential population geneticist, over the role played by chance in the determination of future gene frequencies. (Curiously enough, it was the biologist Wright and not the statistician Fisher who championed cause as a key factor in long-term evolutionary developments.)

---

of the validity of this claim, Glaxo conducted an experiment on a randomly chosen set of migraine sufferers. To prove the company's claim, what should Glaxo have taken as the null and the alternative hypotheses?

3. Suppose a test of

$$H_0: \mu = 0 \qquad \text{against} \qquad H_1: \mu \neq 0$$

resulted in H_0 being rejected at the 5 percent level of significance. Which of the following statements is (are) accurate?
   (a) The data proved that μ is significantly different from 0, meaning that it is far away from 0.
   (b) The data were significantly strong enough to conclude that μ is not equal to 0.

(c) The probability that μ is equal to 0 is less than .05.

(d) The hypothesis that μ is equal to 0 was rejected by a procedure that would have resulted in rejection only 5 percent of the time when μ is equal to 0.

*4. Let μ denote the mean value of some population. Suppose that in order to test

$$H_0: \mu \leq 1.5$$

against the alternative hypothesis

$$H_1: \mu > 1.5$$

a sample is chosen from the population.

(a) Suppose this sample resulted in H_0 not being rejected. Does this imply that the sample data would have resulted in rejection of the null hypothesis if we had been testing the following?

$$H_0: \mu > 1.5 \quad \text{against} \quad H_1: \mu \leq 1.5$$

(b) Suppose this sample resulted in the rejection of H_0. Does this imply that the same sample data would have resulted in not rejecting the null hypothesis if we had been testing the following?

$$H_0: \mu > 1.5 \quad \text{against} \quad H_1: \mu \leq 1.5$$

Assume that all tests are at the 5 percent level of significance, and explain your answers!

9.3   TESTS CONCERNING THE MEAN OF A NORMAL POPULATION: CASE OF KNOWN VARIANCE

Suppose that $X_1, \ldots, X_n$ are a sample from a normal distribution having an unknown mean μ and a known variance σ^2, and suppose we want to test the null hypothesis that the mean μ is equal to some specified value against the alternative that it is not. That is, we want to test

$$H_0: \mu = \mu_0$$

against the alternative hypothesis

$$H_1: \mu \neq \mu_0$$

for a specified value μ_0.

Since the natural point estimator of the population mean μ is the sample mean

$$\overline{X} = \frac{\sum\limits_{i=1}^{n} X_i}{n}$$

it would seem reasonable to reject the hypothesis that the population mean is equal to μ_0 when $\overline{X}$ is far away from μ_0. That is, the critical region of the test should be of the form

$$C = \{X_1, \ldots, X_n: \ |\overline{X} - \mu_0| \geq c\}$$

for a suitable value of c.

Suppose we want the test to have significance level α. Then c must be chosen so that the probability, when μ_0 is the population mean, that $\overline{X}$ differs from μ_0 by c or more is equal to α. That is, c should be such that

$$P\{|\overline{X} - \mu_0| \geq c\} = \alpha \qquad \text{when } \mu = \mu_0 \tag{9.1}$$

However, when μ is equal to μ_0, $\overline{X}$ is normally distributed with mean μ_0 and standard deviation $\sigma/\sqrt{n}$; and so the standardized variable Z, defined by

$$Z = \frac{\overline{X} - \mu_0}{\sigma/\sqrt{n}} = \frac{\sqrt{n}}{\sigma}(\overline{X} - \mu_0)$$

will have a standard normal distribution. Now since the inequality

$$|\overline{X} - \mu_0| \geq c$$

is equivalent to

$$\frac{\sqrt{n}}{\sigma}|\overline{X} - \mu_0| \geq \frac{\sqrt{n}}{\sigma}c$$

it follows that the probability statement 9.1 is equivalent to

$$P\left\{|Z| \geq \sqrt{n}\,\frac{c}{\sigma}\right\} = \alpha$$

Since the probability that the absolute value of a standard normal exceeds some value is equal to twice the probability that a standard normal exceeds that value (see Fig. 9.1), we see from the above that

$$2P\left\{Z \geq \sqrt{n}\,\frac{c}{\alpha}\right\} = \alpha$$

or

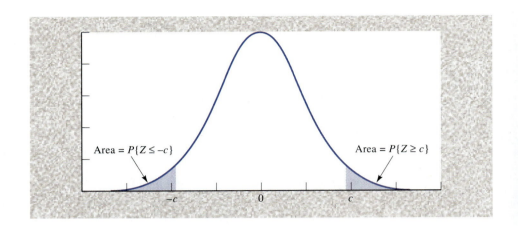

Figure 9.1
$P\{|Z| \geq c\} =$
$P\{Z \geq c\} +$
$P\{Z \leq -c\} =$
$2P\{Z \geq c\}$

Area = $P\{Z \leq -c\}$ Area = $P\{Z \geq c\}$

$-c$ 0 c

$$P\left\{ Z \geq \sqrt{n}\, \frac{c}{\sigma} \right\} = \frac{\alpha}{2}$$

Since $z_{\alpha/2}$ is defined to be such that

$$P\left\{ Z \geq z_{\alpha/2} \right\} = \frac{\alpha}{2}$$

it follows from the preceding that

$$\sqrt{n}\, \frac{c}{\sigma} = z_{\alpha/2}$$

or

$$c = z_{\alpha/2}\, \frac{\sigma}{\sqrt{n}}$$

Therefore, the significance level α test of the null hypothesis that the population mean is equal to the specified value μ_0 against the alternative that it is not equal to μ_0, is to reject the null hypothesis if

$$|\bar{X} - \mu_0| \geq z_{\alpha/2}\, \frac{\sigma}{\sqrt{n}}$$

or, equivalently, to

Reject H_0 if $\dfrac{\sqrt{n}}{\sigma}\, |\bar{X} - \mu_0| \geq z_{\alpha/2}$

Not reject H_0 otherwise

This test is pictorially depicted in Fig. 9.2. Note that in Fig. 9.2 we have superimposed the standard normal density function over the real line since that is the density of the test statistic $\sqrt{n}(\overline{X} - \mu_0)/\sigma$ when H_0 is true. Also, because of this fact, the preceding test is often called the *Z test*.

Example 9.1   Suppose that if a signal of intensity μ is emitted from a particular star, then the value received at an observatory on earth is a normal random variable with mean μ and standard deviation 4. In other words, the value of the signal emitted is altered by *random noise* which is normally distributed with mean 0 and standard deviation 4. It is suspected that the intensity of the signal is equal to 10. Test whether this hypothesis is plausible if the same signal is independently received 20 times and the average of the 20 values received is 11.6. Use the 5 percent level of significance.

Solution   If μ represents the actual intensity of the signal emitted, then the null hypothesis we want to test is

$$H_0: \mu = 10$$

against the alternative

$$H_1: \mu \neq 10$$

Suppose we are interested in testing this at significance level .05. To begin, we compute the value of the statistic

$$\frac{\sqrt{n}}{\sigma}|\overline{X} - \mu_0| = \frac{\sqrt{20}}{4}|11.6 - 10| = 1.79$$

Since this value is less than $z_{.025} = 1.96$, the null hypothesis is not rejected. In other words, we conclude that the data are not inconsistent with the null hypothesis that the

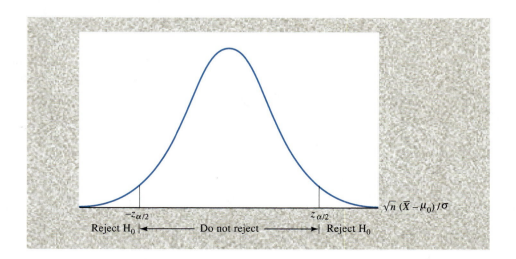

Figure 9.2
Test of $H_0: \mu = \mu_0$ against $H_1: \mu \neq \mu_0$.

value of the signal is equal to 10. The reason for this is that a sample mean as far from the value 10 as the one observed would occur, when H_0 is true, over 5 percent of the time. Note, however, that if the significance level were chosen to be $\alpha = .1$, as opposed to $\alpha = .05$, then the null hypothesis would be rejected (since $z_{\alpha/2} = z_{.05} = 1.645$).

It is important to note that the "correct" level of significance to use in any given hypothesis-testing situation depends on the individual circumstances of that situation. If rejecting H_0 resulted in a large cost that would be wasted if H_0 were indeed true, then we would probably elect to be conservative and choose a small significance level. For instance, suppose that H_1 is the hypothesis that a new method of production is superior to the one presently in use. Since a rejection of H_0 would result in a change of methods, we would want to make certain that the probability of rejection when H_0 is true is quite small; that is, we would want a small value of α. Also, if we initially felt quite strongly that the null hypothesis was true, then we would require very strong data evidence to the contrary for us to reject H_0, and so we would again choose a very small significance level.

The hypothesis test given above can be described as follows: The value, call it v, of the test statistic $\sqrt{n}(\overline{X} - \mu_0)/\sigma$ is determined. The test now calls for rejection of H_0 if the probability that the absolute value of the test statistic will be as large as $|v|$ is, when H_0 is true, less than or equal to α. It therefore follows that the test can be performed by computing, first, the value v of the test statistic and, secondly, the probability that the absolute value of a standard normal will exceed $|v|$. This probability, called the *p value*, gives the critical significance level in the sense that H_0 will be rejected if the *p* value is less than or equal to the significance level α and will not be rejected otherwise.

> The *p value* is the smallest significance level at which the data lead to rejection of the null hypothesis. It gives the probability that data as unsupportive of H_0 as those observed will occur when H_0 is true. A small *p* value (say, .05 or less) is a strong indicator that the null hypothesis is not true. The smaller the *p* value, the greater the evidence for the falsity of H_0.

In practice, the significance level is often not set in advance; rather, the data are used to determine the *p* value. This value is often either so large that it is clear that the null hypothesis should not be rejected or so small that it is clear that the null hypothesis should be rejected.

Example 9.2 Suppose that the average of the 20 values in Example 9.1 is equal to 10.8. In this case the absolute value of the test statistic is

$$\frac{\sqrt{n}}{\sigma}|\overline{X} - \mu_0| = \frac{\sqrt{20}}{4}|10.8 - 10| = .894$$

Since

$$P\{|Z| \geq .894\} = 2P\{Z \geq .894\}$$
$$= .371 \quad \text{(from Table D.1)}$$

it follows that the p value is .371. Therefore, the null hypothesis that the signal value is 10 will not be rejected at any significance level less than .371. Since we never want to use a significance level as high as that, H_0 will not be rejected.

On the other hand, if the value of the sample mean were 7.8, then the absolute value of the test statistic would be

$$\frac{\sqrt{20}}{4}(2.2) = 2.46$$

and so the p value would be

$$p \text{ value} = P\{|Z| \geq 2.46\}$$
$$= 2P\{Z \geq 2.46\}$$
$$= .014$$

Thus, H_0 would be rejected at all significance levels above .014 and would not be rejected for lower significance levels.

The next example is concerned with determining the probability of not rejecting the null hypothesis when it is false.

*Example 9.3 In Example 9.1, assuming a .05 significance level, what is the probability that the null hypothesis (that the signal intensity is equal to 10) will not be rejected when the actual signal value is 9.2?

Solution In Example 9.1, $\sigma = 4$ and $n = 20$. Therefore, the significance level .05 test of

$$H_0: \mu = 10 \quad \text{against} \quad H_1: \mu \neq 10$$

is to reject H_0 if

$$\frac{\sqrt{20}}{4}|\bar{X} - 10| \geq z_{.025}$$

or, equivalently, if

$$|\bar{X} - 10| \geq \frac{4z_{.025}}{\sqrt{20}}$$

Since $4z_{.025}/\sqrt{20} = 4 \times 1.96/\sqrt{20} = 1.753$, this means that H_0 is to be rejected if the distance between $\bar{X}$ and 10 is at least 1.753. That is, H_0 will be rejected if either

$$\bar{X} \geq 10 + 1.753$$

or

$$\bar{X} \leq 10 - 1.753$$

That is, if

$$\bar{X} \geq 11.753 \quad \text{or} \quad \bar{X} \leq 8.247$$

then H_0 will be rejected.

Now if the population mean is 9.2, then $\bar{X}$ will be normal with mean 9.2 and standard deviation $4/\sqrt{20} = .894$; and so the standardized variable

$$Z = \frac{\bar{X} - 9.2}{.894}$$

will be a standard normal random variable. Thus, when the true value of the signal is 9.2, we see that

$$
\begin{aligned}
P\{\text{rejection of } H_0\} &= P\{\bar{X} \geq 11.753\} + P\{\bar{X} \leq 8.247\} \\
&= P\left\{\frac{\bar{X} - 9.2}{.894} \geq \frac{11.753 - 9.2}{.894}\right\} + P\left\{\frac{\bar{X} - 9.2}{.894} \leq \frac{8.247 - 9.2}{.894}\right\} \\
&= P\{Z \geq 2.856\} + P\{Z \leq -1.066\} \\
&= .0021 + .1432 \\
&= .1453
\end{aligned}
$$

That is, when the true signal value is 9.2, there is an 85.47 percent chance that the .05 significance level test will not reject the null hypothesis that the signal value is equal to 10.

Problems

In all problems, assume that the relevant distribution is normal.

1. The device that an astronomer utilizes to measure distances results in measurements that have a mean value equal to the actual distance of the object being surveyed and a standard deviation of .5 light year. Present theory indicates that the actual distance from earth to the asteroid phyla is 14.4 light years. Test this hypothesis, at the 5 percent level of significance, if six independent measurements yielded the data

    15.1, 14.8, 14.0, 15.2, 14.7, 14.5

2. A previous sample of fish in Lake Michigan indicated that the mean polychlorinated biphenyl (PCB) concentration per fish was 11.2 parts per million with a standard deviation of 2 parts per million. Suppose a new random sample of 10 fish has the following concentrations:

$$11.5, \ 12.0, \ 11.6, \ 11.8, \ 10.4, \ 10.8, \ 12.2, \ 11.9, \ 12.4, \ 12.6$$

Assume that the standard deviation has remained equal to 2 parts per million, and test the hypothesis that the mean PCB concentration has also remained unchanged at 11.2 parts per million. Use the 5 percent level of significance.

3. To test the hypothesis

$$H_0\colon \mu = 105 \qquad \text{against} \qquad H_1\colon \mu \neq 105$$

a sample of size 9 is chosen. If the sample mean is $\bar{X} = 100$, find the p value if the population standard deviation is known to be

(a) $\sigma = 5$ (b) $\sigma = 10$
(c) $\sigma = 15$

In which cases would the null hypothesis be rejected at the 5 percent level of significance? What about at the 1 percent level?

4. Repeat Prob. 3 when the sample mean is the same but the sample size is 36.

5. A colony of laboratory mice consists of several thousand mice. The average weight of all the mice is 32 grams with a standard deviation of 4 grams. A laboratory assistant was asked by a scientist to select 25 mice for an experiment. However, before performing the experiment, the scientist decided to weigh the mice as an indicator of whether the assistant's selection constituted a random sample or whether it was made with some unconscious bias (perhaps the mice selected were the ones that were slowest in avoiding the assistant, which might indicate some inferiority about this group). If the sample mean of the 25 mice was 30.4, would this be significant evidence, at the 5 percent level of significance, against the hypothesis that the selection constituted a random sample?

6. It is known that the value received at a local receiving station is equal to the value sent plus a random error that is normal with mean 0 and standard deviation 2. If the same value is sent 7 times, compute the p value for the test of the null hypothesis that the value sent is equal to 14, if the values received are

$$14.6, \ 14.8, \ 15.1, \ 13.2, \ 12.4, \ 16.8, \ 16.3$$

7. Historical data indicate that household water use tends to be normally distributed with a mean of 360 gallons and a standard deviation of 40 gallons per day. To see if this is still the situation, a random sample of 200 households was chosen. The average daily water use in these households was then seen to equal 374 gallons per day.

(a) Are these data consistent with the historical distribution? Use the 5 percent level of significance.

(b) What is the p value?

8. When a certain production process is operating properly, it produces items that each have a measurable characteristic with mean 122 and standard deviation 9. However, occasionally the process goes out of control, and this results in a change in the mean of the items produced. Test the hypothesis that the process is presently in control if a random sample of 10 recently produced items had the following values:

$$123, 120, 115, 125, 131, 127, 130, 118, 125, 128$$

Specify the null and alternative hypotheses, and find the p value.

9. A leasing firm operates on the assumption that the annual number of miles driven in its leased cars is normally distributed with mean 13,500 and standard deviation 4000 miles. To see whether this assumption is valid, a random sample of 36 one-year-old cars has been checked. What conclusion can you draw if the average mileage on these 36 cars is 15,233?

10. A population distribution is known to have standard deviation 20. Determine the p value of a test of the hypothesis that the population mean is equal to 50, if the average of a sample of 64 observations is

(a) 52.5

(b) 55.0

(c) 57.5

11. Traffic authorities claim that traffic lights are red for a time that is normal with mean 30 seconds and standard deviation 1.4 seconds. To test this claim, a sample of 40 traffic lights was checked. If the average time of the 40 red lights observed was 32.2 seconds, can we conclude, at the 5 percent level of significance, that the authorities are incorrect? What about at the 1 percent level of significance?

12. The number of cases of childhood cancer occurring within a 3-year span in communities of a specified size approximately has a normal distribution with mean 16.2 and standard deviation 4.7. To see whether this distribution changes when the community is situated near a high-level electromagnetic field, a researcher chose such a community and subsequently discovered that there had been a total of 32 cases of childhood cancers within the last 3 years. Using these data, find the p value of the test of the hypothesis that the distribution of the number of childhood cancers in communities near high-level electromagnetic fields remains normal with mean 16.2 and standard deviation 4.7.

13. The following data are known to come from a normal population having standard deviation 2. Use them to test the hypothesis that the population mean is equal to 15. Determine the significance levels at which the test would reject and those at which it would not reject this hypothesis.

$$15.6, \ 16.4, \ 14.8, \ 17.2, \ 16.9, \ 15.3, \ 14.0, \ 15.9$$

*14. Suppose, in Prob. 1, that current theory is wrong and that the actual distance to the asteroid phyla is 14.8 light years. What is the probability that a series of 10 readings, each of which has a mean equal to the actual distance and a standard deviation of .8 light year, will result in a rejection of the null hypothesis that the distance is 14 light years? Use a 1 percent level of significance.

*15. In Prob. 6 compute the probability that the null hypothesis that the value 14 is sent will be rejected, at the 5 percent level of significance, when the actual value sent is
   (a) 15
   (b) 13
   (c) 16

9.3.1   One-Sided Tests

So far we have been considering two-sided hypothesis-testing problems in which the null hypothesis is that μ is equal to a specified value μ_0 and the test is to reject this hypothesis if $\bar{X}$ is either too much larger or too much smaller than μ_0. However, in many situations, the hypothesis we are interested in testing is that the mean is less than or equal to some specified value μ_0 versus the alternative that it is greater than that value. That is, we are often interested in testing

$$H_0: \mu \leq \mu_0$$

against the alternative

$$H_1: \mu > \mu_0$$

Since we would want to reject H_0 only when the sample mean $\bar{X}$ is much larger than μ_0 (and no longer when it is much smaller), it can be shown, in exactly the same fashion as was done in the two-sided case, that the significance level α test is to

Reject H_0	if $\sqrt{n} \, \dfrac{\bar{X} - \mu_0}{\sigma} \geq z_\alpha$
Not reject H_0	otherwise

A pictorial depiction of this test is shown in Fig. 9.3.

The above test can be carried out alternatively by first computing the value of the test statistic $\sqrt{n}(\bar{X} - \mu_0)/\sigma$. The p value is then equal to the probability that a standard normal random variable is at least as large as this value. That is, if the value of the test statistic is v, then

$$p \text{ value} = P\{Z \geq v\}$$

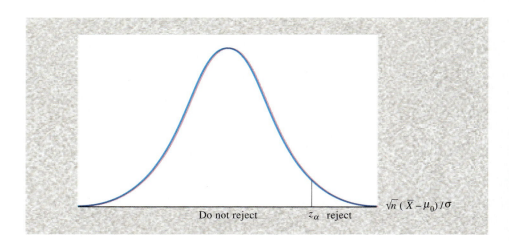

Figure 9.3
Testing H_0: $\mu \leq \mu_0$ against H_1: $\mu > \mu_0$.

The null hypothesis is then rejected at any significance level greater than or equal to the p value.

In similar fashion, we can test the null hypothesis

$$H_0: \mu \geq \mu_0$$

against the alternative

$$H_1: \mu < \mu_0$$

by first computing the value of the test statistic $\sqrt{n}(\bar{X} - \mu_0)/\sigma$. The p value then equals the probability that a standard normal is less than or equal to this value, and the null hypothesis is rejected if the significance level is at least as large as the p value.

Example 9.4   All cigarettes presently being sold have an average nicotine content of at least 1.5 milligrams per cigarette. A firm that produces cigarettes claims that it has discovered a new technique for curing tobacco leaves that results in the average nicotine content of a cigarette being less than 1.5 milligrams. To test this claim, a sample of 20 of the firm's cigarettes was analyzed. If it were known that the standard deviation of a cigarette's nicotine content was .7 milligram, what conclusions could be drawn, at the 5 percent level of significance, if the average nicotine content of these 20 cigarettes were 1.42 milligrams?

Solution To see if the results establish the firm's claim, let us see if they would lead to rejection of the hypothesis that the firm's cigarattes do not have an average nicotine content lower than 1.5 milligrams. That is, we should test

$$H_0: \mu \geq 1.5$$

against the firm's claim of

$$H_1: \mu < 1.5$$

Since the value of the test statistic is

$$\sqrt{n} \, \frac{\bar{X} - \mu_0}{\sigma} = \sqrt{20} \, \frac{1.42 - 1.5}{.7} = -.511$$

it follows that the p value is

$$p \text{ value} = P\{Z \le -.511\} = .305$$

Since the p value exceeds .05, the foregoing data do not enable us to reject the null hypothesis and conclude that the mean content per cigarette is less than 1.5 milligrams. In other words, even though the evidence supports the cigarette producer's claim (since the average nicotine content of those cigarettes tested was indeed less than 1.5 milligrams), that evidence is not strong enough to *prove* the claim. This is because a result at least as supportive of the alternative hypothesis H_1 as that obtained would be expected to occur 30.5 percent of the time when the mean nicotine content was 1.5 milligrams per cigarette.

Statistical hypothesis tests in which either the null or the alternative hypothesis states that a parameter is greater (or less) than a certain value are called *one-sided* tests.

We have assumed so far that the underlying population distribution is the normal distribution. However, we have only used this assumption to conclude that $\sqrt{n}(\bar{X} - \mu)/\sigma$ has a standard normal distribution. But by the central limit theorem this result will approximately hold, no matter what the underlying population distribution, as long as n is reasonably large. A rule of thumb is that a sample size of $n \ge 30$ will almost always suffice. Indeed, for many population distributions, a value of n as small as 4 or 5 will result in a good approximation. Thus, all the hypothesis tests developed so far can often be used even when the underlying population distribution is not normal.

Table 9.1 summarizes the tests presented in this section.

Problems

1. The weights of salmon grown at a commercial hatchery are normally distributed with a standard deviation of 1.2 pounds. The hatchery claims that the mean weight of this year's crop is at least 7.6 pounds. Suppose a random sam-

ple of 16 fish yielded an average weight of 7.2 pounds. Is this strong enough evidence to reject the hatchery's claims at the

(a) 5 percent   (b) 1 percent level of significance?

(c) What is the p value?

2. Consider a test of H_0: $\mu \leq 100$ versus H_1: $\mu > 100$. Suppose that a sample of size 20 has a sample mean of $\bar{X} = 105$. Determine the p value of this outcome if the population standard deviation is known to equal

(a) 5

(b) 10

(c) 15

3. Repeat Prob. 2, this time supposing that the value of the sample mean is 108.

4. It is extremely important in a certain chemical process that a solution which is to be used as a reactant have a pH level greater than 8.40. A method for determining pH that is available for solutions of this type is known to give measurements that are normally distributed with a mean equal to the actual pH and with a standard deviation of .05. Suppose 10 independent measurements yielded the following pH values:

$$8.30, \ 8.42, \ 8.44, \ 8.32, \ 8.43, \ 8.41, \ 8.42, \ 8.46, \ 8.37, \ 8.42$$

Suppose it is a very serious mistake to run the process with a reactant having a pH level less than or equal to 8.40.

| **Table 9.1** | **Hypothesis Tests Concerning the Mean μ of a Normal Population with Known Variance σ^2** |

$X_1, \ldots, X_n$ are sample data, and

$$\bar{X} = \frac{\sum_{i=1}^{n} X_i}{n}$$

H_0	H_1	Test statistic TS	Significance level α test	p Value if TS $= v$
$\mu = \mu_0$	$\mu \neq \mu_0$	$\sqrt{n}\,\dfrac{\bar{X} - \mu_0}{\sigma}$	Reject H_0 if $\|TS\| \geq z_{\alpha/2}$ Do not reject H_0 otherwise	$2P\{Z \geq \|v\|\}$
$\mu \leq \mu_0$	$\mu > \mu_0$	$\sqrt{n}\,\dfrac{\bar{X} - \mu_0}{\sigma}$	Reject H_0 if TS $\geq z_a$ Do not reject H_0 otherwise	$P\{Z \geq v\}$
$\mu \geq \mu_0$	$\mu < \mu_0$	$\sqrt{n}\,\dfrac{\bar{X} - \mu_0}{\sigma}$	Reject H_0 if TS $\leq -z_\alpha$ Do not reject H_0 otherwise	$P\{Z \leq v\}$

Three Mile Island

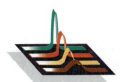

A still unsettled question is whether the nuclear accident at Three Mile Island, which released low-level nuclear radiation in the areas surrounding it, is responsible for an increase in the number of cases of hyperthyroidism. Hyperthyroidism, which results when the thyroid gland is malfunctioning, can lead to mental retardation if it is not treated quickly. It has been reported that 11 babies suffering from hyperthyroidism were born in the surrounding areas between March 28, 1979 (the day of the accident), and December 28, 1979 (nine months later). In addition, it was reported that the normal number of such babies to be born in the surrounding areas over a 9-month period is approximately normally distributed with a mean approximately equal to 3 and a standard deviation approximately equal to 2. Given this information, let us start by determining the probability that such a large number of cases of hyperthyroidism as 11 could have occurred by chance.

To begin, note that if the accident did not have any health effects and if the 9 months following the accident were ordinary months, then the number of newborn babies suffering from hyperthyroidism should approximately have a normal distribution with mean 3 and standard deviation 2. On the other hand, if the accident had a deleterious effect on hyperthyroidism, then the mean of the distribution would be larger than 3. Hence, let us suppose that the data come from a normal distribution with standard deviation 2 and use them to test

$$H_0: \mu \le 3 \quad \text{against} \quad H_1: \mu > 3$$

where μ is the mean number of newborns who suffer from hyperthyroidism.

Since the observed number is 11, the p value of these data is

$$
\begin{aligned}
p \text{ value} &= P\{X \ge 11\} \\
&= P\{X \ge 10.5\} \quad \text{continuity} \\
&\qquad\qquad\qquad\qquad \text{correction} \\
&= P\left\{\frac{X-3}{2} \ge \frac{10.5-3}{2}\right\} \\
&\approx P\{Z \ge 3.75\} \\
&< .0001
\end{aligned}
$$

(a) What null hypothesis should be tested?
(b) What is the alternative hypothesis?
(c) Using the 5 percent level of significance, what would you advise—to use or not to use the solution?
(d) What is the p value of the hypothesis test?

5. An advertisement for a toothpaste claims that use of this product significantly reduces the number of cavities of children in their cavity-prone years. Cavities per year for this age group are normal with mean 3 and standard deviation 1. A study of 2500 children who used this toothpaste found an average of 2.95 cavities per child. Assume that the standard deviation of the number of cavities of a child using this new toothpaste remains equal to 1.

Thus the null hypothesis would be rejected at the 1 percent (or even at the .1 percent) level of significance.

It is important to note that the above test does *not* prove that the nuclear accident was the cause of the increase in hyperthyroidism; and in fact it *does not even prove that there was an increase in this disease.* Indeed, it is hard to know what can be concluded from this test without having a great deal more information. For instance, one difficulty results from our not knowing why the particular hypothesis considered was chosen to be studied. That is, was there some prior scientific reason for believing that a release of nuclear radiation might result in increased hyperthyroidism in newborns, or did someone just check all possible diseases he could think of (and possibly for a variety of age groups) and then test whether there was a significant change in its incidence after the accident? The trouble with such an approach (which is often called *data mining,* or *going on a fishing expedition*) is that even if no real changes resulted from the accident, just by chance some of the many tests might yield a significant result. [For instance, if 20 independent hypothesis tests are run, then even if all the null hypotheses

are true, at least one of them will be rejected at the 1 percent level of significance with probability $1 - (.99)^{20} = .18$.]

Another difficulty in interpreting the results of our hypothesis test concerns the confidence we have in the numbers given to us. For instance, can we really be certain that under normal conditions the mean number of newborns suffering from hyperthyroidism is equal to 3? Is it not more likely that whereas on average 3 newborns would normally be diagnosed to be suffering from this disease, other newborn sufferers may go undetected? Would there not be a much smaller chance that a sufferer would fail to be diagnosed as being such in the period following the accident, given that everyone was alert for such increases in that period? Also, perhaps there are degrees of hyperthyroidism, and a newborn diagnosed as being a sufferer in the tense months following the accident would not have been so diagnosed in normal times.

Note that we are not trying to argue that there was not a real increase in hyperthyroidism following the accident at Three Mile Island. Rather, we are trying to make the reader aware of the potential difficulties in correctly evaluating a statistical study.

(a) Are these data strong enough, at the 5 percent level of significance, to establish the claim of the toothpaste advertisement?

(b) Is this a significant enough reason for your children to switch to this toothpaste?

6. A farmer claims to be able to produce larger tomatoes. To test this claim, a tomato variety that has a mean diameter size of 8.2 centimeters with a standard deviation of 2.4 centimeters is used. If a sample of 36 tomatoes yielded a sample mean of 9.1 centimeters, does this prove that the mean size is indeed larger? Assume that the population standard deviation remains equal to 2.4, and use the 5 percent level of significance.

7. Suppose that the cigarette firm is now, after the test described in Example 9.4, even more convinced about its claim that the mean nicotine content of its cigarettes is less than 1.5 milligrams per cigarette. Would you suggest another test? With the same sample size?

8. The following data come from a normal population having standard deviation 4.

    105, 108, 112, 121, 100, 105, 99, 107, 112, 122, 118, 105

    Use them to test the null hypothesis that the population mean is less than or equal to 100 at the
    (a) 5 percent level of significance
    (b) 1 percent level of significance
    (c) What is the *p* value?

9. A soft drink company claims that its machines dispense, on average, 6 ounces per cup with a standard deviation of .14 ounce. A consumer advocate is skeptical of this claim, believing that the mean amount dispensed is less than 6 ounces. To gain information, a sample of size 100 is chosen. If the average amount per cup is 5.6 ounces, what conclusions can be drawn? State the null and alternative hypotheses, and give the *p* value.

10. The significance level α test of

    $$H_0: \mu = \mu_0 \quad \text{against} \quad H_1: \mu > \mu_0$$

    is the same as the one for testing

    $$H_0: \mu \leq \mu_0 \quad \text{against} \quad H_1: \mu > \mu_0$$

    Does this seem reasonable to you? Explain!

9.4 THE *t* TEST FOR THE MEAN OF A NORMAL POPULATION: CASE OF UNKNOWN VARIANCE

We have previously assumed that the only unknown parameter of the normal population distribution is its mean. However, by far the more common case is when the standard deviation σ is also unknown. In this section we will show how to perform hypothesis tests of the mean in this situation.

To begin, suppose that we are about to observe the results of a sample of size n from a normal population having an unknown mean μ and an unknown standard deviation σ; and suppose that we are interested in using the data to test the null hypothesis

$$H_0: \mu = \mu_0$$

against the alternative

$$H_1: \mu \neq \mu_0$$

As in the previous section, it again seems reasonable to reject H_0 when the point estimator of the population mean μ—that is, the sample mean $\bar{X}$—is far from μ_0. However, how far away it needs to be to justify rejection of H_0 was shown in Sec.

9.3 to depend on the standard deviation σ. Specifically, we showed that a significance level α test called for rejecting H_0 when $|\bar{X} - \mu_0|$ was at least $z_{\alpha/2}\sigma/\sqrt{n}$ or, equivalently, when

$$\frac{\sqrt{n}\,|\bar{X} - \mu_0|}{\sigma} \geq z_{\alpha/2}$$

Now, when σ is no longer assumed to be known, it is reasonable to estimate it by the sample standard deviation S, given by

$$S = \sqrt{\frac{\sum_{i=1}^{n}(X_i - \bar{X})^2}{n-1}}$$

and to employ a test that calls for rejecting H_0 when the absolute value of the test statistic T is large, where

$$T = \sqrt{n}\,\frac{\bar{X} - \mu_0}{S}$$

To determine how large $|T|$ needs to be to justify rejection at the α level of significance, we need to know its probability distribution when H_0 is true. However, as noted in Sec. 8.6, when $\mu = \mu_0$, the statistic T has a t distribution with $n-1$ degrees of freedom. Since the absolute value of such a random variable will exceed $t_{n-1,\,\alpha/2}$ with probability α (see Fig. 9.4), it follows that a significance level α test of

$$H_0\colon \mu = \mu_0 \qquad \text{versus} \qquad H_1\colon \mu \neq \mu_0$$

is, when σ is unknown, to

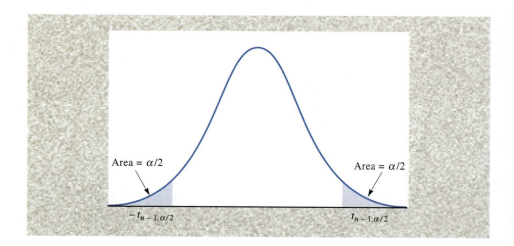

Figure 9.4
$P\{|T_{n-1}| \geq t_{n-1,\,\alpha/2}\} = \alpha$

Reject H_0 if $|T| \geq t_{n-1,\ \alpha/2}$
Not reject H_0 otherwise

The above test, which is pictorially illustrated in Fig. 9.5, is called a *two-sided t test*.

If we let v denote the value of the test statistic $T = \sqrt{n}(\bar{X} - \mu_0)/S$, then the p value of the data is the probability that the absolute value of a t random variable having $n - 1$ degrees of freedom will be as large as $|v|$, which is equal to twice the probability that a t random vairable with $n - 1$ degrees of freedom will be as large as $|v|$. (That is, the p value is the probability that a value of the test statistic at least as large as the one obtained would have occurred if the null hypothesis were true.) The test then calls for rejection at all significance levels that are at least as large as the p value.

If the value of the test statistic is v, then

$$p \text{ value} = P\{|T_{n-1}| \geq |v|\}$$
$$= 2P\{T_{n-1} \geq |v|\}$$

where T_{n-1} is a t random variable with $n - 1$ degrees of freedom.

Example 9.5   Among a clinic's patients having high blood cholesterol levels of at least 240 milliliters per deciliter of blood serum, volunteers were recruited to test a new drug designed to reduce blood cholesterol. A group of 40 volunteers were given

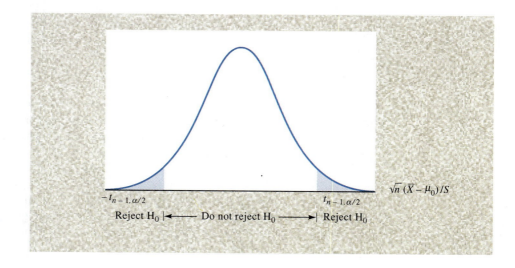

Figure 9.5
The significance level α two-sided t test.

the drug for 60 days, and the changes in their blood cholesterol levels were noted. If the average change was a decrease of 6.8 with a sample standard deviation of 12.1, what conclusions can we draw? Use the 5 percent level of significance.

Solution Let us begin by testing the hypothesis that any changes in blood cholesterol levels were due purely to chance. That is, let us use the data to test the null hypothesis

$$H_0: \mu = 0 \quad \text{versus} \quad H_1: \mu \neq 0$$

where μ is the mean decrease in cholesterol. The value of the test statistic T is

$$T = \frac{\sqrt{n}(\bar{X} - \mu_0)}{S} = \frac{\sqrt{40}(6.8)}{12.1} = 3.554$$

Since, from Table D.2, $t_{39, .025} = 2.02$, the null hypothesis is rejected at the 5 percent level of significance. In fact the p value of the data is given by

$$p \text{ value} = 2P\{T_{39} > 3.554\}$$
$$= .0001 \quad \text{from Program 8-2}$$

Thus, at any significance level greater than .0001, we reject the hypothesis that the change in levels is due solely to chance.

However, note that we would not be justified at this point in concluding that the changes in cholesterol levels are due to the specific drug used and not to some other possibility. For instance, it is well known that any medication received by a patient (whether or not this medication is directly relevant to the patient's suffering) often leads to an improvement in the patient's condition (the *placebo effect*). Also other factors might be involved that could have caused the reduction in blood cholesterol levels; for instance, weather conditions during the testing period might conceivably have affected these levels.

Indeed, it must be concluded that the above testing scheme was very poorly designed for learning about the effectiveness of the drug. For in order to test whether a particular treatment has an effect on a disease that may be affected by many things, it is necessary to design an experiment that neutralizes all other possible causes of change except for the drug. The accepted approach for accomplishing this is to divide the volunteers at random into two groups: one group is to receive the drug, and the other group (the *control* group) is to receive a placebo (that is, a tablet that looks and tastes like the actual drug but which has no physiological effect). The volunteers should not be told whether they are in the actual or in the control group. Indeed, it is best if even the clinicians do not have this information (such tests are called *double-blind*) so as not to allow their own hopes and biases to play a role in their before-and-after evaluations of the patients. Since the two groups are chosen at random from the volunteers, we can now hope that on average all factors affecting the two groups will be the same except that one group received the actual drug and the other received a placebo. Hence, any difference in performance between the two groups can be attributed to the drug.

Program 9-1 computes the value of the test statistic T and the corresponding p value. It can be applied for both one- and two-sided tests. (The one-sided tests will be presented shortly.)

Example 9.6 Historical data indicate that the mean acidity (pH) level of rain in a certain industrial region in West Virginia is 5.2. To see whether there has been any recent change in this value, the acidity levels of 12 rainstorms over the past year have been measured, with the following results.

$$6.1, 5.4, 4.8, 5.8, 6.6, 5.3, 6.1, 4.4, 3.9, 6.8, 6.5, 6.3$$

Are these data strong enough, at the 5 percent level of significance, for us to conclude that the acidity of the rain has changed from its historical value?

Solution To test the hypothesis of no change in acidity, that is, to test

$$H_0: \mu = 5.2 \qquad \text{versus} \qquad H_1: \mu \neq 5.2$$

first we compute the value of the test statistic T. Now, a simple calculation using the above data yields for the values of the sample mean and sample standard deviation

$$\bar{X} = 5.667 \qquad \text{and} \qquad S = .921$$

Thus, the value of the test statistic is

$$T = \sqrt{12} \, \frac{5.667 - 5.2}{.921} = 1.76$$

Since, from Table D.2 of App. D, $t_{11, .025} = 2.20$, the null hypothesis is not rejected at the 5 percent level of significance. That is, the data are not strong enough to enable us to conclude, at the 5 percent level of significance, that the acidity of the rain has changed.

We could also have solved this problem by computing the p value by running Program 9-1.

The value of mu-zero is 5.2

The sample size is 12

The data values are 6.1, 5.4, 4.8, 5.8, 6.6, 5.3, 6.1, 4.4, 3.9, 6.8, 6.5, and 6.3

The program computes the value of the t-statistic as 1.755621

The p-value is .1069365

Thus, the p value is .107, and so the null hypothesis would not be rejected even at the 10 percent level of significance.

Suppose now that we want to test the null hypothesis

$$H_0: \mu \leq \mu_0$$

against the alternative

$$H_1: \mu > \mu_0$$

In this situation, we want to reject the null hypothesis that the population mean is less than or equal to μ_0 only when the test statistic

$$T = \sqrt{n}\, \frac{\bar{X} - \mu_0}{S}$$

is significantly large (for this will tend to occur when the sample mean is significantly larger than μ_0). Therefore, we obtain the following significance level α test:

Reject H_0 if $T \geq t_{n-1,\,\alpha}$
Do not reject H_0 otherwise

A pictorial depiction of the test is shown in Fig. 9.6.
   Equivalently, the preceding test can be performed by first computing the value of the test statistic T, say its value is v, and then computing the p value which is equal to the probability that a t random variable with $n - 1$ degrees of freedom will be at least as large as v. That is, if $T = v$, then

$$p \text{ value} = P\{T_{n-1} \geq v\}$$

If we want to test the hypothesis

$$H_0: \mu \geq \mu_0 \qquad \text{versus} \qquad H_1: \mu < \mu_0$$

then the test is analogous. The significance level α test is again based on the test statistic

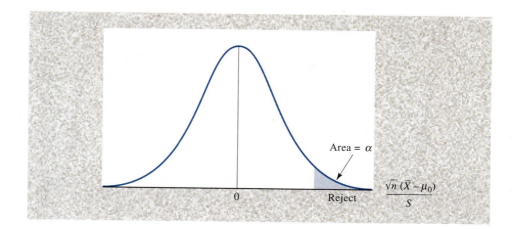

Figure 9.6
Testing H_0: $\mu \leq \mu_0$ against H_1: $\mu > \mu_0$.

$$T = \sqrt{n}\,\frac{\bar{X} - \mu_0}{S}$$

and the test is as follows.

Reject H_0	if $T \leq -t_{n-1,\,\alpha}$
Do not reject H_0	otherwise

In addition, the *p* value equals the probability that a *t* random variable with $n - 1$ degrees of freedom is less than or equal to the observed value of *T*.

Program 9-1 will compute the value of the test statistic *T* and the resulting *p* value. If only summary data are provided, then Program 8-2, which computes probabilities concerning *t* random variables, can be employed.

Example 9.7 The manufacturer of a new fiberglass tire claims that the average life of a set of its tires is at least 50,000 miles. To verify this claim, a sample of 8 sets of tires was chosen, and the tires subsequently were tested by a consumer agency. If the resulting values of the sample mean and sample variance were, respectively, 47.2 and 3.1 (in 1000 miles), test the manufacturer's claim.

Solution To determine whether the foregoing data are consistent with the hypothesis that the mean life is at least 50,000 miles, we will test

$$H_0: \mu \geq 50 \qquad \text{versus} \qquad H_1: \mu < 50$$

A rejection of the null hypothesis H_0 would then discredit the claim of the manufacturer. The value of the test statistic *T* is

What Is the Appropriate Null Hypothesis?

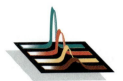

Suppose that the television tubes produced by a certain manufacturer are known to have mean lifetimes of 3000 hours of use. An outside consultant claims that a new production method will lead to a greater mean life. To check this, a pilot program is designed to produce a sample of tubes by the newly suggested approach. How should the manufacturer use the resulting data?

At first glance, it might appear that the data should be used to test

$$H_0: \mu \leq 3000 \quad \text{against} \quad H_1: \mu > 3000$$

Then a rejection of H_0 would be strong evidence that the newly proposed approach resulted in an improved tube. However, the trou-

ble with testing this hypothesis is that if the sample size is large enough, then there is a reasonable chance of rejecting H_0 even in cases where the new mean life is only, say, 3001 hours, and it might not be economically feasible to make the change-over for such a small increase in mean life. Indeed, the data should be used to test

$$H_0: \mu \leq 3000 + c$$

against

$$H_1: \mu > 3000 + c$$

where c is the smallest increase in mean life which would make it economically feasible to make the production change.

$$T = \sqrt{8} \; \frac{47.2 - 50}{3.1} = -2.55$$

Since $t_{7, .05} = 1.895$ and the test calls for rejecting H_0 when T is less than or equal to $-t_{7, \alpha}$, it follows that H_0 is rejected at the 5 percent level of significance. On the other hand, since $t_{7, .01} = 2.998$, H_0 would not be rejected at the 1 percent level. Running Program 8-2 shows that the p value is equal to .019, illustrating that the data strongly indicate that the manufacturer's claim is invalid.

The t test can be used even when the underlying distribution is not normal, provided the sample size is reasonably large. This is true because, by the central limit theorem, the sample mean $\overline{X}$ will be approximately normal no matter what the population distribution, and because the sample standard deviation S will approximately equal σ. Indeed, since for large n the t distribution with $n - 1$ degrees of freedom is almost identical to the standard normal, the above is equivalent to noting that $\sqrt{n}(\overline{X} - \mu_0)/S$ will approximately have a standard normal distribution when μ_0 is the population mean and the sample size n is large.

Table 9.2 summarizes the tests presented in this section.

| Table 9.2 | **Hypothesis Tests Concerning the Mean μ of a Normal Population with Unknown Variance σ^2** |

$X_1, \ldots, X_n$ are sample data;

$$\bar{X} = \frac{\sum\limits_{i=1}^{n} X_i}{n} \qquad S^2 = \frac{\sum\limits_{i=1}^{n} (X_i - \bar{X})^2}{n - 1}$$

H_0	H_1	Test statistic TS	Significance level α test	p Value if TS = v
$\mu = \mu_0$	$\mu \neq \mu_0$	$\sqrt{n}\,\dfrac{\bar{X} - \mu_0}{S}$	Reject H_0 if $\lvert TS \rvert \geq t_{n-1,\,\alpha/2}$ Do not reject otherwise	$2P\{T_{n-1} \geq \lvert v \rvert\}$
$\mu \leq \mu_0$	$\mu > \mu_0$	$\sqrt{n}\,\dfrac{(\bar{X} - \mu_0)}{S}$	Reject H_0 if $TS \geq t_{n-1,\,\alpha}$ Do not reject otherwise	$P\{T_{n-1} \geq v\}$
$\mu \geq \mu_0$	$\mu < \mu_0$	$\sqrt{n}\,\dfrac{(\bar{X} - \mu_0)}{S}$	Reject H_0 if $TS \leq -t_{n-1,\,\alpha}$ Do not reject otherwise	$P\{T_{n-1} \leq v\}$

T_{n-1} is a t random variable with $n - 1$ degrees of freedom, and $t_{n-1,\,\alpha}$ and $t_{n-1,\,\alpha/2}$ are such that $P\{T_{n-1} \geq t_{n-1,\,\alpha}\} = \alpha$ and $P\{T_{n-1} \geq t_{n-1,\,\alpha/2}\} = \alpha/2$.

Problems

1. There is some variability in the amount of phenobarbital in each capsule sold by a manufacturer. However, the manufacturer claims that the mean value is 20.0 milligrams. To test this, a sample of 25 pills yielded a sample mean of 19.7 with a sample standard deviation of 1.3. What inference would you draw from these data? In particular, are the data strong enough evidence to discredit the claim of the manufacturer? Use the 5 percent level of significance.

2. A fast-food establishment has been averaging about $2000 of business per weekday. To see whether business is changing due to a deteriorating economy (which may or may not be good for the fast-food industry), management has decided to carefully study the figures for the next 8 days. Suppose the figures are

    2050, 2212, 1880, 2121, 2205, 2018, 1980, 2188

    (a) What are the null and the alternative hypotheses?
    (b) Are the data significant enough, at the 5 percent level, to prove that a change has occurred?

Ronald A. Fisher

William S. Gosset

In 1908, William Seeley Gosset, writing under the name *Student,* published the distribution of the *t* statistic $\sqrt{n}(\bar{X} - \mu)/S$. It was an important result, for it enabled one to make tests of population means when only small samples were available, as was often the case at the Guinness brewery, where Gosset was employed. Its importance, however, was not noted, and it was mainly ignored by the statistical community at the time. This was primarily because the idea of learning from small samples went against the prevailing scientific beliefs, which were that "if your sample was sufficiently large, then substitute *S* for σ and use the normal distribution, and if your sample was not sufficiently large, then do not apply statistics." One of the few to realize its importance was R. A. Fisher, who in a later paper refined and fixed some technical errors in Gosset's work. However, it was not until Fisher's book *Statistical Methods for Research Workers* appeared in 1925 that the *t* test became widely used and appreciated. Fisher's book was a tremendous success, and it went through 11 editions in its first 25 years of existence. While it was extremely influential, it was, like Fisher's other writings, not easy to read. Indeed, it was said by a coworker at the time that "No student should attempt to read it who has not read it before."

(*Note:* Photograph of Gosset from *Student: A Statistical Biography of William Sealy Gosset.* Based on writings by E. S. Pearson, edited and augmented by R. L. Plackett with the assistance of G. A. Barnard. Clarendon Press, Oxford, 1990. Photograph from *Annals of Eugenics,* 1939, vol. 9.)

(c) What about at the 1 percent level?

(d) If you can run Program 9-1 or some equivalent software, find the *p* value.

3. To test the hypothesis that a normal population has mean 100, a random sample of size 10 is chosen. If the sample mean is 110, will you reject the null hypothesis if the following is known?

(a) The population standard deviation is known to equal 15.

(b) The population standard deviation is unknown, and the sample standard deviation is 15.

Use the 5 percent level of significance.

4. The number of lunches served daily at a school cafeteria last year was normally distributed with mean 300. The menu has been changed this year to healthier foods, and the administration wants to test the hypothesis that the mean number of lunches sold is unchanged. A sample of 12 days yielded the following number of lunches sold:

312, 284, 281, 295, 306, 273, 264, 258, 301, 277, 280, 275

Is the hypothesis that the mean is equal to 300 rejected at the

(a) 10 percent

(b) 5 percent

(c) 1 percent

level of significance?

5. An oceanographer wants to check whether the average depth of the ocean in a certain region is 55 fathoms, as had been previously reported. He took soundings at 36 randomly chosen locations in the region and obtained a sample mean of 56.4 fathoms with s sample standard deviation of 5.1 fathoms. Are these data significant enough to reject the null hypothesis that the mean depth is 55 fathoms, at the
   (a) 10 percent
   (b) 5 percent
   (c) 1 percent
   level of significance?

6. Twenty years ago, entering first-year high school students could do an average of 24 push-ups in 60 seconds. To see whether this remains true today, a random sample of 36 first-year students was chosen. If their average was 22.5 with a sample standard deviation of 3.1, can we conclude that the mean is no longer equal to 24? Use the 5 percent level of significance.

7. The mean response time of a species of pigs to a stimulus is .8 second. Twenty-eight pigs were given 2 ounces of alcohol and then tested. If their average response time was 1.0 second with a standard deviation of .3 second, can we conclude that alcohol affects the mean response time? Use the 5 percent level of significance.

8. Previous studies have shown that mice gain, on average, 5 grams of weight during their first 10 days after being weaned. A group of 36 mice had the artificial sweetener aspartame added to their food. Their average gain was 4.5 grams, with a sample standard deviation of .9 gram. Can we conclude, at the 1 percent level, that the addition of aspartame had an effect?

9. Use the results on last Sunday's National Football League (NFL) professional games to test the hypothesis that the average number of points scored by winning teams is 26.2. Use the 5 percent level of significance.

10. Use the results on last Sunday's major league baseball scores to test the hypothesis that the average number of runs scored by winning teams is 5.6. Use the 5 percent level of significance.

11. A bakery was taken to court for selling loaves of bread that were underweight. These loaves were advertised as weighing 24 ounces. In its defense, the bakery claimed that the advertised weight was meant to imply not that each loaf weighed exactly 24 ounces, but rather that the average value over all loaves was 24 ounces. The prosecution in a rebuttal produced evidence that a randomly chosen sample of 20 loaves had an average weight of 22.8 ounces with a sample standard deviation of 1.4 ounces. In her ruling, the judge stated that advertising a weight of 24 ounces would be acceptable if the mean weight were at least 23 ounces.
    (a) What hypothesis should be tested?
    (b) For the 5 percent level of significance, what should the judge rule?

12. A recently published study claimed that the average academic year salary of full professors at colleges and universities in the United States is $57,800. Students at a certain private school guess that the average salary of their professors is higher than this figure and so have decided to test the null hypothesis

$$H_0: \mu \leq 57{,}800 \qquad \text{against} \qquad H_1: \mu > 57{,}800$$

where μ is the average salary of full professors at their school. A random sample of 10 professors elicited the following salaries (in units of $1000):

61.0, 49.8, 72.0, 63.5, 52.0, 58.6, 60.0, 68.6, 71.0, 54.0

   (a) Is the null hypothesis rejected at the 10 percent level of significance?
   (b) What about at the 5 percent level?

13. A car is advertised as getting at least 31 miles per gallon in highway driving on trips of at least 100 miles. Suppose the miles per gallon obtained in 8 independent experiments (each consisting of a nonstop highway trip of 100 miles) are

28, 29, 31, 27, 30, 35, 25, 29

   (a) If we want to check if these data disprove the advertising claim, what should we take as the null hypothesis?
   (b) What is the alternative hypothesis?
   (c) Is the claim disproved at the 5 percent level of significance?
   (d) What about at the 1 percent level?

14. A manufacturer claims that the mean lifetime of the batteries it produces is at least 250 hours of use. A sample of 20 batteries yielded the following data:

237, 254, 255, 239, 244, 248, 252, 255, 233, 259, 236, 232, 243, 261, 255, 245, 248, 243, 238, 246

   (a) Are these data consistent, at the 5 percent level, with the claim of the manufacturer?
   (b) What about at the 1 percent level?

15. A water official insists that the average daily household water use in a certain county is at least 400 gallons. To check this claim, a random sample of 25 households was checked. The average of those sampled was 367 with a sample standard deviation of 62. Is this consistent with the official's claim?

16. A company supplies plastic sheets for industrial use. A new type of plastic has been produced, and the company would like to prove to an independent assessor that the average stress resistance of this new product is greater than 30.0, where stress resistance is measured in pounds per square inch necessary to

crack the sheet. A random sample of size 12 yielded the following values of stress resistance:

30.1, 27.8, 32.2, 29.4, 24.8, 31.6, 28.8, 29.4, 30.5, 27.6, 33.9, 31.4

(a) Do these data establish that the mean stress resistance is greater than 30.0 pounds per square inch, at the 5 percent level of significance?
(b) What was the null hypothesis in part (a)?
(c) If the answer to (a) is no, do the data establish that the mean stress resistance is less than 30 pounds per square inch?

17. A medical scientist believes that the average basal temperature of (outwardly) healthy individuals has increased over time and is now greater than 98.6°F (37°C). To prove this, she has randomly selected 100 healthy individuals. If their mean temperature is 98.74°F with a sample standard deviation of 1.1°F, does this prove her claim at the 5 percent level? What about at the 1 percent level?

18. In 1981, entering students at a certain university had an average score of 542 on the verbal part of the SAT. A random sample of the scores of 20 students in the 1993 class resulted in these scores:

542, 490, 582, 511, 515, 564, 500, 602, 488, 512, 518, 522, 505, 569, 575, 515, 520, 528, 533, 515

Do the above data prove that the average score has decreased to below 542? Use the 5 percent level of significance.

9.5   HYPOTHESIS TESTS CONCERNING POPULATION PROPORTIONS

In this section we will consider tests concerning the proportion of members of a population that possess a certain characteristic. We suppose that the population is very large (in theory, of infinite size), and we let p denote the unknown proportion of the population with the characteristic. We will be interested in testing the null hypothesis

$$H_0: p \le p_0$$

against the alternative

$$H_1: p > p_0$$

for a specified value p_0.

If a random selection of n elements of the population is made, then X, the number with the characteristic, will have a binomial distribution with parameters n and p. Now it should be clear that we want to reject the null hypothesis that the proportion is less than or equal to p_0 only when X is sufficiently large. Hence, if the observed value of X is x, then the p value of these data will equal the probability that at least as large a value would have been obtained if p had been equal to p_0 (which is the largest possible value of p under the null hypothesis). That is, if we observe that X is equal to x, then

$$p \text{ value} = P\{X \geq x\}$$

where X is a binomial random variable with parameters n and p_0.

The p value can now be computed either by using the normal approximation or by running Program 5-1 which computes the binomial probabilities. The null hypothesis should then be rejected at any significance level that is greater than or equal to the p value.

Example 9.8   A noted educator claims that over half the adult U.S. population is concerned about the lack of educational programs shown on television. To gather data about this issue, a national polling service randomly chose and questioned 920 individuals. If 478 (52 percent) of those surveyed stated that they are concerned at the lack of educational programs on television, does this prove the claim of the educator?

Solution   To prove the educator's claim, we must show that the data are strong enough to reject the hypothesis that at most 50 percent of the population is concerned about the lack of educational programs on television. That is, if we let p denote the proportion of the population that is concerned about this issue, then we should use the data to test

$$H_0\colon p \leq .50 \qquad \text{versus} \qquad H_1\colon p > .50$$

Since 478 people in the sample were concerned, it follows that the p value of these data is

$$\begin{aligned} p \text{ value} &= P\{X \geq 478\} \qquad \text{when X is binomial (920, .50)} \\ &= .1243 \qquad\qquad\quad \text{from Program 5-1} \end{aligned}$$

For such a large p value we cannot conclude that the educator's claim has been proved. Although the data are certainly in support of that claim, since 52 percent of those surveyed were concerned by the lack of educational programs on television, such a result would have had a reasonable chance of occurring even if the claim were incorrect, and so the null hypothesis is not rejected.

If Program 5-1 were not available to us, then we could have approximated the p value by using the normal approximation to binomial probabilities. Since $np = 920(.50) = 460$ and $np(1 - p) = 460(.5) = 230$, this would have yielded the following:

$$p \text{ value} = P\{X \geq 478\}$$
$$= P\{X \geq 477.5\} \qquad \text{continuity correction}$$
$$= P\left\{\frac{X - 460}{\sqrt{230}} \geq \frac{477.5 - 460}{\sqrt{230}}\right\}$$
$$\approx P\{Z \geq 1.154\} = .1242$$

Thus the p value obtained by the normal approximation is quite close to the exact p value obtained by running Program 5-1.

For another type of example in which we are interested in an hypothesis test of a binomial parameter, consider a process that produces items that are classified as being either acceptable or defective. A common assumption is that each item produced is independently defective with a certain probability p; and so the number of defective items in a batch of size n will have a binomial distribution with parameters n and p.

Example 9.9   A computer chip manufacturer claims that at most 2 percent of the chips it produces are defective. An electronics company, impressed by that claim, has purchased a large quantity of chips. To determine if the manufacturer's claim is plausible, the company has decided to test a sample of 400 of these chips. If there are 13 defective chips (3.25 percent) among these 400, does this disprove (at the 5 percent level of significance) the manufacturer's claim?

Solution If p is the probability that a chip is defective, then we should test the null hypothesis

$$H_0: p \leq .02 \qquad \text{against} \qquad H_1: p > .02$$

That is, to see if the data disprove the manufacturer's claim, we must take that claim as the null hypothesis. Since 13 of the 400 chips were observed to be defective, the p value is equal to the probability that such a large number of defectives would have occurred if p were equal to .02 (its largest possible value under H_0). Therefore,

$$p \text{ value} = P\{X \geq 13\} \qquad \text{where } X \text{ is binomial } (400, .02)$$
$$= .0619 \qquad \text{from Program 5-1}$$

and so the data, though clearly not in favor of the manufacturer's claim, are not quite strong enough to reject that claim at the 5 percent level of significance.

If we had used the normal approximation, then we would have obtained the following result for the p value:

$$
\begin{aligned}
p \text{ value} &= P\{X \geq 13\} && \text{where } X \text{ is binomial } (400, .02) \\
&= P\{X \geq 12.5\} && \text{continuity correction} \\
&= P\left\{\frac{X - 8}{\sqrt{8(.98)}} \geq \frac{12.5 - 8}{\sqrt{8(.98)}}\right\} \\
&\approx P\{Z \geq 1.607\} && \text{where } Z \text{ is standard normal} \\
&= .054
\end{aligned}
$$

Thus, the approximate p value obtained by using the normal approximation, though not as close to the actual p value of .062 as we might have liked, is still accurate enough to lead to the correct decision that the data are not quite strong enough to reject the null hypothesis at the 5 percent level of significance.

Once again, let p denote the proportion of members of a large population who possess a certain characteristic, but suppose that we now want to test

$$H_0: p \geq p_0$$

against

$$H_1: p < p_0$$

for some specified value p_0. That is, we want to test the null hypothesis that the proportion of the population with the characteristic is at least p_0 against the alternative that it is less than p_0. If a random sample of n members of the population results in x of them having the characteristic, then the p value of these data is given by

$$p \text{ value} = P\{X \leq x\}$$

where X is a binomial random variable with parameters n and p_0.

That is, when the null hypothesis is that p is at least as large as p_0, then the p value is equal to the probability that a value as small as or smaller than the one observed would have occurred if p were equal to p_0.

9.5.1   Two-Sided Tests of p

Computation of the p value of the test data becomes slightly more involved when we are interested in testing the hypothesis

$$H_0: p = p_0$$

against the two-sided alternative

$$H_1: p \neq p_0$$

for a specified value p_0.

Again suppose that a sample of size n is chosen, and let X denote the number of members of the sample who possess the characteristic of interest. We will want to reject H_0 when X/n, the proportion of the sample with the characteristic, is either much smaller or much larger than p_0; or, equivalently, when X is either very small or very large in relation to np_0. Since we want the total probability of rejection to be less than or equal to α when p_0 is indeed the true proportion, we can attain these objectives by rejecting for both large and small values of X with probability, when H_0 is true, $\alpha/2$. That is, if we observe a value such that the probability is less than or equal to $\alpha/2$ that X would be either that large or that small when H_0 is true, then H_0 should be rejected.

Therefore, if the observed value of X is x, then H_0 will be rejected if either

$$P\{X \leq x\} \leq \frac{\alpha}{2}$$

or

$$P\{X \geq x\} \leq \frac{\alpha}{2}$$

when X is a binomial random variable with parameters n and p_0. Hence, the significance level α test will reject H_0 if

$$\text{Min}\{P\{X \leq x\}, P\{X \geq x\}\} \leq \frac{\alpha}{2}$$

or, equivalently, if

$$2 \, \text{Min}\{P\{X \leq x\}, P\{X \geq x\}\} \leq \alpha$$

where X is binomial (n, p_0). From this, it follows that if x members of a random sample of size n have the characteristic, then the p value for the test of

$$H_0: p = p_0 \qquad \text{versus} \qquad H_1: p \neq p_0$$

is as follows:

$$p \text{ value} = 2 \text{ Min}\{P\{X \leq x\}, P\{X \geq x\}\}$$

where X is a binomial random variable with parameters n and p_0.

Since it will usually be evident which of the two probabilities in the expression for the p value will be smaller (if $x \leq np_0$, then it will almost always be the first, and otherwise the second, probability), Program 5-1 or the normal approximation is needed only once to obtain the p value.

Example 9.10   Historical data indicate that 4 percent of the components produced at a certain manufacturing facility are defective. A particularly acrimonious labor dispute has recently been concluded, and management is curious about whether it will result in any change in this figure of 4 percent. If a random sample of 500 items indicated 16 defectives (3.2 percent), is this significant evidence, at the 5 percent level of significance, to conclude that a change has occurred?

Solution   To be able to conclude that a change has occurred, the data need to be strong enough to reject the null hypothesis when you are testing

$$H_0: p = .04 \qquad \text{versus} \qquad H_1: p \neq .04$$

where p is the probability that an item is defective. The p value of the observed data of 16 defectives in 500 items is

$$p \text{ value} = 2 \text{ Min}\{P\{X \leq 16\}, P\{X \geq 16\}\}$$

where X is a binomial (500, .04) random variable. Since $500 \times .04 = 20$, we see that

$$p \text{ value} = 2P\{X \leq 16\}$$

Since X has mean 20 and standard deviation $\sqrt{20(.96)} = 4.38$, it is clear that twice the probability that X will be less than or equal to 16—a value less than 1 standard deviation lower than the mean—is not going to be small enough to justify rejection. Indeed, it can be shown that

$$p \text{ value} = 2P\{X \leq 16\} = .432$$

and so there is not sufficient evidence to reject the hypothesis that the probability of a defective item has remained unchanged.

Table 9.3 sums up the tests concerning the population proportion p.

Table 9.3	**Hypothesis Tests concerning p, the Proportion of a Large Population That Has a Certain Characteristic**

The number of population members in a sample of size n that have the characteristic is X, and B is a binomial random variable with parameters n and p_0.

H_0	H_1	Test statistic TS	p Value if TS $= x$
$p \leq p_0$	$p > p_0$	X	$P\{B \geq x\}$
$p \geq p_0$	$p < p_0$	X	$P\{B \leq x\}$
$p = p_0$	$p \neq p_0$	X	$2 \, \mathrm{Min}\{P\{B \leq x\}, P\{B \geq x\}\}$

Problems

In solving the following problems, either make use of Program 5-1 or equivalent software to compute the relevant binomial probabilities, or use the normal approximation.

1. A standard drug is known to be effective in 72 percent of cases in which it is used to treat a certain infection. A new drug has been developed, and testing has found it to be effective in 42 cases out of 50. Is this strong enough evidence to prove that the new drug is more effective than the old one? Find the relevant p value.

2. An economist thinks that at least 60 percent of recently arrived immigrants who have been working in the health profession in the United States for more than 1 year feel that they are underemployed with respect to their training. Suppose a random sample of size 450 indicated that 294 individuals (65.3 percent) felt they were underemployed. Is this strong enough evidence, at the 5 percent level of significance, to prove that the economist is correct? What about at the 1 percent level of significance?

3. Shoplifting is a serious problem for retailers. In the past, a large department store found that 1 out of every 14 people entering the store engaged in some form of shoplifting. To help alleviate this problem, 3 months ago the store hired additional security guards. This additional hiring was widely publicized. To assess its effect, the store recently chose 300 shoppers at random and closely followed their movements by camera. If 18 of these 300 shoppers were involved in shoplifting, does this prove, at the 5 percent level of significance, that the new policy is working?

4. Let p denote the proportion of voters in a large city who are in favor of restructuring the city government; and consider a test of the hypothesis

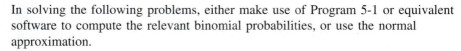

$$H_0: p \geq .60 \quad \text{against} \quad H_1: p < .60$$

Historical Perspective

The First Published Hypothesis Test "Proved" the Existence of God

Remarkably enough, the first published paper in which a statistical test was made of a null hypothesis was used to claim the existence of God. In a paper published in the *Philosophical Transactions of the Royal Society* in 1710, John Arbuthnot looked at the number of males and females born in each of the 82 years from 1629 to 1710; and he discovered that in each of these years the number of male births exceeded the number of female births. Arbuthnot argued that this could not have been due solely to chance, for if each birth were equally likely to be either a boy or a girl (and so each year would be equally

likely to have either more male births or more female births), then the probability of the observed outcome would equal $(1/2)^{82}$. Thus, he argued, the hypothesis that the event occurred solely by chance must be rejected [in our language, the p value of the test of $H_0: p = 1/2$ versus $H_1: p \neq 1/2$ was $2(1/2)^{82}$]. Arbuthnot then argued that the result must have been due to planning; and as he believed it was beneficial to initially have an excess of male babies, since males tend to do more hazardous work than females and thus tend to die earlier, he concluded that it was the work of God. (For reasons not totally understood, it appears that the probability of a newborn's being male is closer to .51 than it is to .50.)

A random sample of n voters indicated that x are in favor of restructuring. In each of the following cases, would a significance level α test result in H_0 being rejected?

(a) $n = 100$, $x = 50$, $\alpha = .10$
(b) $n = 100$, $x = 50$, $\alpha = .05$
(c) $n = 100$, $x = 50$, $\alpha = .01$
(d) $n = 200$, $x = 100$, $\alpha = .01$

5. A politician claims that over 50 percent of the population is in favor of her candidacy. To prove this claim, she has commissioned a polling organization to do a study. This organization chose a random sample of individuals in the population and asked each member of the sample if he or she was in favor of the politician's candidacy.
   (a) To prove the politician's claim, what should be the null and alternative hypotheses?
   Consider the following three alternative results, and give the relevant p values for each one.
   (b) A random sample of 100 voters indicated that 56 (56 percent) are in favor of her candidacy.
   (c) A random sample of 200 voters indicated that 112 (56 percent) are in favor of her candidacy.
   (d) A random sample of 500 indicated that 280 (56 percent) are in favor.
   Give an intuitive explanation for the discrepancy in results, if there are any,

even though in each of cases (b), (c), and (d) the same percentage of the sample was in favor.

6. A revamped television news program has claimed to its advertisers that at least 24 percent of all television sets that are on when the program runs are tuned in to it. This figure of 24 percent is particularly important because the advertising rate increases at that level of viewers. Suppose a random sample indicated that 50 out of 200 televisions were indeed tuned in to the program.
   (a) Is this strong enough evidence, at the 5 percent significance level, to establish the accuracy of the claim?
   (b) Is this strong enough evidence, at the 5 percent significance level, to prove that the claim is unfounded?
   (c) Would you say that the results of this sample are evidence for or against the claim of the news program?
   (d) What do you think should be done next?

7. Three independent news services are running a poll to determine if over half the population supports an initiative concerning limitations on driving automobiles in the downtown area. Each news service wants to see if the evidence indicates that over half the population is in favor. As a result, all three services will be testing

$$H_0: p \leq .5 \quad \text{against} \quad H_1: p > .5$$

where p is the proportion of the population in favor of the initiative.
   (a) Suppose the first news organization samples 100 people, of whom 56 are in favor of the initiative. Is this strong enough evidence, at the 5 percent level of significance, to reject the null hypothesis and in doing so establish that over half the population favors the initiative?
   (b) Suppose the second news organization samples 120 people, of whom 68 are in favor of the initiative. Is this strong enough evidence, at the 5 percent level of significance, to reject the null hypothesis?
   (c) Suppose the third news organization samples 110 people, of whom 62 are in favor of the initiative. Is this strong enough evidence, at the 5 percent level of significance, to reject the null hypothesis?
   (d) Suppose the news organizations combine their samples, to come up with a sample of 330 people, of whom 186 support the initiative. Is this strong enough evidence, at the 5 percent level of significance, to reject the null hypothesis?

8. An ambulance service claims that at least 45 percent of its calls involve life-threatening emergencies. To check this claim, a random sample of 200 calls was selected from the service's files. If 70 of these calls involved life-threatening emergencies, is the service's claim believable
   (a) at the 5 percent
   (b) at the 1 percent
   level of significance?

9. A retailer has received a large shipment of items of a certain type. If it can be established that over 4 percent of the items in the shipment are defective, then the shipment will be returned. Suppose that 5 defectives are found in a random sample of 90 items. Should the shipment be returned to its sender? Use the 10 percent level of significance. What about at the 5 percent level?

10. A campus newspaper editorial claims that at least 75 percent of the students favor traditional course grades rather than a pass/fail option. To gain information, a dean randomly sampled 50 students and learned that 32 of them favor traditional grades. Are these data consistent with the claim made in the editorial? Use the 5 percent level of significance.

11. A recent survey published by the Higher Educational Research Institute stated that 22 percent of entering college students classified themselves as politically liberal. If 65 out of a random sample of 264 entering students at the University of California at Berkeley classified themselves as liberals, does this establish, at the 5 percent level of significance, that the percentage at Berkeley is higher than the national figure?

12. It has been "common wisdom" for some time that 22 percent of the population have a firearm at home. In a recently concluded poll, 54 out of 200 randomly chosen people were found to have a firearm in their homes. Is this strong enough evidence, at the 5 percent level of significance, to disprove common wisdom?

13. The average length of a red light is 30 seconds. Because of this, a certain individual feels lucky whenever he has to wait less than 15 seconds when encountering a red light. This individual assumes that the probability that he is lucky is .5. To test this hypothesis, he timed himself at 30 red lights. If he had to wait more than 15 seconds a total of 19 times, should he reject the hypothesis that p is equal to .5?
(a) Use the 10 percent level of significance.
(b) Use the 5 percent level of significance.
(c) What is the p value?

14. A statistics student wants to test the hypothesis that a certain coin is equally likely to land on either heads or tails when it is flipped. The student flips the coin 200 times, obtaining 116 heads and 84 tails.
(a) For the 5 percent level of significance, what conclusion should be drawn?
(b) What are the null and the alternative hypotheses?
(c) What is the p value?

15. Forty-two percent of women of child-bearing age smoke. A scientist wanted to test the hypothesis that this is also the proportion of smokers in the population of women who suffer ectopic pregnancies. To do so, the scientist chose a random sample of 120 women who had recently suffered an ectopic pregnancy. If 68 of these women turn out to be smokers, what is the p value of the test of the hypothesis

$$H_0: p = .42 \quad \text{against} \quad H_1: p \neq .42$$

where p is the proportion of smokers in the population of women who have suffered an ectopic pregnancy?

Key Terms

Statistical hypothesis: A statement about the nature of a population. It is often stated in terms of a population parameter.

Null hypothesis: A statistical hypothesis that is to be tested.

Alternative hypothesis: The alternative to the null hypothesis.

Test statistic: A function of the sample data. Depending on its value, the null hypothesis will be either rejected or not rejected.

Critical region: If the value of the test statistic falls in this region, then the null hypothesis is rejected.

Significance level: A small value set in advance of the testing. It represents the maximal probability of rejecting the null hypothesis when it is true.

Z test: A test of the null hypothesis that the mean of a normal population having a known variance is equal to a specified value.

p value: The smallest significance level at which the null hypothesis is rejected.

One-sided tests: Statistical hypothesis tests in which either the null or the alternative hypothesis is that a population parameter is less than or equal to (or greater than or equal to) some specified value.

t test: A test of the null hypothesis that the mean of a normal population having an unknown variance is equal to a specified value.

Summary

A *statistical hypothesis* is a statement about the parameters of a population distribution.

The hypothesis to be tested is called the *null hypothesis* and is denoted by H_0. The alternative hypothesis is denoted by H_1.

A hypothesis test is defined by a test statistic, which is a function of the sample data, and a critical region. The null hypothesis is rejected if the value of the test statistic falls within the critical region and is not rejected otherwise. The critical region is chosen so that the probability of rejecting the null hypothesis, when it is true, is no greater than a predetermined value α, called the *significance level* of the test. The significance level is typically set equal to such values as .10, .05, and .01. The 5 percent level of significance, that is, $\alpha = .05$, has become the most common in practice.

Since the significance level is set to equal some small value, there is only a small chance of rejecting H_0 when it is true. Thus a statistical hypothesis test is basically trying to determine whether the data are consistent with a given null hypothesis. Therefore, rejecting H_0 is a strong statement that the null hypothesis does not appear to be consistent with the data, whereas not rejecting H_0 is a much weaker statement to the effect that H_0 is not inconsistent with the data. For this reason, the hypothesis that one is trying to establish should generally be designated as the alternative hypothesis so that it can be "statistically proved" by a rejection of the null hypothesis.

Often in practice a significance level is not set in advance, but rather the test statistic is observed to determine the minimal significance level that would result in a rejection of the null hypothesis. This minimal significance level is called the p *value.* Thus, once the p value is determined, the null hypothesis will be rejected at any significance level which is at least as large as the p value. The following rules of thumb concerning the p value are in rough use:

p value $> .1$ Data are reasonably consistent with H_0.

p value $\approx .05$ Data provide moderate evidence against H_0.

p value $< .01$ Data provide strong evidence against H_0.

---

1. *Testing* H_0: $\mu = \mu_0$ *against* H_1: $\mu \neq \mu_0$ *in a normal population having known standard deviation* σ: The significance level α test is based on the test statistic

$$\sqrt{n} \, \frac{\bar{X} - \mu_0}{\sigma}$$

and it is to

Reject H_0 if $\sqrt{n} \, \dfrac{|\bar{X} - \mu_0|}{\sigma} \geq z_{\alpha/2}$

Not reject H_0 otherwise

If the observed value of the test statistic is v, then the p value is given by

$$\begin{aligned} p \text{ value} &= P\{|Z| \geq |v|\} \\ &= 2P\{Z \geq |v|\} \end{aligned}$$

where Z is a standard normal random variable.

---

2. *Testing*

(1) H_0: $\mu \leq \mu_0$ *against* H_1: $\mu > \mu_0$

or

(2) H_0: $\mu \geq \mu_0$ *against* H_1: $\mu < \mu_0$

in a normal population having known standard deviation σ: These are called one-sided tests. The significance level α test in both situations is based on the test statistic $\sqrt{n}(\bar{X} - \mu_0)/\sigma$. The test in situation (1) is to

Reject H_0	if $\sqrt{n} \dfrac{(\bar{X} - \mu_0)}{\sigma} \geq z_\alpha$
Not reject H_0	otherwise

Alternatively the test in (1) can be performed by first determining the p value of the data. If the value of the test statistic is v, then the p value is

$$p \text{ value} = P\{Z \geq v\}$$

where Z is a standard normal random variable. The null hypothesis will now be rejected at any significance level at least as large as the p value.

In situation (2), the significance level α test is to

Reject H_0	if $\sqrt{n} \dfrac{(\bar{X} - \mu_0)}{\sigma} \leq -z_\alpha$
Not reject H_0	otherwise

Alternatively, if the value of the test statistic $\sqrt{n}(\bar{X} - \mu_0)/\sigma$ is v, then the p value is given by

$$p \text{ value} = P\{Z \leq v\}$$

where Z is a standard normal random variable.

---

3. Two-sided t test of

$$H_0: \mu = \mu_0 \quad against \quad H_1: \mu \neq \mu_0$$

in a normal population whose variance is unknown: This test is based on the test statistic

$$T = \frac{\sqrt{n}(\bar{X} - \mu_0)}{S}$$

where n is the sample size and S is the sample standard deviation. The significance level α test is to

Reject H_0	if $|T| \geq t_{n-1, \alpha/2}$
Not reject H_0	otherwise

The value $t_{n-1, \alpha/2}$ is such that

$$P\{T_{n-1} > t_{n-1,\ \alpha/2}\} = \frac{\alpha}{2}$$

when T_{n-1} is a t random variable having $n - 1$ degrees of freedom. This is called the t test.

    The above t test can be alternatively run by first calculating the value of the test statistic T. If it is equal to v, then the p value is given by

$$\begin{aligned} p \text{ value} &= P\{|T_{n-1}| \geq |v|\} \\ &= 2P\{T_{n-1} \geq |v|\} \end{aligned}$$

where T_{n-1} is a t random variable with $n - 1$ degrees of freedom.

---

4. *One-sided t tests of*

    (1) H_0: $\mu \leq \mu_0$ *against* H_1: $\mu > \mu_0$

or

    (2) H_0: $\mu \geq \mu_0$ *against* H_1: $\mu < \mu_0$

in a normal population having an unknown variance: These tests are again based on the test statistic

$$T = \sqrt{n}\,\frac{\bar{X} - \mu_0}{S}$$

where n is the sample size and S is the sample standard deviation.

    The significance level α test of (1) is to

    Reject H_0 if $T \geq t_{n-1,\ \alpha}$

    Not reject H_0 otherwise

Alternatively, the p value may be derived. If the value of the test statistic T is v, the p value is obtained from

$$p \text{ value} = P\{T_{n-1} \geq v\}$$

where T_{n-1} is a t random variable having $n - 1$ degrees of freedom.

    The significance level α test of (2) is to

    Reject H_0 if $T \leq -t_{n-1,\ \alpha}$

    Not reject H_0 otherwise

If the value of T is v, then the p value of the test of (2) is

$$p \text{ value} = P\{T_{n-1} \leq v\}$$

5. *Hypothesis tests concerning proportions*: If p is the proportion of a large population that has a certain characteristic, then to test

$$H_0: p \le p_0 \qquad \text{versus} \qquad H_1: p > p_0$$

a random sample of n elements of the population should be drawn. The test statistic is X, the number of members of the sample with the characteristic. If the value of X is x, then the p value is given by

$$p \text{ value} = P\{B \ge x\}$$

where B is a binomial random variable with parameters n and p_0.
    Suppose we had wanted to test

$$H_0: p \ge p_0 \qquad \text{versus} \qquad H_1: p < p_0$$

If the observed value of the test statistic is x, then the p value is given by

$$p \text{ value} = P\{B \le x\}$$

where again B is binomial with parameters n and p_0.
    The binomial probabilities can be calculated by using Program 5-1 or can be approximated by making use of the normal approximation to the binomial.
    Suppose now that the desired test is two-sided; that is, we want to test

$$H_0: p = p_0 \qquad \text{versus} \qquad H_1: p \ne p_0$$

If the number of members of the sample with the characteristic is x, then the p value is

$$p \text{ value} = 2 \text{ Min}\{P\{B \le x\}, P\{B \ge x\}\}$$

where B is binomial with parameters n and p_0.

Review Problems and Proposed Case Studies

1. Suppose you were to explain to a person who has not yet studied statistics that a statistical test has just resulted in the rejection of the null hypothesis that a population mean μ is equal to 0. That is, $H_0: \mu = 0$ has been rejected, say, at the 5 percent level of significance. Which of the following is a more accurate statement?
    (a) The evidence of the data indicated that the population mean differs significantly from 0.
    (b) The evidence of the data was significant enough to indicate that the population mean differs from 0.
    What is misleading about the less accurate of these two statements?

2. Suppose that the result of a statistical test was that the *p* value was equal to .11.
   (a) Would the null hypothesis be rejected at the 5 percent level of significance?
   (b) Would you say that this test provided evidence for the truth of the null hypothesis? Briefly explain your answer.

3. Suppose you happened to read the following statement in your local newspaper. "A recent study provided significant evidence that the mean heights of women have increased over the past twenty years."
   (a) Do you regard the above as a precise statement?
   (b) What interpretation would you give to the statement?

4. A fact that has been long known but little understood is that in their early years twins tend to have lower IQ levels and tend to be slower in picking up language skills than nontwins. Recently, some psychologists have speculated that this may be due to the fact that parents spend less time with a twin child than they do with a single child. The reason for this is possibly that a twin always has to share the parent's attention with her or his sibling; and the reason is also possibly economic in nature, since twins place a greater economic burden on parents than do single children, and so parents of twins may have less time in general to spend with their offspring.

   Devise a study that could be used to test the hypothesis that twins obtain less parental time than single children.

   Assuming that the above hypothesis is correct, devise a study that might enable you to conclude that this is the reason for the long known but little understood fact.

5. An individual's present route to work results in, on average, 40 minutes of travel time per trip. An alternate route has been suggested by a friend, who claims that it will reduce the travel time. Suppose that the new route was tried on 10 randomly chosen occasions with the following times resulting:

   44, 38.5, 37.5, 39, 38.2, 36, 42, 36.5, 36, 34

   Do these data establish the claim that the new route is shorter, at the
   (a) 1 percent
   (b) 5 percent
   (c) 10 percent
   level of significance?

6. To test the null hypothesis

   $$H_0: \mu = 15 \quad \text{versus} \quad H_1: \mu \neq 15$$

   a sample of size 12 is taken. If the sample mean is 14.4, find the *p* value if the population standard deviation is known to equal
   (a) .5
   (b) 1.0
   (c) 2.0

7. It has been claimed that over 30 percent of entering college students have blood cholesterol levels of at least 200. Use the last 20 students in the list in App. A to test this hypothesis. What conclusion do you draw at the 5 percent level of significance?

8. Psychologists who consider themselves disciples of Alfred Adler believe that birth order has a strong effect on personality. Adler believed that firstborn (including only) children tend to be more self-confident and success-oriented than later-born children. For instance, of the first 102 appointments to the U.S. Supreme Court, 55 percent have been firstborn children, whereas only 37 percent of the population at large are firstborn.
   (a) Using the above data about the Supreme Court, test the hypothesis that the belief of Adlerians is wrong and being firstborn does not have a statistical effect on one's personality.
   (b) Is the result of (a) a convincing proof of the validity of the Adlerian position? (*Hint:* Recall *data mining*.)
   (c) Construct your own study to try to prove or disprove Adler's belief. Choose some sample of successful people (perhaps sample 200 major league baseball players), and find out what percentage of them are firstborn.

9. An individual named Nicholas Caputo was the clerk of Essex County, New Jersey, for an extended period. One of his duties as clerk was to hold a drawing to determine whether Democratic or Republican candidates would be listed first on county ballots. During his reign as clerk, the Democrats won the drawing on 40 of 41 occasions. As a result, Caputo, a Democrat, acquired the nickname *the man with the golden arm.* In 1985 Essex County Republicans sued Caputo, claiming that he was discriminating against them. If you were the judge, how would you rule? Explain!

10. A recent theory claims that famous people are more likely to die in the 6-month period after their birthday than in the 6-month period preceding it. That is, the claim is that a famous person born on July 1 would be more likely to die between July 1 and December 31 than between January 1 and July 1. The reasoning is that a famous person would probably look forward to all the attention and affection lavished on the birthday, and this anticipation would strengthen the person's "will to live." A countertheory is that famous people are less likely to die in the 6-month period following their birthdays due to their increased strength resulting from their birthday celebration. Still others assert that both theories are wrong.

    Let p denote the probability that a famous person will die within a 6-month period following his or her birthday, and consider a test of

$$H_0: p = \frac{1}{2} \quad \text{versus} \quad H_1: p \neq \frac{1}{2}$$

   (a) Suppose someone compiled a list of 200 famous dead people in each of 25 separate fields, and then ran 25 separate tests of the above null hypothesis.

Even if H_0 is always true, what is the probability that at least one of the tests will result in a rejection of H_0 at the 5 percent level of significance?

(b) Compile a list of between 100 and 200 famous dead people, and use it to test the above hypothesis.

11. Choose a random sample of 16 women from the list provided in App. A, and use their weights to test the null hypothesis that the average weight of all the women on the list is not greater than 110 pounds. Use the 5 percent level of significance.

12. Suppose that team A and team B are to play a National Football League game and team A is favored by f points. Let $S(A)$ and $S(B)$ denote, respectively, the scores of teams A and B, and let $X = S(A) - S(B) - f$. That is, X is the amount by which team A beats the point spread. It has been claimed that the distribution of X is normal with mean 0 and standard deviation 14. Use data concerning randomly chosen football games to test this hypothesis.

13. The random walk model for the price of a stock or commodity assumes that the successive differences in the logarithms of the closing prices of a given commodity constitute a random sample from a normal population. The following data give the closing prices of gold on 17 consecutive trading days in 1994. Use it to test the hypothesis that the mean daily change is equal to 0.

Closing prices					
387.10	391.00	389.50	391.00	395.00	396.25
388.00	391.95	390.25	390.50	393.50	395.45
389.65	391.05	388.00	394.00	396.25	

Note: The data are ordered by columns. The first value is 387.10, the second 388.00, the third 389.65, the fourth 391.00, and so on.

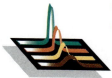

Estimation and Testing Statistical Hypotheses for a Single Population

Purpose

Use Minitab to

1. Construct confidence intervals for μ when σ is known.

2. Test for μ when σ is known.

3. Construct confidence intervals for μ when σ is unknown.

4. Test for μ when σ is unknown.

5. Test and construct confidence intervals for a population proportion.

Procedures

First, load the Minitab (Windows version) software as in the Minitab lab for Chap. 1.

1. CONSTRUCTING CONFIDENCE INTERVALS FOR THE POPULATION MEAN μ WHEN THE POPULATION STANDARD DEVIATION σ IS KNOWN

Example 1   Use Minitab to work Prob. 3 in Sec. 8.5 of the text.

Here we are given that the measurement of the PCB concentration of a fish caught in Lake Michigan is normally distributed with a standard deviation σ of .08 part per million. Ten independent measurements (sample size $n = 10$) of this fish were taken, and we are asked to construct a 95 percent confidence interval estimate for the (average, μ) PCB level of this fish.

To use Minitab, first you need to enter the 10 observed values in the Data window. Enter them in column C1, and rename it PCB. Next, select Stat→Basic Statistics→1-Sample z. Note we are using the 1-Sample z since the standard deviation σ for the population is known. The

1-Sample z dialog box will be displayed. In the Variables text box, select C1 (PCB) or type it in the text box. Use the mouse to select the Confidence interval check box. In the Level text box, type 95 since we are constructing a 95 percent confidence interval. In the Sigma text box type .08, the value of the known population standard deviation. Fig. M9.1 shows the 1-Sample z dialog box with the appropriate selection and entries.

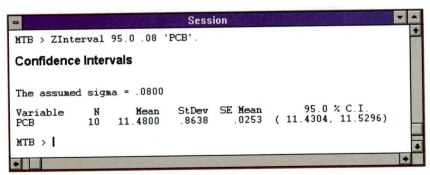

Figure M9.1

Click on the OK button, and the results will be displayed in the Session window. Fig. M9.2 shows this output with the 95 percent confidence interval estimate being (11.4304, 11.5296). That is, we can assert with 95 percent confidence that the average PCB level of the fish lies between 11.4304 and 11.5296 parts per million.

```
┌─────────────────────────────── Session ─────────────────────────────┐
│ MTB > ZInterval 95.0 .08 'PCB'.                                      │
│                                                                      │
│ Confidence Intervals                                                 │
│                                                                      │
│ The assumed sigma = .0800                                            │
│                                                                      │
│ Variable      N       Mean    StDev   SE Mean        95.0 % C.I.     │
│ PCB          10    11.4800    .8638     .0253   ( 11.4304, 11.5296)  │
│                                                                      │
│ MTB > |                                                              │
└──────────────────────────────────────────────────────────────────────┘
```

Figure M9.2

Example 2   Use Minitab to work Prob. 5 in Sec. 8.5 of the text.

Here we are given that the life of a particular brand of television picture tube is normally distributed with a standard deviation σ of 400 hours. A sample n of 20 tubes has an average life $\bar{x}$ of 9000 hours, and we are asked to construct a 90 and a 95 percent confidence interval estimate for the mean lifetime μ of such a tube.

In this case, the sample values are not given. However, we can circumvent this by entering 20 values of 9000 in Minitab. That is, we will trick Minitab into believing that we have 20 sample values each of size 9000. Note the sample mean $\bar{x}$ in this case will still be 9000 but Minitab will now know the sample size as well; and since we know the population standard deviation, we can use the procedures in Example 1. Verify that the 90 and 95 percent confidence interval estimates for the mean lifetime are (8852.8, 9147.2) and (8824.6, 9175.4), respectively.

Note If you have a large sample size *n*, then it would be impractical to type in the value of the sample mean *n* times in a specified column. Suppose that the sample size is $n = 100$ and the sample mean is 9000. To generate 100 values of size 9000 and store them in column C1, select Calc→Set Patterned Data, and the Set Patterned Data dialog box will be displayed. In the Store result in column, type C1. Select the Arbitrary list of constants box and type 9000. In the Repeat each value box, type 100, and in the Repeat the whole list box, type 1. When you click on the OK button, a set of 100 values of size 9000 each will be generated and stored in column C1.

2. PERFORMING HYPOTHESIS TESTS FOR THE POPULATION MEAN μ WHEN THE POPULATION STANDARD DEVIATION σ IS KNOWN

Example 3 Use Minitab to work Prob. 1 in Sec. 9.3.1 of the text.

Here we are given that the weights of salmon are normally distributed with a standard deviation σ of 1.2 pounds. A sample *n* of 16 fish yielded an average weight $\bar{x}$ of 7.2 pounds, and we are required to test at the 5 and 1 percent level of significance the null hypothesis H_0: $\mu \geq 7.6$ pounds. Also, we are required to determine the *p* value.

We know that we are doing a test for the mean (population) weight when the population standard deviation σ is known. Thus we have to perform a *z* test. The alternative hypothesis in this case is H_1: $\mu < 7.6$. Again, since we are not given the 16 observed values, we can enter the 7.2 value 16 times in a given column. The 16 values of 7.2 were entered in C1, which was renamed FISH. Follow the procedures in Example 1; but in the 1-Sample *z* dialog box, select the Test mean check box and type 7.6 in the corresponding text box. In the Alternative drop-down box, click on the arrow and select Less than. Type in 1.2 in the Sigma text box. This dialog box is shown in Fig. M9.3.

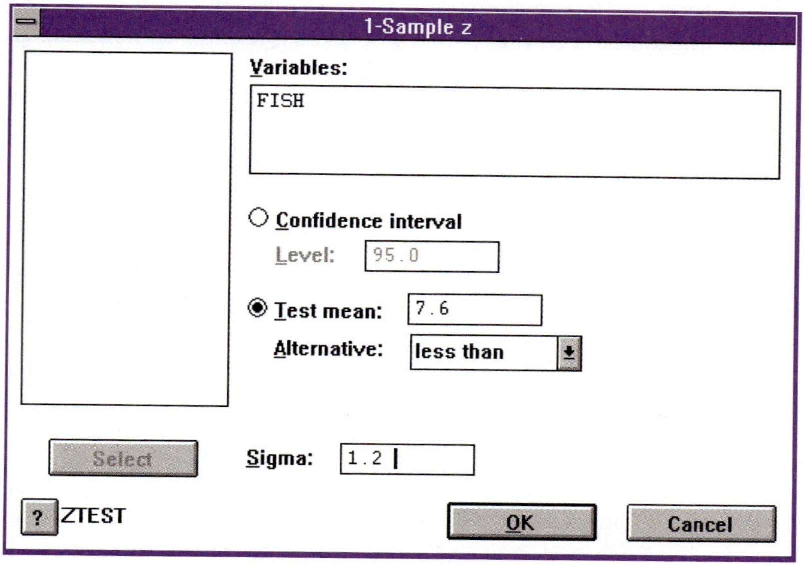

Figure M9.3

Click on the OK button, and the Session window will display the results of the test. It will give you the computed test statistic value and the p value of the test. Now to use Minitab to compute the critical (table) z values such that you can compare them with the computed test statistic, select Calc→Probability Distributions→Normal. In the Normal distribution dialog box, select the Inverse cumulative probability check box; make sure that the Mean text box has a value of 0 and that the Standard deviation text box has a value of 1. Select the Input constant check box, and type .05 in the corresponding text box. The reason for using .05 is that we are doing a left-tail test, and so we need the critical value in the left tail of the standard normal distribution. Click on the OK button, and the critical z value of -1.6449 will be displayed in the Session window. Repeat for .01 and observe that the critical z value will be -2.3263. From Fig. M9.4 the computed test statistic value is $z = -1.33$ which is not less than either critical value. Thus, at both levels of significance, there is insufficient evidence to reject the null hypothesis. Observe that the computed p value for the test is .091.

Note

1. If the alternative hypothesis were H$_1$: $\mu > 7.6$, then we would have used .95 and .99 in the Input constant text box to obtain the critical z values.

2. In Fig. M9.4, the other two options in the Alternative drop-down box are Greater than (right-tail test) and Not equal to (two-tail test).

3. If data values are given in the problem, enter the data set and follow the above procedures.

4. These procedures help to ease the computations, but you will have to write up your hypothesis tests as presented by your instructors.

3. CONSTRUCTING CONFIDENCE INTERVALS FOR THE POPULATION MEAN μ WHEN THE POPULATION STANDARD DEVIATION σ IS UNKNOWN

Example 4   Use Minitab to rework Example 8.13 in Sec. 8.6 of the text.
Enter the data values in column C1, and rename it PCB. Follow the procedures in Sec. 1, but select the 1-Sample t in place of the 1-Sample z option. We have to use the 1-Sample t since the population standard deviation is unknown. Verify that the 95 and 99 percent confidence intervals are those shown in the Session window of Fig. M9.5.

4. PERFORMING HYPOTHESIS TESTS FOR THE POPULATION MEAN μ WHEN THE POPULATION STANDARD DEVIATION σ IS UNKNOWN

Example 5   Use Minitab to work Prob. 4 in Sec. 9.4 of the text.
Follow the procedures in Sec. 2, but select the 1-Sample t in place of the 1-Sample z option. In this example, the sample size was 12, so in the degrees of freedom text box you should type 11. The Session window with the results is shown in Fig. M9.6. This window includes two of the critical t values. At the 1 percent level of significance, the critical t value is 3.1058. To obtain these critical values, follow the procedures in Sec. 2, but use the t distribution rather than the Normal. Also do not forget to divide the level of significance by 2 since you are doing

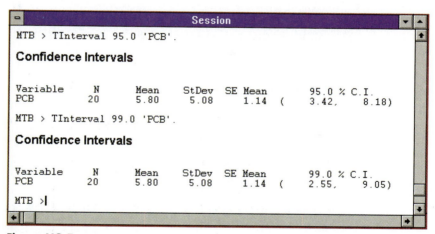

Figure M9.4

Figure M9.5

a two-tailed test. Thus in the Input constant text box in the *t* distribution dialog box, you should use .95, .975, and .995. In all three cases, $|-3.37|$ is greater than 1.7959, 2.2010, and 3.1058. Thus you will reject the null hypothesis of $H_0: \mu = 300$ at the 10, 5, and 1 percent level of significance. Note that the *p* value of .0062 supports this conclusion as well.

Note

1. If you are not given the data values to do the *t* test, you cannot use the procedure of Example 2, since the variance of the constant values will be zero. To do the *t* test, Minitab needs to compute the sample variance and use it.

2. Currently, there are no macros saved in Minitab to help you do a *t* test for the population mean when the data values are not given.

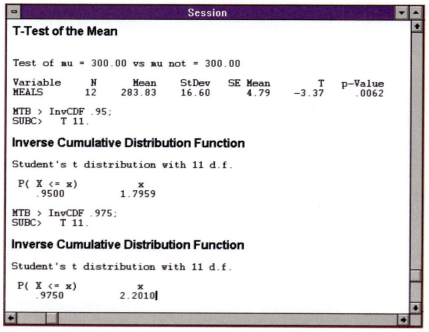

Figure M9.6

Computer Exercises

1. Where possible, use Minitab or any other statistical software to help work the problems at the end of Chap. 8.

2. Where possible, use Minitab or any other statistical software to help work the problems at the end of Chap. 9.

3. Use Minitab or any other statistical software to do Review Prob. 8 in Chap. 9. Present a report and discuss your findings.

4. Use Minitab or any other statistical software to do Review Prob. 10 in Chap. 9. Present a report and discuss your findings.

5. This exercise is set up as if you were using Minitab to do it. You can, of course, use any other statistical software to do the exercise.
   (a) Generate a set of 10 values from an *exponential distribution* with a mean of 12.
   (b) Set up these 10 values as if they were from a discrete distribution with probability .1 each. You can let the 10 generated values be placed in column C1 and the probability values be placed in column C2.
   (c) Generate 1000 values, each of size 10, and save in columns C3 through C12.
   (d) Find the means of the 1000 samples generated in part (c), and save in column C13.
   (e) Construct a histogram for the means in column C13.
   (f) Rank the 1000 means, and save the ranked means in column C14.

(g) Find the 2.5 and 97.5 percentiles of the ranked means in column C14. These two values will be the lower and upper limits, respectively, for the 95 percent confidence interval for the mean of the exponential distribution.

(h) Use column C14 to find the 90 and 99 percent confidence intervals for the mean.

(i) Repeat (a) through (h) 10 times.

(j) Present these results in a report with your observations and discussions.

Note In this exercise we started with a nonnormal distribution (exponential), but the methods of the text required that you sample from a normal distribution. This method of sampling from a nonnormal distribution repeatedly, which enabled us to construct the confidence intervals, is often referred to as the *bootstrap method.*

6. Repeat Exercise 5 with the information that was given in Example 2 in the Minitab lab for Chap. 7.

Hypothesis Tests Concerning Two Populations

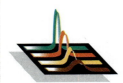

The importance of using a control in the testing of a new drug or a new procedure is discussed, and we see how this often results in comparisons between parameters of two different populations. We show how to test that two normal populations have the same population mean, both when the population variances are known and when they are unknown. We show how to test the equality of two population proportions.

> *Statistics are like therapists—they will testify for both sides.*
> Fiorello La Guardia, former mayor of New York City

> *Numbers don't lie, and they don't forgive.*
> Harry Angstrom in *Rabbit Is Rich* by John Updike

10.1   INTRODUCTION

An ongoing debate of great importance turns on whether megadoses—of the order of 25,000 to 30,000 milligrams daily—of vitamin C can be effective in treating patients suffering from cancerous tumors. On one side of the controversy was the great U.S. chemist Linus Pauling, who was a strong advocate of vitamin C therapy, and a growing number of researchers, and on the other side are the majority of mainstream cancer therapists. While many experiments have been set up to test whether vitamin C is therapeutically effective, there has been controversy surrounding many of them. Some of these experiments which reported negative results have been attacked by vitamin C proponents as utilizing too small dosages of the vitamin. Others of the experiments, reported by Prof. Pauling and his associates, have been met with skepticism by some in the medical community. To settle all doubts, a definitive experiment was planned and carried out in recent years at the Mayo Clinic. In this famous study, part of a group of terminally ill cancer patients was given, in addition to the regular medication, large doses of vitamin C for 3 months. The remainder of the group received a placebo along with the regular medication. After the 3-month period, the experiment was discontinued. These patients were then monitored until death to determine if the life span of those who had received vitamin C was longer than that of the control group. A summary statement was issued at the end of the experiment. This statement, which was widely disseminated by the news media, reported that there was no significant difference in the life span of those patients who had received the vitamin C treatment. This experiment, regarded by some in the medical community as being of seminal importance in discrediting vitamin C cancer therapy, was attacked by Pauling as being irrelevant to the claims of the proponents of vitamin C. According to the theory developed by Pauling and others, vitamin C would be expected to have protective value only while it was being taken and would not be expected to have any ongoing effect once it was discontinued. Indeed, according to earlier writings of Pauling, an immediate (as was done in the study) rather than a gradual stopping of vitamin C could have some potential negative effects. The controversy continues.

It is important to note that it would not have been sufficient for the Mayo Clinic to have given megadoses of vitamin C to all the volunteer patients. Even if there were significant increases in the additional life spans of these patients in comparison with the known life span distribution of patients suffering from this cancer, it would not be possible to attribute the cause of this increase to vitamin C. For one thing, the placebo effect—in which any type of "extra" treatment gives additional hope to a patient, and this in itself can have beneficial effects—could not be ruled out. For another, the additional life span could be due to factors totally unconnected to the experiment. Thus, to be able to draw a valid conclusion from the experiment, it was necessary to have a second group of volunteer patients, treated in all manners the same as the first, except that they did not receive additional vitamin C but rather only medication that looks and tastes like it. (Of course, to ensure that the two groups are as alike as possible, with the exception of their vitamin C intake, the group of volunteer patients was randomly divided into the two groups—the treated group whose members received the vitamin C and the control group whose members received the placebo.) This resulted in two separate samples, and the resultant data were used to test the hypothesis that the mean additional lifetimes of these two groups are identical.

Indeed, in all situations in which one is trying to study the effect of a given factor, such as the administration of vitamin C, one wants to hold all other factors constant so that any change from the norm can be attributed solely to the factor under study. However, as this is often impossible to achieve outside experiments in the physical sciences, it is usually necessary to consider two samples—one of which is to receive the factor under study and the other of which is a control group that will not receive the factor—and then determine whether there is a statistically significant difference in the responses of these two samples. For this reason, tests concerning two sampled populations are of great importance in a variety of applications.

In this chapter we will show how to test the hypothesis that two population means are equal when a sample from each population is available. In Sec. 10.2 we will suppose that the underlying population distributions are normal with known variances. Although it is rarely the case that the population variances will be known, the analysis presented in this section will be useful in showing us how to handle the more important cases where this assumption is no longer made. In fact, we show in Sec. 10.3 how to test the hypothesis that the two population means are equal when the variances are unknown, provided that the sample sizes are large. The case where the sample sizes are not large is considered in Sec. 10.4. To be able to test the hypothesis in this case, it turns out to be necessary to assume that the unknown population variances are equal.

In Sec. 10.5 we consider situations in which the two samples are related because of a natural pairing between the elements of the two data sets. For instance, one of the data values in the first sample might refer to an individual's blood pressure before receiving any medication, whereas one of the data values in the second sample might refer to that same person's blood pressure after receiving medication.

In Sec. 10.6 we consider tests concerning the equality of two binomial proportions.

10.2   TESTING EQUALITY OF MEANS OF TWO NORMAL POPULATIONS: CASE OF KNOWN VARIANCES

Suppose that $X_1, \ldots, X_n$ are a sample from a normal population having mean μ_x and variance σ_x^2; and suppose that $Y_1, \ldots, Y_m$ are an independent sample from a normal population having mean μ_y and variance σ_y^2. Assuming that the population variances σ_x^2 and σ_y^2 are known, let us consider a test of the null hypothesis that the two population means are equal; that is, let us consider a test of

$$H_0: \mu_x = \mu_y$$

against the alternative

$$H_1: \mu_x \neq \mu_y$$

Since the estimators of μ_x and μ_y are the respective sample means

$$\bar{X} = \frac{\sum\limits_{i=1}^{n} X_i}{n} \quad \text{and} \quad \bar{Y} = \frac{\sum\limits_{i=1}^{m} Y_i}{m}$$

it seems reasonable that H_0 should be rejected when $\bar{X}$ and $\bar{Y}$ are far apart. That is, for an appropriate constant c, it would seem that the test should be to

Reject H_0 if $|\bar{X} - \bar{Y}| \geq c$

Not reject H_0 otherwise

To specify the appropriate value of c, say, for a significance level α test, first we need to determine the probability distribution of $\bar{X} - \bar{Y}$. Now $\bar{X}$ is normal with mean μ_x and variance σ_x^2/n. and similarly $\bar{Y}$ is normal with mean μ_y and variance σ_y^2/m. Since the difference of independent normal random variables remains normally distributed, it follows that $\bar{X} - \bar{Y}$ is normal with mean

$$E[\bar{X} - \bar{Y}] = E[\bar{X}] - E[\bar{Y}] = \mu_x - \mu_y$$

and variance

$$\begin{aligned}
\text{Var}\,(\bar{X} - \bar{Y}) &= \text{Var}\,(\bar{X}) + \text{Var}\,(-\bar{Y}) \\
&= \text{Var}\,(\bar{X}) + (-1)^2\,\text{Var}\,(\bar{Y}) \\
&= \text{Var}\,(\bar{X}) + \text{Var}\,(\bar{Y}) \\
&= \frac{\sigma_x^2}{n} + \frac{\sigma_y^2}{m}
\end{aligned}$$

Hence, the standardized variable

$$\frac{\bar{X} - \bar{Y} - (\mu_x - \mu_y)}{\sqrt{\sigma_x^2/n + \sigma_y^2/m}}$$

has a standard normal distribution. Therefore, when the null hypothesis H_0: $\mu_x = \mu_y$ is true, the test statistic TS, given by

$$\text{TS} = \frac{\bar{X} - \bar{Y}}{\sqrt{\sigma_x^2/n + \sigma_y^2/m}} \tag{10.1}$$

will have a standard normal distribution. Now a standard normal random variable Z will, in absolute value, exceed $z_{\alpha/2}$ with probability α; that is,

$$P\{|Z| \geq z_{\alpha/2}\} = 2P\{Z \geq z_{\alpha/2}\} = \alpha$$

Thus, since we want to reject H_0 when $|\text{TS}|$ is large, it follows that the appropriate significance level α test of

$$H_0: \mu_x = \mu_y \qquad \text{against} \qquad H_1: \mu_x \neq \mu_y$$

is to

Reject H_0 if $|\text{TS}| \geq z_{\alpha/2}$
Not reject H_0 otherwise

where the test statistic TS is given by Eq. (10.1).

An alternative way of carrying out this test is to first compute the value of the test statistic TS; say that the data yield the value v. The resulting p value for the test of H_0 versus H_1 is the probability that the absolute value of a standard normal random variable is at least as large as $|v|$. That is, if TS is v, then

$$p \text{ value} = P\{|Z| \geq |v|\} = 2P\{Z \geq |v|\}$$

where Z is a standard normal random variable.

Example 10.1   Two new methods for producing a tire have been proposed. The manufacturer believes that there will be no appreciable difference in the lifetimes of tires produced by these methods. To test the plausibility of such an hypothesis, a sample of 9 tires are produced by method 1 and a sample of 7 tires by method 2. The first sample of tires is to be road-tested at location A and the second at location B. It is known, from previous experience, that the lifetime of a tire that is tested at either of these locations is a normal random variable with a mean life due to the tire but with a variance that is due to the location. Specifically, it is known that the lifetimes of tires tested at location A are normal with a standard deviation equal to 3000 kilometers, whereas those tested at location B have lifetimes that are normal with a standard deviation of 4000 kilometers.

Should the data in Table 10.1 cause the manufacturer to reject the hypothesis that the mean lifetime is the same for both types of tires? Use a 5 percent level of significance.

Solution   Call the tires tested at location A the X sample and those tested at B the Y sample. To test

$$H_0: \mu_x = \mu_y \qquad \text{against} \qquad H_1: \mu_x \neq \mu_y$$

Table 10.1   **Tire Lives in Units of 1000 Kilometers**

Tires tested at A	Tires tested at B	Tires tested at A	Tires tested at B
66.4	58.2	61.4	58.7
61.6	60.4	62.5	56.1
60.5	55.2	64.4	
59.1	62.0	60.7	
63.6	57.3		

we need to compute the value of the test statistic TS. Now, the sample means are given by

$$\bar{X} = 62.2445 \qquad \bar{Y} = 58.2714$$

Since $n = 9$, $m = 7$, $\sigma_x = 3$, and $\sigma_y = 4$, we see that the value of the test statistic is

$$TS = \frac{62.2445 - 58.2714}{\sqrt{9/9 + 16/7}} = 2.192$$

Thus the p value is equal to

$$p \text{ value} = 2P\{Z \geq 2.192\} = .0284$$

and so the hypothesis of equal means is rejected at any significance level greater than or equal to .0284. In particular, it is rejected at the 5 percent ($\alpha = .05$) level of significance.

If we were interested in testing the null hypothesis

$$H_0: \mu_x \leq \mu_y$$

against the one-sided alternative

$$H_1: \mu_x > \mu_y$$

then the null hypothesis will be rejected only when the test statistic TS is large. In this case, therefore, the significance level α test is to

Reject H_0	if TS $\geq z_\alpha$
Not reject H_0	otherwise

where

$$TS = \frac{\bar{X} - \bar{Y}}{\sqrt{\sigma_x^2/n + \sigma_y^2/m}}$$

Equivalently, if the observed value of TS is v, then the p value is

$$p \text{ value} = P\{Z \geq v\}$$

Example 10.2   Suppose the purpose of the experiment in Example 10.1 was to attempt to prove the hypothesis that the mean life of the first set of tires exceeded that of the second set by more than 1000 kilometers. Are the data strong enough to estab-lish this at, say, the 5 percent level of significance?

Solution Let Y_i denote the life of the ith tire of the second set, $i = 1, \ldots, 7$. If we set $W_i = Y_i + 1$, then we are interested in determining whether the data will enable us to conclude that $\mu_x > \mu_w$, where μ_x is the mean life of tires in the first set and μ_w is the mean of W_i. To decide this, we should take this conclusion to be the alter-native hypothesis. That is, we should test

$$H_0: \mu_x \le \mu_w \quad \text{against} \quad H_1: \mu_x > \mu_w$$

In other words, a rejection of H_0 would be strong evidence for the validity of the hypothesis that the mean life of the first set of tires exceeds that of the second set by more than 1000 kilometers.

To test the above, we compute the value of the test statistic TS, being careful to add 1 to the values given in Table 10.1 for tires tested at location B. This yields

$$\bar{X} = 62.2445 \qquad \bar{W} = 59.2714$$

and

$$\text{TS} = \frac{62.2445 - 59.2714}{\sqrt{9/9 + 16/7}} = 1.640$$

Since we want to reject H_0 when TS is large, the p value is the probability that a standard normal will exceed 1.640. That is,

$$p \text{ value} = P\{Z \ge 1.640\} = .0505$$

| Table 10.2 | **Tests of Means of Two Normal Populations Having Known Variances When Samples Are Independent** |

The sample mean of a sample of size n from a normal population having mean μ_x and known variance σ_x^2 is $\bar{X}$. The sample mean of a sample of size m from a second normal population having mean μ_y and known variance σ_y^2 is $\bar{Y}$. The two samples are independent.

H_0	H_1	Test statistic TS	Significance level α test	p Value if TS $= v$
$\mu_x = \mu_y$	$\mu_x \ne \mu_y$	$\dfrac{\bar{X} - \bar{Y}}{\sqrt{\sigma_x^2/n + \sigma_y^2/m}}$	Reject H_0 if $\lvert\text{TS}\rvert \ge z_{\alpha/2}$ Do not reject otherwise	$2P\{Z \ge \lvert v \rvert\}$
$\mu_x \le \mu_y$	$\mu_x > \mu_y$	$\dfrac{\bar{X} - \bar{Y}}{\sqrt{\sigma_x^2/n + \sigma_y^2/m}}$	Reject H_0 if $\text{TS} \ge z_{\alpha}$ Do not reject otherwise	$P\{Z \ge v\}$

Thus, even though the evidence is strongly in favor of the alternative hypothesis, it is not quite strong enough to cause us to reject the null hypothesis at the 5 percent level of significance.

Table 10.2 details both the two-sided test and the one-sided test presented in this section.

Problems

1. An experiment is performed to test the difference in effectiveness of two methods of cultivating wheat. A total of 12 patches of ground are treated with shallow plowing and 14 with deep plowing. The average yield per ground area of the first group is 45.2 bushels, and the average yield for the second group is 48.6 bushels. Suppose it is known that shallow plowing results in a ground yield having a standard deviation of .8 bushel, while deep plowing results in a standard deviation of 1.0 bushel.
   (a) Are the above data consistent, at the 5 percent level of significance, with the hypothesis that the mean yield is the same for both methods?
   (b) What is the p value for this hypothesis test?

2. A method for measuring the pH level of a solution yields a measurement value that is normally distributed with a mean equal to the actual pH of the solution and with a standard deviation equal to .05. An environmental pollution scientist claims that two different solutions come from the same source. If this is so, then the pH level of the solutions will be equal. To test the plausibility of this claim, 10 independent measurements were made of the pH level for both solutions, with the following data resulting:

Measurements of solution A	Measurements of solution B	Measurements of solution A	Measurements of solution B
6.24	6.27	6.26	6.31
6.31	6.25	6.24	6.28
6.28	6.33	6.29	6.29
6.30	6.27	6.22	6.34
6.25	6.24	6.28	6.27

   (a) Do the above data disprove the scientist's claim? Use the 5 percent level of significance.
   (b) What is the p value?

3. Two machines used for cutting steel are calibrated to cut exactly the same lengths. To test this hypothesis, each machine is used to cut 10 pieces of steel.

These pieces are then measured (with negligible measuring error). Suppose the resulting data are as follows:

Machine 1	Machine 2	Machine 1	Machine 2
122.40	122.36	121.76	122.40
123.12	121.88	122.31	122.12
122.51	122.20	123.20	121.78
123.12	122.88	122.48	122.85
122.55	123.43	121.96	123.04

Assume that it is known that the standard deviation of the length of a cut (made by either machine) is equal to .50.

(a) Test the hypothesis that the machines are set at the same value, that is, that the mean lengths of their cuttings are equal. Use the 5 percent level of significance.

(b) Find the p value.

4. The following are the values of independent samples from two different populations.

Sample 1: 122, 114, 130, 165, 144, 133, 139, 142, 150

Sample 2: 108, 125, 122, 140, 132, 120, 137, 128, 138

Let μ_1 and μ_2 be the respective means of the two populations. Find the p value of the test of the null hypothesis

$$H_0: \mu_1 \leq \mu_2$$

against the alternative

$$H_1: \mu_1 > \mu_2$$

when the population variances are $\sigma_1 = 10$ and
(a) $\sigma_2 = 5$
(b) $\sigma_2 = 10$
(c) $\sigma_2 = 20$

5. In this section, we presented the test of

$$H_0: \mu_x \leq \mu_y \qquad \text{against} \qquad H_1: \mu_x > \mu_y$$

Explain why it was not necessary to separately present the test of

$$H_0: \mu_x \geq \mu_y \qquad \text{against} \qquad H_1: \mu_x < \mu_y$$

6. The device used by astronomers to measure distances results in measurements that have a mean value equal to the actual distance of the object being surveyed and a standard deviation of .5 light year. An astronomer is interested in testing the widely held hypothesis that asteroid *A* is at least as close to the earth as is asteroid *B*. To test this hypothesis, the astronomer made 8 independent measurements on asteroid *A* and 12 on asteroid *B*. If the average of the measurements for asteroid *A* was 22.4 light years and the average of those for asteroid *B* was 21.3, will the hypothesis be rejected at the 5 percent level of significance? What is the *p* value?

7. The value received at a certain message receiving station is equal to the value sent plus a random error that is normal with mean 0 and standard deviation 2. Two messages, each consisting of a single value, are to be sent. Because of the random error, each message will be sent 9 times. Before reception, the receiver is fairly certain that the first message value will be less than or equal to the second. Should this hypothesis be rejected if the average of the values relating to message 1 is 5.6 whereas the average of those relating to message 2 is 4.1? Use the 1 percent level of significance.

8. A large industrial firm has its manufacturing operations at one end of a large river. A public health official thinks that the firm is increasing the polychlorinated biphenyl (PCB) level of the river by dumping toxic wastes. To gain information, the official took 12 readings of water from the part of the river situated by the firm and 14 readings near the other end of the river. The sample mean of the 12 readings of water near the firm was 32 parts per billion, and the sample mean of the other set of 14 readings was 22 parts per billion. Assume that the value of each reading of water is equal to the actual PCB level at that end of the river where the water is collected plus a random error due to the measuring device that is normal with mean 0 and standard deviation 8 parts per billion.
   (a) Using the above data and the 5 percent level of significance, can we reject the hypothesis that the PCB level at the firm's end of the river is no greater than the PCB level at the other end?
   (b) What is the *p* value?

10.3 TESTING EQUALITY OF MEANS: UNKNOWN VARIANCES AND LARGE SAMPLE SIZES

In the previous section we supposed that the population variances were known to the experimenter. However, it is by far more common that these parameters are unknown. That is, if the mean of a population is unknown, then it is likely that the variance will also be unknown.

Let us again suppose that we have two independent samples $X_1, \ldots, X_n$ and $Y_1, \ldots, Y_m$ and are interested in testing a hypothesis concerning their means μ_x and μ_y. Although we do not assume that the population variances σ_x^2 and σ_y^2 are known, we will suppose that the sample sizes n and m are large.

To determine the appropriate test in this situation, we will make use of the fact that for large sample sizes the sample variances will approximately equal the population variances. Thus, it seems reasonable that we can substitute the sample variances S_x^2 and S_y^2 for the population variances and make use of the analysis developed in the previous section. That is, analogous with the result that

$$\frac{\bar{X} - \bar{Y} - (\mu_x - \mu_y)}{\sqrt{\sigma_x^2/n + \sigma_y^2/m}}$$

has a standard normal distribution, it would seem that for large values of n and m, the random variable

$$\frac{\bar{X} - \bar{Y} - (\mu_x - \mu_y)}{\sqrt{S_x^2/n + S_y^2/m}}$$

will approximately have a standard normal distribution. Since this result is indeed true, it follows that we can utilize the same tests developed in Sec. 10.2 except that the sample variances are now utilized in place of the population variances. For instance, the significance level α test of

$$H_0: \mu_x = \mu_y$$

against

$$H_1: \mu_x \neq \mu_y$$

is to reject when $|TS| \geq z_{\alpha/2}$, where the test statistic TS is now given by

$$TS = \frac{\bar{X} - \bar{Y}}{\sqrt{S_x^2/n + S_y^2/m}}$$

An equivalent way of determining the outcome is to first determine the value of the test statistic TS, say it is v, and then calculate the p value, given by

$$p \text{ value} = P\{|Z| \geq |v|\} = 2P\{Z \geq |v|\}$$

Also, if we want to test the one-sided hypothesis

$$H_0: \mu_x \leq \mu_y$$

against

$$H_1: \mu_x > \mu_y$$

then we use the same test statistic as above. The test is to

Reject H_0 if $TS \geq z_\alpha$

Not reject H_0 otherwise

Equivalently, if the observed value of TS is v, then the p value is

$$p \text{ value} = P\{Z \geq v\}$$

Remarks: We have not yet specified how large n and m should be for the preceding to be valid. A general rule of thumb is for both sample sizes to be at least 30, although values of 20 or more will usually suffice.

Even when the underlying population distributions are themselves not normal, the central limit theorem implies that the sample means $\bar{X}$ and $\bar{Y}$ will be approximately normal. For this reason the preceding tests of population means can be used for arbitrary underlying distributions provided that the sample sizes are large. (Again, sample sizes of at least 20 should suffice.)

Example 10.3   To test the effectiveness of a new cholesterol-lowering medication, 100 volunteers were randomly divided into two groups of size 50 each. Members of the first group were given pills containing the new medication while members of the second, or *control,* group were given pills containing lovastatin, one of the standard medications for lowering blood cholesterol. All the volunteers were instructed to take a pill every 12 hours for the next 3 months. None of the volunteers knew which group they were in.

Suppose that the result of the above experiment was that there was an average reduction of 8.2 with a sample variance of 5.4 in the blood cholesterol levels of those taking the old medication, and an average reduction of 8.8 with a sample variance of 4.5 of those taking the newer medication. Do these results prove, at the 5 percent level, that the new medication is more effective than the old one?

Solution Let μ_x denote the mean cholesterol reduction of a volunteer who is given the new medication, and let μ_y be the equivalent value for one given the control. If we want to see if the data were sufficient to prove that $\mu_x > \mu_y$, then we should use them to test

$$H_0: \mu_x \leq \mu_y \qquad \text{against} \qquad H_1: \mu_x > \mu_y$$

The value of the test statistic is

$$TS = \frac{8.8 - 8.2}{\sqrt{4.5/50 + 5.4/50}} = 1.3484$$

Since this is a one-sided test where the null hypothesis will be rejected when TS is large, the p value equals the probability that a standard normal (which would be the approximate distribution of TS if $\mu_x = \mu_y$) is as large as 1.3484. That is, the p value of these data is

p value $= P\{Z \geq 1.3484\} = .089$

Since the p value is greater than .05, the evidence is not strong enough to establish, at the 5 percent level of significance, that the new medication is more effective than the old.

In Example 10.3, note that we compared the new drug to a standard medication rather than to a placebo. Now, when a new drug is tested in situations where there is no accepted treatment, the drug should always be tested against a placebo. However, if there is a viable treatment already in place, then the new drug should be tested against it. This is obvious in very serious diseases where there may be ethical questions related to prescribing a placebo. Also, in general, one always hopes to conclude that a new drug is better than the previous state-of-the-art drug as opposed to concluding that it is "better than nothing."

Example 10.4 A phenomenon that is quite similar to the placebo effect is often observed in industrial human factor experiments. It has been noted that a worker's productivity usually increases when that worker becomes aware that she or he is being monitored. As this phenomenon was documented and widely publicized after some studies on increasing productivity carried out at the Hawthorne plant of the Western Electric company, it is sometimes referred to as the *Hawthorne effect*. To counter this effect, industrial experiments often make use of a control group.

An industrial consultant has suggested a modification of the existing method for producing semiconductors. She claims that this modification will increase the number of semiconductors that a worker can produce in a day. To test the effectiveness of her ideas, management has set up a small study. A group of 50 workers have been randomly divided into two groups. One of the groups, consisting of 30 workers, has been trained in the modification proposed by the consultant. The other group, acting as a control, has been trained in a different modification. These two modifications are considered by management to be roughly equal in complexity of learning and in time of implementation. In addition, management is quite certain that the alternative (to the one proposed by the consultant) modification would not have any real effect on productivity. Neither group was told whether it was learning the consultant's proposal or not.

The workers were then monitored for a period of time with the following results.

For those trained in the technique of the consultant:

The average number of semiconductors produced per worker was 242.

The sample variance was 62.2.

For those workers in the control group:

The average number of semiconductors produced per worker was 234.

The sample variance was 58.4.

Are the above data sufficient to prove that the consultant's modification will increase productivity?

Solution Let μ_x denote the mean number of semiconductors that would be produced over the period of the study by workers trained in the method of the consultant. Also let μ_y denote the mean number produced by workers given the alternative technique. To prove the consultant's claim that $\mu_x > \mu_y$, we need to test

$$H_0: \mu_x \le \mu_y \qquad \text{against} \qquad H_1: \mu_x > \mu_y$$

The data are

$$
\begin{array}{ll}
n = 30 & m = 20 \\
\bar{X} = 242 & \bar{Y} = 234 \\
S_x^2 = 62.2 & S_y^2 = 58.4
\end{array}
$$

Thus the value of the test statistic is

$$TS = \frac{242 - 234}{\sqrt{62.2/30 + 58.4/20}} = 3.58$$

Hence, the p value of these data is

$$p \text{ value} = P\{Z \ge 3.58\} = .0002$$

Thus, the data are significant enough to prove that the consultant's modification was more effective than the one used by the control group.

When we are given raw data, rather than summary statistics, the sample means and sample variances can be calculated by a manual computation or by using a calculator or by a computer program such as Program 3-1. These quantities should then

---

H i s t o r i c a l   P e r s p e c t i v e

The Hawthorne effect illustrates that the presence of an observer may affect the behavior of those being observed. As noted in Example 10.4, the recognition of this phenomenon grew out of research conducted during the 1920s at the Hawthorne plant of Western Electric. Investigators set out to determine how the productivity of workers at this plant could be improved. Their initial studies were designed to examine the ef- fects of changes in lighting on the pro- ductivity of workers assembling tele- phone components. Gradual increases in lighting were made, and each change led to increased productivity. Productiv- ity, in fact, continued to increase even when the lighting was made abnormally bright. Even more surprising was the fact that when the lighting was reduced, productivity still continued to rise.

---

be used to determine the value of the test statistic TS. Finally, the p value can then be obtained by using the normal probability table (Table D.1 in App. D).

Example 10.5   Test

$$H_0: \mu_x \leq \mu_y \qquad \text{against} \qquad H_1: \mu_x > \mu_y$$

for the following data:

X: 22, 21, 25, 29, 31, 18, 28, 33, 28, 26, 32, 35, 27, 29, 26

Y: 14, 17, 22, 18, 19, 21, 24, 33, 28, 22, 27, 18, 21, 19, 33, 31

Solution   A simple calculation yields that

$$
\begin{array}{ll}
n = 15 & m = 16 \\
\bar{X} = 27.333 & \bar{Y} = 22.938 \\
S_x^2 = 21.238 & S_y^2 = 34.329
\end{array}
$$

Hence the value of TS is

$$\text{TS} = \frac{4.395}{\sqrt{21.238/15 + 34.329/16}} = 2.33$$

Since this is a one-sided test which will call for rejection only at large values of TS, we have

$$p \text{ value} = P\{Z \geq 2.33\} = .01$$

Table 10.3	**Tests of Means of Two Normal Populations Having Unknown Variances When Samples Are Independent and Sample Sizes Are Large**

The sample mean and sample variance of a sample of size n from a normal population having mean μ_x and unknown variance σ_x^2 are, respectively, $\bar{X}$ and S_x^2. The sample mean and sample variance of a sample of size m from a second normal population having mean μ_y and unknown variance σ_y^2 are, respectively, $\bar{Y}$ and S_y^2. The two samples are independent, and both n and m are at least 20.

H_0	H_1	Test statistic TS	Significance level α test	p Value if TS = v
$\mu_x = \mu_y$	$\mu_x \neq \mu_y$	$\dfrac{\bar{X} - \bar{Y}}{\sqrt{S_x^2/n + S_y^2/m}}$	Reject H_0 if $\lvert TS \rvert \geq z_{\alpha/2}$ Do not reject otherwise	$2P\{Z \geq \lvert v \rvert\}$
$\mu_x \leq \mu_y$	$\mu_x > \mu_y$	$\dfrac{\bar{X} - \bar{Y}}{\sqrt{S_x^2/n + S_y^2/m}}$	Reject H_0 if $TS \geq z_\alpha$ Do not reject otherwise	$P\{Z \geq v\}$

Therefore, the hypothesis that the mean of the X population is no greater than that of the Y population would be rejected at all significance levels greater than or equal to .01.

Table 10.3 details both the two-sided and the one-sided tests presented in this section.

Problems

1. A high school is interested in determining whether two of its instructors are equally able to prepare students for a statewide examination in geometry. Seventy students taking geometry this semester were randomly divided into two groups of 35 each. Instructor 1 taught geometry to the first group, and instructor 2 to the second. At the end of the semester, the students took the statewide examination, with the following results:

Class of instructor 1	Class of instructor 2
$\bar{X} = 72.6$	$\bar{Y} = 74.0$
$S_x^2 = 6.6$	$S_y^2 = 6.2$

Can we conclude from these results that the instructors are not equally able in preparing students for the examinations? Use the 5 percent level of significance. Give the null and alternative hypotheses and the resulting p value.

2. Sample weights (in pounds) of newborn babies born in two adjacent counties in western Pennsylvania yielded the following data:

$$n = 53 \qquad m = 44$$
$$\bar{X} = 6.8 \qquad \bar{Y} = 7.2$$
$$S^2 = 5.2 \qquad S^2 = 4.9$$

Consider a test of the hypothesis that the mean weight of newborns is the same in both counties. What is the resulting p value? How would you express your conclusions to an intelligent person who has not yet studied statistics?

3. An administrator of a large exercise spa is curious as to whether women members younger than 40 years old use the spa with the same frequency as do women members over age 40. Random samples of 30 women younger than 40 years of age and 30 women older than age 40 were chosen and the women tracked for the following month. The result was that the younger group had a sample mean of 3.6 visits with a sample standard deviation of 1.3 visits, while the older group had a sample mean of 3.8 visits with a sample standard deviation of 1.4 visits. Use these data to test the hypothesis that the mean number of visits of the population of older women is the same as that of younger women.

4. You are interested in testing the hypothesis that the mean travel time from your home to work in the morning is the same as the mean travel time from work back to home in the evening. To check this hypothesis, you recorded the times for 40 workdays. It turned out that the sample mean for the trip to work was 38 minutes with a sample standard deviation of 4 minutes, and the sample mean of the return trip home was 42 minutes with a sample standard deviation of 7 minutes.
(a) What conclusion can you draw at the 5 percent level of significance?
(b) What is the p value?

5. The following experiment was conducted to compare the yields of two varieties of tomato plants. Thirty-six plants of each variety were randomly selected and planted in a field. The first variety produced an average yield of 12.4 kilograms per plant with a sample standard deviation of 1.6 kilograms. The second variety produced an average yield of 14.2 kilograms per plant with a sample standard deviation of 1.8 kilograms. Does this provide sufficient evidence to conclude that there is a difference in the mean yield for the two varieties? At what level of significance?

6. Data were collected to determine if there is a difference between the mean IQ scores of urban and rural students in upper Michigan. A random sample of 100 urban students yielded a sample mean score of 102.2 and a sample standard deviation of 11.8. A random sample of 60 rural students yielded a sample mean score of 105.3 with a sample standard deviation of 10.6. Are the data significant enough, at the 5 percent level, for us to reject the hypothesis that the mean scores of urban and rural students are the same?

7. In Prob. 6, are the data significant enough, at the 1 percent level, to conclude that the mean score of rural students in upper Michigan is greater than that of urban students? What are the null and the alternative hypotheses?

8. Suppose in Prob. 5 that the experimenter wanted to prove that the average yield of the second variety was greater than that of the first. What conclusion would have been drawn? Use a 5 percent level of significance.

9. A firm must decide between two different suppliers of lightbulbs. Management has decided to order from supplier A unless it can be "proved" that the mean lifetime of lightbulbs from supplier B is superior. A test of 28 lightbulbs from A and 32 lightbulbs from B yielded the following data as to the number of hours of use given by each lightbulb:
   A: 121, 76, 88, 103, 96, 89, 100, 112, 105, 101, 92, 98, 87, 75, 111, 118, 121, 96, 93, 82, 105, 78, 84, 96, 103, 119, 85, 84
   B: 127, 133, 87, 91, 81, 122, 115, 107, 109, 89, 82, 90, 81, 104, 109, 110, 106, 85, 93, 90, 100, 122, 117, 109, 98, 94, 103, 107, 101, 99, 112, 90
   At the 5 percent level of significance, which supplier should be used? Give the hypothesis to be tested and the resulting p value.

10. An administrator of a business school claims that the average salary of its graduates is, after 10 years, at least $5000 higher than that of comparable graduates of a rival institution. To study this claim, a random sample of 50 students who had graduated 10 years ago was selected, and the salaries of the graduates were determined. A similar sample of students from the rival institute was also chosen. Suppose the following data resulted:

College	Rival institution
$n = 50$	$m = 50$
$\bar{X} = 55.2$	$\bar{Y} = 44.8$
$S_x^2 = 26.4$	$S_y^2 = 24.5$

   (a) To determine whether the above data prove the administrator's claim, what should be the null and the alternative hypotheses?
   (b) What is the resulting p value?
   (c) What conclusions can you draw?

11. An attempt was recently made to verify whether women are being discriminated against, as far as wages are concerned, in a certain industry. To study this claim, a court-appointed researcher obtained a random sample of employees with 8 or more years' experience and with a history of regular employment during that time. With the unit of wages being $1, the following data on hourly pay resulted:

Female workers		Male workers	
Sample size:	55	Sample size:	72
Sample mean:	8.80	Sample mean:	9.40
Sample variance:	.90	Sample variance:	1.1

(a) What hypothesis should be tested? Give the null and the alternative.
(b) What is the resulting p value?
(c) What does this prove?

12. The following data summary was obtained from a comparison of the lead content of human hair removed from adult individuals who had died between 1880 and 1920 with the lead content of present-day adults. The data were in units of micrograms, equal to one-millionth of a gram.

	1880–1920	Today
Sample size	30	100
Sample mean	48.5	26.6
Sample standard deviation	14.5	12.3

(a) Do the above data establish, at the 1 percent level of significance, that the mean lead content of human hair is less today than it was in the years between 1880 and 1920? Clearly state what the null and alternative hypotheses are.
(b) What is the p value for the hypothesis tested in part (a)?

13. Forty workers were randomly divided into two sets of 20 each. Each set spent 2 weeks in a self-training program that was designed to teach a new production technique. The first set of workers was accompanied by a supervisor whose only job was to check that the workers were all paying attention. The second group was left on its own. After the program ended, the workers were tested. The results were as follows:

	Sample mean	Sample standard deviation
Supervised group	70.6	8.4
Unsupervised group	77.4	7.4

(a) Test the null hypothesis that supervision had no effect on the performance of the workers. Use the 1 percent level of significance.
(b) What is the p value?
(c) What would you conclude was the result of the supervision?

10.4 TESTING EQUALITY OF MEANS: SMALL-SAMPLE TESTS WHEN THE UNKNOWN POPULATION VARIANCES ARE EQUAL

Suppose again that we have independent samples from two normal populations:

$$X_1, \ldots, X_n \quad \text{and} \quad Y_1, \ldots, Y_m$$

and we are interested in testing hypotheses concerning the respective population means μ_x and μ_y. Unlike in the previous sections, we will suppose neither that the population variances are known nor that the sample sizes n and m are necessarily large.

In many situations, even though they are unknown, it is reasonable to suppose that the population variances σ_x^2 and σ_y^2 are approximately equal. So let us assume they are equal and denote their common value by σ^2. That is, suppose that

$$\sigma_x^2 = \sigma_y^2 = \sigma^2$$

To obtain a test of the null hypothesis

$$H_0: \mu_x = \mu_y \qquad \text{against} \qquad H_1: \mu_x \neq \mu_y$$

when the population variances are equal, we start with the fact, shown in Sec. 10.2, that

$$\frac{\bar{X} - \bar{Y} - (\mu_x - \mu_y)}{\sqrt{\sigma_x^2/n + \sigma_y^2/m}}$$

has a standard normal distribution.

Thus, since $\sigma_x^2 = \sigma_y^2 = \sigma^2$, we see that when H_0 is true (and so $\mu_x - \mu_y = 0$), then $(\bar{X} - \bar{Y})/\sqrt{\sigma^2/n + \sigma^2/m}$ has a standard normal distribution. That is,

When H_0 is true,

$$\frac{\bar{X} - \bar{Y}}{\sqrt{\sigma^2/n + \sigma^2/m}} \qquad (10.2)$$

has a standard normal distribution.

The preceding result cannot be directly employed to test the null hypothesis of equal means since it involves the unknown parameter σ^2. As a result, we will first obtain an estimator of σ^2 and then determine the effect on the distribution of the above quantity when σ^2 is replaced by its estimator.

To obtain an estimator for σ^2, we make use of the fact that the sample variances S_x^2 and S_y^2 are both estimators of the common population variance σ^2. It is thus natural to combine, or *pool,* these two estimators. In other words, it is natural to consider a weighted average of the two sample variances. To determine the appropriate weights to attach to each one, recall that the sample variance from a sample of size, say, k has $k - 1$ degrees of freedom associated with it. From this we see that S_x^2 has $n - 1$ degrees of freedom associated with it, and S_y^2 has $m - 1$ degrees of freedom. Thus, we will use a pooled estimator which weights S_x^2 by the factor $(n - 1)/(n - 1 + m - 1)$ and weights S_y^2 by the factor $(m - 1)/(n - 1 + m - 1)$.

Definition

The estimator S_p^2 defined by

$$S_p^2 = \frac{n - 1}{n + m - 2} S_x^2 + \frac{m - 1}{n + m - 2} S_y^2$$

is called the *pooled estimator* of σ^2.

Note that the larger the sample size, the greater the weight given to its sample variance in estimating σ^2. Also note that the pooled estimator will have $n - 1 + m - 1 = n + m - 2$ degrees of freedom attached to it.

If, in Eq. (10.2), we replace σ^2 by its pooled estimator S_p^2, then the resultant statistic can be shown, when H_0 is true, to have a t distribution with $n + m - 2$ degrees of freedom. [This is directly analogous to what happens to the distribution of $\sqrt{n}(\bar{X} - \mu)/\sigma$ when the population variance σ^2 is replaced by the sample variance S^2—namely, this replacement changes the standard normal random variable $\sqrt{n}(\bar{X} - \mu)/\sigma$ to $\sqrt{n}(\bar{X} - \mu)/S$ which is a t random variable with $n - 1$ degrees of freedom.]

From the preceding we see that to test

$$H_0: \mu_x = \mu_y \qquad \text{against} \qquad H_1: \mu_x \neq \mu_y$$

one should first compute the value of the test statistic

$$TS = \frac{\bar{X} - \bar{Y}}{\sqrt{S_p^2\,(1/n + 1/m)}}$$

The significance level α test is then to

Reject H_0	if $|TS| \geq t_{n+m-2,\,\alpha/2}$
Not reject H_0	otherwise

Alternatively the test can be run be determining the p value. If TS is observed to equal v, then the resulting p value of the test of H_0 against H_1 is given by

$$p \text{ value} = P\{|T_{n+m-2}| \geq |v|\}$$
$$= 2P\{T_{n+m-2} \geq |v|\}$$

where T_{n+m-2} is a t random variable having $n + m - 2$ degrees of freedom.

If we are interested in testing the one-sided hypothesis

$$H_0: \mu_x \leq \mu_y \qquad \text{against} \qquad H_1: \mu_x > \mu_y$$

then H_0 will be rejected at large values of TS. Thus the significance level α test is to

Reject H_0 if $\text{TS} \geq t_{n+m-2, \alpha}$

Not reject H_0 otherwise

If the value of the test statistics TS is v, then the p value is given by

$$p \text{ value} = P\{T_{n+m-2} \geq v\}$$

Program 10-1 will compute both the value of the test statistic and the corresponding p value for either a one-sided or a two-sided test.

Example 10.6   Twenty-two volunteers at a cold research institute caught a cold after having been exposed to various cold viruses. A random selection of 10 volunteers were given tablets containing 1 gram of vitamin C. These tablets were taken 4 times a day. The control group, consisting of the other 12 volunteers, was given placebo tablets that looked and tasted exactly like the vitamin C ones. This was continued for each volunteer until a doctor, who did not know whether the volunteer was receiving vitamin C or the placebo, decided that the volunteer was no longer suffering from the cold. The length of time the cold lasted was then recorded.

At the end of this experiment, the following data resulted:

Treated with vitamin C	Treated with placebo	Treated with vitamin C	Treated with placebo
5.5	6.5	7.5	7.5
6.0	6.0	5.5	6.5
7.0	8.5	7.0	7.5
6.0	7.0	6.5	6.0
7.5	6.5		8.5
6.0	8.0		7.0

Do the above data prove that taking 4 grams of vitamin C daily reduces the time that a cold lasts? At what level of significance?

Solution To prove the above hypothesis, we need to reject the null hypothesis in a test of

$$H_0: \mu_p \leq \mu_c \qquad \text{against} \qquad H_1: \mu_p > \mu_c$$

where μ_c is the mean time a cold lasts when the vitamin C tablets are taken and μ_p is the mean time when the placebo is taken. Assuming that the variance of the length of the cold is the same for the vitamin C patients and the placebo patients, we test the above by running Program 10-1. This program computes the p value when testing that two normal populations having equal but unknown variances have a common mean.

Enter the size of sample 1, which is 12. The sample 1 values are a follows: 6.5, 6, 8.5, 7, 6.5, 8, 7.5, 6.5, 7.5, 6, 8.5, and 7.

Enter the size of sample 2, which is 10. The sample 2 values are as follows: 5.5, 6, 7, 6, 7.5, 6, 7.5, 5.5, 7, and 6.5.

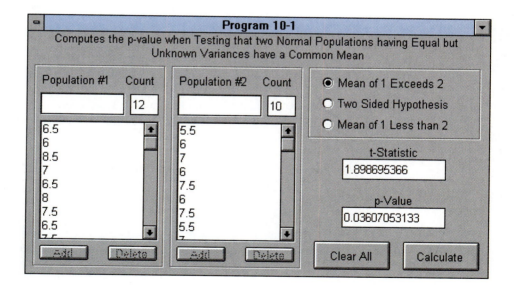

Table 10.4   **Tests of Means of Two Normal Populations Having Unknown Though Equal Variances When Samples Are Independent**

The sample mean and sample variance, respectively, of a sample size n from a normal population having mean μ_x and variance σ^2 are $\overline{X}$ and S_x^2. And the sample mean and sample variance of a sample of size m from a second normal population having mean μ_y and variance σ^2 are $\overline{Y}$ and S_y^2. The two samples are independent.

$$S_p^2 = \frac{n-1}{n+m-2} S_x^2 + \frac{m-1}{n+m-2} S_y^2$$

H_0	H_1	Test statistic TS	Significance level α test	p Value if TS = v
$\mu_x = \mu_y$	$\mu_x \neq \mu_y$	$\dfrac{\overline{X} - \overline{Y}}{\sqrt{S_p^2(1/n + 1/m)}}$	Reject H_0 if $\|TS\| \geq t_{n+m-2,\alpha/2}$ Do not reject otherwise	$2P\{T_{n+m-2} \geq \|v\|\}$
$\mu_x \leq \mu_y$	$\mu_x > \mu_y$	$\dfrac{\overline{X} - \overline{Y}}{\sqrt{S_p^2(1/n + 1/m)}}$	Reject H_0 if $TS \geq t_{n+m-2,\alpha}$ Do not reject otherwise	$P\{T_{n+m-2} \geq v\}$

Program 10-1 computes the value of the t statistic as 1.898695366.

When we enter the values into Program 10-1, we make sure to note that the alternative hypothesis is not two-sided but rather is that the mean of sample 1 exceeds that of sample 2.

Consequently, the program computes the p value as .03607053133.

Thus H_0 would be rejected at the 5 percent level of significance.

Of course, if it was not convenient to run Program 10-1, then we could perform the test by first computing the values of the statistics $\overline{X}$, $\overline{Y}$, S_x^2, S_y^2, and S_p^2, where the X sample corresponds to those receiving a placebo and the Y sample to those receiving vitamin C. These computations give the values

$$\overline{X} = 7.125 \qquad \overline{Y} = 6.450$$
$$S_x^2 = .778 \qquad S_y^2 = .581$$

Therefore,

$$S_p^2 = \frac{11}{20} S_x^2 + \frac{9}{20} S_y^2 = .689$$

and the value of the test statistic is

$$TS = \frac{.675}{\sqrt{.689(1/12 + 1/10)}} = 1.90$$

Since, from Table D.2, $t_{20, .05} = 1.725$, the null hypothesis is rejected at the 5 percent level of significance. That is, the above evidence is significant, at the 5 percent level, in establishing that vitamin C reduces the mean time that a cold persists.

Table 10.4 details both the two-sided test and the one-sided test presented in this section.

Problems

In the following problems, assume that the population distributions are normal and have equal variances.

1. Twenty-five males between the ages of 25 and 30 who were participating in a well-known heart study carried out in Framingham, Massachusetts were randomly selected. Of these 11 were smokers and 14 were not. The following data refer to readings of their systolic blood pressure:

Smokers	Nonsmokers	Smokers	Nonsmokers
124	130	131	127
134	122	133	135
136	128	125	120
125	129	118	122
133	118		120
127	122		115
135	116		123

Do the data indicate, at the 1 percent level of significance, a difference in mean systolic blood pressure levels for the populations represented by the two groups? If not, what about at the 5 percent level?

2. A study was instituted to learn how the diets of women changed during the winter and the summer. A random group of 12 women were observed during the month of July, and the percentage of each woman's calories that came from fat was determined. Similar observations were made on a different randomly selected group of size 12 during the month of January. Suppose the results were as follows:

July: 32.2 27.4 28.6 32.4 40.5 26.2 29.4 25.8 36.6 30.3 28.5 32.0
January: 30.5 28.4 40.2 37.6 36.5 38.8 34.7 29.5 29.7 37.2 41.5 37.0

Test the hypothesis that the mean fat intake is the same for both months. Use the
(a) 5 percent
(b) 1 percent
level of significance.

3. A consumer organization has compared the time it takes a generic pain reliever tablet to dissolve with the time it takes a name-brand tablet. Nine tablets of each were checked. The following data resulted:

   Generic: 14.2 14.7 13.9 15.3 14.8 13.6 14.6 14.9 14.2
   Name: 14.3 14.9 14.4 13.8 15.0 15.1 14.4 14.7 14.9

   (a) Do the above data establish, at the 5 percent level of significance, that the name-brand tablet is quicker to dissolve?
   (b) What about at the 10 percent level of significance?

4. To learn about the feeding habits of bats, a collection of 22 bats were tagged and tracked by radio. Of these 22 bats, 12 were female and 10 were male. The distances flown (in meters) between feedings were noted for each of the 22 bats, and the following summary statistics were obtained.

Female bats	Male bats
$n = 12$	$m = 10$
$\bar{X} = 180$	$\bar{Y} = 136$
$S_x = 92$	$S_y = 86$

   Test the hypothesis that the mean distance flown between feedings is the same for the populations of male and female bats. Use the 5 percent level of significance.

5. To determine the effectiveness of a new method of teaching reading to young children, a group of 20 nonreading children were randomly divided into two groups of 10 each. The first group was taught by a standard and the second group by an experimental method. At the end of the school term, a reading examination was given to each of the students, with the following summary statistics resulting.

Students using standard	Students using experimental
Average score = 65.6	Average score = 70.4
Standard deviation = 5.4	Standard deviation = 4.8

   Are the above data strong enough to prove, at the 5 percent level of significance, that the experimental method results in a higher mean test score?

6. Redo Prob. 2 of Sec. 10.3, assuming that the population variances are equal.
   (a) Would you reject the null hypothesis at the 5 percent level of significance?
   (b) How does the p value compare with the one previously obtained?

7. To learn about how diet affects the chances of getting diverticular disease, 20 vegetarians, 6 of whom had the disease, were studied. The total daily dietary fiber consumed by each of these individuals was determined with the following results:

With disease	Without disease
$n = 6$	$m = 14$
$\bar{X} = 26.8$ grams	$\bar{Y} = 42.5$ grams
$S_x = 9.2$ grams	$S_y = 9.5$ grams

Test the hypothesis that the mean dietary fiber consumed daily is the same for the population of vegetarians having diverticular disease and the population of vegetarians who do not have this disease. Use the 5 percent level of significance.

8. It is "well known" that the average automobile commuter in the Los Angeles area drives more miles daily than does a commuter in the San Francisco Bay area. To see whether this "fact" is indeed true, a random sample of 20 Los Angeles area commuters and 20 San Francisco Bay area commuters were randomly chosen and their driving habits monitored. The following data relating to the average number and standard deviation of miles driven resulted.

Los Angeles commuter	San Francisco commuter
$\bar{X} = 57.4$	$\bar{Y} = 52.8$
$S_x = 12.4$	$S_y = 13.8$

Do the above data prove the hypothesis that the mean distance driven by Los Angeles commuters exceeds that of San Francisco commuters? Use the
(a) 10
(b) 5
(c) 1
percent level of significance.

9. The following are the results of independent samples of two different populations.

X: 10.3, 10.4, 11.3, 13.5, 12.7, 11.1, 10.9, 9.7, 14.5, 13.3
Y: 12.4, 11.7, 13.5, 12.9, 13.4, 15.5, 16.3, 13.7, 14.3

Test the null hypothesis that the two population means are equal against the alternative that they are unequal, at the
(a) 10 percent
(b) 5 percent
(c) 1 percent
level of significance.

10. A manager is considering instituting an additional 15-minute coffee break if it can be shown to decrease the number of errors that employees commit. The manager divided a sample of 20 employees into two groups of 10 each. Members of one group followed the same work schedule as before, but the members of the other group were given a 15-minute coffee break in the middle of the day. The following data give the total number of errors committed by each of the 20 workers over the next 20 working days.

Coffee break group: 8, 7, 5, 8, 10, 9, 7, 8, 4, 5
No-break group: 7, 6, 14, 12, 13, 8, 9, 6, 10, 9

Test the hypothesis, at the 5 percent level of significance, that instituting a coffee break does not reduce the mean number of errors. What is your conclusion?

10.5 PAIRED-SAMPLE *t* TEST

Suppose that $X_1, \ldots, X_n$ and $Y_1, \ldots, Y_n$ are samples of the same size from different normal populations having respective means μ_x and μ_y. In certain situations there will be a relationship between the data values X_i and Y_i. Because of this relationship, the pairs of data values $X_i, Y_i, i = 1, \ldots, n$, will not be independent; so we will not be able to use the methods of previous sections to test hypotheses concerning μ_x and μ_y.

Example 10.7   Suppose we are interested in learning about the effect of a newly developed gasoline detergent additive on automobile mileage. To gather information, seven cars have been assembled, and their gasoline mileages (in units of miles per gallon) have been determined. For each car this determination is made both when gasoline without the additive is used and when gasoline with the additive is used. The data can be represented as follows:

Car	Mileage without additive	Mileage with additive
1	24.2	23.5
2	30.4	29.6
3	32.7	32.3
4	19.8	17.6
5	25.0	25.3
6	24.9	25.4
7	22.2	20.6

For instance, car 1 got 24.2 miles per gallon by using gasoline without the additive and only 23.5 miles per gallon by using gasoline with the additive, whereas car 4 obtained 19.8 miles per gallon by using gasoline without the additive and 17.6 miles per gallon by using gasoline with the additive.

Now it is easy to see that two factors will determine a car's mileage per gallon. One factor is whether the gasoline includes the additive, and the second factor is the car itself. For this reason we should not treat the two samples as being independent; rather, we should consider paired data.

Suppose we want to test

$$H_0: \mu_x = \mu_y \quad \text{against} \quad H_1: \mu_x \neq \mu_y$$

where the two samples consist of the paired data X_i, Y_i, $i = 1, \ldots, n$. We can test this null hypothesis that the population means are equal by looking at the differences between the data values in a pairing. That is, let

$$D_i = X_i - Y_i \quad i = 1, \ldots, n$$

Now

$$E[D_i] = E[X_i] - E[Y_i]$$

or, with $\mu_d = E[D_i]$,

$$\mu_d = \mu_x - \mu_y$$

The hypothesis that $\mu_x = \mu_y$ is therefore equivalent to the hypothesis that $\mu_d = 0$. Thus we can test the hypothesis that the population means are equal by testing

$$H_0: \mu_d = 0 \quad \text{against} \quad H_1: \mu_d \neq 0$$

Assuming that the random variables $D_1, \ldots, D_n$ constitute a sample from a normal population, we can test the above null hypothesis by using the *t* test described in Sec. 9.4. That is, if we let $\bar{D}$ and S_d denote, respectively, the sample mean and sample standard deviation of the data $D_1, \ldots, D_n$, then the test statistic TS is given by

$$\text{TS} = \sqrt{n}\, \frac{\bar{D}}{S_d}$$

The significance level α test will be to

Reject H_0 if $|\text{TS}| \geq t_{n-1, \alpha/2}$
Not reject H_0 otherwise

where the value of $t_{n-1, \alpha/2}$ can be obtained from Table D.2.

Equivalently, the test can be performed by computing the value of the test statistic TS, say it is equal to v, and then computing the resulting p value, given by

$$p \text{ value} = P\{|T_{n-1}| \geq |v|\} = 2P\{T_{n-1} \geq |v|\}$$

where T_{n-1} is a t random variable with $n - 1$ degrees of freedom. If a personal computer is available, then Program 9-1 can be used to determine the value of the test statistic and the resulting p value. The successive data values entered in this program should be $D_1, D_2, \ldots, D_n$, and the value of μ_0 (the null hypothesis value for the mean of D) entered should be 0.

Example 10.8   Using the data of Example 10.7, test, at the 5 percent level of significance, the null hypothesis that the additive does not change the mean number of miles obtained per gallon of gasoline.

Solution If it is not convenient to run Program 9-1, we can use the data to compute first the differences D_i and then the summary statistics $\bar{D}$ and S_d. Using the data differences

$$.7, .8, .4, 2.2, -.3, -.5, 1.6$$

results in the values

$$\bar{D} = .7 \qquad S_d = .966$$

Therefore, the value of the test statistic is

$$TS = \frac{\sqrt{7}(.7)}{.966} = 1.917$$

Since, from Table D.2, $t_{6, .025} = 2.447$, the null hypothesis that the mean mileage is the same whether or not the gasoline used contains the additive is not rejected at the 5 percent level of significance.

   If a personal computer is available, then we can solve the above by running Program 9-1. This yields the following:

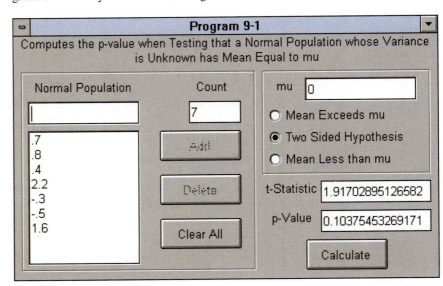

Thus the null hypothesis will not even be rejected at the 10 percent level of significance. ◼ ◼

One-sided tests concerning the two population means are similarly obtained. For instance, to test

$$H_0: \mu_x \leq \mu_y \quad \text{against} \quad H_1: \mu_x > \mu_y$$

we use the data $D_1, \ldots, D_n$ and test

$$H_0: \mu_d \leq 0 \quad \text{against} \quad H_1: \mu_d > 0$$

Again with the test statistic

$$\text{TS} = \sqrt{n}\,\frac{\bar{D}}{S_d}$$

the significance level α test is to

Reject H_0 if TS $> t_{n-1,\alpha}$
Not reject H_0 otherwise

Equivalently, if the value of TS is v, then the p value is

$$p \text{ value} = P\{T_{n-1} \geq v\}$$

Program 9-1 can be used again to determine the value of the test statistic and the resulting p value. (If summary statistics $\bar{D}$ and S_d are given, then the p value can be obtained by calculating v, the value of the test statistic, and then running Program 8-1 to determine $P\{T_{n-1} \geq v\}$.)

Example 10.9   The management of a chain of stores wanted to determine whether advertising tended to increase its sales of women's shoes. To do so, management determined the number of shoe sales at six stores during a 2-week period. While there were no advertisements in the first week, advertising was begun at the beginning of the second week. Assuming that any change in sales is due solely to the advertising, do the resulting data prove that advertising increases the mean number of sales? Use the 1 percent level of significance.

Store	First-week sales	Second-week sales
1	46	54
2	54	60
3	74	96
4	60	75
5	63	80
6	45	50

Solution Letting D_i denote the increase in sales at store i, we need to check if the data are significant enough to establish that $\mu_d > 0$. Hence, we should test

$$H_0: \mu_d \leq 0 \qquad \text{against} \qquad H_1: \mu_d > 0$$

Using the data values 8, 6, 22, 15, 17, 5, we run Program 9-1 to obtain the following:

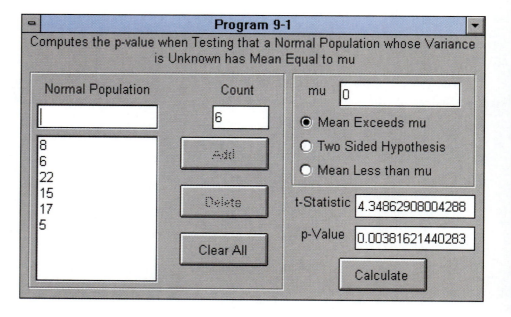

Thus the hypothesis that advertising does not result in increased sales is rejected at any significance level greater than or equal to .0038. Therefore, it is rejected at the 1 percent level of significance.

Problems

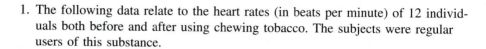

1. The following data relate to the heart rates (in beats per minute) of 12 individuals both before and after using chewing tobacco. The subjects were regular users of this substance.

Subject	Heart rate before use	Heart rate after use
1	73	77
2	67	69
3	68	73
4	60	70
5	76	74
6	80	88
7	73	76
8	77	82
9	66	69
10	58	61
11	82	84
12	78	80

(a) Test the hypothesis, at the 5 percent level of significance, that chewing smokeless tobacco does not result in a change in the mean heart rate of the population of regular users of chewing tobacco.

(b) What is the resulting p value?

2. A shoe salesman claims that using his company's running shoes will result, on average, in faster times. To check this claim, a track coach assembled a team of 10 sprinters. The coach randomly divided the runners into two groups of size 5. The members of the first group then ran 100 yards, using their usual running shoes; and the members of the second group ran 100 yards, using the company's shoes. After time was given for rest, the group who ran with their usual shoes changed into the company's shoes and members of the other group changed to their usual shoes. Then they all ran another dash of 100 yards. The following data resulted:

Racer	1	2	3	4	5	6	7	8	9	10
Time (old shoes)	10.5	10.3	11.0	10.9	11.3	9.9	10.1	10.7	12.2	11.1
Time (new shoes)	10.3	10.0	10.6	11.1	11.0	9.8	10.2	10.5	11.8	10.5

Do the above data prove the claim of the salesman that the company's new shoes result, on average, in lower times? Use the 10 percent level of significance. What about at the 5 percent level?

3. Use the t test on the following paired data to test

$$H_0: \mu_x = \mu_y \qquad \text{against} \qquad H_1: \mu_x \neq \mu_y$$

at the 5 percent level of significance.

i	1	2	3	4	5	6	7	8	9	10	11
X_i	122	132	141	127	141	119	124	131	145	140	135
Y_i	134	126	133	122	155	116	118	137	140	133	142

4. A question of medical interest is whether jogging leads to a reduction in systolic blood pressure. To learn about this question, eight nonjogging volunteers have agreed to begin a 1-month jogging program. At the end of the month their blood pressures were determined and compared with earlier values, with the following data resulting.

Subject	1	2	3	4	5	6	7	8
Blood pressure before	134	122	118	130	144	125	127	133
Blood pressure after	130	120	123	127	138	121	132	135

(a) Suppose you want to see if the above data are significant enough to prove that jogging for 1 month will tend to reduce the systolic blood pressure. Give the null and alternative hypotheses.

(b) Do the data prove the hypothesis in (a) at the 5 percent level of significance?

(c) Do the data prove that the hypothesis is false?

(d) How would you present the results of this experiment to a medical person who is not trained in statistics?

5. The following table gives the scores on a test of intelligence for 14 pairs of monozygotic (commonly called *identical*) twins who were separated at birth. One member of each pair was raised by a biological parent while the other was raised in a home that did not contain either of its biological parents. The IQ test used is known in the psychological literature as the "dominoes" IQ test.

Twin raised by mother or father	Twin raised by neither parent	Twin raised by mother or father	Twin raised by neither parent
23	18	22	15
30	25	31	23
25	28	29	27
18	22	24	26
19	14	28	19
25	34	31	30
28	36	27	28

(a) Test the hypothesis that the mean IQ test score of a twin is not affected by whether he or she is raised by a biological parent. Use the 5 percent level of significance.

(b) What conclusions, if any, can be drawn from the above?

6. Consider Prob. 2 of Sec. 10.4. Suppose that the same women were used for both months and that the data in each of the columns referred to the same woman's fat intake during the summer and winter.

(a) Test the hypothesis that there is no difference in fat intake during summer and winter. Use the 5 percent level of significance.

(b) Repeat (a), this time using the 1 percent level.

7. The following are scores on two IQ tests of 12 university students. One of the tests was taken before, and the other was taken after, the student had a course in statistics.

Student	IQ score before course	IQ score after course
1	104	111
2	125	120
3	127	138
4	102	113
5	140	142
6	122	130
7	118	114
8	110	121
9	126	135
10	138	145
11	116	118
12	125	125

Use these data to test the hypothesis that a student's score on an IQ test will not tend to be any different after the student takes a statistics course. Use the 5 percent level of significance.

8. To see whether there are any differences in starting salaries for women and men law school graduates, a set of eight law firms was selected. For each of these firms a recently hired woman and a recently hired man were randomly chosen. The following starting salary information resulted from interviewing those chosen.

Company	1	2	3	4	5	6	7	8
Woman's salary	52	53.2	78	75	62.5	72	39	49
Man's salary	54	55.5	78	81	64.5	70	42	51

Use the above data to test the hypothesis, at the 10 percent level of significance, that the starting salary is the same for both sexes.

9. To study the effectiveness of a certain commercial liquid protein diet, the Food and Drug Administration sampled nine individuals who were entering the program. Their weights both immediately before they entered and 6 months after they completed the 2-week program were recorded. The following data resulted.

Person	Weight before	Weight after
1	197	185
2	212	220
3	188	180
4	226	217
5	170	185
6	194	197
7	233	219
8	166	170
9	205	202

Suppose we want to determine if the above data prove that the diet is effective in the sense that the expected weight loss after 6 months is positive.
(a) What is the null hypothesis to be tested, and what is the alternative?
(b) Do the data prove that the diet works? Use the 5 percent level.

10. The following are the motor vehicle death rates per 100 million vehicle miles for a random selection of states in both 1985 and 1989.

State	1985 Rate	1989 Rate
Arkansas	3.4	3.3
Colorado	2.4	1.9
Indiana	2.6	1.9
Kentucky	2.6	2.4
Massachusetts	1.9	1.7
Ohio	2.1	2.1
Tennessee	3.4	2.3
Wyoming	2.7	2.3

Source: Accident Facts, National Safety Council, Chicago, 1990.

(a) Do the above data establish, at the 5 percent level of significance, that the motor vehicle death rate was lower in 1989 than in 1985?
(b) What is the p value for the hypothesis test in part (a)?

11. The following data give the marriage rates per 1000 population in a random sample of countries for 1987 and 1989.

Country	1987 Rate	1989 Rate
Belgium	5.8	6.4
Finland	5.4	5.1
Greece	6.3	6.0
Israel	6.9	7.0
New Zealand	6.0	6.1
Norway	5.0	4.9
Switzerland	6.6	6.8
United States	9.9	9.7
Yugoslavia	7.0	6.7

Source: United Nations, *Monthly Bulletin of Statistics*, June 1990.

Test the hypothesis that the worldwide marriage rates in 1987 are no greater than those in 1989.

10.6 TESTING EQUALITY OF POPULATION PROPORTIONS

Consider two large populations, and let p_1 and p_2 denote, respectively, the proportions of the members of these two populations that have a certain characteristic of interest. Suppose that we are interested in testing the hypothesis that these proportions are equal against the alternative that they are unequal. That is, we are interested in testing

$$H_0: p_1 = p_2 \quad \text{against} \quad H_1: p_1 \neq p_2$$

To test this null hypothesis, suppose that independent random samples, of respective sizes n_1 and n_2, are drawn from the populations. Let X_1 and X_2 represent the number of elements in the two samples that have the characteristic.

Let $\hat{p}_1$ and $\hat{p}_2$ denote, respectively, the proportions of the members of the two samples that have the characteristic. That is, $\hat{p}_1 = X_1/n_1$ and $\hat{p}_2 = X_2/n_2$. Since $\hat{p}_1$ and $\hat{p}_2$ are the respective estimators of p_1 and p_2, it is evident that we want to reject H_0 when $\hat{p}_1$ and $\hat{p}_2$ are far apart, that is, when $|\hat{p}_1 - \hat{p}_2|$ is sufficiently large. To see how far apart they need be to justify rejection of H_0, first we need to determine the probability distribution of $\hat{p}_1 - \hat{p}_2$.

Recall from Sec. 7.5 that the mean and variance of the proportion of the first sample that has the characteristic is given by

$$E[\hat{p}_1] = p_1 \quad \text{Var}(\hat{p}_1) = \frac{p_1(1 - p_1)}{n_1}$$

and, similarly, for the second sample

$$E[\hat{p}_2] = p_2 \quad \text{Var}(p_2) = \frac{p_2(1 - p_2)}{n_2}$$

Thus we see that

$$E[\hat{p}_1 - \hat{p}_2] = E[\hat{p}_1] - E[\hat{p}_2]$$
$$= p_1 - p_2$$
$$\text{Var}(\hat{p}_1 - \hat{p}_2) = \text{Var}(\hat{p}_1) + \text{Var}(\hat{p}_2)$$
$$= \frac{p_1(1 - p_1)}{n_1} + \frac{p_2(1 - p_2)}{n_2}$$

In addition, if we suppose that n_1 and n_2 are reasonably large, then $\hat{p}_1$ and $\hat{p}_2$ will approximately have a normal distribution, and thus so will their difference $\hat{p}_1 - \hat{p}_2$. As a result, the standardized variable

$$\frac{\hat{p}_1 - \hat{p}_2 - (p_1 - p_2)}{\sqrt{p_1(1 - p_1)/n_1 + p_2(1 - p_2)/n_2}}$$

will have a distribution that is approximately that of a standard normal random variable.

Now suppose that H_0 is true, and so the proportions are equal. Let p denote their common value; that is, $p_1 = p_2 = p$. In this case $p_1 - p_2 = 0$, and so the quantity

$$W = \frac{\hat{p}_1 - \hat{p}_2}{\sqrt{p(1 - p)/n_1 + p(1 - p)/n_2}} \tag{10.3}$$

will approximately have a standard normal distribution. We cannot, however, base our test directly on W for it depends on the unknown quantity p. However, we can estimate p by noting that of the combined sample of size $n_1 + n_2$ there are a total of $X_1 + X_2 = n_1\hat{p}_1 + n_2\hat{p}_2$ elements that have the characteristic of interest. Therefore, when H_0 is true and each population has the same proportion of its members with the characteristic, the natural estimator of that common proportion p is

$$\hat{p} = \frac{n_1\hat{p}_1 + n_2\hat{p}_2}{n_1 + n_2} = \frac{X_1 + X_2}{n_1 + n_2}$$

The estimator $\hat{p}$ is called the *pooled* estimator of p.

We will now substitute the estimator $\hat{p}$ for the unknown parameter p in Eq. (10.3) for W and base our test on the resulting expression. That is, we will use the test statistic

$$\text{TS} = \frac{\hat{p}_1 - \hat{p}_2}{\sqrt{\hat{p}(1-\hat{p})/n_1 + \hat{p}(1-\hat{p})/n_2}} = \frac{\hat{p}_1 - \hat{p}_2}{\sqrt{(1/n_1 + 1/n_2)\,\hat{p}(1 - \hat{p})}}$$

It can be shown that for reasonably large values of n_1 and n_2 (both being at least 30 should suffice) TS will, when H_0 is true, have a distribution that is approximately equal to the standard normal distribution. Thus, the significance level α test of

$$H_0: p_1 = p_2 \qquad \text{against} \qquad H_1: p_1 \neq p_2$$

is to

Reject H_0 if $|TS| \geq z_{\alpha/2}$
Not reject H_0 otherwise

The test can also be performed by determining the value of the test statistic, say it is equal to v, and then determining the p value given by

$$p \text{ value} = P\{|Z| \geq |v|\} = 2P\{Z \geq |v|\}$$

where Z is (as always) a standard normal random variable.

Example 10.10   In criminal proceedings a convicted defendant is sometimes sent to prison by the presiding judge and is sometimes not. A question has arisen in legal circles as to whether a judge's decision is affected by (1) whether the defendant pleaded guilty or (2) whether he or she pleaded innocent but was subsequently found guilty. The following data refer to individuals, all having previous prison records, convicted of second-degree robbery.

74, out of 142 who pleaded guilty, went to prison

61, out of 72 who pleaded not guilty, went to prison

Do the above data indicate that a convicted individual's chance of being sent to prison depends on whether she or he had pleaded guilty?

Solution Let p_1 denote the probability that a convicted individual who pleaded guilty will be sent to prison, and let p_2 denote the corresponding probability for one who pleaded innocent but was adjudged guilty. To see if the data are significant enough to prove that $p_1 \neq p_2$, we need to test

$$H_0: p_1 = p_2 \qquad \text{against} \qquad H_1: p_1 \neq p_2$$

The data yield

$$n_1 = 142 \qquad p_1 = \frac{74}{142} = .5211$$

$$n_2 = 72 \qquad p_2 = \frac{61}{72} = .8472$$

The value of the pooled estimator $\hat{p}$ is

$$\hat{p} = \frac{74 + 61}{142 + 72} = .6308$$

and the value of the test statistic is

$$\text{TS} = \frac{.5211 - .8472}{\sqrt{1/142 + 1/72}\,(.6308)(1 - .6308)} = -4.66$$

The p value is given by

$$p \text{ value} = 2P\{Z \geq 4.66\} \approx 0$$

For such a small p value the null hypothesis will be rejected. That is, we can conclude that the decision of a judge, with regards to whether a convicted defendant should be sent to prison, is indeed affected by whether that defendant pleaded guilty or innocent. (We cannot, however, conclude that pleading guilty is a good strategy for a defendant as far as avoiding prison is concerned. The reason we cannot is that a defendant who pleads innocent has a chance of being acquitted.)

Our next example illustrates the difficulties that abound in modeling real phenomena.

Example 10.11: Predicting a child's gender
Suppose we are interested in determining a model for predicting the gender of future children in families. The simplest model would be to suppose that each new birth, no matter what the present makeup of the family, will be a boy with some probability p_0. (Interestingly enough, existing data indicate that p_0 would be closer to .51 than to .50.)

Somewhat surprisingly, the above simple model does not hold up when real data are considered. For instance, data on the gender of members of French families were given by Malinvaud in 1955. Consider families having four or more children. Malinvaud reported 36,694 such families whose first three children were girls (that is, the three eldest children were girls), and he reported 42,212 such families whose first three children were all boys. Malinvaud's data indicated that in those families whose first three children were all girls, their next child was a boy 49.6 percent of the time, whereas in the families whose first three children were all boys, the next one was a boy 52.3 percent of the time.

Let p_1 denote the probability that the next child of a family presently composed of three girls is a boy, and let p_2 denote the corresponding probability for a family presently composed of three boys. If we use the above data to test

$$\text{H}_0: p_1 = p_2 \quad \text{against} \quad \text{H}_1: p_1 \neq p_2$$

then we have

$$n_1 = 36{,}694 \qquad n_2 = 42{,}212$$
$$\hat{p}_1 = .496 \qquad \hat{p}_2 = .523$$

and so

$$\hat{p} = \frac{36{,}694(.496) + 42{,}212(.523)}{36{,}694 + 42{,}212} = .51044$$

Therefore, the value of the test statistic is

$$\text{TS} = \frac{.496 - .523}{\sqrt{(1/36{,}694 + 1/42{,}212)(.5104)(1 - .5104)}} = -7.567$$

Since $|\text{TS}| \geq z_{.005} = 2.58$, the null hypothesis that the probability that the next child is a boy is the same regardless of whether the present family is made up of three girls or of three boys is rejected at the 1 percent level of significance. Indeed, the p value of these data is

$$p \text{ value} = P\{|Z| \geq 7.567\} = 2P\{Z \geq 7.567\} \approx 0$$

This shows that any model which assumes that the probability of the gender of an unborn does not depend on the present makeup of the family is not consistent with existing data. (One model that is consistent with the above data is to suppose that each family has its own probability that a newborn will be a boy, with this probability remaining the same no matter what the present makeup of the family. This probability, however, differs from family to family.)   ◼ ◼

If we are interested in verifying the one-sided hypothesis that p_1 is larger than p_2, then we should take that to be the alternative hypothesis and so test

$$H_0: p_1 \leq p_2 \qquad \text{against} \qquad H_1: p_1 > p_2$$

The same test statistic TS as used before is still employed, but now we reject H_0 only when TS is large (since this occurs when $\hat{p}_1 - \hat{p}_2$ is large). Thus, the one-sided significance level α test is to

| Reject H_0 | if $\text{TS} \geq z_\alpha$ |
| Not reject H_0 | otherwise |

Alternatively, if the value of the test statistic TS is v, then the resulting p value is

$$p \text{ value} = P\{Z \geq v\}$$

where Z is a standard normal.

Remark: The test of

$$H_0: p_1 \leq p_2 \qquad \text{against} \qquad H_1: p_1 > p_2$$

is the same as

$$H_0: p_1 = p_2 \qquad \text{against} \qquad H_1: p_1 > p_2$$

This is so because in both cases we want to reject H_0 when $\hat{p}_1 - \hat{p}_2$ is so large that such a large value would have been highly unlikely if p_1 were not greater than p_2.

Example 10.12   A manufacturer has devised a new method for producing computer chips. He feels that this new method will reduce the proportion of chips that turn out to have defects. To verify this, 320 chips were produced by the new method and 360 by the old. The result was that 76 of the former and 94 of the latter were defective. Is this significant enough evidence for the manufacturer to conclude that the new method will produce a smaller proportion of defective chips? Use the 5 percent level of significance.

Solution Let p_1 denote the probability that a chip produced by the old method will be defective, and let p_2 denote the corresponding probability for a chip produced by the new method. To conclude that $p_1 > p_2$, we need to reject H_0 when testing

$$H_0: p_1 \leq p_2 \qquad \text{against} \qquad H_1: p_1 > p_2$$

The data are

$$n_1 = 360 \qquad\qquad n_2 = 320$$
$$\hat{p}_1 = \frac{94}{360} = .2611 \qquad\qquad \hat{p}_2 = \frac{76}{320} = .2375$$

The value of the pooled estimator is thus

$$\hat{p} = \frac{94 + 76}{360 + 320} = .25$$

Hence, the value of the test statistic is

$$TS = \frac{.2611 - .2375}{\sqrt{(1/360 + 1/320)(.25)(.75)}} = .7094$$

Since $z_{.05} = 1.96$, we cannot reject the null hypothesis at the 5 percent level of significance. That is, the evidence is not significant enough for us to conclude that the new method will produce a smaller percentage of defective chips than the old method. The p value for the data is

Do Not Misinterpret a Rejection

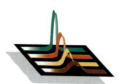

We must be careful when deciding what a rejection of the null hypothesis really means, for often interpretations are given that are not warranted by the available data. For instance, suppose a hypothesis test was performed to study whether the probabilities that a patient does not survive an operation are the same at hospitals A and B. Suppose that a random sample of the operations performed at hospital A yielded that 72 out of 480 patients operated on did not survive; whereas a sample at hospital B yielded that 30 of 360 did not survive. While we can certainly conclude from these data that the survival probabilities are unequal, we cannot conclude that hospital A is not doing as good a job as hospital B; for without additional data we cannot rule out such possibilities as that hospital A is performing more high-risk operations than is B, and that is the reason it has a lower survival rate.

For another example that indicates how careful we must be when interpreting the meaning of a rejected hypothesis, consider a hypothetical study of the salaries of male and female salespeople at a large corporation. Suppose that a random sample of 50 male and 50 female employees indicated that the average salary of the men was $40,000 per year whereas that of the women was $36,000. Assuming that the sample variances were small, a test of the hypothesis that the mean salary was the same for both populations would be rejected. But what could we conclude from this? For instance, would we be

justified in concluding that the women are being discriminated against? The answer is that we cannot come to such a conclusion with the information presented, for there are many possible explanations for the apparent differences in mean salary.

One possibility might be that the mix of experienced and inexperienced workers is different for the two sexes. For instance, taking into account whether an employee had worked for more or less than 5 years might have produced the following data.

Years of employment	Number	Average salary ($)
Men:		
Less than 5	10	34,000
More than 5	40	41,500
Total	50	40,000
Women:		
Less than 5	40	34,500
More than 5	10	42,000
Total	50	36,000

For instance, a total of 10 of the 50 women have been employed more than 5 years, and their average salary is $42,000 per year. Thus, we see from the above that even though the average salary of the men is higher than that of the women, when time of employment is taken into account, the female employees are actually receiving higher salaries than their male counterparts.

$$p \text{ value} = P\{Z \geq .7094\} = .239$$

indicating that a value of TS at least as large as the one observed will occur 24 percent of the time when the two probabilities are equal.

| Table 10.5 | **Tests Concerning Two Binomial Probabilities** |

The proportions of members of two populations that have a certain characteristic are p_1 and p_2. A random sample of size n_1 is chosen from the first population, and an independent random sample of size n_2 is chosen from the second population. The numbers of members of the two samples with the characteristic are X_1 and X_2, respectively.

$$\hat{p}_1 = \frac{X_1}{n_1} \qquad \hat{p}_2 = \frac{X_2}{n_2}$$

$$\hat{p} = \frac{X_1 + X_2}{n_1 + n_2}$$

H_0	H_1	Test statistic TS	Significance level α test	p Value if TS $= v$
$p_1 = p_2$	$p_1 \neq p_2$	$\dfrac{\hat{p}_1 - \hat{p}_2}{\sqrt{(1/n_1 + 1/n_2)\hat{p}(1 - \hat{p})}}$	Reject H_0 if $|TS| \geq z_{\alpha/2}$ Do not reject otherwise	$2P\{Z \geq |v|\}$
$p_1 \leq p_2$	$p_1 > p_2$	$\dfrac{\hat{p}_1 - \hat{p}_2}{\sqrt{(1/n_1 + 1/n_2)\hat{p}(1 - \hat{p})}}$	Reject H_0 if TS $\geq z_{\alpha}$ Do not reject otherwise	$P\{Z \geq v\}$

Table 10.5 details the tests considered in this section.

Problems

1. Two methods have been proposed for producing transistors. If method 1 resulted in 20 unacceptable transistors out of a total of 100 produced, and method 2 resulted in 12 unacceptable transistors out of a total of 100 produced, can we conclude that the proportion of unacceptable transistors that will be produced by the two methods are different?
   (a) Use the 5 percent level of significance.
   (b) What about at the 10 percent level of significance?

2. A random sample of 220 women and 210 men coffee drinkers were questioned. The result was that 71 of the women and 58 of the men indicated a preference for decaffeinated coffee. Do these data establish, at the 5 percent level of significance, that the proportion of women coffee drinkers who prefer decaffeinated coffee differs from the corresponding proportion for men? What is the p value?

3. An automobile insurance company selected random samples of 300 single male policyholders and 300 married male policyholders, all between the ages of 25 and 30. It recorded the number who had reported accidents at some time within the past 3 years. The resulting data were that 19 percent of the single policyholders and 12 percent of the married ones had reported an accident.
   (a) Does this establish, at the 10 percent level of significance, that there is a difference in these two types of policyholders?
   (b) What is the p value for the test in part (a)?

4. A large swine flu vaccination program was instituted in 1976. Approximately 50 million of the roughly 220 million North Americans received the vaccine. Of the 383 persons who subsequently contracted swine flu, 202 had received the vaccine.
   (a) Test the hypothesis, at the 5 percent level, that the probability of contracting swine flu is the same for the vaccinated portion of the population as for the unvaccinated.
   (b) Do the results of part (a) indicate that the vaccine itself was causing the flu? Can you think of any other possible explanations?

5. Two insect sprays are to be compared. Two rooms of equal size are sprayed, one with spray 1 and the other with spray 2. Then 100 insects are released in each room, and after 2 hours the dead insects are counted. Suppose the result is 64 dead insects in the room sprayed with spray 1 and 52 dead insects in the other room.
   (a) Is the above evidence significant enough for us to reject, at the 5 percent level, the hypothesis that the two sprays have equal ability to kill insects?
   (b) What is the p value of the test in part (a)?

6. Random samples of 100 residents from San Francisco and 100 from Los Angeles were chosen, and the residents were questioned about whether they favored raising the driving age. The result was that 56 of those from San Francisco and 45 of those from Los Angeles were in favor.
   (a) Are these data strong enough to establish, at the 10 percent level of significance, that the proportions of the population in the two cities that are in favor are different?
   (b) What about at the 5 percent level?

7. In 1983, a random sample of 1000 scientists included 212 female scientists. On the other hand, a random sample of 1000 scientists drawn in 1990 included 272 women. Use these data to test the hypothesis, at the 5 percent level of significance, that the proportion of scientists who are female was the same in 1983 as in 1990. Also find the p value.

8. Example 10.11 considered a model for predicting a child's gender. One generalization of the model considered would be to suppose that a child's gender depends only on the number of previous children in the family and on the number of these who are boys. If this were so, then the gender of the third

child in families whose children presently consist of one boy and one girl would not depend on whether the order of the first two children was boy-girl or girl-boy. The following data give the gender of the third child in families whose first two children were a boy and a girl. It distinguishes whether the boy or the girl was older. (Boy-girl means, for instance, that the older child was a boy.)

Boy-girl families	Girl-boy families
412 boys	560 boys
418 girls	544 girls

Use the above data to test the hypothesis that the sex of a third child in a family presently having one boy and one girl does not depend upon the gender birth order of the two older siblings. Use the 5 percent level of significance.

9. According to the National Center for Health Statistics, there were a total of 330,535 African American females and 341,441 African American males born in 1988. Also in that year, 1,483,487 white females and 1,562,675 white males were born. Use these data to test the hypothesis that the proportion of all African American babies who are female is equal to the proportion of all white babies who are female. Use the 5 percent level of significance. Also find the p value.

10. Suppose a random sample of 480 heart-bypass operations at hospital A showed that 72 patients did not survive, whereas a random sample of 360 operations at hospital B showed that 30 patients did not survive. Find the p value of the test of the hypothesis that the survival probabilities are the same at the two hospitals.

11. A birthing class run by the University of California has recently added a lecture on the importance of the use of automobile car seats for children. This decision was made after a study of the results of an experiment in which the lecture was given in some of the birthing classes and not in others. A follow-up interview, carried out 1 year later, questioned 82 couples who had heard the lecture and 120 who had not. A total of 78 of the couples who had heard the lecture stated that they always used an infant car seat, whereas a total of 90 of those couples not attending the lecture made the same claim.
   (a) Assuming the accuracy of the above, is the above difference significant enough to conclude that instituting the lecture will result in increased use of car seats? Use the 5 percent level of significance.
   (b) What is the p value?

12. In a recently conducted poll, 52 out of 200 people surveyed claimed to have a handgun at home. In an earlier survey, 28 out of 150 people made that claim. Does this prove that more people now have, or at least claim to have, handguns in their homes?

(a) Use the 5 percent level of significance.

(b) What is the p value?

13. To see how effective a newly developed vaccine is against the common cold, 204 workers at a ski resort were randomly divided into two groups of size 102 each. Members of the first group were given the vaccine throughout the winter months while members of the second group were given a placebo. By the end of the winter season, it turned out that 29 individuals who had been receiving the vaccine caught at least one cold, compared to 34 of those receiving the placebo. Does this prove, at the 5 percent level of significance, that the vaccine is effective in preventing colds?

14. The American Cancer Society recently sampled 2500 adults and determined that 738 of them were smokers. A similar poll of 2000 adults carried out in 1986 yielded a total of 640 smokers. Do these figures prove that the proportion of adults who smoke has decreased since 1986?

(a) Use the 5 percent level of significance.

(b) Use the 1 percent level of significance.

15. In a recent study of 22,000 male physicians, half were given a daily dose of aspirin while the other half were given a placebo. The study was continued for a period of 6 years. During this time 104 of those taking the aspirin, and 189 of those taking the placebo, suffered heart attacks. Does this result indicate that taking a daily dose of aspirin decreases the risk of suffering a heart attack? Give the null hypothesis and the resulting p value.

Key Terms

Two-sample tests: Tests concerning the relationships of parameters from two separate populations.

Paired-sample tests: Tests where the data consist of pairs of dependent variables.

Summary

I. Testing Equality of Population Means: Independent Samples

Suppose that $X_1, \ldots, X_n$ and $Y_1, \ldots, Y_m$ are independent samples from normal populations having respective parameters μ_x, σ_x^2 and μ_y, σ_y^2.

Case 1: σ_x^2 and σ_y^2 are known:

To test

$$H_0: \mu_x = \mu_y \quad \text{against} \quad H_1: \mu_x \neq \mu_y$$

use the test statistic

$$TS = \frac{\bar{X} - \bar{Y}}{\sqrt{\sigma_x^2/n + \sigma_y^2/m}}$$

The significance level α test is to

Reject H_0	if $|TS| \geq z_{\alpha/2}$
Not reject H_0	otherwise

If the value of TS is v, then

$$p \text{ value} = P\{|Z| \geq |v|\} = 2P\{Z \geq |v|\}$$

where Z is a standard normal random variable.

The significance level α test of

$$H_0: \mu_x \leq \mu_y \qquad \text{against} \qquad H_1: \mu_x > \mu_y$$

uses the same test statistic. The test is to

Reject H_0	if TS $\geq z_\alpha$
Not reject H_0	otherwise

If TS $= v$, then the p value is

$$p \text{ value} = P\{Z \geq v\}$$

Case 2: σ_x^2 and σ_y^2 are unknown and n and m are large:

To test

$$H_0: \mu_x = \mu_y \qquad \text{against} \qquad H_1: \mu_x \neq \mu_y$$

or

$$H_0: \mu_x \leq \mu_y \qquad \text{against} \qquad H_1: \mu_x > \mu_y$$

use the test statistic

$$TS = \frac{\bar{X} - \bar{Y}}{\sqrt{S_x^2/n + S_y^2/m}}$$

where S_x^2 and S_y^2 are the respective sample variances. The test and the p value are then exactly the same as in case 1.

Case 3: σ_x^2 and σ_y^2 are assumed to be unknown but equal:

To test

$$H_0: \mu_x = \mu_y \qquad \text{against} \qquad H_1: \mu_x \neq \mu_y$$

use the test statistic

$$\text{TS} = \frac{\bar{X} - \bar{Y}}{\sqrt{S_p^2(1/n + 1/m)}}$$

where S_p^2, called the *pooled estimator* of the common variance, is given by

$$S_p^2 = \frac{n-1}{n+m-2} S_x^2 + \frac{m-1}{n+m-2} S_y^2$$

The significance level α test is to

Reject H_0	if $|\text{TS}| \geq t_{n+m-2, \alpha/2}$
Not reject H_0	otherwise

If $\text{TS} = v$, then the p value is

$$p \text{ value} = 2P\{T_{n+m-2} \geq |v|\}$$

In the preceding, T_{n+m-2} is a t random variable having $n+m-2$ degrees of freedom, and $t_{n+m-2, \alpha}$ is such that

$$P\{T_{n+m-2} \geq t_{n+m-2,\alpha}\} = \alpha$$

To test

$$H_0: \mu_x \leq \mu_y \qquad \text{against} \qquad H_1: \mu_x > \mu_y$$

use the same test statistic. The significance level α test is to

Reject H_0	if $\text{TS} \geq t_{n+m-2, \alpha}$
Not reject H_0	otherwise

If $\text{TS} = v$, then

$$p \text{ value} = P\{T_{n+m-2} \geq v\}$$

II. Testing Equality of Population Means: Paired Samples

Suppose $X_1, \ldots, X_n$ and $Y_1, \ldots, Y_n$ are samples from populations having respective means of μ_x and μ_y. Suppose also that these samples are not independent, but that the n pairs of random variables X_i and Y_i are dependent, $i = 1, \ldots, n$. Let, for each i,

$$D_i = X_i - Y_i$$

and suppose that $D_1, \ldots, D_n$ constitute a sample from a normal population. Let

$$\mu_d = E[D_i] = \mu_x - \mu_y$$

To test

$$H_0: \mu_x = \mu_y \qquad \text{against} \qquad H_1: \mu_x \neq \mu_y$$

test the equivalent hypothesis

$$H_0: \mu_d = 0 \qquad \text{against} \qquad H_1: \mu_d \neq 0$$

Testing that the two samples have equal means is thus equivalent to testing that a normal population has mean 0. This latter hypothesis is tested by using the t test presented in Sec. 9.4. The test statistic is

$$\text{TS} = \sqrt{n}\, \frac{\bar{D}}{S_d}$$

and the significance level α test is to

Reject H_0 if $|\text{TS}| \geq t_{n-1, \alpha/2}$
Not reject H_0 otherwise

If $\text{TS} = v$, then

$$p \text{ value} = 2P\{T_{n-1} \geq |v|\}$$

To test the one-sided hypothesis

$$H_0: \mu_x \leq \mu_y \qquad \text{against} \qquad H_1: \mu_x > \mu_y$$

use the test statistic

$$TS = \sqrt{n}\,\frac{\bar{D}}{S_d}$$

The significance level α test is

Reject H_0	if TS $\geq t_{n-1,\alpha}$
Not reject H_0	otherwise

If TS $= v$, then

$$p \text{ value} = P\{T_{n-1} \geq v\}$$

III. Testing Equality of Population Proportions

Consider two large populations and a certain characteristic possessed by some members of these populations. Let p_1 and p_2 denote, respectively, the proportions of the members of the first and second populations that possess this characteristic. Suppose that a random sample of size n_1 is chosen from population 1, and one of size n_2 is chosen from population 2. Let X_1 and X_2 denote, respectively, the numbers of members of these samples that possess the characteristic.

Let

$$p_1 = \frac{X_1}{n_1} \quad \text{and} \quad p_2 = \frac{X_2}{n_2}$$

denote the proportions of the samples that have the characteristic; and let

$$\hat{p} = \frac{X_1 + X_2}{n_1 + n_2}$$

denote the proportion of the combined samples with the characteristic.

To test

$$H_0: p_1 = p_2 \quad \text{against} \quad H_1: p_1 \neq p_2$$

use the test statistic

$$TS = \frac{\hat{p}_1 - \hat{p}_2}{\sqrt{(1/n_1 + 1/n_2)\hat{p}(1 - \hat{p})}}$$

The significance level α test is to

Reject H_0 if $|\text{TS}| \geq z_{\alpha/2}$
Not reject H_0 otherwise

If the values of TS is v, then

$$p \text{ value} = 2P\{Z \geq |v|\}$$

To test

$$H_0: p_1 \leq p_2 \quad \text{against} \quad H_1: p_1 > p_2$$

use the same test statistic. The significance level α test is to

Reject H_0 if $\text{TS} \geq z_\alpha$
Not reject H_0 otherwise

If $\text{TS} = v$, then

$$p \text{ value} = P\{Z \geq v\}$$

Remark: In the above, as in all the text, Z always refers to a standard normal random variable, and z_α is such that

$$P\{Z \geq z_\alpha\} = \alpha$$

Review Problems

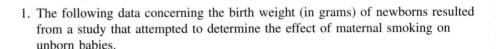

1. The following data concerning the birth weight (in grams) of newborns resulted from a study that attempted to determine the effect of maternal smoking on unborn babies.

Nonsmokers	Smokers
$n = 1820$	$m = 1340$
$\bar{X} = 3480$ grams	$\bar{Y} = 3260$ grams
$S_x = 9.2$ grams	$S_y = 10.4$ grams

(a) Test the hypothesis, at the 5 percent level of significance, that the mean weight of a newborn is the same whether or not the mother is a smoker.

(b) What is the p value in part (a)?

2. A study was initiated to compare two treatments for reducing the possibility of rejection in heart transplants. The first treatment involves giving the patient sodium salicylate, and the second calls for this drug to be given in conjunction with a second drug, azathioprine. The study was conducted on male rats with one type of rat being used as heart donor and a second type being used as recipient. (The use of different types of rats ensured that recipients would not survive too long.) The variable of interest is the survival time in days after receipt of the transplanted heart. The following summary statistics were obtained.

Sodium salicylate	Sodium salicylate with azathioprine
$n = 14$	$m = 12$
$\bar{X} = 15.2$ days	$\bar{Y} = 14$ days
$S_x = 9.2$ days	$S_y = 9.0$ days

Test the hypothesis, at the 5 percent level, that both treatments are equally effective in the population of rats.

3. A recent study concerning knee injuries of football players compared two types of football shoes. Out of a randomly chosen group of 1440 players, 240 used multicleated shoes, and 1200 used more conventional football shoes. All played on natural grass. Of those using the multicleated shoes, 13 suffered knee injuries. Of those using conventional shoes, 78 suffered knee injuries.
   (a) Test the hypothesis that the probability of a knee injury is the same for both groups of players. Use the 5 percent level of significance.
   (b) What is the p value in part (a)?
   (c) Are the above data strong enough to establish that the multicleated shoes are superior to the conventional ones in terms of reducing the probability of a knee injury?
   (d) In part (c), at what levels of significance would the evidence be strong enough?

4. Use the first 60 data values in App. A. Test, at the 5 percent level of significance, the hypothesis that men's and women's mean
   (a) Cholesterol
   (b) Blood pressure
   are equal.

5. The following data come from an experiment performed by Charles Darwin and reported in his 1876 book *The Effects of Cross and Self-Fertilization in the Vegetable Kingdom*. The data were first analyzed by Darwin's cousin Francis Galton. Galton's analysis was, however, in error. A correct analysis was eventually done by R. A. Fisher.

Darwin's experiment dealt with 15 pairs of *Zea mays,* a type of corn plant. One plant in each pair had been cross-fertilized while the other plant had been self-fertilized. The pairs were grown in the same pot, and their heights were measured. The data were as follows:

Pair	Cross-fertilized plant	Self-fertilized plant
1	23.5	17.375
2	12	20.375
3	21	20
4	22	20
5	19.125	18.375
6	21.5	18.625
7	22.125	18.625
8	20.375	15.25
9	18.25	16.5
10	21.625	18
11	23.25	16.25
12	21	18
13	22.125	12.75
14	23	15.5
15	12	18

(a) Test, at the 5 percent level of significance, the hypothesis that the mean height of cross-fertilized *Zea mays* corn plants is equal to that of self-fertilized *Zea mays* plants.

(b) Determine the p value for the test of the hypothesis in part (a).

6. A continuing debate in public health circles concerns the dangers of being exposed to dioxin, an environmental contaminant. A recent German study, published in the October 19, 1991, issue of *The Lancet,* a British medical journal, considered records of workers at a herbicide manufacturing plant that made use of dioxin. A control group consisted of workers at a nearby gas supply company who had similar medical profiles. The following data relating to the number of workers who had died from cancer were obtained.

	Control group	Dioxin-exposed group
Sample size	1583	1242
Number dying from cancer	113	123

(a) Test the hypothesis that the probability of dying from cancer is the same for the two groups. Use the 1 percent level of significance.

(b) Find the p value for the test of part (a).

7. Consider Prob. 6. Of the 1583 gas company workers whose records were studied, there were a total of 1184 men and 399 women. Of these individuals, 93

men and 20 women died of cancer. Test the hypothesis, at the 5 percent level, that the probability of dying from cancer is the same for workers of both sexes.

8. A random sample of 56 women revealed that 38 were in favor of gun control. A random sample of 64 men revealed that 32 were in favor. Use these data to test the hypothesis that the proportion of men and the proportion of women in favor of gun control are the same. Use the 5 percent level of significance. What is the p value?

9. Use the data presented in Review Prob. 19 of Chap. 8 to test the hypothesis that the chances of scoring a run are the same when there is one out and a player on second base and when there are no outs and a player on first base.

10. The following data concern 100 randomly chosen professional baseball games and 100 randomly chosen professional football games in the 1990–91 season. The data present, for the two sports, the number of games in which the team leading at the three-quarter mark (end of the seventh inning in baseball and end of the third quarter in football) ended up losing the game.

Sport	Number of games	Number of games lost by leader
Baseball	92	6
Football	93	21

Find the p value of the test of the hypothesis that the probability of the leading team's losing the game is the same in both sports. (*Note:* The number of games is not 100 because 8 of the baseball games and 7 of the football games were tied at the three-quarter point.)

11. The following relates to the same set of sample games reported in Prob. 10. It details the number of games in which the home team was the winner.

Sport	Number of games	Number of games that home team won
Baseball	100	53
Football	99	57

Test the hypothesis, at the 5 percent level of significance, that the proportion of games won by the home team is the same in both sports.

12. Suppose that a test of H_0: $\mu_x = \mu_y$ against H_1: $\mu_x \neq \mu_y$ results in rejecting H_0 at the 5 percent level of significance. Which of the following statements is (are) true?
   (a) The difference in sample means was statistically significant at the 1 percent level of significance.
   (b) The difference in sample means was statistically significant at the 10 percent level of significance.
   (c) The difference in sample means is equal to the difference in population means.

Hypothesis Tests
Concerning Two Populations

Purpose

Use Minitab to

1. Test two normal population means when the variances are known.

2. Test two normal population means when the variances are unknown but the sample sizes are large.

3. Test two normal population means when the variances are unknown but equal and the sample sizes are small.

4. Perform the paired-sample t test.

Procedures

First, load the Minitab (Windows version) software as in the Minitab lab for Chap. 1.

1. TESTING TWO NORMAL POPULATION MEANS WHEN THE POPULATION VARIANCES ARE KNOWN

Example 1 Use Minitab to work Example 10.1 in Sec. 10.2 of the text.

Minitab currently does not have a feature that will enable us to do the problem directly. However, we can use Minitab to help with some of the computations. Enter the values in columns C1 and C2. Rename C1; call it *A*. Rename C2; call it *B*. Type commands in the Session window, or use the Calc→Mathematical Expressions sequence to compute the test statistic. The Minitab commands that are used to compute the test statistic are shown in the Session window in Fig. M10.1. Note that K1/K2 is equivalent to Eq. 10.1 in Sec. 10.2. The computed test statistic value is 2.19182.

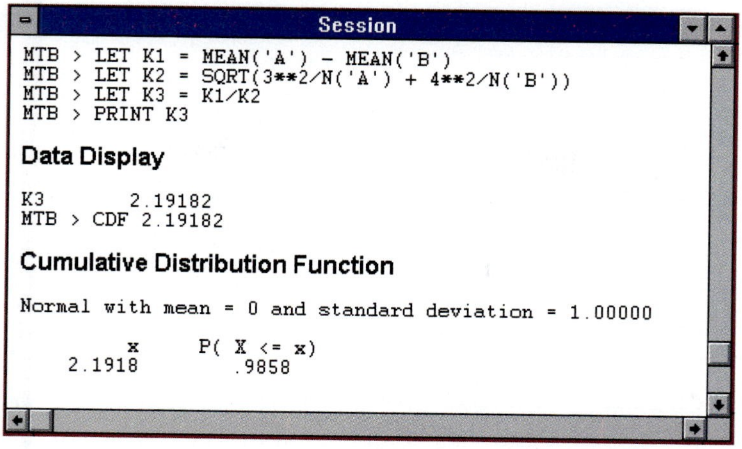

```
Session
MTB > LET K1 = MEAN('A') - MEAN('B')
MTB > LET K2 = SQRT(3**2/N('A') + 4**2/N('B'))
MTB > LET K3 = K1/K2
MTB > PRINT K3

Data Display

K3        2.19182
MTB > CDF 2.19182

Cumulative Distribution Function

Normal with mean = 0 and standard deviation = 1.00000

         x      P( X <= x)
    2.1918         .9858
```

Figure M10.1

To find the p value for the test at the MTB prompt in the Session window you can type CDF 2.19182; or you can use the sequence Calc→Probability Distributions→Normal, and the Normal distribution dialog box will be displayed. Select the Cumulative probability check box. Also select the Input constant check box, and select K3 for the corresponding text box. The results are displayed in Fig. M10.1. Thus the p value for the test will be twice the value to the right of 2.19182, since the test is two-sided. At the 5 percent level of significance, p value $= 2(1 - .9858) = .0284 < 5$ percent $= .05$. Thus we reject H_0: $\mu_x = \mu_y$.

Note You can use the same procedures as in this section to compare two means when the variances are unknown but the sample sizes are large. Keep in mind that you would have to replace the population variances with the corresponding sample variances.

2. TESTING TWO NORMAL POPULATION MEANS WHEN THE POPULATION VARIANCES ARE UNKNOWN

In this section *we assume that the unknown population variances are equal and that the sample sizes are small.* Thus, in this case we will have to use the t test statistic.

Example 2 Use Minitab to work Prob. 1 in Sec. 10.4 of the text.

Here we are asked to do a two-sample t, test on the mean systolic blood pressure for smokers and nonsmokers at the 1 percent level of significance. To use Minitab, enter the values for smokers in column C1 and for nonsmokers in C2. Rename C1; call it SMOKE. Rename C2; call it NSMOKE. Next select Stat→Basic Statistics→2-Sample t, and the 2-Sample t dialog box will be displayed. Choose the Samples in different columns check box, and select the columns for the First and Second text boxes. From the information given in the problem, we need to do a two-sided test; so select Not equal to in the Alternative drop-down box. Since the level of significance is at 1 percent, type 99 in the Confidence level text box. Make sure that the Assume equal variances option is selected by clicking on the box until an × is highlighted in it. The dialog box with all the appropriate selections and entries is shown in Fig. M10.2.

Click on the OK button, and the Session window will display the results. This is shown in Fig. M10.3. Observe that the computed test statistic is $T = 2.52$, the p value $= .019$, degrees of freedom DF $= 23$, and the pooled standard deviation $= 5.73$. Since the p value $= .019 >$

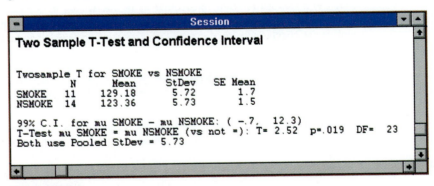

Figure M10.2

level of significance = .01, we fail to reject the null hypothesis of equal mean systolic blood pressures for smokers and nonsmokers. Also observe that a 99 percent confidence interval was generated for the *difference of the two means* (SMOKE − NSMOKE).

```
─                          Session                        ▼ ▲
                                                           ▲
 Two Sample T-Test and Confidence Interval


 Twosample T for SMOKE vs NSMOKE
             N       Mean      StDev    SE Mean
 SMOKE      11      129.18      5.72      1.7
 NSMOKE     14      123.36      5.73      1.5

 99% C.I. for mu SMOKE - mu NSMOKE: ( -.7,  12.3)
 T-Test mu SMOKE = mu NSMOKE (vs not =): T= 2.52  p=.019  DF=  23
 Both use Pooled StDev = 5.73
                                                           ▼
 ◄     ▐▌                                             ► ◄──►
```

Figure M10.3

Note These procedures help to ease in the computations, but you will have to write up the hypothesis tests in a manner presented by your instructor.

3. THE PAIRED-SAMPLE *t* TEST

When the samples are dependent, use Minitab to find the *difference* between the paired values, and do a one-sample *t* test for these differences as you did in the Minitab lab for Chap. 9. Suppose that the paired data set was stored in columns C1 and C2. To use Minitab to compute the difference C1 − C2 and save the differences in C3, select Calc→Mathematical Expressions and the Expressions dialog box will be displayed. In the Variable (new or modified) text box, type C3, and in the Expressions box type $C1 - C2$. When the OK button is selected, the differences will be computed and saved in column C3. You will observe these results in the Data window.

Note   Currently Minitab does not have a feature that will enable you to test two population proportions (two binomial probabilities) directly. However, check with your instructors to see whether a macro is available from them to help with the computations.

Computer Exercises

1. Where possible, use Minitab or any other statistical software to help work the problems at the end of each section in Chap. 10.

2. Where possible, use Minitab or any other statistical software to help work the review problems at the end of Chap. 10.

3. Use your library or any other resource to collect a *large* data set for two *independent* random variables. Use Minitab or any other statistical software to test for the equality of their population means. Establish whether these samples came from normal populations as in previous labs. Write up a report with your discussion and findings. If you select data from published research, include relevant factors about the validity and reliability of the study.

4. Use your library or any other resource to collect a *small* data set for two *independent* random variables. Use Minitab or any other statistical software to test for the equality of their population means. Establish whether these samples came from normal populations as in previous labs. Write up a report with your discussion and findings. If you select data from published research, include relevant factors about the validity and reliability of the study. State any assumptions you make.

5. Use your library or any other resource to collect a data set for two *dependent* random variables. Use Minitab or any other statistical software to test for the equality of their population means. Establish whether these samples came from normal populations as in previous labs. Write up a report with your discussion and findings. If you select data from published research, include relevant factors about the validity and reliability of the study.

6. Use your library or any other resource to collect a *large* data set for two independent binomial random variables. Use Minitab or any other statistical software to test for the equality of their population proportions. Write up a report with your discussion and findings. If you select data from published research, include relevant factors about the validity and reliability of the study.

CHAPTER 11

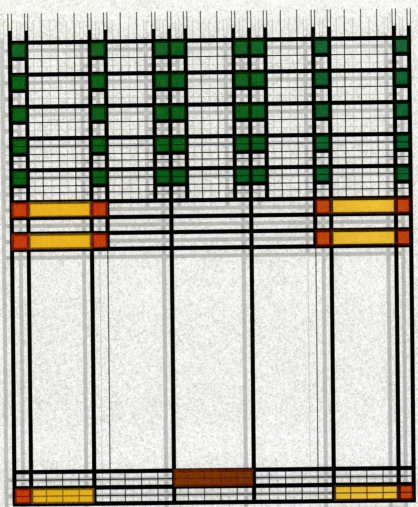

Analysis of Variance

> Statistics will prove anything, even the truth.
> N. Moynihan
> (British writer)

We present a general approach, called the *analysis of variance* (ANOVA), for making inferences about the mean values of a variety of random variables. In one-factor ANOVA, the mean of a variable depends on only a single factor, namely, the sample to which it belongs. In two-factor ANOVA, the random variables are thought of as being arrayed in a rectangular arrangement, and the mean of a variable depends on both its row and its column factor. We show how to test the hypothesis that the mean of a random variable does not depend on which row the random variable is in, as well as the analogous hypothesis that the mean does not depend on which column it is in.

11.1 INTRODUCTION

In recent years, many people have expressed the fear that large parts of U.S. industry are becoming increasingly unable to compete effectively in the world economy. For instance, U.S. public opinion has shifted to the belief that Japanese-made automobiles are of higher quality than their U.S.-made counterparts. Japan, and not the United States, is now considered by many to be the world leader in applying statistical techniques to improve quality.

Statistical quality control methods were developed by U.S. industrial statisticians in the 1920s and 1930s. These early methods were primarily concerned with surveillance of existing manufacturing processes. They relied to a large degree on the use of statistical sampling procedures to enable statisticians to quickly detect when something had gone wrong with the manufacturing process. In recent years, however, the emphasis in statistical quality control has shifted from overseeing a manufacturing process to designing that process. That is, led by some Japanese quality control experts, a feeling has developed that the primary contribution of statistics should be in determining effective ways of manufacturing a product.

For instance, when producing computer chips, the manufacturer needs to decide upon the raw materials to be used, the temperature at which to fuse the parts, the shape and the size of the chip, and other factors. For a given set of choices of these factors, the manufacturer wants to know the mean quality value of the resulting chip. This will enable her or him to determine the choices of the factors of production that would be most appropriate for obtaining a quality product.

In this chapter we introduce the statistical technique used for analyzing the above type of problem. It is a general method for making inferences about a multitude of parameters relating to population means. Its use will enable us, for instance, to determine the mean quality level of a manufactured item for a variety of choices of factor settings. The statistical technique was invented by R. A. Fisher and is known as the *analysis of variance* (ANOVA).

Whereas the previous chapter was concerned with hypothesis tests of two population means, this chapter considers tests of multiple population means. For instance, in Sec. 11.2 we will suppose that we have data from m populations and are interested in testing the hypothesis that all the population means are equal. This scenario is said to constitute a *one-factor* analysis of variance since the model assumes that the mean of a variable depends on only one factor, namely, the sample from which the observation is taken.

In Sec. 11.3 we consider models in which it is assumed that two factors determine the mean value of a variable. In such cases the variables to be observed can be thought of as being arranged in a rectangular array, and the mean value of a specified variable depends on both the row and the column in which it is located. For this *two-factor* analysis of variance problem we show how to estimate the mean values. In addition we show how to test the hypothesis that a specified factor does not affect the mean. For instance, we might have data of the yearly rainfall in various desert locations over a series of years. Two factors would affect the yearly amount of rainfall in a region—the location of the region and the year considered—and we might be interested in testing whether it is only the location and not the year that makes a difference in the mean yearly rainfall.

In all the models considered in this chapter, we assume that the data are normally distributed with the same (though unknown) variance σ^2. The analysis of variance approach for testing a null hypothesis H_0 concerning multiple parameters is based on deriving two estimators of the common variance σ^2. The first estimator is a valid estimator of σ^2 whether the null hypothesis is true or not, while the second one is a valid estimator only when H_0 is true. In addition, when H_0 is not true, this latter estimator will overestimate σ^2 in that the estimator will tend to exceed it. The test compares the values of these two estimators and rejects H_0 when the ratio of the second estimator to the first is sufficiently large. In other words, since the two estimators should be close to each other when H_0 is true (as they both estimate σ^2 in this case) whereas the second estimator should tend to be larger than the first when H_0 is not true, it is natural to reject H_0 when the second estimator is significantly larger than the first.

11.2   ONE-FACTOR ANALYSIS OF VARIANCE

Consider m samples, each of size n. Suppose that these samples are independent and that sample i comes from a population that is normally distributed with mean μ_i and variance σ^2, $i = 1, \ldots, m$. We will be interested in testing the null hypothesis

$$H_0: \mu_1 = \mu_2 = \cdots = \mu_m$$

against

$$H_1: \text{not all the means are equal}$$

That is, we will be testing the null hypothesis that all the population means are equal against the alternative that at least two population means differ.

Let $\bar{X}_i$ and S_i^2 denote the sample mean and sample variance, respectively, for the data of the ith sample, $i = 1, \ldots, m$. Our test of the above null hypothesis will be carried out by comparing the values of two estimators of the common variance σ^2. Our first estimator, which will be a valid estimator of σ^2 whether or not the null hypothesis is true, is obtained by noting that each of the sample variances S_i^2 is an unbiased estimator of its population variance σ^2. Since we have m of these estimators, namely, $S_1^2, \ldots, S_m^2$, we will combine them into a single estimator by taking their average. That is, our first estimator of σ^2 is given by

$$\frac{1}{m} \sum_{i=1}^{m} S_i^2$$

Note that this estimator was obtained without assuming anything about the truth or falsity of the null hypothesis.

Our second estimator of σ^2 will be a valid estimator only when the null hypothesis is true. So let us assume that H_0 is true, and thus all the population means μ_i are equal, say, $\mu_i = \mu$ for all i. Under this condition it follows that the m sample means $\overline{X}_1, \overline{X}_2, \ldots, \overline{X}_m$ will all be normally distributed with the same mean μ and the same variance σ^2/n. In other words, when the null hypothesis is true, the data $\overline{X}_1, \overline{X}_2, \ldots, \overline{X}_m$ constitute a sample from a normal population having variance σ^2/n. As a result, the sample variance of these data will, when H_0 is true, be an estimator of σ^2/n. Designate this sample variance by $\overline{S}^2$. That is,

$$\overline{S}^2 = \frac{\sum_{i=1}^{m} (\overline{X}_i - \overline{\overline{X}})^2}{m - 1}$$

where

$$\overline{\overline{X}} = \frac{1}{m} \sum_{i=1}^{m} \overline{X}_i$$

Since $\overline{S}^2$ is an unbiased estimator of σ^2/n when H_0 is true, it follows in this case that $n\overline{S}^2$ is an estimator of σ^2. That is, our second estimator of σ^2 is $n\overline{S}^2$. Hence, we have shown that

$$\sum_{i=1}^{m} \frac{S_i^2}{m} \qquad \text{always estimates } \sigma^2$$

$$n\overline{S}^2 \qquad \text{estimates } \sigma^2 \text{ when } H_0 \text{ is true}$$

Since it can be shown that $n\overline{S}^2$ will tend to be larger than σ^2 when H_0 is not true, it is reasonable to let the test statistic TS be given by

$$\text{TS} = \frac{n\overline{S}^2}{\sum_{i=1}^{m} S_i^2/m}$$

and to reject H_0 when TS is sufficiently large.

To determine how large TS need be to justify rejecting H_0, we use the fact that when H_0 is true, TS will have what is known as an *F distribution* with $m - 1$ numerator and $m(n - 1)$ denominator degrees of freedom. Let $F_{m-1,m(n-1),\alpha}$ denote the α critical value of this distribution. That is, the probability that an F random variable having numerator and denominator degrees of freedom $m - 1$ and $m(n - 1)$, respectively, will exceed $F_{m-1,m(n-1),\alpha}$ is equal to α (see Fig. 11.1). The significance level α test of H_0 is as follows:

Reject H_0 $\qquad$ if $\dfrac{n\overline{S}^2}{\sum_{i=1}^{m} \dfrac{S_i^2}{m}} \geq F_{m-1,m(n-1),\alpha}$

Do not reject H_0 $\qquad$ otherwise

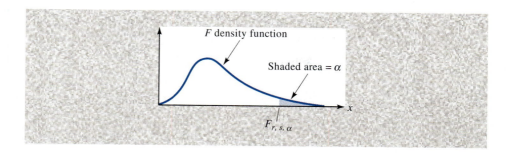

Figure 11.1
Random variable F with degrees of freedom r, s:
$P\{F \geq F_{r,s,\alpha}\} = \alpha$.

Values of $F_{r,s,.05}$ for various values of r and s are presented in App. Table D.4. Part of this table is presented now in Table 11.1.

For instance, from Table 11.1, we see that there is a 5 percent chance that an F random variable having 3 numerator and 10 denominator degrees of freedom will exceed 3.71.

A Remark on the Degrees of Freedom

The numerator degrees of freedom of the F random variable are determined by the numerator estimator $n\bar{S}^2$. Since $\bar{S}^2$ is the sample variance from a sample of size m, it follows that it has $m - 1$ degrees of freedom. Similarly, the denominator estimator is based on the statistic $\sum_{i=1}^{m} S_i^2$. Since each of the sample variances S_i^2 is based on a sample of size n, it follows that they each have $n - 1$ degrees of freedom. Summing the m sample variances then results in a statistic with $m(n - 1)$ degrees of freedom.

Example 11.1   An investigator for a consumer cooperative organized a study of the mileages obtainable from three different brands of gasoline. Using 15 identical motors set to run at the same speed, the investigator randomly assigned each brand of gasoline to 5 of the motors. Each of the motors was then run on 10 gallons of gasoline, with the total mileages obtained as given below.

Table 11.1   **Values of $F_{r,s,.05}$**

s = Degrees of freedom for denominator	r = degrees of freedom for numerator			
	1	2	3	4
4	7.71	6.94	6.59	6.39
5	6.61	5.79	5.41	5.19
⋮	...	...	...	...
10	4.96	4.10	3.71	3.48

Gas 1	Gas 2	Gas 3
220	244	252
251	235	272
226	232	250
246	242	238
260	225	256

Test the hypothesis that the average mileage obtained is the same for all three types of gasoline. Use the 5 percent level of significance.

Solution Since there are three samples, each of size 5, we see that $m = 3$ and $n = 5$. The sample means are

$$\bar{X}_1 = \frac{1203}{5} = 240.6$$

$$\bar{X}_2 = \frac{1178}{5} = 235.6$$

$$\bar{X}_3 = \frac{1268}{5} = 253.6$$

The average of the three sample means is

$$\bar{\bar{X}} = \frac{240.6 + 235.6 + 253.6}{3} = 243.2667$$

Therefore, the sample variance of the data $\bar{X}_i$, $i = 1, 2, 3$, is

$$\bar{S}^2 = \frac{(240.6 - 243.2667)^2 + (235.6 - 243.2667)^2 + (253.6 - 243.2667)^2}{2}$$

$$= 86.33335$$

The numerator estimate is thus

$$5\bar{S}^2 = 431.667$$

Computing the sample variances from the three samples yields $S_1^2 = 287.8$, $S_2^2 = 59.3$, and $S_3^2 = 150.8$, and so the denominator estimate is

$$\sum_{i=1}^{3} \frac{S_i^2}{3} = 165.967$$

Therefore, the value of the test statistic is

Table 11.2 **One-Factor ANOVA Table**

Variables $\bar{X}_i$ and S_i^2, $i = 1, \ldots, m$, are the sample means and sample variances, respectively, of independent samples of size n from normal populations having means μ_i and a common variance σ^2.

Source of estimator	Estimator of σ^2	Value of test statistic
Between samples	$n\bar{S}^2 = \dfrac{n\sum_{i=1}^{m}(\bar{X}_i - \bar{\bar{X}})^2}{m-1}$	$TS = \dfrac{n\bar{S}^2}{\sum_{i=1}^{m}\dfrac{S_i^2}{m}}$
Within samples	$\sum_{i=1}^{m}\dfrac{S_i^2}{m}$	

Significance level α test of H_0: all μ_i values are equal:

Reject H_0 if $TS \geq F_{m-1,m(n-1),\,\alpha}$
Do not reject H_0 otherwise

If $TS = v$, then p value $= P\{F_{m-1,m(n-1)} \geq v\}$

where $F_{m-1,m(n-1)}$ is an F random variable with $m-1$ numerator and $m(n-1)$ denominator degrees of freedom.

$$TS = \frac{431.667}{165.967} = 2.60$$

Since $m - 1 = 2$ and $m(n - 1) = 12$, we must compare the value of the TS with the value of $F_{2,12,.05}$. Now, from App. Table D.4, we see that $F_{2,12,.05} = 3.89$. Since the value of the test statistic does not exceed 3.89, it follows that at the 5 percent level of significance we cannot reject the null hypothesis that the gasolines give equal mileage.

Another way of doing the computations for the hypothesis test that all the population means are equal is by computing the p value. If the value of the test statistic TS is v, then the p value will be given by

$$p \text{ value} = P\{F_{m-1,m(n-1)} \geq v\}$$

where $F_{m-1,m(n-1)}$ is an F random variable with $m - 1$ numerator and $m(n - 1)$ denominator degrees of freedom.

Program 11-1 will compute the value of the test statistic TS and the resulting p value.

Example 11.2   Let us do the computations of Example 11.1 by using Program 11-1. After the data have been entered, we get the following output.

The denominator estimate is 165.9668
The numerator estimate is 431.6675
The value of the f-statistic is 2.600927
The p-value is .112408

Table 11.2 summarizes the results of this section.

Remark: When $m = 2$, the preceding is a test of the null hypothesis that two independent samples, having a common population variance, have the same mean. The reader might wonder how this compares with the one presented in Chap. 10. It turns out that the tests are exactly the same. That is, assuming the same data are used, they always give rise to exactly the same p value.

Problems

1. Consider the data from three samples, each of size 4. (That is, $m = 3$, $n = 4$.)

Sample 1	5	9	12	6
Sample 2	13	12	20	11
Sample 3	8	12	16	8

(a) Determine the three sample means $\bar{X}_i$, $i = 1, 2, 3$.
(b) Find $\bar{\bar{X}}$, the average of the three sample means.
(c) Show that $\bar{\bar{X}}$ is equal to the average of the 12 data values.

2. Use the data in Prob. 1 to test the hypothesis that the three population means are equal. Use the 5 percent level of significance.

3. A nutritionist randomly divided 15 bicyclists into three groups of five each. Members of the first group were given vitamin supplements to take with each of their meals over the next 3 weeks. The second group was instructed to eat a particular type of high-fiber whole-grain cereal for the next 3 weeks. Members of the third group were instructed to eat as they normally do. After the 3-week period elapsed, the nutritionist had each bicyclist ride 6 miles. The following times were recorded:

Vitamin group	15.6	16.4	17.2	15.5	16.3
Fiber cereal group	17.1	16.3	15.8	16.4	16.0
Control group	15.9	17.2	16.4	15.4	16.8

Are the above data consistent with the hypothesis that neither the vitamin nor fiber cereal affect the speed of a bicyclist? Use the 5 percent level of significance.

4. To determine whether the percentage of calories in a person's diet that is due to fat is the same across the country, random samples of 20 volunteers each were chosen in the three different regions. Each volunteer's percentage of total calories due to fat was determined, with the following summarized data resulting.

Region i	$\bar{X}_i$	S_i^2
$i = 1$	32.4	102
$i = 2$	36.4	112
$i = 3$	37.1	138

Test the null hypothesis that the percentage of calories due to fat does not vary for individuals living in the three regions. Use the 5 percent level of significance.

5. Six servings each of three different brands of processed meat were tested for fat content. The following data (in fat percentage per gram of weight) resulted:

Brand	Fat content					
1	32	34	31	35	33	30
2	40	36	33	29	35	32
3	37	30	28	33	37	39

Do the above data enable us to reject, at the 5 percent level of significance, the hypothesis that the average fat content is the same for all three brands?

6. An important factor in the sales of a new golf ball is how far it will travel when hit. Four different types of balls were hit by an automatic driving machine 25 times each, and their distances (in yards) were recorded. The following data, referring to the sample means and sample variances obtained with each type of ball, resulted.

Ball i	$\bar{X}_i$	S_i^2
1	212	26
2	220	23
3	198	25
4	214	24

Using the 5 percent level of significance, test the null hypothesis that the mean distance traveled is the same for each type of ball.

7. Three standard chemical procedures are used to determine the magnesium content in a certain chemical compound. Each procedure was used 4 times on a given compound with the following data resulting:

Method 1	76.43	78.61	80.40	78.22
Method 2	80.40	82.24	72.70	76.04
Method 3	82.16	84.14	80.20	81.33

Test the hypothesis that the mean readings are the same for all three methods. Use the 5 percent level of significance.

8. An emergency room physician wanted to learn whether there were any differences in the time it takes for three different inhaled steroids to clear a mild asthmatic attack. Over a period of weeks, she randomly administered these steroids to asthma sufferers and noted the number of minutes it took for the patient's lungs to become clear. Afterward, she discovered that 12 patients had been treated with each type of steroid, with the following sample means and sample variances resulting.

Steroid	$\bar{X}_i$	S_i^2
A	32	145
B	40	138
C	30	150

Test the hypothesis that the mean time to clear a mild asthmatic attack is the same for all three steroids. Use the 5 percent level of significance.

9. The following data refer to the numbers of deaths per 10,000 adults in a large eastern city in different seasons of the years from 1982 to 1986.

Year	Winter	Spring	Summer	Fall
1982	33.6	31.4	29.8	32.1
1983	32.5	30.1	28.5	29.9
1984	35.3	33.2	29.5	28.7
1985	34.4	28.6	33.9	30.1
1986	37.3	34.1	28.5	29.4

Test the hypothesis that death rates do not depend on the season. Use the 5 percent level of significance.

10. A nutrition expert claims that the amount of running a person does relates to that person's blood cholesterol level. Six runners from each of three running categories were randomly chosen to have their blood cholesterol levels checked. The sample means and sample variances were as follows.

Weekly miles run	$\bar{X}_i$	S_i^2
Less than 15	188	190
Between 15 and 30	181	211
More than 30	174	202

Do these data prove the nutritionist's claim? Use the 5 percent level of significance.

11. A college administrator claims that there is no difference in first-year grade-point averages for students entering the college from any of three different local high schools. The following data give the first-year grade-point averages of 15 randomly chosen students—5 from each of the three high schools. Are they strong enough, at the 5 percent level, to disprove the claim of the administrator?

School *A*	School *B*	School *C*
3.2	2.8	2.5
2.7	3.0	2.8
3.0	3.3	2.4
3.3	2.5	2.2
2.6	3.1	3.0

12. A psychologist conducted an experiment concerning maze test scores of a strain of laboratory mice trained under different laboratory conditions. A group of 24 mice was randomly divided into three groups of 8 each. Members of the first group were given a type of cognitive training, those in the second group were given a type of behavioral training, and those in the third group were not trained at all. The maze test scores (judged by someone who did not know which training particular mice received) were summarized as follows.

Group	$\bar{X}_i$	S_i^2
1	74.2	111.4
2	78.5	102.1
3	80.0	124.0

Is this sufficient evidence to conclude that the different types of training have an effect on maze test scores? Use the 5 percent level of significance.

11.3 TWO-FACTOR ANALYSIS OF VARIANCE: INTRODUCTION AND PARAMETER ESTIMATION

Whereas the model of Sec. 11.2 enabled us to study the effect of a single factor on a data set, we can also study the effects of several factors. In this section we suppose that each data value is affected by two factors.

Example 11.3 Four different standardized reading achievement tests were administered to each of five students. Their scores were as follows:

	Student				
Examination	1	2	3	4	5
1	75	73	60	70	86
2	78	71	64	72	90
3	80	69	62	70	85
4	73	67	63	80	92

Each value in this set of 20 data points is affected by two factors, namely, the examination and the student whose score on that examination is being recorded. The examination factor has four possible values, or *levels*, and the student factor has five possible levels.

In general, let us suppose that there are m possible levels of the first factor and n possible levels of the second. Let X_{ij} denote the value of the data obtained when the first factor is at level i and the second factor is at level j. We often portray the data set in the following array of rows and columns:

$$
\begin{array}{ccccc}
X_{11} & X_{12} & \cdots & X_{1j} & \cdots & X_{1n} \\
X_{21} & X_{22} & \cdots & X_{2j} & \cdots & X_{2n} \\
\cdots & \cdots & \cdots & \cdots & \cdots & \cdots \\
X_{i1} & X_{i2} & \cdots & X_{ij} & \cdots & X_{in} \\
\cdots & \cdots & \cdots & \cdots & \cdots & \cdots \\
X_{m1} & X_{m2} & \cdots & X_{mj} & \cdots & X_{mn}
\end{array}
$$

Because of this we refer to the first factor as the *row* factor and the second factor as the *column* factor. Also, the data value X_{ij} is the value in row i and column j.

As in Sec. 11.2, we suppose that all the data values X_{ij}, $i = 1, \ldots, m$, $j = 1, \ldots, n$, are independent normal random variables with common variance σ^2. However, whereas in Sec. 11.2 we supposed that only a single factor affected the mean value of a data point—namely, the sample to which it belonged—in this section we will suppose that the mean value of the data point depends on both its row and its column. However, before specifying this model, we first recall the model of Sec. 11.2. If we let X_{ij} represent the value of the jth member of sample i, then this model supposes that

$$E[X_{ij}] = \mu_i$$

If we now let μ denote the average value of the μ_i, that is,

$$\mu = \frac{\displaystyle\sum_{i=1}^{m} \mu_i}{m}$$

then we can write the preceding as

$$E[X_{ij}] = \mu + \alpha_i$$

where $\alpha_i = \mu_i - \mu$. With this definition of α_i equal to the deviation of μ_i from the average of the means μ, it is easy to see that

$$\sum_{i=1}^{m} \alpha_i = 0$$

In the case of two factors, we write our model in terms of row and column deviations. Specifically, we suppose that the expected value of variable X_{ij} can be expressed as follows:

$$E[X_{ij}] = \mu + \alpha_i + \beta_j$$

The value μ is referred to as the *grand mean*, α_i is the *deviation from the grand mean due to row i*, and β_j is the *deviation from the grand mean due to column j*. In addition, these quantities satisfy the following equalities.

$$\sum_{i=1}^{m} \alpha_i = \sum_{j=1}^{n} \beta_j = 0$$

Let us start by determining estimators for parameters μ, α_i, and β_j, $i = 1, \ldots, m$, $j = 1, \ldots, n$. To do so, we will find it convenient to introduce the following "dot" notation. Let

$$X_{i\cdot} = \frac{\sum_{j=1}^{n} X_{ij}}{n} = \text{average of all values in row } i$$

$$X_{\cdot j} = \frac{\sum_{i=1}^{m} X_{ij}}{m} = \text{average of all values in column } j$$

$$X_{\cdot\cdot} = \frac{\sum_{i=1}^{m} \sum_{j=1}^{n} X_{ij}}{nm} = \text{average of all } nm \text{ data values}$$

It is not difficult to show that

$$E[X_{i\cdot}] = \mu + \alpha_i$$

$$E[X_{\cdot j}] = \mu + \beta_j$$

$$E[X_{\cdot\cdot}] = \mu$$

Since the above is equivalent to

$$E[X_{..}] = \mu$$

$$E[X_{i.} - X_{..}] = \alpha_i$$

$$E[X_{.j} - X_{..}] = \beta_j$$

we see that unbiased estimators of μ, α_i, and β_j—call them $\hat{\mu}$, $\hat{\alpha}_i$, and $\hat{\beta}_j$—are given by

$$\hat{\mu} = X_{..}$$

$$\hat{\alpha}_i = X_{i.} - X_{..}$$

$$\hat{\beta}_j = X_{.j} - X_{..}$$

Example 11.4   The following data from Example 11.3 give the scores obtained when four different reading tests were given to each of five students. Use it to estimate the parameters of the model.

Examination	Student					Row totals	$X_{i.}$
	1	2	3	4	5		
1	75	73	60	70	86	364	72.8
2	78	71	64	72	90	375	75
3	80	69	62	70	85	366	73.2
4	73	67	63	80	92	375	75
Column totals	306	280	249	292	353	1480 ← grand total	
$X_{.j}$	76.5	70	62.25	73	88.25		

$$X_{..} = \frac{1480}{20} = 74$$

The estimators are

$$\hat{\mu} = 74$$

$$\hat{\alpha}_1 = 72.8 - 74 = -1.2 \qquad \hat{\beta}_1 = 76.5 - 74 = 2.5$$

$$\hat{\alpha}_2 = 75 - 74 = 1 \qquad \hat{\beta}_2 = 70 - 74 = -4$$

$$\hat{\alpha}_3 = 73.2 - 74 = -.8 \qquad \hat{\beta}_3 = 62.25 - 74 = -11.75$$

$$\hat{\alpha}_4 = 75 - 74 = 1 \qquad \hat{\beta}_4 = 73 - 74 = -1$$

$$\hat{\beta}_5 = 88.25 - 74 = 14.25$$

Therefore, for instance, if one of the students is randomly chosen and then given a randomly chosen examination, then our estimate of the mean score that will be obtained is $\hat{\mu} = 74$. If we were told that examination i was taken, then this would increase our estimate of the mean score by the amount $\hat{\alpha}_i$; and if we were told that the student chosen was number j, then this would increase our estimate of the mean score by the amount $\hat{\beta}_j$. Thus, for instance, we would estimate that the score obtained on examination 1 by student 2 is the value of a random variable whose mean is $\hat{\mu} + \hat{\alpha}_1 + \hat{\beta}_2 = 74 - 1.2 - 4 = 68.8$.

Remark: In the preceding we defined $X_{..}$ by using the double-summation notation. That is, we used notation of the form

$$\sum_{i=1}^{m}\sum_{j=1}^{n} X_{ij}$$

This expression is meant to be the sum of the terms X_{ij} for all nm possible values of the pair i, j.

Equivalently, suppose that the data values X_{ij} are arranged in rows and columns as given at the beginning of this section. Let T_i denote the sum of the values in row i. That is,

$$T_i = \sum_{j=1}^{n} X_{ij}$$

Then the double summation notation is defined by

$$\sum_{i=1}^{m}\sum_{j=1}^{n} X_{ij} = \sum_{i=1}^{m} T_i$$

In words, the double summation is equal to the sum of all the row sums; that is, it is just the sum of all the nm data values X_{ij}. (It is easy to see that it is also equal to the sum of the n column sums.)

Problems

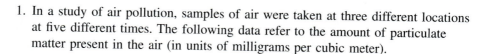

1. In a study of air pollution, samples of air were taken at three different locations at five different times. The following data refer to the amount of particulate matter present in the air (in units of milligrams per cubic meter).

	Location		
Time	1	2	3
1. January 1988	78	84	87
2. July 1988	75	69	82
3. January 1989	66	60	70
4. July 1989	71	64	61
5. January 1990	58	55	52

Assuming the model

$$E[X_{ij}] = \mu + \alpha_i + \beta_j$$

estimate the unknown parameters.

2. Using the data of Prob. 1, verify that

$$X_{..} = \frac{\sum\limits_{i=1}^{m} X_{i.}}{m} = \frac{\sum\limits_{j=1}^{n} X_{.j}}{n}$$

Express in words what the above equation states.

3. The following data refer to the numbers of boxes packed by each of three men during three different shifts.

	Man		
Shift	1	2	3
1. 9–11 a.m.	32	27	29
2. 1–3 p.m.	31	26	22
3. 3–5 p.m.	33	30	25

Assuming the model of this section, estimate the unknown parameters.

4. Use the results of Example 11.4 to estimate $E[X_{ij}] = \mu + \alpha_i + \beta_j$ for all the possible values of i and j, $i = 1, 2, 3, 4$, $j = 1, 2, 3, 4, 5$. Compare the estimated values $E[X_{ij}]$ with the observed values of X_{ij} as given in that example.

5. The following table gives the birth rates per 1000 population for four different countries in four different years.

	Birth rate			
Country	1975	1980	1985	1989
Australia	16.9	15.3	15.7	14.9
France	14.1	14.8	13.9	13.6
Japan	17.2	13.7	11.9	10.1
United States	14.0	16.2	15.7	16.0

Source: United Nations, *Monthly Bulletin of Statistics,* May 1991.

Assuming the model of this section, estimate the
(a) Grand mean of the birth rates
(b) Deviation from the grand mean of U.S. birth rates
(c) Deviation from the grand mean of the 1989 birth rates

6. The following table provides the unemployment rates for three levels of educational attainment in three different years.

Level of education	1980	1984	1988
Did not graduate from high school	8.4	12.1	9.6
High school graduate	5.1	7.2	5.4
College graduate	1.9	2.7	1.7

Source: U.S. Bureau of Labor Statistics, *Labor Force Statistics.*

Assuming the model of this section, estimate
(a) The grand mean μ
(b) The row deviations α_i, $i = 1, 2, 3$
(c) The column deviations β_j, $j = 1, 2, 3$

7. The following table provides the unemployment rates for five different industries in four different years.

Industry	1975	1980	1984	1987
Agriculture	10.3	11.0	13.5	10.5
Mining	4.1	6.4	10.0	10.0
Construction	18.0	14.1	14.3	11.6
Manufacturing	10.9	8.5	7.5	6.0
Services	7.1	5.9	6.6	5.4

Source: U.S. Bureau of Labor Statistics, *Employment and Earnings,* monthly.

Assuming the model of this section, estimate the unknown parameters.

8. Suppose that $x_{ij} = i + 4j$. (So, for instance, $x_{11} = 1 + 4 = 5$, and $x_{23} = 2 + 4 \cdot 3 = 14$.) Write out in a rectangular array of rows and columns all the 12 values of x_{ij} where i is 1 or 2 or 3 and j is 1 or 2 or 3 or 4. Put the value of x_{ij} in the location joining row i and column j.

9. In Prob. 8, determine

(a) $\displaystyle\sum_{j=1}^{4} x_{1j}$ (b) $\displaystyle\sum_{j=1}^{4} x_{2j}$

(c) $\displaystyle\sum_{j=1}^{4} x_{3j}$ (d) $\displaystyle\sum_{i=1}^{3} \sum_{j=1}^{4} x_{ij}$

11.4 TWO-FACTOR ANALYSIS OF VARIANCE: TESTING HYPOTHESES

Consider the two-factor model in which one has data values X_{ij}, $i = 1, \ldots, m$ and $j = 1, \ldots, n$. These data are assumed to be independent normal random variables with a common variance σ^2 and with mean values satisfying

$$E[X_{ij}] = \mu + \alpha_i + \beta_j$$

where

$$\sum_{i=1}^{m} \alpha_i = \sum_{j=1}^{n} \beta_j = 0$$

In this section we will test the hypothesis

$$H_0: \text{all } \alpha_i = 0$$

against

$$H_1: \text{not all } \alpha_i \text{ are } 0$$

This null hypothesis states that there is no row effect, in that the value of a datum is not affected by its row factor level.

We will also test the analogous hypothesis for columns, namely,

$$H_0: \text{all } \beta_j \text{ are } 0$$

against

$$H_1: \text{not all } \beta_j \text{ are } 0$$

To obtain tests for the above null hypotheses, we will apply the analysis of variance approach in which two different estimators are derived for the variance σ^2. The first will always be a valid estimator whereas the second will be a valid estimator only when the null hypothesis is true. In addition, the second estimator will tend to overestimate σ^2 when the null hypothesis is not true.

To obtain our first estimator of σ^2, we recall that the sum of the squares of N standard normal random variables is a chi-squared random variable with N degrees of freedom. Since the nm standardized variables

$$\frac{X_{ij} - E[X_{ij}]}{\sigma}$$

$i = 1, \ldots, m, j = 1, \ldots, n$ are all standard normal, it follows that

$$\frac{\sum_{i=1}^{m} \sum_{j=1}^{n} (X_{ij} - E[X_{ij}])^2}{\sigma^2} = \frac{\sum_{i=1}^{m} \sum_{j=1}^{n} (X_{ij} - \mu - \alpha_i - \beta_j)^2}{\sigma^2}$$

is chi squared with nm degrees of freedom. If in the above expression we now replace the unknown parameters $\mu, \alpha_1, \alpha_2, \ldots, \alpha_m, \beta_1, \beta_2, \ldots, \beta_n$ by their estimators $\hat{\mu}, \hat{\alpha}_1, \hat{\alpha}_2, \ldots, \hat{\alpha}_m, \hat{\beta}_1, \hat{\beta}_2, \ldots, \hat{\beta}_n$, then it turns out that the resulting expression will remain chi squared but will lose 1 degree of freedom for each parameter that is estimated. To determine how many parameters are to be estimated, we must be careful to remember that $\sum_{i=1}^{m} \alpha_i = \sum_{j=1}^{n} \beta_j = 0$. Since the sum of all the α_i is 0, it follows that once we have estimated $m - 1$ of the α_i, then we have also estimated the final one. Hence, only $m - 1$ parameters are to be estimated in order to determine all the estimators $\hat{\alpha}_i$. For the same reason only $n - 1$ of the β_j need to be estimated to determine estimators for all n of them. Since μ is also to be estimated, we see that the number of parameters to be estimated is

$$1 + (m - 1) + (n - 1) = m + n - 1$$

As a result, it follows that

$$\frac{\sum_{i=1}^{m} \sum_{j=1}^{n} (X_{ij} - \hat{\mu} - \hat{\alpha}_i - \hat{\beta}_j)^2}{\sigma^2}$$

is a chi-squared random variable with $nm - (n + m - 1) = (n - 1)(m - 1)$ degrees of freedom.

Since

$$\hat{\mu} = X_{..}$$
$$\hat{\alpha}_i = X_{i.} - X_{..}$$
$$\hat{\beta}_j = X_{.j} - X_{..}$$

we see that

$$\hat{\mu} + \hat{\alpha}_i + \hat{\beta}_j = X_{..} + X_{i.} - X_{..} + X_{.j} - X_{..}$$
$$= X_{i.} + X_{.j} - X_{..}$$

Thus, the statistic

$$\frac{\sum_{i=1}^{m} \sum_{j=1}^{n} (X_{ij} - X_{i.} - X_{.j} + X_{..})^2}{\sigma^2} \tag{11.1}$$

is chi squared with $(n - 1)(m - 1)$ degrees of freedom.

The sum of squares SS_e defined by

$$SS_e = \sum_{i=1}^{m} \sum_{j=1}^{n} (X_{ij} - X_{i\cdot} - X_{\cdot j} + X_{\cdot\cdot})^2$$

is called the *error sum of squares*.

If we think of the difference between a random variable and its estimated mean as being an "error," then SS_e is equal to the sum of the squares of the errors. Since SS_e/σ^2 is just the expression in Eq. (11.1), we see that SS_e/σ^2 is chi squared with $(n - 1)(m - 1)$ degrees of freedom. As the expected value of a chi squared random variable is equal to its number of degrees of freedom, we have

$$E\left[\frac{SS_e}{\sigma^2}\right] = (n - 1)(m - 1)$$

or

$$E\left[\frac{SS_e}{(n - 1)(m - 1)}\right] = \sigma^2$$

That is, letting $N = (n - 1)(m - 1)$, we have shown the following.

$\dfrac{SS_e}{N}$ is an unbiased estimator of σ^2.

Suppose now that we want to test the null hypothesis that there is no row effect; that is, we want to test

H_0: all the α_i are 0

against

H_1: not all the α_i are 0

To obtain a second estimator of σ^2, consider the row averages $X_{i\cdot}$, $i = 1, \ldots, m$. Note that when H_0 is true, each α_i is equal to 0, and so

$$E[X_{i.}] = \mu + \alpha_i = \mu$$

Since each $X_{i.}$ is the average of n random variables, each having variance σ^2, it follows that

$$\text{Var }(X_{i.}) = \frac{\sigma^2}{n}$$

Thus, we see that when H_0 is true,

$$\frac{\sum\limits_{i=1}^{m} (X_{i.} - E[X_{i.}])^2}{\text{Var }(X_{i.})} = \frac{n\sum\limits_{i=1}^{m} (X_{i.} - \mu)^2}{\sigma^2}$$

will be chi squared with m degrees of freedom. If we now substitute $X_{..}$ (the estimator of μ) for μ in the above, the resulting expression will remain chi squared but with one less degree of freedom. That is, it will have $m - 1$ degrees of freedom. We thus have the following:

When H_0 is true, then

$$\frac{n\sum\limits_{i=1}^{m} (X_{i.} - X_{..})^2}{\sigma^2}$$

is chi squared with $m - 1$ degrees of freedom.
The statistic SS_r defined by

$$SS_r = n \sum\limits_{i=1}^{m} (X_{i.} - X_{..})^2$$

is called the *row sum of squares*.

We have seen above that when H_0 is true, SS_r/σ^2 is chi squared with $m - 1$ degrees of freedom. As a result, when H_0 is true,

$$E\left[\frac{SS_r}{\sigma^2}\right] = m - 1$$

or, equivalently,

$$E\left[\frac{SS_r}{m-1}\right] = \sigma^2$$

In addition, it can be shown that $SS_r/(m-1)$ will tend to be larger than σ^2 when H_0 is not true. Thus, once again we have obtained two estimators of σ^2. The first estimator SS_e/N, where $N = (n-1)(m-1)$, is a valid estimator whether or not the null hypothesis is true. The second estimator $SS_r/(m-1)$ is a valid estimator of σ^2 only when H_0 is true and tends to be larger than σ^2 when H_0 is not true.

The test of the null hypothesis H_0 that there is no row effect involves comparing the two estimators given above, and it calls for rejection when the second is significantly larger than the first. Specifically, we use the test statistic

$$TS = \frac{SS_r/(m-1)}{SS_e/N}$$

and the significance level α test is to

Reject H_0 if $TS \geq F_{m-1,N,\alpha}$
Not reject H_0 otherwise

Alternatively the test can be performed by calculating the p value. If the value of the test statistic is v, then the p value is given by

$$p \text{ value} = P\{F_{m-1,N} \geq v\}$$

where $F_{m-1,N}$ is an F random variable with $m-1$ numerator and N denominator degrees of freedom.

A similar test can be derived to test the null hypothesis that there is no column effect, that is, that all the β_j are equal to 0. The results of both tests are summarized in Table 11.3.

Program 11-2 will do the computations and give the p value.

Example 11.5   The following are the numbers of defective items produced by four workers using, in turn, three different machines.

	Worker			
Machine	1	2	3	4
1	41	42	40	35
2	35	42	43	36
3	42	39	44	47

Test whether there are significant differences between the machines and the workers.

Solution   Since there are three rows and four columns, we see that $m = 3$ and $n = 4$. Computing the row and column averages gives the following results:

Table 11.3   **Two-Factor ANOVA**

	Sum of squares	Degrees of freedom
Row	$SS_r = n \sum_{i=1}^{m} (X_{i\cdot} - X_{\cdot\cdot})^2$	$m - 1$
Column	$SS_c = m \sum_{j=1}^{n} (X_{\cdot j} - X_{\cdot\cdot})^2$	$n - 1$
Error	$SS_e = \sum_{i=1}^{m} \sum_{j=1}^{n} (X_{ij} - X_{i\cdot} - X_{\cdot j} + X_{\cdot\cdot})^2$	$(n - 1)(m - 1)$

Let $N = (n - 1)(m - 1)$

Null hypothesis	Test statistic	Significance level α test	p Value if TS = v
No row effect (all $\alpha_i = 0$)	$\dfrac{SS_r/(m - 1)}{SS_e/N}$	Reject if TS $\geq F_{m-1,N,\alpha}$	$P\{F_{m-1,N} \geq v\}$
No column effect (all $\beta_j = 0$)	$\dfrac{SS_c/(n - 1)}{SS_e/N}$	Reject if TS $\geq F_{n-1,N,\alpha}$	$P\{F_{n-1,N} \geq v\}$

$$X_{1\cdot} = \frac{41 + 42 + 40 + 35}{4} = 39.5 \qquad X_{\cdot 1} = \frac{41 + 35 + 42}{3} = 39.33$$

$$X_{2\cdot} = \frac{35 + 42 + 43 + 36}{4} = 39 \qquad X_{\cdot 2} = \frac{42 + 42 + 39}{3} = 41$$

$$X_{3\cdot} = \frac{42 + 39 + 44 + 47}{4} = 43 \qquad X_{\cdot 3} = \frac{40 + 43 + 44}{3} = 42.33$$

$$X_{\cdot 4} = \frac{35 + 36 + 47}{3} = 39.33$$

Also

$$X_{\cdot\cdot} = \frac{39.5 + 39 + 43}{3} = 40.5$$

Thus,

$$SS_r = n\sum_{i=1}^{m} (X_{i\cdot} - X_{\cdot\cdot})^2$$
$$= 4[1^2 + (1.5)^2 + (2.5)^2]$$
$$= 38$$

and

$$SS_c = m \sum_{j=1}^{n} (X_{\cdot j} - X_{\cdot\cdot})^2$$
$$= 3[(1.17)^2 + (.5)^2 + (1.83)^2 + (1.17)^2]$$
$$= 19.010$$

The calculation of SS_e is more involved because we must add the sum of the squares of the terms $X_{ij} - X_{i\cdot} - X_{\cdot j} + X_{\cdot\cdot}$ as i ranges from 1 to 3 and j from 1 to 4. The first term in this sum, when $i = 1$ and $j = 1$, is

$$(41 - 39.5 - 39.33 + 40.5)^2$$

Adding all 12 terms gives

$$SS_e = 94.05$$

Since $m - 1 = 2$ and $N = 2 \cdot 3 = 6$, the test statistic for the hypothesis that there is no row effect is

$$TS(\text{row}) = \frac{38/2}{94.05/6} = 1.21$$

From App. Table D.4 we see that $F_{2, 6, .05} = 5.14$, and so the hypothesis that the mean number of defective items is unaffected by which machine is used is not rejected at the 5 percent level of significance.

The test statistic for the hypothesis that there is no column effect is

$$TS(\text{col.}) = \frac{19.010/3}{94.06/6} = .40$$

From App. Table D.4 we see that $F_{3, 6, .05} = 4.76$, and so the hypothesis that the mean number of defective items is unaffected by which worker is used is also not rejected at the 5 percent level of significance.

We could also have solved the above by running a program such as Program 11-2. Running Program 11-2 yields the following output:

The value of the f-statistic for testing that there is no row effect is 1.212766

The p-value for testing that there is no row effect is .3571476

The value of the f-statistic for testing that there is no column effect is .4042554

The p-value for testing that there is no column effect is .7555629

Since both p values are greater than .05, we cannot reject at the 5 percent significance level the hypothesis that the machine used does not affect the mean number of defective items produced; nor can we reject the hypothesis that the worker employed does not affect the mean number of defective items produced.

Problems

1. An experiment was performed to determine the effect of three different fuels and three different types of launchers on the range of a certain missile. The following data, in the number of miles traveled by the missile, resulted.

	Fuel 1	Fuel 2	Fuel 3
Launcher 1	70.4	71.7	78.5
Launcher 2	80.2	82.8	76.4
Launcher 3	90.4	85.7	84.8

Find out whether the above implies, at the 5 percent level of significance, that there are differences in the mean mileages obtained by using
(a) Different launchers
(b) Different fuels

2. An important consideration in deciding which database management system to employ is the mean time required to learn how to use the system. A test was designed involving three systems and four users. Each user took the following amount of time (in hours) in training with each system:

	User			
	1	2	3	4
System 1	20	23	18	17
System 2	20	21	17	16
System 3	28	26	23	22

(a) Using the 5 percent level of significance, test the hypothesis that the mean training time is the same for all the systems.
(b) Using the 5 percent level of significance, test the hypothesis that the mean training time is the same for all the users.

3. Five different varieties of oats were planted in each of four separated fields. The following yields resulted.

Oat variety	Field 1	2	3	4
1	296	357	340	348
2	402	390	420	335
3	345	342	358	308
4	360	322	336	270
5	324	339	357	308

Find out whether the data are consistent with the hypothesis that the mean yield does not depend on
(a) The field
(b) The oat variety
Use the 5 percent level of significance.

4. In Example 11.3, test the hypothesis that the mean score of a student does not depend on which test is taken.

5. In Prob. 1 of Sec. 11.3, test the hypothesis that the mean air pollution level
(a) Is unchanging in time
(b) Does not depend on the location
Use the 5 percent level of significance.

6. In Prob. 3 of Sec. 11.3, test the hypothesis that the mean number of boxes packed does not depend on
(a) The worker doing the packing
(b) The shift
Use the 5 percent level of significance.

7. The following data give the percentages of randomly sampled women in different age groups who smoked in the years 1980, 1982, 1984, 1986, 1988, and 1990.

Age group	Percentage of women who smoked in: 1980	1982	1984	1986	1988	1990
16–19	32	30	32	30	28	28
20–24	40	40	36	38	37	36
25–34	44	37	36	38	35	33
35–49	43	38	36	34	35	31
50–59	44	40	39	35	34	34
≥ 60	24	23	23	22	21	22

(a) Test the hypothesis that the percentages of women who smoke do not depend on the year considered.

(b) Test the hypothesis that there is no effect due to age group.

(c) Repeat (a) and (b) when the data referring to women over age 60 are deleted.

8. In Prob. 5 of Sec. 11.3, test the hypothesis that

(a) The mean birth rates do not depend on the particular country being considered.

(b) The mean birth rates do not depend on the particular year being considered.

9. In Prob. 7 of Sec. 11.3, test the hypothesis that

(a) The mean unemployment rates do not depend on the particular industry being considered.

(b) The mean unemployment rates do not depend on the particular year being considered.

11.5   FINAL COMMENTS

This chapter presented a brief introduction to a powerful statistical technique known as the *analysis of variance* (ANOVA). This technique enables statisticians to draw inferences about population means when these mean values are affected by many different factors. For instance, whereas we have considered only one- and two-factor ANOVA problems, any number of factors could affect the value of an outcome. In addition, there could be interactions between some of these factors. For instance, in two-factor ANOVA, it might be the case that the combination of a particular row and a particular column greatly affects a mean value. For example, while individually each of two carcinogens may be relatively harmless, perhaps in conjunction they are devastating. The general theory of ANOVA shows how to deal with the above and a variety of other situations.

ANOVA was developed by R. A. Fisher, who applied it to a large number of agricultural problems during his tenure as chief scientist at the Rothamstead Experimental Laboratories. ANOVA has since been widely applied in a variety of fields. For instance, in education one might want to study how a student's learning of algebra is affected by such factors as the instructor, the syllabus of the algebra course, the time spent on each class, the number of classes, the number of students in each class, and the textbook used. ANOVA has also been widely applied in studies in psychology, social science, manufacturing, biology, and many other fields. Indeed, ANOVA is probably the most widely used of all the statistical techniques.

Key Terms

One-factor analysis of variance: A model concerning a collection of normal random variables. It supposes that the variances of these random variables are equal and that their mean values depend on only a single factor, namely, the sample to which the random variable belongs.

***F* statistic:** A test statistic that is, when the null hypothesis is true, a ratio of two estimators of a common variance.

Two-factor analysis of variance: A model in which a set of normal random variables having a common variance is arranged in an array of rows and columns. The mean value of any of them depends on two factors, namely, the row and the column in which the variable lies.

Summary

One-Factor Analysis of Variance

Consider m independent samples, each of size n. Let $\mu_1, \mu_2, \ldots, \mu_m$ be the respective means of these m samples, and consider a test of

H_0: all the means are equal

against

H_1: not all the means are equal

Let $\bar{X}_i$ and S_i^2 denote the sample mean and sample variance, respectively, from sample i, $i = 1, \ldots, m$. Also, let $\bar{S}^2$ be the sample variance of the data set $\bar{X}_1, \ldots, \bar{X}_m$.

To test H_0 against H_1, use the test statistic

$$\text{TS} = \frac{n\bar{S}^2}{\displaystyle\sum_{i=1}^{m} S_i^2/m}$$

The significance level α test is to

Reject H_0 if $\text{TS} \geq F_{m-1,m(n-1),\alpha}$
Not reject H_0 otherwise

If the value of TS is v, then

$$p \text{ value} = P\{F_{m-1,m(n-1)} \geq v\}$$

Program 11-1 can be used both to compute the value of TS and to obtain the resulting p value.

Note: Variable $F_{r,s}$ represents an F random variable having r numerator and s denominator degrees of freedom. Also, $F_{r,s,\alpha}$ is defined to be such that

$$P\{F_{r,s} \geq F_{r,s,\alpha}\} = \alpha$$

Two-Factor Analysis of Variance

The Model

Suppose that each data value is affected by two factors, and suppose that there are m possible values, or levels, of the first factor and n of the second factor. Let X_{ij} denote the datum obtained when the first factor is at level i and the second factor is at level j. The data set can be arranged in the following array of rows and columns:

$$
\begin{array}{ccccccc}
X_{11} & X_{12} & \cdots & X_{1j} & \cdots & X_{1n} \\
X_{21} & X_{22} & \cdots & X_{2j} & \cdots & X_{2n} \\
\cdots & \cdots & \cdots & \cdots & \cdots & \cdots \\
X_{i1} & X_{i2} & \cdots & X_{ij} & \cdots & X_{in} \\
\cdots & \cdots & \cdots & \cdots & \cdots & \cdots \\
X_{m1} & X_{m2} & \cdots & X_{mj} & \cdots & X_{mn}
\end{array}
$$

The two-factor ANOVA model supposes that the X_{ij} are normal random variables having means given by

$$E[X_{ij}] = \mu + \alpha_i + \beta_j$$

and a common variance

$$\text{Var}(X_{ij}) = \sigma^2$$

The above parameters satisfy

$$\sum_{i=1}^{m} \alpha_i = \sum_{j=1}^{m} \beta_j = 0$$

Estimating the Parameters

Let

$$X_{i\cdot} = \frac{\sum_{j=1}^{n} X_{ij}}{n}$$

$$X_{\cdot j} = \frac{\sum_{i=1}^{m} X_{ij}}{m}$$

$$X_{\cdot\cdot} = \frac{\sum_{i=1}^{m} \sum_{j=1}^{n} X_{ij}}{nm}$$

The estimators of the parameters are as follows:

$$\hat{\mu} = X_{..}$$

$$\hat{\alpha}_i = X_{i.} - X_{..}$$

$$\hat{\beta}_j = X_{.j} - X_{..}$$

Testing Hypotheses

Let

$$SS_e = \sum_{i=1}^{m} \sum_{j=1}^{n} (X_{ij} - X_{i.} - X_{.j} + X_{..})^2$$

$$SS_r = n \sum_{i=1}^{m} (X_{i.} - X_{..})^2$$

$$SS_c = m \sum_{j=1}^{n} (X_{.j} - X_{..})^2$$

and SS_e, SS_r, and SS_c are called, respectively, the error sum of squares, the row sum of squares, and the column sum of squares. Also let $N = (n - 1)(m - 1)$.

To test H_0: all $\alpha_i = 0$ against H_1: not all $\alpha_i = 0$, use test statistic

$$TS = \frac{SS_r/(m - 1)}{SS_e/N}$$

The significance level α test is to

Reject H_0	if TS $\geq F_{m-1,N,\alpha}$
Not reject H_0	otherwise

If TS $= v$, then the p value is given by

$$p \text{ value} = P\{F_{m-1,N} \geq v\}$$

To test H_0: all $\beta_j = 0$ versus H_1: not all $\beta_j = 0$, use test statistic

$$TS = \frac{SS_c/(n - 1)}{SS_e/N}$$

The significance level α test is to

Reject H_0	if TS $\geq F_{n-1,N,\alpha}$
Not reject H_0	otherwise

If TS $= v$, then the p value is given by

$$p \text{ value} = P\{F_{n-1,N} \geq v\}$$

Program 11-2 can be used for the above hypothesis tests. It will compute the values of the two test statistics and give the resulting p values.

Review Problems

1. A corporation has three apparently identical manufacturing plants. Wanting to see if these plants are equally effective, management randomly chose 30 days. On 10 of these days it determined the daily output at plant 1. On another 10 days, it determined the daily output at plant 2, and on the final 10 days management determined the daily output at plant 3. The following summary data give the sample means and sample variances of the daily numbers of items produced at the three plants over those days.

Plant i	$\bar{X}_i$	S_i^2
$i = 1$	325	450
$i = 2$	413	520
$i = 3$	366	444

Test the hypothesis that the mean number of items produced daily is the same for all three plants. Use the 5 percent level of significance.

2. Sixty nonreading preschool students were randomly divided into four groups of 15 each. Each group was given a different type of course in learning how to read. Afterward, the students were tested with the following results.

Group	$\bar{X}_i$	S_i^2
1	65	224
2	62	241
3	68	233
4	61	245

Test the null hypothesis that the reading courses are equally effective. Use the 5 percent level of significance.

3. Preliminary studies indicate a possible connection between one's natural hair color and threshold for pain. A sample of 12 women were classified as to having light, medium, or dark hair. Each was then given a pain sensitivity test, with the following scores resulting.

Light	Medium	Dark
63	60	45
72	48	33
52	44	57
60	53	40

Are the above data sufficient to establish that hair color affects the results of a pain sensitivity test? Use the 5 percent level of significance.

4. Three different washing machines were employed to test four different detergents. The following data give a coded score of the effectiveness of each washing.

	Machine		
Detergent	**1**	**2**	**3**
1	53	50	59
2	54	54	60
3	56	58	62
4	50	45	57

   (a) Estimate the improvement in mean value with detergent 1 over detergent (i) 2, (ii) 3, (iii) 4.
   (b) Estimate the improvement in mean value when machine 3 is used as opposed to machine (i) 1 and (ii) 2.
   (c) Test the hypothesis that the detergent used does not affect the score.
   (d) Test the hypothesis that the machine used does not affect the score.
   In both (c) and (d), use the 5 percent level of significance.

5. Suppose in Prob. 4 that the 12 applications of the detergents were all on different randomly chosen machines. Test the hypothesis, at the 5 percent significance level, that the detergents are equally effective.

6. In Example 11.3 test the hypothesis that the mean test score depends only on the test taken and not on which student is taking the test.

7. A manufacturer of women's beauty products is considering four new variations of a hair dye. An important consideration in a hair dye is its lasting power, defined as the number of days until treated hair becomes indistinguishable from untreated hair. To learn about the lasting power of its new variations, the company hired three long-haired women. Each woman's hair was divided into four sections, and each section was treated by one of the dyes. The following data concerning the lasting power resulted.

	Dye			
Woman	**1**	**2**	**3**	**4**
1	15	20	27	21
2	30	33	25	27
3	37	44	41	46

   (a) Test, at the 5 percent level of significance, the hypothesis that the four variations have the same mean lasting power.

(b) Estimate the mean lasting power obtained when woman 2 uses dye 2.

(c) Test, at the 5 percent level of significance, the hypothesis that the mean lasting power does not depend on which woman is being tested.

8. Use the following data to test the hypotheses of (a) no row effect and (b) no column effect.

17	23	35	39	5
42	28	19	40	14
36	23	31	44	13
27	40	25	50	17

9. Problem 9 of Sec. 11.2 implicitly assumes that the number of deaths is not affected by the year under consideration. However, consider a two-factor ANOVA model for this problem.

(a) Test the hypothesis that there is no effect due to the year.

(b) Test the hypothesis that there is no seasonal effect.

10. The following data relate to the ages at death of a certain species of rats that were fed one of three types of diet. The rats chosen were of a type having a short life span, and they were randomly divided into three groups. The data are the sample means and sample variances of the ages of death (measured in months) of the three groups. Each group is of size eight.

	Very low-calorie	Moderate-calorie	High-calorie
Sample mean	22.4	16.8	13.7
Sample variance	24.0	23.2	17.1

Test the hypothesis, at the 5 percent level of significance, that the mean lifetime of a rat is not affected by its diet. What about at the 1 percent level?

MINITAB LAB

Analysis of Variance

Purpose

Use Minitab to

1. Do one-factor analysis of variance.

2. Do two-factor analysis of variance.

Procedures

First, load the Minitab (Windows version) software as in the Minitab lab for Chapter 1.

1. ONE-FACTOR ANALYSIS OF VARIANCE

Example 1 Use Minitab to work Example 11.1 in Sec. 11.2 of the text.

First, enter the values for GAS1, GAS2, and GAS3 in columns C1, C2, and C3, respectively. Note that this is a one-factor analysis of variance, with the factor being gasoline, and the three populations (levels) from which the samples were taken are GAS1, GAS2, and GAS3. To use Minitab, select Stat→ANOVA→Oneway (Unstacked) and the One-way analysis of variance dialog box will be displayed. Select GAS1–GAS3 for the Responses (in separate columns) text box, as shown in Fig. M11.1.

Click on the OK button, and the Session window will display the results of the computations for the test. This is shown in Fig. M11.2. Observe that the degrees of freedom for the *between-samples* (factor) sum of squares (SS) is 2, and this corresponds to the numerator degrees of freedom for the critical (table) F value. For the *within-samples* (error) sum of squares, there are 12 degrees of freedom, and this corresponds to the denominator degrees of freedom for the critical F value. The between-samples SS is 863, and the within-samples SS is 1992. The mean square (MS) values are the respective sum of squares divided by their degrees of freedom; and the computed F value is calculated from the ratio of these two mean square values.

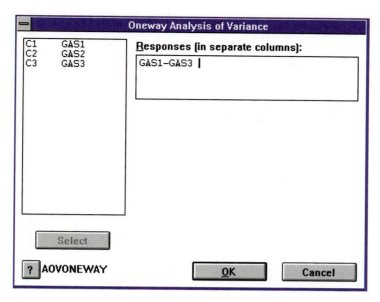

Figure M11.1

The calculated test statistic value is given by $F = 2.60$ with the corresponding p value for the hypothesis test equal to .115. In addition, individual 95 percent confidence intervals for the population means (GAS1, GAS2, and GAS3) are displayed in the window. *With this output you can now write up the hypothesis test for the example.*

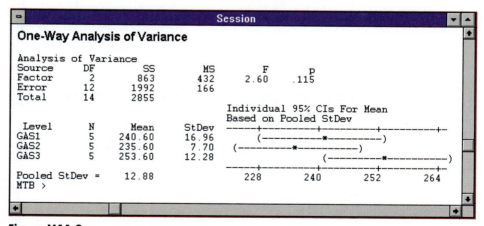

Figure M11.2

To use Minitab to compute the critical (table) F value, select Calc→Probability Distribution→F, and the F distribution dialog box will be displayed. Select the Inverse cumulative probability check box. In the Numerator degrees of freedom text box, type 2; and in the Denominator degrees of freedom text box, type 12. Select the Input constant check box and type .95 in its text box.

Select the OK button, and the critical value of 3.8853 will be computed and displayed in the Session window.

Note   The sequence Stat→ANOVA→Oneway (Unstacked) will also work for unequal sample sizes.

2. TWO-FACTOR ANALYSIS OF VARIANCE

Example 2   Use Minitab to work Example 11.5 in Sec. 11.4 of the text.

To use Minitab to help solve this problem, enter the data values in the Data window. Take care in entering the values. Figure M11.3 shows one way of entering the values.

	C1	C2	C3	
↓	MACHINES	WORKERS	DEF	
1	1	1	41	
2	1	2	42	
3	1	3	40	
4	1	4	35	
5	2	1	35	
6	2	2	42	
7	2	3	43	
8	2	4	36	
9	3	1	42	
10	3	2	39	
11	3	3	44	
12	3	4	47	
13				

Figure M11.3

Once the data have been entered, select Stat→ANOVA→Twoway, and the Two-way analysis of variance dialog box will be displayed. For the Response text box, select DEF (number of defective items). For the Row factor and Column factor text boxes, select MACHINES and WORKERS, respectively. In this example, the means are to be displayed by the Display means boxes. The appropriate selections are shown in Fig. M11.4.

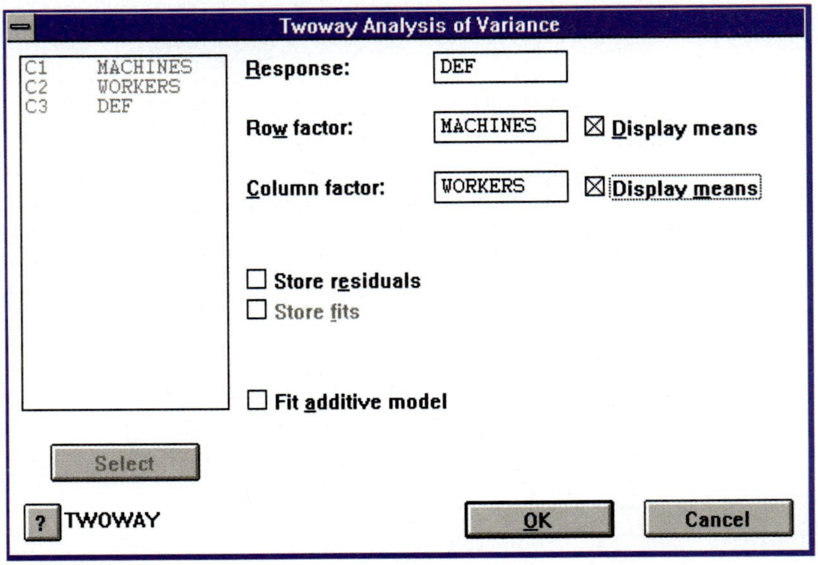

Twoway Analysis of Variance

C1	MACHINES
C2	WORKERS
C3	DEF

Response: DEF

Row factor: MACHINES   ⊠ **Display means**

Column factor: WORKERS   ⊠ **Display means**

☐ **Store residuals**
☐ **Store fits**

☐ **Fit additive model**

Select

? TWOWAY   **OK**   **Cancel**

Figure M11.4

Click on the OK button, and the Session window will display some of the required results. The test statistic for the hypothesis of no row (machines) effect is obtained from the ratio MS(machines)/MS(error) = 19.0/15.7 = 1.2102; and the test statistic for the hypothesis of no column (workers) effect is obtained from MS(workers)/MS(error) = 6.3/15.7 = .4013. You can find the critical (table) values for the two tests by using degrees of freedom 2 (numerator) and 6 (denominator), and 3 (numerator) and 6 (denominator), as in Example 1. If the significancce level is 5 percent, these values are 5.1433 and 4.7571.

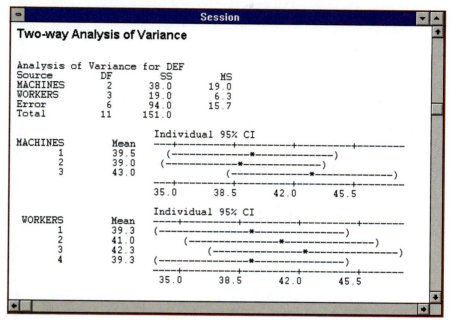

Figure M11.5

The p value for the test of row effect is computed from $P(F \geq 1.2102)$, where F has an F distribution with 2 and 6 degrees of freedom. For the column effect test, the p value is computed from $P(F \geq .4013)$, where F has an F distribution with 3 and 6 degrees of freedom. To compute these P values, select Calc→Probability Distributions→F, and the F distribution dialog box will be displayed. Select Cumulative probability with degrees of freedom 2 and 6 and with Input constant of 1.2102. Click on the OK button, and $P(F \leq 1.2102)$ will be computed and displayed in the Session window. This value is .6382. Thus the p value for the test of row effect will be $P(F \geq 1.2102) = 1 - P(F \leq 1.2102) = 1 - .6382 = .3618$. The p value for the test of column effect is .7573 (verify). Thus you have enough information now to write up the appropriate hypothesis tests.

Computer Exercises

1. Use Minitab or any other statistical software to help work Probs. 3, 5, 7, 9, and 11 at the end of Sec. 11.2.

2. Use Minitab or any other statistical software to help work Probs. 1, 2, 3, and 7 at the end of Sec. 11.4.

3. Use Minitab or any other statistical software to help work Review Probs. 3, 4, and 7 at the end of Chap. 11.

4. Use your library or any other resource to collect a data set for a *one-factor experiment.* Use Minitab or any other statistical software to test for the equality of the population means for the different levels. Establish whether these samples came from normal populations as in previous labs. Write up a report with your discussion and findings. If you select data from published research, include relevant factors about the validity and reliability of the study.

5. Use your library or any other resource to collect a data set for a *two-factor experiment.* Use Minitab or any other statistical software to test for the equality of the population means for the different levels (for rows as well as for columns). Establish whether these samples came from normal populations as in previous labs. Write up a report with your discussion and findings. If you select data from published research, include relevant factors about the validity and reliability of the study.

6. Design your own experiment, and collect data to do a one-factor analysis of variance. Use Minitab or any other statistical software to present your findings in a report.

7. Design your own experiment, and collect data to do a two-factor analysis of variance. Use Minitab or any other statistical software to present your findings in a report.

CHAPTER 12

Linear Regression

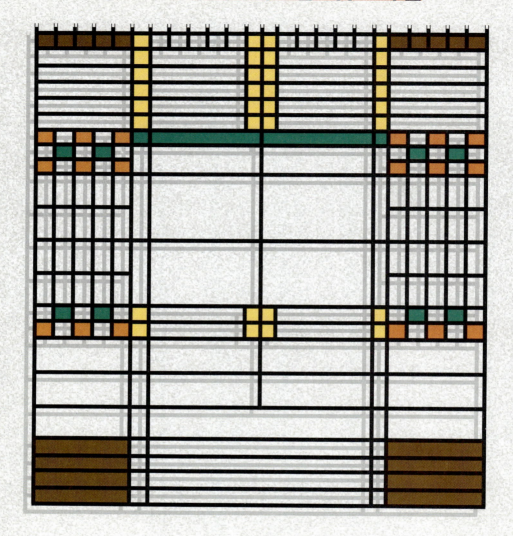

Linear Regression

We know a thing when we understand it.
George Berkeley (British philosopher for whom the California city was named)

You can observe a lot just by watching.
Yogi Berra

We study the simple linear regression model which, except for random error, assumes a straight-line relationship between a response and an input variable. We use the method of least squares to estimate the parameters of this model. Assuming that the random error is normal with mean 0 and variance σ^2, we show how to test hypotheses concerning the parameters of the model. The concept of regression to the mean is introduced; we explain when it arises and how one must be careful to avoid the regression fallacy in its presence. We explain the coefficient of determination. Finally, we introduce the multiple linear regression model which relates a response variable to a set of input variables.

It was a spring day in 1888, and Francis Galton was out for a stroll in the countryside. While walking, he considered a question that had concerned him for some time. What was the relationship between a child's physical and mental characteristics and those of the child's parents? For instance, simplifying his ideas somewhat, Galton believed that the height of a child at adulthood should have an expected value equal to the height of his or her (same-sex) parent. But if this were so, then it would follow that about one-half of the offspring of very tall (short) people would be even taller (shorter) than their parents. Thus each new generation should produce taller (as well as shorter) people than the previous generation. However, on the contrary, data indicated a stability in the heights of the population from generation to generation. How could this apparent contradiction be explained?

It came to Galton in a flash. In his own words, "A temporary shower drove me to seek refuge by a recess in the rock by the side of a pathway. There the idea flashed across me and I forgot everything else for a moment in my great delight."

Galton's flash of insight was that the mean value of a child's characteristic (such as height) was not equal to his or her parent's height but rather was between this value and the average value of the entire population. Thus, for instance, the heights of the offspring of very tall people (called, by Galton, people "taller than mediocrity") would tend to be shorter than their parents. Similarly, the offspring of those shorter than mediocrity would tend to be taller than their parents. Galton called this insight "regression to mediocrity"; we call it *regression to the mean*.

12.1 INTRODUCTION

We are often interested in trying to determine the relationship between a pair of variables. For instance, how does the amount of money spent in advertising a new product relate to the first month's sales figures for that product? Or how does the amount of catalyst employed in a scientific experiment relate to the yield of that experiment? Or how does the height of a father relate to that of his son?

In many situations the values of the variables are not determined simultaneously in time; rather, one of the variables will be set at some value, and this will, in turn, affect the value of the second variable. For instance, the advertising budget would be set before the sales figures are determined, and the amount of catalyst to be used would be set before the resulting yield could be determined. The variable whose value is determined first is called the *input* or *independent* variable and the other is called the *response* or *dependent* variable.

Suppose that the value of the independent variable is set to equal x. Let Y denote the resulting value of the dependent variable. The simplest type of relationship between this pair of variables is a straight-line, or *linear,* relation of the form

$$Y = \alpha + \beta x \tag{12.1}$$

This model, however, supposes that (once the values of the parameters α and β are determined) it would be possible to predict exactly the response for any value of the input variable. In practice, however, such precision is almost never attainable, and the most that one can expect is that the preceding equation is valid *subject to random error.*

In Sec. 12.2 we explain precisely the meaning of the *linear regression* model which assumes that Eq. (12.1) is valid subject to random error. In Sec. 12.3 we show how data can be used to estimate the regression parameters α and β. The estimators presented are based on the least-squares approach to finding the best straight-line fit for a set of data pairs. Section 12.4 deals with the *error* random variable, which will be taken to be a normal random variable having mean 0 and variance σ^2. The problem of estimating σ^2 will be considered.

In Sec. 12.5 we consider tests of the statistical hypothesis that there is no linear relationship between the response variable Y and the input value x. Section 12.6 deals with the concept of *regression to the mean.* It is shown that this phenomenon arises when the value of the regression parameter β is between 0 and 1. We explain how this phenomenon will often occur in testing-retesting situations and how a careless analysis of such data can often lead one into the *regression fallacy.* In addition, we indicate in this section how regression to the mean, in conjunction with the central limit theorem and the passing of many generations, can be used to explain why biological data sets are so often approximately normally distributed.

Section 12.7 is concerned with determining an interval which, with a fixed degree of confidence, will contain a future response corresponding to a specified input. These intervals, which make use of previously obtained data, are known as *prediction intervals.* Sections 12.8 and 12.9 present, respectively, the coefficient of determination and the correlation coefficient. Both quantities can be used to indicate the degree of fit of the linear regression model to the data.

In Sec. 12.10 we consider the multiple linear regression model where one tries to predict a response not on the basis of the value of a single input variable but on the basis of the values of two or more such variables.

12.2 SIMPLE LINEAR REGRESSION MODEL

Consider a pair of variables, one of which is called the *input variable* and the other the *response variable.* Suppose that for a specified value x of the input variable the value of the response variable Y can be expressed as

$$Y = \alpha + \beta x + e$$

The quantities α and β are parameters. The variable e, called the *random error,* is assumed to be a random variable having mean 0.

Definition

The relationship between the response variable Y and the input variable x specified in the preceding equation is called a *simple linear regression.*

The simple linear regression relationship can also be expressed by stating that for any value x of the input variable, the response variable Y is a random variable with mean given by

$$E[Y] = \alpha + \beta x$$

Thus a simple linear regression model supposes a straight-line relationship between the mean value of the response and the value of the input variable. Parameters α and β will almost always be unknown and will have to be estimated from data.

To see if a simple linear regression might be a reasonable model for the relationship between a pair of variables, one should first collect and then plot data on the paired values of the variables. For instance, suppose there is available a set of data pairs (x_i, y_i), $i = 1, \ldots, n$, meaning that when the input variable was set to equal x_i, then the observed value of the response variable was y_i. These points should then be plotted to see if, subject to random error, a straight-line relationship between x and y appears to be a reasonable assumption. The resulting plot is called a *scatter diagram.*

Example 12.1   A new type of washing machine was recently introduced in 11 department stores. These stores are of roughly equal size and are located in similar types of communities. The manufacturer varied the price charged in each store, and the following data giving the number of units sold in 1 month for each of the different prices resulted.

Price ($)	Number sold
280	44
290	41
300	34
310	38
320	33
330	30
340	32
350	26
360	28
370	23
380	20

A plot of the number of units sold y versus the price x for these 11 data pairs is given in Fig. 12.1. The resulting scatter diagram indicates that, subject to random error, the assumption of a straight-line relationship between the number of units sold and the price appears to be reasonable. That is, a simple linear regression model appears to be appropriate.

Remark: The input variable x in the linear regression model is not usually thought of as being a random variable. Rather it is regarded as a constant that can be set at various values. The resulting response Y, on the other hand, is regarded as a random variable

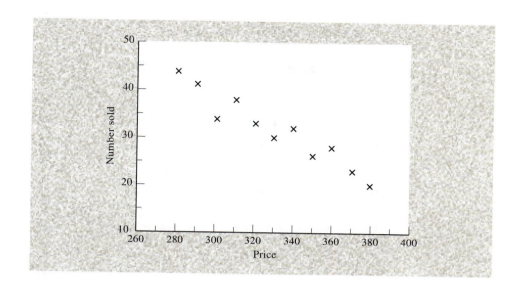

Figure 12.1
A scatter diagram for the data of Example 12.1.

whose mean value depends in a linear way on the input x. It is for this reason that we use the capital, or uppercase, letter Y to represent the response. We use a small, or lowercase, y to denote an observed value of Y, and so y will represent the observed value of the response at the input value x.

Problems

1. The following 12 data pairs relate y, the percentage yield of a laboratory experiment, to x, the temperature at which the experiment was conducted.

i	x_i	y_i	i	x_i	y_i
1	100	45	7	150	69
2	110	51	8	160	74
3	120	54	9	170	78
4	125	53	10	180	86
5	130	59	11	190	89
6	140	63	12	200	94

(a) Represent these data in a scatter diagram.
(b) Do you think a simple linear regression model would be appropriate for describing the relationship between percentage yield and temperature?

2. An area manager in a department store wants to study the relationship between the number of workers on duty and the value of merchandise lost to shoplifters. To do so, she assigned a different number of clerks for each of 10 weeks. The results were as follows:

Week	Number of workers	Loss
1	9	420
2	11	350
3	12	360
4	13	300
5	15	225
6	18	200
7	16	230
8	14	280
9	12	315
10	10	410

(a) Which variable should be the input variable and which should be the response?

(b) Plot the above in a scatter diagram.

(c) Does a simple linear regression model appear reasonable?

3. The following data relate the traffic density, described in the number of automobiles per mile, to the average speed of traffic on a moderately large city thoroughfare. The data were collected at the same location at 10 different times within a span of 3 months.

Density	Speed
69	25.4
56	32.5
62	28.6
119	11.3
84	21.3
74	22.1
73	22.3
90	18.5
38	37.2
22	44.6

(a) Which variable is the input and which is the response?

(b) Draw a scatter diagram.

(c) Does a simple linear regression model appear to be reasonable?

4. Repeat Prob. 3, but now let the square root of the speed, rather than the speed itself, be the response variable.

5. The use that can be obtained from a tire is affected by the air pressure in the tire. A new type of tire was tested for wear at different pressures, with the following results.

Pressure (pounds per square inch)	Mileage (thousands of miles)
30	29.4
31	32.2
32	35.9
33	38.4
34	36.6
35	34.8
36	35.0
37	32.2
38	30.5
39	28.6
40	27.4

(a) Plot the data in a scatter diagram.

(b) Does a simple linear regression model appear appropriate for describing the relation between tire pressure and miles of use?

12.3 ESTIMATING THE REGRESSION PARAMETERS

Suppose that the responses Y_i corresponding to the input values x_i, $i = 1, \ldots , n$, are to be observed and used to estimate the parameters α and β in a simple linear regression model

$$Y = \alpha + \beta x + e$$

To determine estimators of α and β, we reason as follows: If A and B were the respective estimators of α and β, then the estimator of the response corresponding to the input value x_i would be $A + Bx_i$. Since the actual response is Y_i, it follows that the difference between the actual response and its estimated value is given by

$$\epsilon_i \equiv Y_i - (A + Bx_i)$$

That is, ϵ_i represents the error that would result from using estimators A and B to predict the response at input value x_i (see Fig. 12.2).

   Now, it is reasonable to choose our estimates of α and β to be the values of A and B that make these errors as small as possible. To accomplish this, we choose A and B to minimize the value of $\sum_{i=1}^{n} \epsilon_i^2$, the sum of the squares of the errors. The resulting estimators of α and β are called *least-square estimators*.

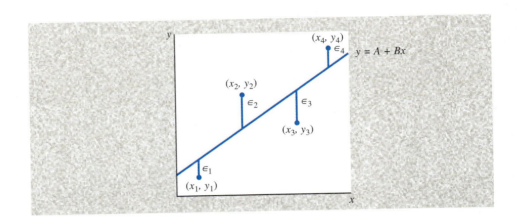

Figure 12.2
The errors.

For given data pairs (x_i, Y_i), $i = 1, \ldots, n$, the least-square estimators of α and β are the values of A and B that make

$$\sum_{i=1}^{n} \epsilon_i^2 = \sum_{i=1}^{n} (Y_i - A - Bx_i)^2$$

as small as possible.

Remark: The reason we want $\sum_{i=1}^{n} \epsilon_i^2$, rather than $\sum_{i=1}^{n} \epsilon_i$, to be small is that the sum of the errors can be small even when individual error terms are large (since large positive and large negative errors cancel). On the other hand, this could not happen with the sum of the *squares* of the errors since none of the terms could be negative.

It can be shown that the least-square estimators of α and β, which we call $\hat{\alpha}$ and $\hat{\beta}$, are given by

$$\hat{\beta} = \frac{\sum_{i=1}^{n} (x_i - \bar{x})(Y_i - \bar{Y})}{\sum_{i=1}^{n} (x_i - \bar{x})^2}$$

$$\hat{\alpha} = \bar{Y} - \hat{\beta}\,\bar{x}$$

where

$$\bar{x} = \frac{\sum_{i=1}^{n} x_i}{n} \qquad \text{and} \qquad \bar{Y} = \frac{\sum_{i=1}^{n} Y_i}{n}$$

The line

$$y = \hat{\alpha} + \hat{\beta}x$$

is called the *estimated regression line:* $\hat{\beta}$ is the slope, and $\hat{\alpha}$ is the intercept of this line.

Notation: If we let

$$S_{xY} = \sum_{i=1}^{n} (x_i - \bar{x})(Y_i - \bar{Y})$$

$$S_{xx} = \sum_{i=1}^{n} (x_i - \bar{x})^2$$

$$S_{YY} = \sum_{i=1}^{n} (Y_i - \bar{Y})^2$$

then the least-square estimators can be expressed as

$$\hat{\beta} = \frac{S_{xY}}{S_{xx}}$$

$$\hat{\alpha} = \bar{Y} - \hat{\beta}\bar{x}$$

The values of $\hat{\alpha}$ and $\hat{\beta}$ can be obtained either by a pencil-and-paper computation or by using a hand calculator. In addition, Program 12-1 will compute the least-squares estimators and the estimated regression line. This program also gives the user the option of computing some other statistics whose values will be needed in the following sections.

Example 12.2   A large midwestern bank is planning on introducing a new word processing system to its secretarial staff. To learn about the amount of training that is needed to effectively implement the new system, the bank chose eight employees of roughly equal skill. These workers were trained for different amounts of time and were then individually put to work on a given project. The following data indicate the training times and the resulting times (both in hours) that it took each worker to complete the project.

Worker	Training time (= x)	Time to complete project (= Y)
1	22	18.4
2	18	19.2
3	30	14.5
4	16	19.0
5	25	16.6
6	20	17.7
7	10	24.4
8	14	21.0

(a) What is the estimated regression line?

(b) Predict the amount of time it would take a worker who receives 28 hours of training to complete the project.

(c) Predict the amount of time it would take a worker who receives 50 hours of training to complete the project.

Solution

(a) Rather than calculating by hand (which you will be asked to do in Prob. 2), we run Program 12-1, which computes the least-squares estimators and related statistics in simple linear regression models. We obtain the following:

First, enter the number of data pairs n, which is 8.

Next, enter the 8 successive pairs, which are:

    22, 18.4
    18, 19.2
    30, 14.5
    16, 19
    25, 16.6
    20, 17.7
    10, 24.4
    14, 21

The program computes the least-squares estimators as follows:

    A = 27.46606   (b)
    B = −.4447002   (a)

The estimated regression line is as follows:

    Y = 27.46606 + −.4447002x

A plot of the scatter diagram and the resulting estimated regression line is given in Fig. 12.3.

(b) The best prediction of the completion time corresponding to the training time of 28 hours is its mean value, namely,

$$\alpha + 28\beta$$

By using the estimates of α and β previously derived, the predicted completion time is

$$27.466 - 28(.445) = 15.006$$

(c) This part asks for the prediction at the input value 50, which is far greater than all the input values in our data set. As a result, even though the scatter diagram indicates that a straight-line fit should be a reasonable approximation for the range of input values considered, one should be extremely cautious about assuming that the relationship will continue to be a straight line for input values as large as 50. Thus, it is prudent not to attempt to answer part (c) on the basis of the available data.

Warning: *Do not use the estimated regression line to predict responses at input values that are far outside the range of the ones used to obtain this line.*

The following formulas can be useful when you are computing by hand.

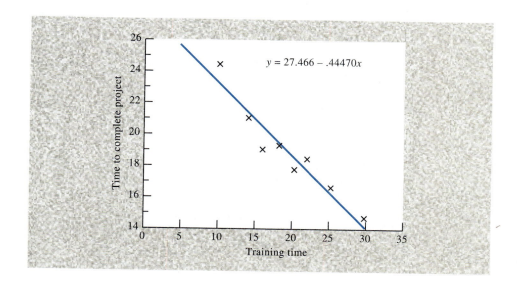

Figure 12.3
A scatter diagram and the estimated regression line.

$$S_{xY} = \sum_{i=1}^{n} x_i Y_i - n\bar{x}\bar{Y}$$

$$S_{xx} = \sum_{i=1}^{n} x_i^2 - n\bar{x}^2$$

Problems

1. Find, by a hand computation, the estimated regression line for the following data:

x	y
1	4
2	7
3	8
5	12

   (a) Plot the scatter diagram, and draw the estimated regression line.
   (b) Double all the data values and repeat part (a).

2. Verify the value given in Example 12.2 for the estimated regression line either by a pencil-and-paper computation or by using a hand calculator.

3. The following pairs of data represent the amounts of damages (in thousands of dollars) in fires at middle-class residences in a certain city and the distances (in miles) from these residences to the nearest fire station.

Distance	Damage
5.2	36.2
4.4	28.8
3.0	22.6
1.2	8.8
7.5	41.5
9.4	25.4

   (a) Draw a scatter diagram.
   (b) Try to approximate the relationship between the distance and damage by drawing a straight line through the data.
   (c) Find the estimated regression line, and compare it to the line drawn in part (b).

4. Consider Prob. 1 of Sec. 12.2.
   (a) Draw a straight line through the data points.
   (b) Determine the estimated regression line, and compare it to the line drawn in part (a).

5. The following table gives the U.S. per capita consumption (in pounds) of fresh fruits for the years 1985 through 1991.

Year	1985	1986	1987	1988	1989	1990	1991
Fresh fruit consumption	86.8	93.1	97.5	97.4	98.8	92.6	90.6

Source: U.S. Department of Agriculture, *Food Consumption, Prices and Expenditures.*

(a) Letting the year be the independent variable and consumption the dependent variable, plot a scatter diagram of the above.
(b) Find the estimated regression line. (To simplify the algebra, you can take the x variable to be the year minus 1985. That is, year 1985 has value 0, 1986 has value 1, and so on.
(c) Plot the estimated regression line on the scatter diagram.
(d) Predict the per capita consumption in 1988.

6. The following table gives the U.S. per capita consumption (in pounds) of canned fruits for the years 1980 through 1987.

Year	1980	1981	1982	1983	1984	1985	1986	1987
Canned fruit consumption	10.7	10.0	9.7	9.2	8.9	8.4	8.4	8.7

Source: U.S. Department of Agriculture, *Food Consumption, Prices and Expenditures.*

(a) Letting the year be the independent variable and consumption the dependent variable, plot a scatter diagram of the above.
(b) Find the estimated regression line.
(c) Predict the 1988 per capita consumption.
(d) Predict the 1990 per capita consumption.

7. Find the estimated regression line for Prob. 2 of Sec. 12.2.

8. The following table gives the percentage of the U.S. labor force that belonged to unions for certain years.

Year	1980	1983	1984	1985	1986	1987	1988	1989	1992
Percentage in union	21.9	20.1	18.8	18.0	17.5	17.0	16.8	16.4	15.8

Source: Bureau of Labor Statistics.

(a) Plot these data in a scatter diagram.
(b) Find the estimated regression line.
(c) Predict what the percentage was in 1981.

(d) Predict what the percentage was in 1982.

(e) Predict what the percentage was in 1990.

9. Find the estimated regression line for Prob. 4 of Sec. 12.2.

10. The following table gives, for certain years, the percentage of graduating high school students who had ever used marijuana.

Year	1980	1983	1984	1985	1986	1987	1988	1989
Percentage of students	60.3	57.0	54.9	54.2	50.9	50.2	47.2	43.7

Source: National Institute of Drug Abuse.

(a) Plot these data in a scatter diagram.

(b) Find the estimated regression line.

(c) Predict what the percentage was in 1981.

(d) Predict what the percentage was in 1982.

(e) Predict what the percentage was in 1990.

11. The following table relates the world production of wood pulp to the world production of newsprint for each of seven different years. The data come from the Statistical Division of the United Nations, New York, *Monthly Bulletin of Statistics* and are in units of 1 million metric tons.

Wood pulp	Newsprint
124.4	25.4
131.3	27.8
133.1	28.3
136.6	29.3
142.0	30.6
150.1	32.3
150.3	33.1

(a) Taking the amount of wood pulp as the independent (or input) variable, find the estimated regression line.

(b) Predict the amount of newsprint produced in a year in which 146.0 million metric tons of wood pulp is produced.

(c) Taking the amount of newsprint as the input variable, find the estimated regression line.

(d) Predict the amount of wood pulp produced in a year in which 32.0 million metric tons of newsprint is produced.

12. It is believed that the more alcohol there is in an individual's bloodstream, the slower is that person's reaction time. To test this, 10 volunteers were given different amounts of alcohol. Their blood alcohol levels were determined as percentages of their body weights. The volunteers were then tested to determine their reaction times to a given stimulus. The following data resulted.

x = amount of alcohol in blood (percent)	y = reaction time (seconds)
.08	.32
.10	.38
.12	.44
.14	.42
.15	.47
.16	.51
.18	.63

(a) Plot a scatter diagram.

(b) Approximate the estimated regression line by drawing a straight line through the data.

(c) What is the estimated regression line?

(d) Compare the lines in parts (b) and (c). Are their slopes nearly equal? How about the intercepts?

Predict the reaction time for an individual (not one of the above) whose blood alcohol content is

(e) .15

(f) .17

13. In Example 12.2, suppose the eight training times had been chosen in advance. How do you think the decision as to the assignment of the eight workers to these training times should have been made?

14. In an experiment designed to study the relationship between the number of alcoholic drinks consumed and blood alcohol concentration, seven individuals having the same body size were randomly assigned a certain number of alcoholic drinks. After a wait of 1 hour, their blood alcohol levels were checked. The results were as follows.

Number of drinks	Blood alcohol level
.5	.01
1	.02
2	.05
3	.09
4	.10
5	.14
6	.20

(a) Draw a scatter diagram.

(b) Find the estimated regression line, and draw it on the scatter diagram.

(c) Predict the blood alcohol level of a person, of the same general size as the people in the experiment, who had 3 drinks 1 hour ago.

(d) What if the person in part (b) had 7 drinks 1 hour ago?

15. The following data relate the per capita consumption of cigarettes in 1930 and men's death rates from lung cancer in 1950, for a variety of countries.

Country	1930 Per capita cigarette consumption	1950 Deaths per million men
Australia	480	180
Canada	500	150
Denmark	380	170
Finland	1100	350
Great Britain	1100	460
Iceland	230	60
The Netherlands	490	240
Norway	250	90
Sweden	300	110
Switzerland	510	250
United States	1300	200

(a) Determine the estimated regression line.
Predict the number of 1950 lung cancer deaths per million men in a country whose 1930 per capita cigarette consumption was
(b) 600
(c) 850
(d) 1000

16. The following are the average scores on the mathematics part of the Scholastic Aptitude Test (SAT) in the years from 1980 to 1989, excluding 1983.

Year	1980	1981	1982	1984	1985	1986	1987	1988	1989
Average score	466	466	467	471	475	475	476	476	476

Source: College Entrance Examination Board.

(a) Predict the average mathematics score in 1983.
(b) Predict the average mathematics score in 1993.

17. Use the data of Prob. 3 in Sec. 3.7 to predict the IQ of the daughter of a woman having an IQ of 130.

18. Use the data of Prob. 6 in Sec. 3.7 to predict the number of adults on parole in a state having 14,500 adults in prison.

19. The following data give the number of foreign students enrolled in U.S. institutions of higher education from 1987 to 1992. The data are in units of 1000.

Year	1987	1988	1989	1990	1991	1992
Foreign students	350	356	366	387	408	420

Source: Open Doors, Institute of International Education, New York, annual.

Use these data to predict the number of foreign students enrolled in (a) 1993, (b) 1994, and (c) 1995.

12.4 ERROR RANDOM VARIABLE

We have defined the linear regression model by the relationship

$$Y = \alpha + \beta x + e$$

where α and β are unknown parameters that will have to be estimated and e is an error random variable having mean 0. To be able to make statistical inferences about the regression parameters α and β, it is necessary to make some additional assumptions concerning the error random variable e. The usual assumption, which we will be making, is that e is a normal random variable with mean 0 and variance σ^2. Thus we are assuming that the variance of the error random variable remains the same no matter what input value x is used.

Put another way, the above assumption is equivalent to assuming that for any input value x, the response variable Y is a random variable that is normally distributed with mean

$$E[Y] = \alpha + \beta x$$

and variance

$$\text{Var}(Y) = \sigma^2$$

An additional assumption we will make is that all response variables are independent. That is, for instance, the response from input value x_1 will be assumed to be independent of the response from input value x_2.

The quantity σ^2 is an unknown which will have to be estimated from the data. To see how this can be accomplished, suppose that we will be observing the response values Y_i corresponding to the input values x_i, $i = 1, \ldots, n$. Now, for each value of i the standardized variable

$$\frac{Y_i - E[Y_i]}{\sqrt{\text{Var}(Y_i)}} = \frac{Y_i - (\alpha + \beta x_i)}{\sigma}$$

will have a standard normal distribution. Thus, since a chi-squared random variable with n degrees of freedom is defined to be the sum of the squares of n independent standard normals, we see that

$$\frac{\sum_{i=1}^{n} (Y_i - \alpha - \beta x_i)^2}{\sigma^2}$$

is chi squared with n degrees of freedom.

If we now substitute the estimators $\hat{\alpha}$ and $\hat{\beta}$ for α and β in the preceding, then the resulting variable will remain chi squared but will now have $n - 2$ degrees of freedom (since 1 degree of freedom will be lost for each parameter that is estimated). That is,

$$\frac{\sum_{i=1}^{n} (Y_i - \hat{\alpha} - \hat{\beta} x_i)^2}{\sigma^2}$$

is chi squared with $n - 2$ degrees of freedom.

The quantities

$$Y_i - \hat{\alpha} - \hat{\beta} x_i \qquad i = 1, \ldots, n$$

are called *residuals*. They represent the differences between the actual and the predicted responses. We will let SS_R denote the sum of the squares of these residuals. That is,

$$SS_R = \sum_{i=1}^{n} (Y_i - \hat{\alpha} - \hat{\beta} x_i)^2$$

From the preceding result, we thus have

$$\frac{SS_R}{\sigma^2}$$

is chi squared with $n - 2$ degrees of freedom.

Since the expected value of a chi-squared random variable is equal to its number of degrees of freedom, we obtain

$$\frac{E[SS_R]}{\sigma^2} = n - 2$$

or

$$E\left[\frac{SS_R}{n - 2} \right] = \sigma^2$$

In other words, $SS_R/(n - 2)$ can be used to estimate σ^2.

$$\frac{SS_R}{n - 2}$$

is the estimator of σ^2.

Program 12-1 can be utilized to compute the value of SS_R.

Example 12.3   Consider Example 12.2 and suppose that we are interested in estimating the value of σ^2. To do so, we could again run Program 12-1, this time asking
for the additional statistics. This would result in the following additional output:

S(x,Y) = -125.3499
S(x,x) = 281.875
S(Y,Y) = 61.08057
SSR = 5.337465
THE SQUARE ROOT OF (n-2)S(x,x)/SSR is 17.80067

The estimate of σ^2 is 5.3375/6 = .8896.

The following formula for SS_R is useful when you are using a calculator or computing by hand.

Computational formula for SS_R:

$$SS_R = \frac{S_{xx}S_{YY} - S_{xY}^2}{S_{xx}}$$

use notes

The easiest way to compute SS_R by hand is to first determine S_{xx}, S_{xY}, and S_{YY} and
then apply the above formula.

Problems

1. Estimate σ^2 in Prob. 1 of Sec. 12.2.

2. Estimate σ^2 in Prob. 2 of Sec. 12.2.

3. The following data relate the speed of a particular typist and the temperature setting of his office. The units are words per minute and degrees Fahrenheit.

Temperature	Typing speed
50	63
60	74
70	79

(a) Compute, by hand, the value of SS_R.
(b) Estimate σ^2.
(c) If the temperature is set at 65, what typing speed would you predict?

4. Estimate σ^2 in Prob. 3 of Sec. 12.2.

5. Estimate σ^2 in Prob. 14 of Sec. 12.3.

6. The following data give, for certain years between 1972 and 1988, the percentages of British women who were cigarette smokers.

Year	1972	1976	1978	1980	1982	1984	1986	1988
Percentage	41.2	38.0	37.3	36.9	33.1	31.8	31.1	30.4

Treat the above data as coming from a linear regression model with the input being the year and the response being the percentage. Take 1972 as the base year, so 1972 has input value $x = 0$, 1976 has input value $x = 4$, and so on.
(a) Estimate the value of σ^2.
(b) Predict the percentage of British women who smoked in 1992.

7. Estimate σ^2 in Prob. 15 of Sec. 12.3.

8. In data relating the ages at which 25 fathers (x) and their respective sons (Y) first began to shave, the following summary statistics resulted.

$$\bar{x} = 13.9 \qquad \bar{Y} = 14.6$$

$$S_{xx} = 46.8 \qquad S_{YY} = 53.3 \qquad S_{xY} = 12.2$$

(a) Determine the estimated regression line.
(b) Predict the age at which a boy will begin to shave if his father began to shave at age 15.1 years.
(c) Estimate σ^2.

12.5 TESTING THE HYPOTHESIS THAT $\beta = 0$

An important hypothesis to consider with respect to the simple linear regression model

$$Y = \alpha + \beta x + e$$

is the hypothesis that $\beta = 0$. Its importance lies in the fact that it is equivalent to stating that a response does not depend on the value of the input; or in other words, there is no regression on the input value.

To derive a test of

$$H_0\colon \beta = 0 \qquad \text{against} \qquad H_1\colon \beta \neq 0$$

first it is necessary to study the distribution of $\hat{\beta}$, the estimator of β. That is, we will clearly want to reject H_0 when $\hat{\beta}$ is far from 0 and not to reject it otherwise. To determine how far away $\hat{\beta}$ need be from 0 to justify rejection of the null hypothesis, it is necessary to know something about its distribution.

It can be shown that $\hat{\beta}$ is normally distributed with mean and variance, respectively, given by

$$E[\hat{\beta}] = \beta$$

and

$$\text{Var}\,(\hat{\beta}) = \frac{\sigma^2}{S_{xx}}$$

Hence, the standard variable

$$\frac{\hat{\beta} - \beta}{\sqrt{\sigma^2/S_{xx}}} = \sqrt{\frac{S_{xx}}{\sigma^2}}\,(\hat{\beta} - \beta)$$

will have a standard normal distribution.

We cannot directly base a test on the preceding fact, however, since the standardized variable involves the unknown parameter σ^2. However, if we replace σ^2 by its estimator $\text{SS}_R/(n - 2)$, which is chi squared with $n - 2$ degrees of freedom, then it can be shown that the resulting quantity will now have a t distribution with $n - 2$ degrees of freedom. That is,

$$\sqrt{\frac{(n - 2)S_{xx}}{\text{SS}_R}}\,(\hat{\beta} - \beta)$$

has a t distribution with $n - 2$ degrees of freedom.

It follows from the preceding that if H_0 is true and so $\beta = 0$, then

$$\sqrt{\frac{(n-2)S_{xx}}{SS_R}}\,\hat{\beta}$$

has a t distribution with $n-2$ degrees of freedom. This gives rise to the following test of H_0.

A significance level γ test of H_0 is to

  Reject H_0 if $|TS| \geq t_{n-2,\ \gamma/2}$
  Not reject H_0 otherwise

where

$$TS = \sqrt{\frac{(n-2)S_{xx}}{SS_R}}\,\hat{\beta}$$

An equivalent way of performing the above test is to first compute the value of the test statistic TS; say, its value is v. The null hypothesis should then be rejected if the desired significance level γ is at least as large as the p value given by

$$p \text{ value} = P\{|T_{n-2}| \geq |v|\}$$
$$= 2P\{T_{n-2} \geq |v|\}$$

where T_{n-2} is a t random variable with $n-2$ degrees of freedom. Program 8-2 can be used to compute this latter probability.

Example 12.4   An individual claims that the fuel consumption of his automobile does not depend on how fast the car is driven. To test the plausibility of this hypothesis, the car was tested at various speeds between 45 and 75 miles per hour. The miles per gallon attained at each of these speeds were determined, with the following data resulting.

Speed	Miles per gallon
45	24.2
50	25.0
55	23.3
60	22.0
65	21.5
70	20.6
75	19.8

Do these data refute the claim that the mileage per gallon of gas is unaffected by the speed at which the car is being driven?

Solution Suppose that a simple linear regression model

$$Y = \alpha + \beta x + e$$

relates Y, the miles per gallon of the car, to x, the speed at which it is being driven. Now, the claim being made is that the regression coefficient β is equal to 0. To see if the data are strong enough to refute this claim, we need to see if they lead to a rejection of the null hypothesis in testing

$$H_0: \beta = 0 \qquad \text{against} \qquad H_1: \beta \neq 0$$

To compute the value of the test statistic, first we will compute the values of S_{xx}, S_{YY}, and S_{xY}. A hand calculation yields

$$S_{xx} = 700 \qquad S_{YY} = 21.757 \qquad S_{xY} = -119$$

The computational formula for SS_R presented at the end of Sec. 12.4 gives

$$SS_R = \frac{S_{xx} S_{YY} - S_{xY}^2}{S_{xx}}$$

$$= \frac{700(21.757) - 119^2}{700} = 1.527$$

Since

$$\hat{\beta} = \frac{S_{xY}}{S_{xx}} = \frac{-119}{700} = -.17$$

the value of the test statistic is

$$TS = \sqrt{\frac{5(700)}{1.527}} (-.17) = -8.139$$

Since from App. Table D.2 $t_{5,\ .005} = 4.032$, it follows that the hypothesis that $\beta = 0$ is rejected at the 1 percent level of significance. Thus, the claim that the mileage does not depend on the speed at which the car is driven is rejected. Indeed, there is clearly strong evidence that increased speeds lead to decreased efficiencies.

Problems

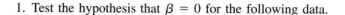

1. Test the hypothesis that $\beta = 0$ for the following data.

x	Y
3	7
8	8
10	6
13	7

   Use the 5 percent level of significance.

2. The following data set presents the heights of 12 male law school classmates whose law school examination scores were roughly equal. It also gives their annual salaries 5 years after graduation. Each went into corporate law. The height is in inches, and the salary is in units of $1000.

Height	Salary
64	91
65	94
66	88
67	103
69	77
70	96
72	105
72	88
74	122
74	102
75	90
76	114

   (a) Do the above data establish the hypothesis that a lawyer's salary is related to his height? Use the 5 percent level of significance.
   (b) What was the null hypothesis in part (a)?

3. The following table relates the number of sunspots that appeared each year from 1970 to 1980 to the number of automobile accident deaths during that year. The data for automobile accident deaths are in units of 1000 deaths.

Year	Sunspots	Automobile deaths
70	165	54.6
71	89	53.3
72	55	56.3
73	34	49.6
74	9	47.1
75	30	45.9
76	59	48.5
77	83	50.1
78	109	52.4
79	127	52.5
80	153	53.2

Test the hypothesis that the number of automobile accident deaths is not related to the number of sunspots. Use the 5 percent level of significance.

4. An electric utility wants to estimate the relationship between the daily summer temperature and the amount of electricity used by its customers. The following data were collected.

Temperature (degrees Fahrenheit)	Electricity (millions of kilowatts)
85	22.5
90	23.7
76	20.3
91	23.4
84	24.2
94	23.5
88	22.9
85	22.4
97	26.1
86	23.1
82	22.5
78	20.9
77	21.0
83	22.6

(a) Find the estimated regression line.
(b) Predict the electricity that will be consumed tomorrow if the predicted temperature for tomorrow is 93.
(c) Test the hypothesis, at the 5 percent level of significance, that the daily temperature has no effect on the amount of electricity consumed.

Problems 5 through 8 refer to the following data relating cigarette smoking and death rates for four types of cancers in 14 states. The data are based in part on records concerning 1960 cigarette tax receipts.

Cigarette Smoking and Cancer Death Rates

State	Cigarettes per person	Deaths per year per 100,000 people			
		Bladder cancer	Lung cancer	Kidney cancer	Leukemia
California	2860	4.46	22.07	2.66	7.06
Idaho	2010	3.08	13.58	2.46	6.62
Illinois	2791	4.75	22.80	2.95	7.27
Indiana	2618	4.09	20.30	2.81	7.00
Iowa	2212	4.23	16.59	2.90	7.69
Kansas	2184	2.91	16.84	2.88	7.42
Kentucky	2344	2.86	17.71	2.13	6.41
Massachusetts	2692	4.69	22.04	3.03	6.89
Minnesota	2206	3.72	14.20	3.54	8.28
New York	2914	5.30	25.02	3.10	7.23
Alaska	3034	3.46	25.88	4.32	4.90
Nevada	4240	6.54	23.03	2.85	6.67
Utah	1400	3.31	12.01	2.20	6.71
Texas	2257	3.21	20.74	2.69	7.02

5. (a) Draw a scatter diagram of cigarettes smoked versus death rate from bladder cancer.
   (b) Find the estimated regression line.
   (c) Test the hypothesis, at the 5 percent level of significance, that cigarette consumption does not affect the death rate from bladder cancer.
   (d) Repeat part (c) at the 1 percent level of significance.

6. (a) Draw a scatter diagram of cigarettes smoked versus death rate from lung cancer.
   (b) Find the estimated regression line.
   (c) Test the hypothesis, at the 5 percent level of significance, that cigarette consumption does not affect the death rate from lung cancer.
   (d) Repeat part (c) at the 1 percent level of significance.

7. (a) Draw a scatter diagram of cigarettes smoked versus death rate from kidney cancer.
   (b) Find the estimated regression line.
   (c) Test the hypothesis, at the 5 percent level of significance, that cigarette consumption does not affect the death rate from kidney cancer.
   (d) Repeat part (c) at the 1 percent level of significance.

8. (a) Draw a scatter diagram of cigarettes smoked versus death rate from leukemia.
   (b) Find the estimated regression line.
   (c) Test the hypothesis, at the 5 percent level of significance, that cigarette consumption does not affect the death rate from leukemia.
   (d) Repeat part (c) at the 1 percent level of significance.

9. In Prob. 3 of Sec. 12.3, test the null hypothesis that the amount of fire damage sustained by a property does not depend on its distance to the nearest fire station. Use the 5 percent level of significance.

10. The following are the average scores on the verbal part of the Scholastic Aptitude Test (SAT) in the years from 1984 to 1992.

Year	1984	1985	1986	1987	1988	1989	1990	1991	1992
Average SAT scores	426	431	431	430	428	427	424	422	423

Source: College Entrance Examination Board.

Assuming a simple linear regression model, test the hypothesis that the average verbal score is not changing over time. Use the 5 percent level of significance.

11. The following table gives the U.S. per capita consumption of bananas, apples, and oranges (in pounds) in seven different years.

Year	Bananas	Apples	Oranges	Year	Bananas	Apples	Oranges
1970	17.4	16.2	15.7	1983	21.2	17.6	15.6
1975	17.6	18.2	15.4	1985	23.4	16.6	12.0
1980	20.8	18.3	15.4	1987	24.9	20.3	13.9
1982	22.5	17.1	12.3				

Source: U.S. Department of Agriculture, *Food Consumption, Prices and Expenditures*.

Test the hypotheses that the yearly amount of bananas consumed is unrelated to the yearly amount of
(a) Apples consumed
(b) Oranges consumed
(c) Test the hypothesis that the yearly per capita amount of oranges consumed is unrelated to the yearly amount of apples consumed.

12. The following are the average scores on the mathematics part of the SAT in the years from 1984 to 1992.

Year	1984	1985	1986	1987	1988	1989	1990	1991	1992
Average score	471	475	475	476	476	476	476	474	476

Source: College Entrance Examination Board.

Assuming a simple linear regression model, test the hypothesis that the average mathematics score is not changing over time. Use the 5 percent level of significance.

12.6 REGRESSION TO THE MEAN

The term *regression* was originally employed by Francis Galton while describing the laws of inheritance. Galton believed that these laws caused population extremes to "regress towards the mean." By this he meant that children of individuals having extreme values of a certain characteristic would tend to have less extreme values of this characteristic than their parents.

If we assume a linear regression relationship between the characteristic of the offspring Y and that of the parent x then a regression to the mean will occur when the regression parameter β is between 0 and 1. That is, if

$$E[Y] = \alpha + \beta x$$

and $0 < \beta < 1$, then $E[Y]$ will be smaller than x when x is large and will be greater than x when x is small. That this statement is true can be easily checked either algebraically or by plotting the two straight lines

$$y = \alpha + \beta x$$

and

$$y = x$$

A plot indicates that when $0 < \beta < 1$, the line $y = \alpha + \beta x$ is above the line $y = x$ for small values of x and is below it for large values of x. Such a plot is given in Fig. 12.4.

Example 12.5 To illustrate Galton's thesis of regression to the mean, the British statistician Karl Pearson plotted the heights of 10 randomly chosen sons versus those of their fathers. The resulting data (in inches) were as follows.

Father's height	Son's height
60	63.6
62	65.2
64	66
65	65.5
66	66.9
67	67.1
68	67.4
70	68.3
72	70.1
74	70

A scatter diagram representing these data is presented in Fig. 12.5.

Note that whereas the data appear to indicate that taller fathers tend to have taller sons, they also appear to indicate that the sons of fathers who are either extremely short or extremely tall tend to be more "average" than their fathers; that is, there is a *regression toward the mean*.

We will determine whether the preceding data are strong enough to prove that there is a regression toward the mean by taking this statement as the alternative hypothesis. That is, we use the above data to test

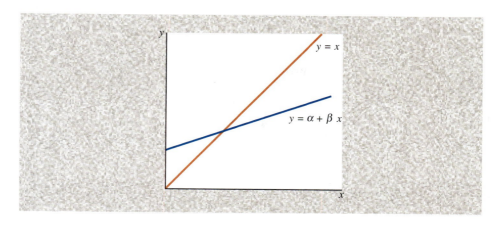

Figure 12.4
Regression to the mean occurs when $0 < \beta < 1$. For x small, $\alpha + \beta x > x$; for x large, $\alpha + \beta x < x$.

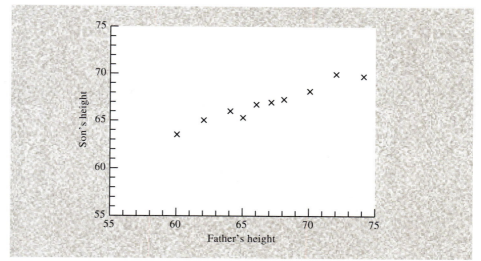

Figure 12.5
Scatter diagram of son's height versus father's height.

$$H_0: \beta \geq 1 \qquad \text{against} \qquad H_1: \beta < 1$$

Now the test of the above is equivalent to a test of

$$H_0: \beta = 1 \qquad \text{against} \qquad H_1: \beta < 1$$

and will be based on the fact that

$$\sqrt{\frac{(n-2)S_{xx}}{SS_R}}\,(\hat{\beta} - \beta)$$

has a t distribution with $n - 2$ degrees of freedom.

Hence, when $\beta = 1$, the test statistic

$$TS = \sqrt{\frac{8S_{xx}}{SS_R}} (\hat{\beta} - 1)$$

has a t distribution with 8 degrees of freedom. The significance level α test will be to reject H_0 when the value of TS is sufficiently small (since this will occur when $\hat{\beta}$, the estimator of β, is sufficiently smaller than 1). Specifically, the test is to

Reject H_0 if TS $\leq -t_{8, \alpha}$

Not reject H_0 otherwise

To determine the value of the test statistic TS, we run Program 12-1 and obtain the following:

The least-squares estimators are as follows

A = 35.97757
B = .4645573

The estimated regression line is
Y = 35.97757 + .4645573 x

S(x,Y) = 79.71875
S(x,x) = 171.6016
S(Y,Y) = 38.53125
SSR = 1.497325
The square root of (n−2)S(x,x)/SSR is 30.27942

From the preceding we see that

$$TS = 30.2794(.4646 - 1) = -16.21$$

Since $t_{8, .01} = 2.896$, we see that

$$TS < -t_{8, .01}$$

and so the null hypothesis that $\beta \geq 1$ is rejected at the 1 percent level of significance. In fact the p value is

$$p \text{ value} = P\{T_8 \leq -16.213\} \approx 0$$

and so the null hypothesis that $\beta \geq 1$ is rejected at almost any significance level, thus establishing a regression toward the mean.

A modern biological explanation for the regression to the mean phenomenon would roughly go along the lines of noting that since an offspring obtains a random selection of one-half of each parent's genes, it follows that the offspring of a very tall parent would, by chance, tend to have fewer "tall" genes than its parent.

While the most important applications of the regression to the mean phenomenon concern the relationship between the biological characteristics of an offspring and those of its parents, this phenomenon also arises in situations where we have two sets of data referring to the same variables. We illustrate it in Example 12.6.

Example 12.6   The following data relate the number of motor vehicle deaths occurring in 12 counties in the northwestern United States in the years 1988 and 1989.

County	Deaths in 1988	Deaths in 1989
1	121	104
2	96	91
3	85	101
4	113	110
5	102	117
6	118	108
7	90	96
8	84	102
9	107	114
10	112	96
11	95	88
12	101	106

The scatter diagram for this data set appears in Fig. 12.6. A glance at Fig. 12.6 indicates that in 1989 there was, for the most part, a reduction in the number of deaths in those counties that had a large number of motor vehicle deaths in 1988. Similarly, there appears to have been an increase in those counties that had a low value in 1988. Thus, we would expect that a regression to the mean is in effect. In fact, running Program 12-1 yields the estimated regression equation

$$y = 74.589 + .276x$$

which shows that the estimated value of β indeed appears to be less than 1.

One must be careful when considering the reason behind the regression to the mean phenomenon in the preceding data. For instance, it might be natural to suppose that those counties that had a large number of deaths caused by motor vehicles in 1988 would have made a large effort—perhaps by improving the safety of their roads or by making people more aware of the potential dangers of unsafe driving—to reduce this number. In addition, we might suppose that those counties that had the fewest number of deaths in 1988 might have "rested on their laurels" and not made much

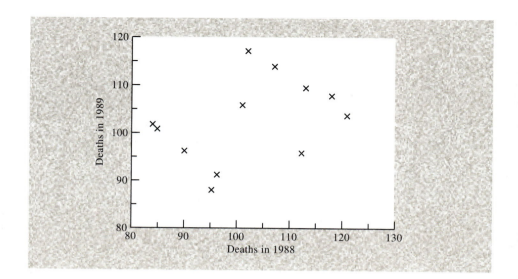

Figure 12.6

Scatter diagram of 1989 deaths versus 1988 deaths.

of an effort to further improve their numbers—and as a result had an increase in the number of casualties the following year.

While the above suppositions might be correct, it is important to realize that a regression to the mean would probably have occurred even if none of the counties had done anything out of the ordinary. Indeed, it could very well be the case that those counties having large numbers of casualties in 1988 were just very unlucky in that year, and thus a decrease in the next year was just a return to a more normal result for them. (For an analogy, if 9 heads result when 10 fair coins are flipped, then it is quite likely that another flip of these 10 coins will result in fewer than 9 heads.) Similarly, those counties having few deaths in 1988 might have been "lucky" that year, and a more normal result in 1989 would thus lead to an increase.

The mistaken belief that regression to the mean is due to some outside influence, when it is in reality just due to "chance," is heard frequently enough that it is often referred to as the *regression fallacy*.

Regression to the mean plays a key role in the explanation of why so many communal biological data sets from a homogeneous population tend to have "normal curve" histograms. For instance, if one plotted the heights of all senior girls in a specified high school, then it is a good bet that the resulting histogram would strongly resemble the bell-shaped normal curve. One possible explanation for this combines the central limit theorem, regression to the mean, and the passing of many generations. We now sketch it.

*12.6.1 Why Biological Data Sets Are Often Normally Distributed

We will present this argument in the context of considering the heights of females in a population. We will follow this population of females over many generations. Suppose that there are initially k women, whom we will refer to as the *initial generation*, and their heights are $x_1, \ldots, x_k$. These k values are considered to be totally arbitrary. Let d denote the largest minus the smallest of these values. For instance, if

$$k = 3 \qquad x_1 = 60 \qquad x_2 = 58 \qquad x_3 = 66$$

then $d = 66 - 58 = 8$.

If Y denotes the height of a female child of a woman whose height is x, then we will assume the linear regression model

$$Y = \alpha + \beta x + e$$

In the above model we will make the usual assumption that e is an error random variable that is normally distributed with mean 0 and variance σ^2. However, whereas this assumption is often made without any real attempt at justification, it seems quite reasonable in the above application because of the central limit theorem. That is, the height of a daughter of a woman of height x can be thought of as being composed of the sum of a large number of approximately independent random variables that relate, among other things, to the random set of genes that she receives as well as to environmental factors. Hence, by the central limit theorem, her height should be approximately normally distributed. We will also assume that regression to the mean is in effect, that is, that $0 < \beta < 1$.

Thus, the heights of the daughters of the k women of the initial population are all normally distributed. However, it is important to note that their mean heights are all different. For instance, a daughter of the woman of height x_1 will have a normally distributed height with mean value $\alpha + \beta x_1$, whereas the daughter of the woman of height x_2 will have a different mean height, namely, $\alpha + \beta x_2$. Thus, the heights of all the daughters do *not* come from the *same* normal distribution, and for that reason a plot of all their heights would not follow the normal curve.

However, if we now consider the difference between the largest and the smallest *mean* height of all the daughters of the initial set of women, then it is not difficult to show that

$$\text{Difference} \leq \beta d$$

(If each of the initial set of women had at least one daughter, then the above inequality would be an equality.) If we now consider the daughters of these daughters, then it can be shown that their heights will be normally distributed with differing means and a common variance. The difference between the largest and the smallest mean height of these second-generation daughters can be shown to satisfy

$$\text{Difference} \leq \beta^2 d$$

Indeed, if we suppose that more and more generations have passed and we consider the women of the nth generation after the initial population, then it can be shown that the heights of the women in this generation are normally distributed with the same variance and with mean values which, while differing, are such that the difference between the largest and smallest of them satisfies

$$\text{Difference} \leq \beta^n d$$

Now, since $0 < \beta < 1$, it follows that as n grows larger, $\beta^n d$ gets closer and closer to 0. Thus, after a large enough number of generations have passed, all the women in the population will have normally distributed heights with approximately the same mean and with a common variance. That is, after many generations have passed, the heights of the women will come from approximately the same normal population, and thus at this point a plot of these heights will approximately follow the bell-shaped normal curve.

Problems

1. The following data come from an experiment performed by Francis Galton. The data relate the diameter of an offspring seed to that of its parent seed in the case of a self-fertilized seed.

Diameter of parent seed	Diameter of offspring seed
15	15.3
16	16.4
17	15.5
18	16.2
19	16.0
20	17.4
21	17.5

   (a) Estimate the regression parameters.
   (b) Does there appear to be a regression to the mean?

2. In Example 12.6 it was shown that the estimated value of β is less than 1. Using the data of this example, test the hypothesis

$$H_0: \beta \geq 1 \qquad \text{against} \qquad H_1: \beta < 1$$

   Would H_0 be rejected at the 5 percent level of significance?

3. Would you be surprised if the following data sets exhibited a regression to the mean? Would you expect them to exhibit this phenomenon? Explain your answers.
   (a) You go to 10 different restaurants which you know nothing about in advance. You eat a meal in each one and give a numerical ranking—anywhere from 0 to 100—to the quality of the meal. You then return to each of these restaurants and again give a ranking to the meal. The data consist of the two scores of each of the 10 restaurants.

(b) At the beginning of an hour, 12 individuals check their pulse rates to determine the number of heartbeats per minute. Call these the x values. After 1 hour, they repeat this to obtain the y values.

(c) The set considers paired data concerning different mutual funds. For each mutual fund, the x variable is the 1995 ranking of the fund, and the corresponding y variable is the 1996 ranking.

(d) The data consist of the paired scores of 20 first-year preschoolers with the first value in the pair being the student's test score on an IQ examination given to all entering students and the second value being the same student's score on an IQ test given at the end of the first month in school.

4. Test

$$H_0: \beta = 1 \qquad \text{against} \qquad H_1: \beta < 1$$

for the following set of data. Use the 5 percent level of significance.

x	y
24	27
21	24
26	20
17	22
15	21
24	20
23	17

5. The following data give the 1988 and 1989 birth rates per 1000 population for certain midwestern states.

State	1988	1989
Iowa	13.5	13.1
Indiana	14.7	14.8
Michigan	15.1	15.4
Minnesota	15.5	15.3
Illinois	15.9	16.0
Idaho	15.7	15.2
Kentucky	13.7	14.1
Kansas	15.5	14.2
Missouri	14.9	15.5

Source: Department of
Health and Human
Services.

(a) Find the estimated regression line.
(b) Does it indicate a regression to the mean?

6. Figure 12.7 presents a histogram of the heights of 8585 men. How well does it appear to fit a normal curve?

7. Figure 12.8 presents a histogram of the weights of 7738 men. How well do these data fit a normal curve?

12.7 PREDICTION INTERVALS FOR FUTURE RESPONSES

Suppose, in the linear regression model, that input values x_i have led to the response values y_i, $i = 1, \ldots, n$. The best prediction of the value of a new response at input x_0 is, of course, $\hat{\alpha} + \hat{\beta}x_0$. However, rather than give a single number as the predicted value, it is often more useful to be able to present an interval which you predict, with a certain degree of confidence, will contain the response value. Such a *prediction interval* is given by the following.

Prediction interval for a response at input value x_0, based on the response values y_i at the input values x_i, $i = 1, \ldots, n$:

With $100(1 - \gamma)$ degree confidence, the response Y at the input value x_0 will lie in the interval

$$\hat{\alpha} + \hat{\beta}x_0 \pm t_{n-2,\, \gamma/2}W$$

where $t_{n-2,\, \gamma/2}$ is the $100(1 - \gamma/2)$ percentile of the t distribution with $n-2$ degrees of freedom, and

$$W = \sqrt{\left[1 + \frac{1}{n} + \frac{(x_0 - \bar{x})^2}{S_{xx}}\right] \frac{\text{SS}_R}{n - 2}}$$

The quantities $\hat{\alpha}$, $\hat{\beta}$, $\bar{x}$, S_{xx}, and SS_R are all computed from the data x_i, y_i, $i = 1, \ldots, n$.

Example 12.7   Using the data of Example 12.6, specify an interval which, with 95 percent confidence, will contain the adult height of a newborn son whose father is 70 inches tall.

Solution   From the output of Program 12-1, we obtain

$$\hat{\alpha} + 70\hat{\beta} = 68.497$$
$$W = .4659$$

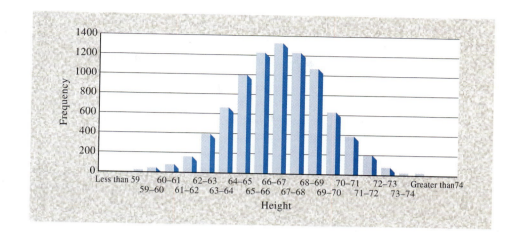

Figure 12.7
A histogram of heights.

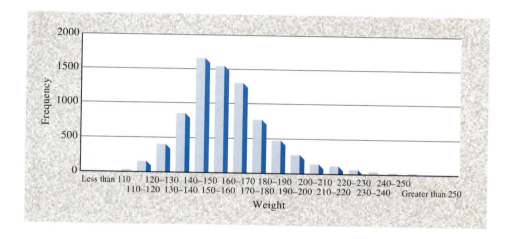

Figure 12.8
A histogram of weights.

Since from Table D.2, $t_{8, .025} = 2.306$, we see that the 95 percent prediction interval of the height of the son of a 70-inch-tall man is

$$68.497 \pm 2.306(.4659) = 68.497 \pm 1.074$$

That is, we can be 95 percent confident that the son's height will be between 67.423 and 69.571 inches.

Example 12.8   A company that runs a hamburger concession at a college football stadium must decide on Monday how much to order for the game that is to be played on the following Saturday. The company bases its order on the number of tickets for the game that have already been sold by Monday. The following data give the advance ticket sales and the number of hamburgers purchased for each game played this year. All data are in units of 1000.

Advance ticket sales	Hamburgers sold
29.4	19.5
21.4	16.2
18.0	15.3
25.2	18.0
32.5	20.4
23.9	16.8

If 26,000 tickets have been sold by Monday for next Saturday's game, determine a 95 percent prediction interval for the amount of hamburgers that will be sold.

Solution Running Program 12-1 gives the following output, if we request predicted future responses and the value of the input is 26.

The predicted response is 18.04578.

W = .3381453

Since $t_{4, .025} = 2.776$, we see from the output that the 95 percent prediction interval is

$$18.046 \pm 2.776(.338) = 18.046 \pm .938$$

That is, with 95 percent confidence, between 17,108 and 18,984 hamburgers will be sold.

Problems

1. Use the data below to
   (a) Predict the response at the input value $x = 4$.
   (b) Determine an interval that contains, with 95 percent confidence, the response in part (a).

x	y
1	5
2	8
5	15

2. An official of a large automobile manufacturing firm wanted to study the relationship between a worker's age and his or her level of absenteeism. The following data concerning 10 randomly chosen employees were collected.

Age	40	28	34	27	21	38	19	55	31	35
Days missed	1	6	6	9	12	4	13	2	5	3

(a) Predict the number of days missed by a worker aged 42.
(b) Determine a 95 percent prediction interval for the quantity in part (a).

3. The following data were recently reported by an economist who wanted to learn about the relationship between a family's income and the proportion of that income spent on food. Each of the families consisted of a married couple with two teenage children.

Income (in $1000)	Percentage spent on food
14	35
18	33
22	32
28	29
35	23
39	19
42	17

(a) Find the estimated regression line.
(b) Predict the amount of money spent on food by a family of size 4 that earns $31,000 annually.
(c) Determine a 95 percent confidence prediction interval for the amount in (b).
(d) Repeat part (c), but this time obtain a prediction interval having 99 percent confidence.

4. The following data relate the scores of 10 students on a college entrance examination to their grade-point average at the end of their first year.

Entrance examination score	Grade-point average
88	3.2
74	2.7
70	2.3
77	2.9
83	2.8
94	3.6
92	3.0
81	2.8
85	3.3
92	3.1

(a) Predict the grade-point average of a student, not listed above, who scored 88 on the entrance examination.

(b) Obtain a 90 percent prediction interval for the score of the student described in part (a).

(c) Test the hypothesis, at the 5 percent level of significance, that a student's grade-point average is independent of her or his score on the entrance examination.

5. Glass plays an important role in criminal investigations, because criminal activity often results in the breakage of windows and other glass objects. Since glass fragments often lodge in the clothing of criminals, it is important to be able to identify such fragments as having come from the crime scene. Two physical properties of glass that are useful for identification purposes are its refractive index, which is relatively easy to measure, and its density, which is much more difficult to measure. The measurement of density is, however, greatly facilitated when one has a good estimate of this value before setting up the laboratory experiment needed to determine it exactly. Thus, it would be quite useful if one could use the refractive index of a glass fragment to estimate its density.

The following data relate the refractive index to the density for 12 selected pieces of glass.

Refractive index	Density	Refractive index	Density
1.514	2.480	1.516	2.484
1.515	2.482	1.517	2.486
1.516	2.480	1.518	2.495
1.517	2.490	1.519	2.498
1.517	2.482	1.522	2.511
1.520	2.505	1.525	2.520

(a) Predict the density of a fragment of glass whose refractive index is 1.520.

(b) Determine an interval which, with 95 percent confidence, will contain the density of a fragment of glass whose refractive index is 1.520.

6. The following summary data relate to the ages of puberty of 20 mother-daughter pairs. The x data refer to the mother's and the Y data to her daughter's age at puberty.

$$\bar{x} = 12.8 \qquad \bar{Y} = 12.9$$
$$S_{xx} = 36.5 \qquad S_{YY} = 42.4 \qquad S_{xY} = 24.4$$

(a) Find the estimated regression line.

(b) Use the computational formula given at the end of Sec. 12.4 to compute SS_R.

(c) Test, at the 5 percent level of significance, the hypothesis that $\beta = 0$.

(d) If a mother reached puberty at age 13.8, determine an interval which, with 95 percent confidence, will contain the age at which her daughter reaches puberty.

7. The following data relate the grade-point average (GPA) in accounting courses to starting annual salary of eight recent accounting graduates.

Accounting GPA	Starting salary (in $1000)
3.4	42
2.5	29
3.0	33
2.8	32
3.7	40
3.5	44
2.7	30
3.1	35

(a) Predict the annual salary of a recent graduate whose grade-point average in accounting courses was 2.9.
(b) Determine an interval which, with 95 percent confidence, will contain the annual salary in part (a).
(c) Repeat parts (a) and (b) for a graduate having a 3.6 GPA.

8. The following data give the average price of all books reviewed in the journal *Science* from 1987 to 1992.

Year	Price ($)
1987	47.37
1988	54.05
1989	54.58
1990	54.43
1991	54.08
1992	57.58

Give an interval which, with 95 percent confidence, will contain the average price of all books reviewed in *Science* in 1993.

12.8 COEFFICIENT OF DETERMINATION

Suppose we want to measure the amount of variation in the set of response values Y_1, ..., Y_n corresponding to the set of input values $x_1, \ldots, x_n$. A standard measure in statistics of the amount of variation in a set of values $Y_1, \ldots, Y_n$ is given by the quantity

$$S_{YY} = \sum_{i=1}^{n} (Y_i - \bar{Y})^2$$

For instance, if all the Y_i are equal—and thus are all equal to $\bar{Y}$—then S_{YY} will equal 0.

The variation in the values of the Y_i arises from two factors. First, since the input values x_i are different, the response variables Y_i all have different mean values, which will result in some variation in their values. Second, the variation also arises from the fact that even when the difference in the input values is taken into account, each of the response variables Y_i has variance σ^2 and thus will not exactly equal the predicted value at its input x_i.

Let us consider now the question of how much of the variation in the values of the response variables is due to the different input values and how much is due to the inherent variance of the responses even when the input values are taken into account. To answer this question, note that the quantity

$$SS_R = \sum_{i=1}^{n} (Y_i - \hat{\alpha} - \hat{\beta}x_i)^2$$

measures the remaining amount of variation in the response values after the different input values have been taken into account. Thus,

$$S_{YY} - SS_R$$

represents the amount of variation in the response variables that is *explained* by the different input values; and so the quantity R^2 defined by

$$R^2 = \frac{S_{YY} - SS_R}{S_{YY}}$$
$$= 1 - \frac{SS_R}{S_{YY}}$$

represents the proportion of variation in the response variables that is explained by the different input values.

Definition
The quantity R^2 is called the *coefficient of determination*.

The coefficient of determination R^2 will have a value between 0 and 1. A value of R^2 near 1 indicates that most of the variation of the response data is explained by the different input values, whereas a value of R^2 near 0 indicates that little of the variation is explained by the different input values.

Example 12.9   In Example 12.5, which relates the height of a son to that of his father, the output from Program 12-1 yielded

$$S_{YY} = 38.521 \qquad SS_R = 1.497$$

Thus,

$$R^2 = 1 - \frac{1.497}{38.531} = .961$$

In other words, 96 percent of the variation of the heights of the 10 individuals is explained by the heights of their fathers. The remaining (unexplained) 4 percent of the variation is due to the variance of a son's height even when the father's height is taken into account. (That is, it is due to σ^2, the variance of the error random variable.)

The value of R^2 is often used as an indicator of how well the regression model fits the data, with a value near 1 indicating a good fit and one near 0 indicating a poor fit. In other words, if the regression model is able to explain most of the variation in the response data, then it is considered to fit the data well.

Example 12.10   In Example 12.8, which relates the number of hamburgers sold at a football game to the advance ticket sales for that game, Program 12-1 yielded

$$S_{YY} = 19.440 \qquad SS_R = .390$$

Thus,

$$R^2 = 1 - \frac{.390}{19.440} = .98$$

and so 98 percent of the variation in the different amounts of hamburgers sold in the six games is explained by the advance ticket sales for these games. (Loosely put, 98 percent of the amount sold is explained by the advance ticket sales.)

Problems

1. A real estate brokerage gathered the following information relating the selling prices of three-bedroom homes in a particular neighborhood to the sizes of these homes. (The square footage data are in units of 1000 square feet whereas the selling price data are in units of $1000.)

Square footage	Selling price
2.3	140
1.8	112
2.6	153
3.0	180
2.4	148
2.3	132
2.7	160

(a) Plot the above in a scatter diagram.

(b) Determine the estimated regression line.

(c) What percentage of the selling price is explained by the square footage?

(d) A house of size 2600 square feet has just come on the market. Determine an interval in which, with 95 percent confidence, the selling price of this house will lie.

2. Determine R^2 for the following data set:

x	y
2	10
3	16
5	22

3. It is difficult and time-consuming to directly measure the amount of protein in a liver sample. As a result, medical laboratories often make use of the fact that the amount of protein is related to the amount of light that would be absorbed by the sample. As a result, a spectrometer that emits light is shined upon a solution that contains the liver sample, and the amount of light absorbed is then used to estimate the amount of protein.

The above procedure was tried on five samples having known amounts of protein, with the following data resulting:

Light absorbed	Amount of protein (mg)
.44	2
.82	16
1.20	30
1.61	46
1.83	55

(a) Determine the coefficient of determination.

(b) Does this appear to be a reasonable way of estimating the amount of protein in a liver sample?

(c) What is the estimate of the amount of protein when the light absorbed is 1.5?

(d) Determine a prediction interval in which we can have 90 percent confidence for the quantity in part (c).

4. Determine the coefficient of determination for the data of Prob. 1 of Sec. 12.2.

5. Determine the coefficient of determination for the data of Example 12.6.

6. A new-car dealer is interested in the relationship between the number of salespeople working on a weekend and the number of cars sold. Data were gathered for six consecutive Sundays:

Number of salespeople	Number of cars sold
5	22
7	20
4	15
2	9
4	17
8	25

(a) Determine the estimated regression line.
(b) What is the coefficient of determination?
(c) How much of the variation in the number of automobiles sold is explained by the number of salespeople?
(d) Test the null hypothesis that the mean number of sales does not depend on the number of salespeople working.

7. Find the coefficient of determination in Prob. 8 of Sec. 12.4.

12.9 SAMPLE CORRELATION COEFFICIENT

Consider a set of data pairs (x_i, Y_i), $i = 1, \ldots, n$. In Sec. 3.7 we defined the *sample correlation coefficient* of this data set by

$$r = \frac{\sum\limits_{i=1}^{n} (x_i - \bar{x})(Y_i - \bar{Y})}{\sqrt{\sum\limits_{i=1}^{n} (x_i - \bar{x})^2 \sum\limits_{i=1}^{n} (Y_i - \bar{Y})^2}}$$

It was noted that r provided a measure of the degree to which high values of x are paired with high values of Y and low values of x with low values of Y. A value of r near $+1$ indicated that large x values were strongly associated with large Y values and small x values were strongly associated with small Y values, whereas a value near -1 indicated that large x values were strongly associated with small Y values and small x values with large Y values.

In the notation of this chapter, r would be expressed as

$$r = \frac{S_{xY}}{\sqrt{S_{xx}S_{YY}}}$$

Upon using the identity

$$SS_R = \frac{S_{xx}S_{YY} - S_{xY}^2}{S_{xx}}$$

that was presented at the end of Sec. 12.4, we can show that the absolute value of the sample correlation coefficient r can be expressed as

$$|r| = \sqrt{1 - \frac{SS_R}{S_{YY}}}$$

That is,

$$|r| = \sqrt{R^2}$$

and so, except for its sign indicating whether it is positive or negative, the sample correlation coefficient is equal to the square foot of the coefficient of determination. The sign of r is the same as that of $\hat{\beta}$.

The above gives additional meaning to the sample correlation coefficient. For instance, if a data set has its sample correlation coefficient r equal to .9, then this implies that a simple linear regression model for these data explains 81 percent (since $R^2 = .9^2 = .81$) of the variation in the response values. That is, 81 percent of the variation in the response values is explained by the different input values.

Problems

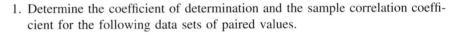

1. Determine the coefficient of determination and the sample correlation coefficient for the following data sets of paired values.

(a)

x	y
2	4
3	5
5	9

(b)

x	y
4	2
5	3
9	5

What does this lead you to conclude?

*2. Show that the sample correlation coefficient of a given set of data pairs (u_i, v_i) is the same regardless of whether the u_i are considered to be the input values or the response values.

3. Find the sample correlation coefficient when the coefficient of determination and the estimated regression line are
(a) $R^2 = .64$, $y = 2x + 4$
(b) $R^2 = .64$, $y = 2x - 4$
(c) $R^2 = .64$, $y = -2x + .4$
(d) $R^2 = .64$, $y = -2x - .4$

4. If the sample correlation coefficient is .95, how much of the variation in the responses is explained by the different input values?

5. The following data relate the ages of wives and husbands when they were married. Before you look at the data, would you expect a positive, negative, or near-zero value of the sample correlation coefficient?

Wife's age	18	24	40	33	30	25
Husband's age	21	29	51	30	36	25

(a) Letting the wife's age be the input, find the estimated regression line for determining the husband's age.

(b) Letting the husband's age be the input, find the estimated regression line for determining the wife's age.

(c) Determine the coefficient of determination and the sample correlation coefficient for the situation described in part (a).

(d) Determine the coefficient of determination and the sample correlation coefficient for the situation described in part (b).

6. Find the sample correlation coefficient in Prob. 6 of Sec. 12.7.

12.10 MULTIPLE LINEAR REGRESSION MODEL

Up to now we have been concerned with predicting the value of a response on the basis of the value of a single input variable. However, in many situations the response is dependent on a multitude of input variables.

Example 12.11   In laboratory experiments two factors that often affect the percentage yield of the experiment are the temperature and the pressure at which the experiment is conducted. The following data detail the results of four independent experiments. For each experiment, we have the temperature (in degrees Fahrenheit) at which the experiment is run, the pressure (in pounds per square inch), and the percentage yield.

Experiment	Temperature	Pressure	Percentage yield
1	140	210	68
2	150	220	82
3	160	210	74
4	130	230	80

Suppose that we are interested in predicting the response value Y on the basis of the values of the k input variables $x_1, x_2, \ldots, x_k$.

Definition

The *multiple linear regression* model supposes that the response Y is related to the input values $x_i, i = 1, \ldots, k$, through the relationship

$$Y = \beta_0 + \beta_1 x_1 + \beta_2 x_2 + \cdots + \beta_k x_k + e$$

In the above expression β_0, β_1, ..., β_k are *regression parameters,* and e is an *error* random variable that has mean 0. The regression parameters will not be initially known and must be estimated from a set of data.

Suppose that we have at our disposal a set of n responses corresponding to n different sets of the k input values. Let y_i denote the ith response, and let the k input values corresponding to this response be x_{i1}, x_{i2}, ..., x_{ik}, $i = 1$, ..., n. Thus, for instance, y_1 was the response when the k input values were x_{11}, x_{12}, ..., x_{1k}. The data set is presented in Fig. 12.9.

Set	Input 1	Input 2	$\cdots$	Input k	Response
1	x_{11}	x_{12}	$\cdots$	x_{1k}	y_1
2	x_{21}	x_{22}	$\cdots$	x_{2k}	y_2
3	x_{31}	x_{32}	$\cdots$	x_{3k}	y_3
$\vdots$					
n	x_{n1}	x_{n2}	$\cdots$	x_{nk}	y_n

Figure 12.9

Data on n experiments.

Example 12.12   In Example 12.11 there are two input variables, the temperature and the pressure, and so $k = 2$. There are four experimental results, and so $n = 4$. The value x_{i1} refers to the temperature and x_{i2} to the pressure of experiment i. The value y_i is the percentage yield (response) of experiment i. Thus, for instance,

$$x_{31} = 160 \qquad x_{32} = 210 \qquad y_3 = 74$$

To estimate the regression parameters again, as in the case of simple linear regression, we use the method of least squares. That is, we start by noting that if B_0, B_1, ..., B_k are estimators of the regression parameters β_0, β_1, ..., β_k, then the estimate of the response when the input values are x_{i1}, x_{i2}, ..., x_{ik} is given by

$$\text{Estimated response} = B_0 + B_1 x_{i1} + B_2 x_{i2} + \cdots + B_k x_{ik}$$

Since the actual response was y_i, we see that the difference between the actual response and what would have been predicted if we had used the estimators B_0, B_1, ..., B_k is

$$\epsilon_i = y_i - (B_0 + B_1 x_{i1} + B_2 x_{i2} + \cdots + B_k x_{ik})$$

Thus, ϵ_i can be regarded as the *error* that would have resulted if we had used the estimators B_i, $i = 0$, ..., k. The estimators that make the sum of the squares of the errors as small as possible are called the *least-squares estimators.*

The least-squares estimators of the regression parameters are the choices of B_i that make

$$\sum_{i=1}^{n} \epsilon_i^2$$

as small as possible.

The actual computations needed to obtain the least-squares estimators are algebraically messy and will not be presented here. Instead we refer to Program 12-2 to do the computations for us. The outputs of this program are the estimates of the regression parameters. In addition, the program provides predicted response values for specified sets of input values. That is, if the user enters the values $x_1, x_2, \ldots, x_k$, then the computer will print out the value of $B(0) + B(1)x_1 + \cdots + B(k)x_k$, where $B(0)$, $B(1), \ldots, B(k)$ are the least-squares estimators of the regression parameters.

Example 12.13   The following data relate the suicide rate y to the population size x_1 and the yearly divorce rate x_2 in eight different cities.

Location	Population (thousands)	Divorce rate per 100,000	Suicide rate per 100,000
Akron, OH	679	30.4	11.6
Anaheim, CA	1420	34.1	16.1
Buffalo, NY	1349	17.2	9.3
Austin, TX	296	26.8	9.1
Chicago, IL	3975	29.1	8.4
Columbia, SC	323	18.7	7.7
Detroit, MI	2200	32.6	11.3
Gary, IN	633	32.5	8.4

(a) Fit a multiple regression model to these data. That is, fit a model of the form

$$Y = \beta_0 + \beta_1 x_1 + \beta_2 x_2 + e$$

where Y is the suicide rate, x_1 is the population, and x_2 is the divorce rate.

(b) Predict the suicide rate in a county having a population of 400,000 people and a divorce rate of 28.4 divorces yearly for every 1000 people.

Solution Running Program 12-2 gives the following output.

The estimates of the regression coefficients are as follows

Historical Perspective

Method of Least Squares

The first publication detailing the method of least squares was due to the French mathematical scientist Adrien Marie Legendre in 1805. Legendre presented the method in the appendix to his book *Nouvelles methodes pour la determination des orbites des cometes (New Methods for Determining the Orbits of Comets)*. After explaining the method, Legendre worked out an example, using data from the 1795 survey of the French meridian arc, an example in which $k = 2$ and $n = 5$. In 1809 Karl Friedrich Gauss published a justification of the method of least squares which highlighted the normal as the distribution of the error term. In his paper Gauss started a controversy by claiming that he had been using the method since 1795. Gauss claimed that he had used the method of least squares in 1801 to locate the missing asteroid Ceres. This asteroid, the largest in the solar system and the first to be discovered, was spotted by the Italian astronomer Giuseppe Piazzi of the Palerno Observatory on January 1, 1801. Piazzi observed it for 40 consecutive days at which time the asteroid, which had a very low luminos-

ity, disappeared from view. In the hope that other scientists would be able to determine its path, Piazzi published the data concerning his observations. Months later the news and data reached the attention of Gauss. In a short time, and without any explanation of his method, Gauss published a predicted orbit for the asteroid. Shortly afterward, Ceres was found in almost the exact position predicted by Gauss.

The ensuing priority dispute between Legendre and Gauss became rather heated. In the 1820 edition of his book, Legendre added an attack on Gauss that he attributed to the anonymous writer Monsieur ***. Gauss, in turn, solicited testimony from colleagues to the effect that he had told them of his method before 1805. Present-day scholars for the most part accept Gauss' claim that he knew and used the method of least squares before Legendre. (Gauss is famous for often letting many years go by before publishing his results.) However, most scholars also feel that priority should be determined by the earliest date of publication and so the credit for the discovery of the method of least squares rightfully belongs to Legendre.

B(0)= 3.686646
B(1)=−2.411092E−04
B(2)= .2485504

If the two input values are 400 and 28.4, the predicted response is 10.64903.

That is, the estimated multiple regression equation is

$$Y = 3.6866 - .00024x_1 + .24855x_2$$

The predicted suicide rate is

$$y = 3.6866 - .00024 \times 400 + .24855 \times 28.4$$
$$= 10.649$$

That is, we predict that in such a county the yearly suicide rate is 10.649 per 100,000 residents. Since the population size is 400,000, this means a prediction of 42.596 suicides per year.

Problems

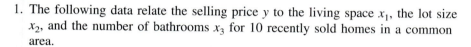

1. The following data relate the selling price y to the living space x_1, the lot size x_2, and the number of bathrooms x_3 for 10 recently sold homes in a common area.

Selling price (thousands of dollars)	House size (square feet)	Lot size (acres)	Number of bathrooms
70	1300	.25	1
77	1450	.30	1.5
91	1600	.30	2
94	1850	.45	2
102	2100	.40	2
110	2000	.40	2.5
114	2100	.50	2
128	2400	.50	2.5
140	2700	.50	2.5
152	2600	.70	3

(a) Fit a multiple linear regression model to the above.
(b) Predict the selling price of a home of 2500 square feet whose lot size is .4 acre and which has two bathrooms.
(c) What if the house in part (b) had three bathrooms?

2. In Example 12.11, predict the yield of an experiment run at a temperature of 150 degrees Fahrenheit and with a pressure of 215 pounds per square inch.

3. Fit a multiple linear regression model to the following data set:

x_1	x_2	x_3	x_4	y
1	3	5	9	121
2	4	4	10	165
1.5	8	2	14	150
3	9	3	8	170
1	11	4	12	140

Predict the value of a response taken at the input values

$$x_1 = 2 \qquad x_2 = 7 \qquad x_3 = 3 \qquad x_4 = 13$$

4. The following data set refers to Stanford heart transplants. It relates the survival time of patients after receiving a heart transplant to their age and to a mismatch score that is used as an indicator of how well the transplanted heart should match the recipient.

Survival time (days)	Mismatch score	Age
624	1.32	51.0
1350	.87	54.1
64	1.89	54.6
46	.61	42.5
1024	1.13	43.4
280	1.12	49.5
10	2.76	55.3
60	.69	64.5
836	1.58	45.0
136	1.62	52.0
730	.96	58.4
39	1.38	42.8

(a) Fit a multiple linear regression model to the above.
(b) Estimate the survival time of a 50-year-old heart transplant patient whose mismatch score is 1.46.

5. A steel company will be producing cold-reduced sheet steel consisting of .15 percent copper and produced at an annealing temperature of 1150 degrees Fahrenheit. The company is interested in estimating the mean hardness of this steel. It collected the following data on 10 different specimens of sheet steel that had been produced at different copper contents and annealing temperatures.

Hardness	Copper content	Annealing temperature
79.2	.02	1050
64.0	.03	1200
55.7	.03	1250
56.3	.04	1300
58.6	.10	1300
49.8	.09	1450
51.1	.12	1400
61.0	.09	1200
70.4	.15	1100
84.3	.16	1000

Estimate the mean hardness of the steel to be produced.

Key Terms

Simple linear regression: A model that relates a response variable Y to an input variable x by the equation

$$Y = \alpha + \beta x + e$$

The quantities α and β are parameters of the regression model, and e is an error random variable.

Dependent variable: Another term for the response variable.

Independent variable: Another term for the input variable.

Method of least squares: A method for obtaining estimators of the regression parameters α and β. It chooses as estimators those values which make the sum of the squares of the differences between the observed and the predicted responses as small as possible.

Regression to the mean: This phenomenon occurs when the regression parameter β is strictly between 0 and 1. It results in the mean response corresponding to the input level x being larger than x when x is small and being smaller than x when x is large. This phenomenon is common in testing-retesting situations.

Regression fallacy: The belief in testing-retesting situations that the regression to the mean phenomenon has a significant cause when it is actually just a by-product of random fluctuations.

Coefficient of determination: A statistic whose value indicates the proportion of the variation in the response values that is caused by the different input values.

Sample correlation coefficient: Its absolute value is the square root of the coefficient of determination. Its sign is the same as that of the estimator of the regression parameter β.

Multiple linear regression: A model that relates a response variable Y to a set of k input variables $x_1, \ldots, x_k$ by the equation

$$Y = \beta_0 + \beta_1 x_1 + \beta_2 x_2 + \cdots + \beta_k x_k + e$$

Summary

The simple linear regression model relates the value of a *response* random variable Y to the value of an *input* variable x by the equation

$$Y = \alpha + \beta x + e$$

The parameters α and β are *regression parameters* that have to be estimated from data. The quantity e is an *error* random variable that has expected value 0.

The *method of least squares* is used to estimate the regression parameters α and β. Suppose that experiments are run at the input levels x_i, $i = 1, \ldots, n$. Let Y_i, $i = 1, \ldots, n$, denote the corresponding outputs. The least-squares approach is to choose as estimators of α and β the values of A and B that make

$$\sum_{i=1}^{n} (Y_i - A - Bx_i)^2$$

as small as possible. The values of A and B that accomplish this—call these values $\hat{\alpha}$ and $\hat{\beta}$—are given by

$$\hat{\beta} = \frac{S_{xY}}{S_{xx}}$$

$$\hat{\alpha} = \bar{Y} - \hat{\beta}\bar{x}$$

where $\bar{x}$ and $\bar{Y}$ are the average values of the x_i's and the Y_i's, respectively, and

$$S_{xY} = \sum_{i=1}^{n} (x_i - \bar{x})(Y_i - \bar{Y}) = \sum_{i=1}^{n} x_i Y_i - n\bar{x}\bar{Y}$$

$$S_{xx} = \sum_{i=1}^{n} (x_i - \bar{x})^2 = \sum_{i=1}^{n} x_i^2 - n\bar{x}^2$$

The straight-line relationship

$$y = \hat{\alpha} + \hat{\beta}x$$

is called the *estimated regression line*.

The error random variable e is assumed to be a normal random variable with expected value 0 and variance σ^2, where σ^2 is unknown and needs to be estimated from the data. The estimator of σ^2 is

$$\frac{SS_R}{n-2}$$

where the quantity SS_R is called the *sum of the squares of the residuals* and is defined by

$$SS_R = \sum_{i=1}^{n} (Y_i - \hat{\alpha} - \hat{\beta}x_i)^2$$

The quantities $Y_i - \hat{\alpha} - \hat{\beta}x_i$, representing the difference between the actual response and its predicted value under the least-squares estimators, are called the *residuals*.

The following is a useful computational formula for finding SS_R when you are using a hand calculator.

$$SS_R = \frac{S_{xx}S_{YY} - S_{xY}^2}{S_{YY}}$$

where

$$S_{YY} = \sum_{i=1}^{n} (Y_i - \bar{Y})^2$$

If the regression parameter β is equal to 0, then the value of a response will not be affected by its input value x. To see if this hypothesis is plausible, we test

$$H_0: \beta = 0 \quad \text{against} \quad H_1: \beta \neq 0$$

The significance level γ test is based on the test statistic

$$TS = \sqrt{\frac{(n-2)S_{xx}}{SS_R}} \, \hat{\beta}$$

and is to

Reject H_0 if $|TS| \geq t_{n-2,\,\gamma/2}$
Not reject H_0 otherwise

Equivalently, if the value of TS is v, then the p value is given by

$$p \text{ value} = 2P\{T_{n-2} \geq |v|\}$$

where T_{n-2} is a t random variable with $n - 2$ degrees of freedom.

The phenomenon of *regression to the mean* is said to occur when the regression parameter β lies between 0 and 1. When this is the case, the expected response corresponding to the input value x will be greater than x when x is small and will be less than x when x is large.

The regression to the mean phenomenon is often seen in testing-retesting situations involving a homogeneous population. This is because some of those being tested will, purely by chance, do significantly better or worse than is their norm. In the repeated test they will often obtain a more normal result. Thus, those scoring high on the first test often come down somewhat on the second while those scoring low on the first test often improve on the second. The belief that something significant has caused the regression to the mean phenomenon (for instance, that the lower-scoring students studied much harder for the retest while the higher-scoring ones were complacent) when in fact it was just due to random fluctuations about the mean value is called the *regression fallacy*.

The input-response data pairs (x_i, y_i), $i = 1, \ldots, n$, can be used to provide a *prediction interval* which, with a prescribed degree of confidence, will contain a future response at the input value x_0. Specifically, we can assert, with $100(1 - \gamma)$ percent confidence, that the response at the input value x_0 will lie in the interval

$$\hat{\alpha} + \hat{\beta} x_0 \pm t_{n-2, \, \gamma/2} W$$

where

$$W = \sqrt{\left[1 + \frac{1}{n} + \frac{(x_0 - \bar{x})^2}{S_{xx}}\right] \frac{SS_R}{n - 2}}$$

The quantities $\hat{\alpha}$, $\hat{\beta}$, $\bar{x}$, S_{xx}, and W are all based on the data pairs (x_i, y_i), $i = 1$, $\ldots, n$, and can be obtained by running Program 12-1.

The quantity R^2 defined by

$$R^2 = 1 - \frac{SS_R}{S_{YY}}$$

is called the *coefficient of determination.* Its value, which will always lie between 0 and 1, can be interpreted as the proportion of the variation in the response values that is explained by the different input values.

The quantity r, defined by

$$r = \frac{S_{xY}}{\sqrt{S_{xx} S_{YY}}}$$

is called the *sample correlation coefficient.* Aside from its sign (either positive or negative) it is equal to the square root of the coefficient of determination. That is,

$$|r| = \sqrt{R^2}$$

The *multiple linear regression* model relates a response random variable Y to a set of input variables $x_1, \ldots, x_k$ according to the equation

$$Y = \beta_0 + \beta_1 x_1 + \beta_2 x_3 + \cdots + \beta_k x_k + e$$

In this equation, $\beta_0, \beta_1, \ldots, \beta_k$ are regression parameters, and e is an error random variable having mean 0.

The regression parameters are estimated from data by using the method of least squares. That is, the estimators are chosen to minimize the sum of the squares of the differences between the actual observed response values and their predicted values. Program 12-2 can be used to obtain these estimates. This program will also return predicted values of responses corresponding to arbitrarily entered input values.

Review Problems

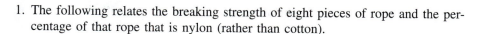

1. The following relates the breaking strength of eight pieces of rope and the percentage of that rope that is nylon (rather than cotton).

Percentage nylon	Breaking strength (pounds)
0	160
10	240
20	325
20	340
30	395
40	450
50	510
50	520

(a) Plot the above in a scatter diagram.
(b) Give the estimated regression line.
(c) Predict the breaking strength of a new piece of rope that is 50 percent nylon.
(d) Give an interval that, with 95 percent confidence, will contain the breaking strength of a piece of rope that is 50 percent nylon.

2. It is generally accepted that by increasing the number of units it produces, a manufacturer can often decrease its cost per unit. The following relates the manufacturing cost per unit to the number of units produced.

Number of units	10	20	50	100	150	200
Cost per unit	9.4	9.2	9.0	8.5	8.1	7.4

(a) Predict the cost per unit when a production run of 125 units is called for.
(b) Estimate the variance of the cost in part (a).
(c) Give an interval which, with 99 percent confidence, will contain the cost per unit when a production run of 110 units is used.

3. Use the data relating to the first 20 women on the data set given in App. A. Let the input variable be the weight and the response variable be the systolic blood pressure.
(a) Estimate the regression parameters.
(b) Give a 95 percent prediction interval for the systolic blood pressure of a female student who weighs 120 pounds.
(c) Find all female students in App. A that weigh between 119 and 121 pounds. What percentage of them have systolic blood pressures that fall within the interval given in (c)?

4. A set of 10 married couples are randomly chosen from a given community, and the 20 individuals are given an IQ test. Number the couples, and let x_i and y_i denote the score of the wife and of the husband of couple i. Do you think that a plot of the resulting scatter diagram will indicate a regression to the mean? Explain.

5. Experienced flight instructors have claimed that praise for an exceptionally fine landing is typically followed by a poorer landing on the next attempt; whereas criticism of a faulty landing is typically followed by an improved landing. Should we thus conclude that verbal praise tends to lower performance levels whereas verbal criticism tends to raise them? Or is some other explanation possible?

6. The following data relate the average number of cigarettes smoked daily to the number of free radicals found in the lungs of eight individuals.

Number of cigarettes	Free radicals
0	94
10	144
14	182
5	120
18	240
20	234
30	321
40	400

(a) Represent this data set in a scatter diagram.
(b) Fit a straight line to the data "by hand."
(c) Determine the estimated regression line, and compare it to the one drawn in part (b).
(d) Predict the amount of free radicals in someone who smokes an average of 26 cigarettes daily.
(e) Determine a prediction interval which, with 95 percent confidence, will contain the amount of free radicals in an individual who smokes an average of 26 cigarettes daily.

7. The following data give the U.S. per capita amount of purchased fresh vegetables, excluding potatoes, in each year from 1980 to 1987. The data are in pounds.

Year	1980	1981	1982	1983	1984	1985	1986	1987
Amount	72.8	71.5	74.2	74.7	78.8	78.8	79.9	78.6

Source: U.S. Department of Agriculture.

(a) Find the estimated regression line.
(b) Test the hypothesis, at the 5 percent level of significance, that $\beta = 0$.

8. The following are the average scores of college-bound high school students on the science and reasoning section of the American College Testing (ACT) examination in the years from 1987 through 1992, excluding 1990.

Year	1987	1988	1989	1991	1992
Average ACT scores	21.4	21.4	21.2	20.7	20.7

Source: *High Schools Profile Report,* ACT Program, Iowa City, IA.

(a) Predict the 1990 average score.
(b) Find a 95 percent prediction interval for that score.

9. The following table gives the percentage of workers in manufacturing who were union members in both 1984 and 1989 for a random sample of nine states.

State	1984	1989
Alabama	27.3	23.8
Colorado	10.9	9.5
Illinois	40.9	29.8
Kentucky	27.0	21.5
Minnesota	25.7	16.4
New Jersey	25.4	24.4
Texas	15.9	13.8
Wisconsin	31.1	23.0
New York	50.4	47.2

Source: *Manufacturing Climates Study,* Grant/Thornton, Chicago, annual.

(a) The percentage of Ohio's manufacturing workers who were union members was 41.6 in 1984. Predict the 1989 percentage.
(b) Oklahoma's union membership percentage was 17.5 in 1984. Construct an interval that, with 95 percent confidence, contains Oklahoma's membership percentage in 1989.

10. The following data give the average price of all technical books in the natural sciences that were reviewed in the journal *Science* in each year from 1987 to 1992.

Year	Price ($)
1987	59.06
1988	71.70
1989	73.73
1990	75.57
1991	73.19
1992	76.78

Give an interval which, with 95 percent confidence, will contain the average price of all books reviewed in *Science* in 1994.

11. The tensile strength of a certain synthetic fiber is thought to be related to the percentage of cotton in the fiber and to the drying time of the fiber. A study of eight pieces of fiber yielded the following results.

Percentage of cotton	Drying time	Tensile strength
13	2.1	212
15	2.2	221
18	2.5	230
20	2.4	219
18	3.2	245
20	3.3	238
17	4.1	243
18	4.3	242

(a) Fit a multiple regression equation with tensile strength being the response and the percentage of cotton and the drying time being the input variables.
(b) Predict the tensile strength of a synthetic fiber having 22 percent cotton whose drying time is 3.5.

12. The following data refer to the seasonal wheat yield per acre at eight different locations, all having roughly the same quality soil. The data relate the wheat yield at each location to the seasonal amount of rainfall and the amount of fertilizer used per acre.

Rainfall (inches)	Fertilizer (pounds per acre)	Wheat yield
15.4	100	46.6
18.2	85	45.7
17.6	95	50.4
18.4	140	66.5
24.0	150	82.1
25.2	100	63.7
30.3	120	75.8
31.0	80	58.9

(a) Estimate the regression parameters.
(b) Estimate the additional yield in wheat for each additional inch of rain.
(c) Estimate the additional yield in wheat for each additional pound of fertilizer.
(d) Predict the wheat yield in a year having 26 inches of rain if the amount of fertilizer used that year was 130 pounds per acre.

13. A recently completed study attempted to relate job satisfaction to income and seniority for a random sample of nine municipal workers. The job satisfaction

value given for each worker is his or her own assessment of such, with a score of 1 being the lowest and 10 being the highest. The following data resulted.

Yearly income (thousands of dollars)	Years on the job	Job satisfaction
27	8	5.6
22	4	6.3
34	12	6.8
28	9	6.7
36	16	7.0
39	14	7.7
33	10	7.0
42	15	8.0
46	22	7.8

(a) Estimate the regression parameters.

(b) What qualitative conclusions can you draw about how job satisfaction changes when income remains fixed and the number of years of service increases?

(c) Predict the job satisfaction of an employee who has spent 5 years on the job and earns a yearly salary of $31,000.

14. Suppose in Prob. 13 that job satisfaction was related to years on the job, and so the following data would have resulted:

Years on the job	Job satisfaction	Years on the job	Job satisfaction
8	5.6	14	7.7
4	6.3	10	7.0
12	6.8	15	8.0
9	6.7	22	7.8
16	7.0		

(a) Estimate the regression parameters α and β.

(b) What is the qualitative relationship between years of service and job satisfaction? That is, based on the above, what appears to happen to job satisfaction as service increases?

(c) Compare your answer to part (b) to the answer you obtained in part (b) of Prob. 13.

(d) What conclusion, if any, can you draw from your answer in part (c)?

15. The correct answer to Prob. 5 of Sec. 12.5 is to reject the hypothesis that cigarette consumption and bladder cancer rates are unrelated. Does this imply that cigarette smoking directly leads to an increased risk of contracting bladder cancer or can you think of another explanation? (*Hint*: Is there another variable you can think of that is statistically associated both with smoking and bladder cancer? What type of data collection and statistical procedure would you recommend to increase our knowledge about the factors affecting bladder cancer rates).

MINITAB LAB

Linear Regression

Purpose

Use Minitab to

1. Discuss properties of the simple linear regression model.

2. Discuss the multiple linear regression model.

Procedures

First, load the Minitab (Windows version) software as in the Minitab lab for Chap. 1.

1. SIMPLE LINEAR REGRESSION MODEL

Review Exercise   Use Minitab to work Prob. 2 in Sec. 12.2 of the text. Refer to the Minitab lab in Chap. 2.

Example 1   Use Minitab to work Prob. 12 in Sec. 12.3 of the text.

Parts (a) and (b) can be done by using the procedures in the Minitab lab for Chap. 2. The lab for Chap. 2 includes the procedures for a scatter diagram and a fitted line. Next, to determine the estimated regression line, enter the x values in column C1 and the y values in column C2. Rename these columns X and Y, respectively. Note that the x values represent the amount of alcohol in the blood (%), and the y values are the reaction time (seconds). Select Stat→Regression→Regression, and the Regression dialog box will be displayed. All we will use at this point in the Regression dialog box are the Response and Predictors text boxes. The response variable here is Y (C2), and the predictor variable is X (C1). So select X and Y for the appropriate text boxes.

Click on the OK button, and the results will be displayed in the Session window, as shown in Fig. M12.1. This output gives several results. However, at this point we are only interested in the estimated regression line. From this output, the estimated regression line (regression equation) is given by $Y = .0993 + 2.66X$. Thus the slope of the line is 2.66, and the intercept is .0993.

```
┌──────────────────────────────────────────────────────────────────┐
│ ▭                          Session                        ▼ ▲   ▲ │
│ Regression Analysis                                            ▲  │
│                                                                   │
│                                                                   │
│ The regression equation is                                        │
│ Y = .0993 + 2.66X                                                 │
│                                                                   │
│ Predictor        Coef       Stdev      t-ratio         p          │
│ Constant        .09926     .06026         1.65      .160          │
│ X               2.6615      .4407         6.04      .002          │
│                                                                   │
│ s = .03776      R-sq = 87.9%      R-sq(adj) = 85.5%               │
│                                                                   │
│ Analysis of Variance                                              │
│                                                                   │
│ SOURCE         DF          SS          MS          F        p     │
│ Regression      1      .052013     .052013     36.47     .002     │
│ Error           5      .007130     .001426                        │
│ Total           6      .059143                                ▼  │
│ ◄  ▯                                                          ►  │
└──────────────────────────────────────────────────────────────────┘
```

Figure M12.1

The predicted reaction time for a blood alcohol content of .15 will be $Y = .0993 + 2.66$ (.15) $= .4983$, and for .17 it will be .5515.

Now find the estimate for the variance of the response variable Y. Recall that this is the same as the variance for the error random variable e.

The estimate of the variance σ^2 is computed from $SS_R/(n - 2)$. That is, it is computed from the ratio of the residual (error) sum of squares and the degrees of freedom for this sum of squares. From Fig. M12.1 this value is given as the mean square (MS) for the error term in the analysis of variance portion of the output. Thus, the estimate for σ^2 is .001426.

Suppose now that we want to determine whether the response (reaction time) does not depend on the value of the input (blood alcohol content). That is, we are asked to test for no regression on the input value. This is equivalent to testing the hypothesis H_0: $\beta = 0$. Minitab does this test automatically for you. In Fig. M12.1, observe that just below the regression equation, the standard deviations (Stdev), t ratios, and p values (p) are computed for the estimators (predictors) of α (constant) and β (X). The p value for this test is .002; thus we can reject the hypothesis H_0: $\beta = 0$ since the p value is rather small. If we were testing at the 5 percent (.05) level of significance, the conclusion would be the same since the p value is smaller than the significance level. Thus we can conclude that there is a regression on the input value. Observe that the p value for the hypothesis H_0: $\alpha = 0$ is .160. Now determine the 95 percent prediction interval for a .17 amount of alcohol in the blood.

Select Stat→Regression→Regression, and enter the X and Y variables in the appropriate text boxes in the Regression dialog box. The Regression options dialog box with the appropriate entries is shown in Fig. M12.2. Next click on the Options button, and the Regression options dialog box will be displayed. In the Prediction intervals for new observations text box, type .17. Type in 95 in the Confidence level text box, and select the Prediction limits box. The Regression options dialog box with the appropriate entries is shown in Fig. M12.2.

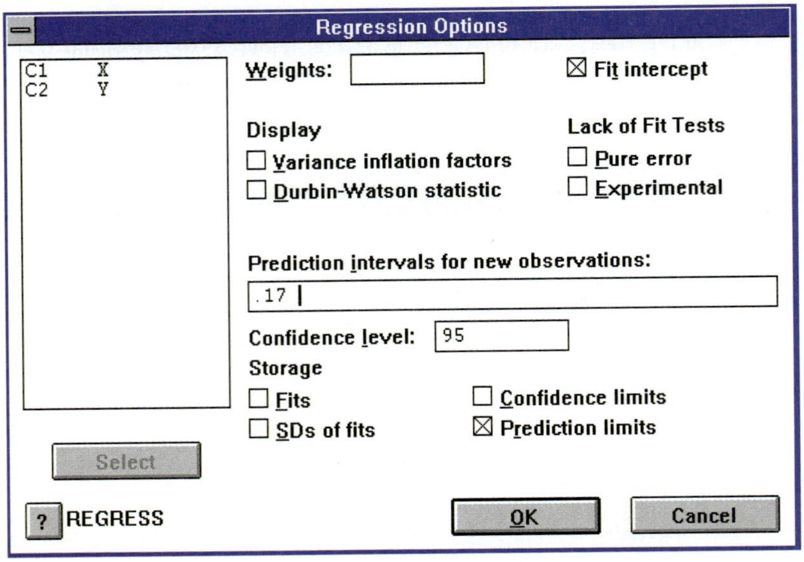

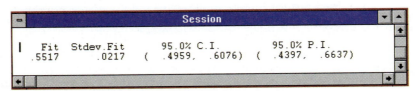

Figure M12.2

Click twice on the OK button, and the prediction interval (P.I.) will be computed and displayed in the Session window. A portion of this window is shown in Fig. M12.3. Note that the 95 percent prediction interval is (.4397, .6637). That is, we can be 95 percent confident that a person with a .17 amount of alcohol in the blood will have a reaction time between .4397 and .6637 second.

```
  ─                         Session                        ▼ ▲
                                                              ↑
│   Fit   Stdev.Fit       95.0% C.I.          95.0% P.I.
  .5517       .0217    (  .4959,   .6076)  (  .4397,   .6637)
                                                              ↓
  ←  ←                                                    →  →
```

Figure M12.3

Now compute R^2, the coefficient of determination. The coefficient of determination is computed automatically by Minitab when you use the Stat→Regression→Regression procedure initially. From Fig. M12.1, $R^2 = 87.9$ percent. That is, approximately 88 percent of the variation of the reaction time of the seven individuals is explained by the amount of alcohol in the blood.

Now we compute r, the sample correlation coefficient. This is obtained by taking the square root of R^2. You need to determine the sign for the square root. This can be obtained by observing the slope of the regression line. From Fig. M12.1, the slope is 2.66, which is positive. Thus r will be positive. That is, $r = \sqrt{.879} = .9375$. This implies that there is a strong positive correlation between the reaction time and amount of alcohol in the blood for these seven individuals.

2. MULTIPLE LINEAR REGRESSION MODEL

Example 2   Use Minitab to work Example 12.13 in Sec. 12.10.

Enter the values in columns C1 through C3. Rename C1, and call it POP for population. Rename C2, and call it DRATE for divorce rate. Rename C3, and call it SRATE for suicide rate. Select Stat→Regression→Regression, and in the Regression dialog box select SRATE for the Response text box; select POP and DRATE for the Predictors text box.

Click on the OK button, and the Session window will display the results, as shown in Fig. M12.4.

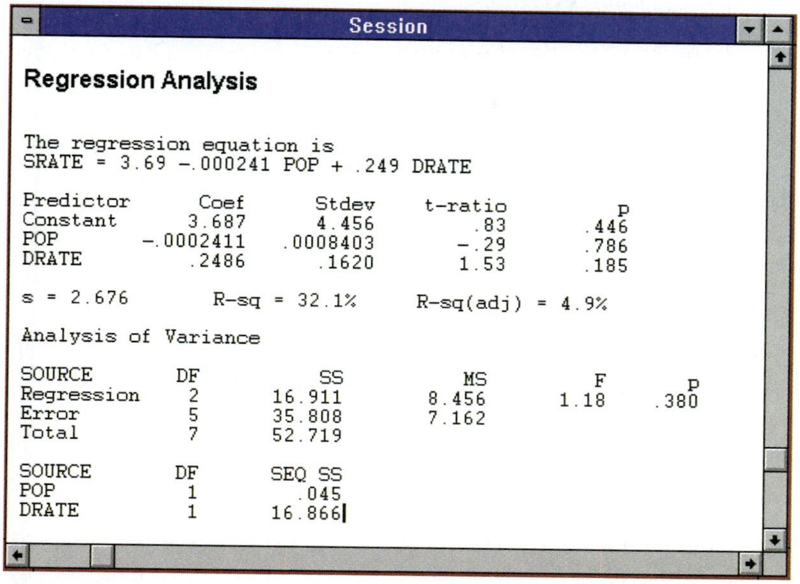

```
┌─────────────────────────────────────────────────────────────────────────┐
│ ⊖                                Session                            ▼  ▲ │
├─────────────────────────────────────────────────────────────────────────┤
│                                                                       ▲ │
│  Regression Analysis                                                    │
│                                                                         │
│  The regression equation is                                            │
│  SRATE = 3.69 −.000241 POP + .249 DRATE                                 │
│                                                                         │
│  Predictor       Coef      Stdev     t-ratio          p                 │
│  Constant        3.687      4.456        .83        .446                 │
│  POP          −.0002411    .0008403     −.29        .786                 │
│  DRATE           .2486      .1620       1.53        .185                 │
│                                                                         │
│  s = 2.676        R-sq = 32.1%      R-sq(adj) = 4.9%                     │
│                                                                         │
│  Analysis of Variance                                                   │
│                                                                         │
│  SOURCE          DF         SS          MS        F       p             │
│  Regression       2      16.911      8.456     1.18    .380             │
│  Error            5      35.808      7.162                              │
│  Total            7      52.719                                         │
│                                                                         │
│  SOURCE          DF      SEQ SS                                         │
│  POP              1        .045                                         │
│  DRATE            1      16.866                                      ▼  │
│  ◄                                                                   ►  │
└─────────────────────────────────────────────────────────────────────────┘
```

Figure M12.4

Computer Exercises

1. Use Minitab or any other statistical software to work Probs. 1 to 5 in Sec. 12.2.

2. Use Minitab or any other statistical software to work Probs. 1, 3, 5, 8 and 11 in Sec. 12.3.

3. Use Minitab or any other statistical software to work Probs. 5 and 6 in Sec. 12.4.

4. Use Minitab or any other statistical software to work Probs. 5 to 8 in Sec. 12.5.

5. Use Minitab or any other statistical software to work Probs. 1 and 5 in Sec. 12.6.

6. Use Minitab or any other statistical software to work Probs. 3, 5, and 7 in Sec. 12.7.

7. Use Minitab or any other statistical software to work Probs. 1, 3, and 6 in Sec. 12.8.

8. Use Minitab or any other statistical software to work Probs. 1 and 5 in Sec. 12.9.

9. Use Minitab or any other statistical software to work Probs. 1 and 4 in Sec. 12.10.

10. From the Session window in Fig. M12.4, write up appropriate hypothesis tests for the parameters associated with the variables POP and DRATE. Discuss the reliability of the model.

11. Use your library or any other resource to collect a data set that could be modeled by a simple linear regression model. Use Minitab or any other statistical software to do a complete analysis of the data. Present any relevant graphs, analyses, and discussion in a report. If you select data from published research, include relevant factors about the validity and reliability of the study.

12. Use your library or any other resource to collect a data set that could be modeled by a multiple linear regression model. Use Minitab or any other statistical software to do a complete analysis of the data. Present any relevant graphs, analyses, and discussion in a report. If you select data from published research, include relevant factors about the validity and reliability of the study.

13. Design your own experiment, and collect data that could be modeled by simple linear regression. Use Minitab or any other statistical software to present your findings in a report.

14. Design your own experiment, and collect data that could be modeled by multiple linear regression. Use Minitab or any other statistical software to present your findings in a report.

CHAPTER 13

Chi-Squared Goodness
of Fit Tests

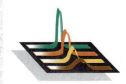

We consider a population in which each member has any one of k possible values. We show how to test the hypothesis that a specified set of probabilities represents the proportions of the members of the population that have each of the different possible values. We consider populations in which each member is classified as having two values, and we show how to test the hypothesis that the two values of a randomly selected member of the population are independent.

13.1   INTRODUCTION

The manipulation of data to make them conform to a particular scientific hypothesis is considered to be an instance of scientific fraud. There have been many such cases of scientific fraud over the years, ranging in severity from slight "fudging" to outright falsification of data. For instance, one of the most egregious examples involved the British educational psychologist Cyril Burt. Burt was highly regarded in his lifetime—indeed, he was eventually knighted by the Queen of England and became Sir Cyril—for his research on the IQs of identical twins who were raised apart. However, it is now widely accepted that in his published work he invented not only the data he published but also the very existence of his supposed research subjects and collaborators.

Perhaps the most puzzling instance of scientific fraud involves the Austrian monk Gregor Mendel (1822–1884), who is regarded as the founder of the theory of genetics. In 1865 Mendel published a paper outlining the results of a series of experiments carried out on garden peas. One of the experiments was concerned with the color—either yellow or green—of the seeds of such peas. Mendel began his experiment by breeding peas of pure yellow strain, which is a strain of peas in which every plant in every generation has only yellow seeds. He also bred a pure green strain. Mendel then crossed peas of the pure yellow strain with those of the pure green strain. The result of this crossing, known as *first-generation hybrid* seeds, was always a yellow seed. That is, there were no green seeds in this generation.

Mendel then crossed these first-generation seeds with themselves, to obtain second-generation seeds. Surprisingly, green seeds reappeared in this generation. In fact, approximately 25 percent of second-generation seeds were green, and 75 percent were yellow.

In his paper, Mendel presented a theory to explain the above results. His theory supposed that each seed contained two entities, which we now call *genes,* which together determine the color of the seed. Each gene is one of two types: type y (for yellow) or type g (for green). Mendel's theory was that the pair of genes in the pure yellow strain seeds is always y,y. That is, both genes in a pea from the pure yellow strain are yellow. Similarly, the pair of genes in seeds from the pure green strain are g,g. Mendel now supposed that when two seeds are crossed, the resulting offspring obtains one gene from each parent. In addition, Mendel supposed that the gene obtained from a parent is equally likely to be either one of the two genes of that parent. Thus, when a pure y,y yellow seed is matched with a pure g,g green seed, the offspring will necessarily have one y and one g seed; that is, the offspring will have the gene pair y,g. Since every offspring resulting from a cross of a pure yellow and a pure green seed was itself yellow, Mendel postulated that the y gene was dominant over the g gene in that a seed having the gene pair y,g would be yellow. See Fig. 13.1.

Consider now what happens when two first-generation seeds are crossed. First note that both seeds are hybrids having the gene pair y,g. Also note that in order for the offspring seed to be green, it must receive the g gene from each parent. Since each parent is equally likely to contribute either its y or its g gene, it follows that the probability that both parents contribute their g genes is $1/2 \times 1/2 = 1/4$. Thus, the result of a large number of crossings of first-generation seeds should be that approximately

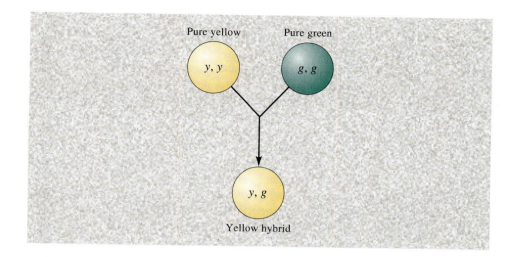

Figure 13.1
Crossing pure
yellow with
pure green
seeds.

25 percent of the next-generation seeds are green. This is exactly the result reported by Mendel. See Fig. 13.2.

Mendel's discovery is first-rate science of the highest order. Although it took some time (as his work was ignored for almost 30 years), his theory of genetics has become a cornerstone of basic science.

It is not difficult to imagine the consternation among geneticists when, in 1936, R. A. Fisher published a paper that analyzed Mendel's data and concluded that they fit the theory too well to be explained by chance. Using the chi-squared goodness of fit test that had been developed by Karl Pearson, Fisher showed that the probability that an overall data fit was at least as good as the one reported by Mendel would have occurred with a probability equal to .00004.

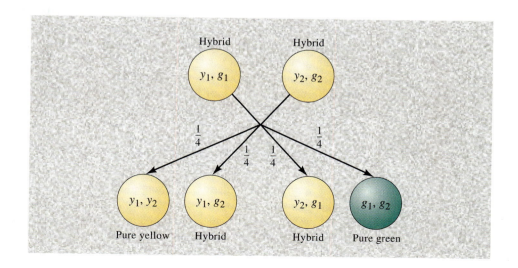

Figure 13.2
Crossing hybrid
first-generation
seeds.

For instance, Mendel reported that of 8023 second-generation peas, 6022 were yellow and 2001 were green. That is, the fraction of the second-generation peas that was green was 2001/8023 = .2494, which is almost exactly equal to the theoretical probability .25. Although such a good fit in itself is not that unlikely (at least as good a fit as this would occur roughly 10 percent of the time), the trouble was that almost all the experiments reported by Mendel resulted in data which were in unusually close agreement with the theoretical probability. By combining the results of all the experiments reported by Mendel, Fisher came up with the (p value) probability of .00004.

Although believing that Mendel's data had been manipulated, Fisher apparently exonerated Mendel himself from direct blame. Indeed, Fisher has gone on record as believing that the data were probably manipulated by an assistant who knew the results that Mendel expected. (Of course, equally plausible is that Mendel himself made mistakes in recording the data. Even honest people can see what is not there, when they believe it should be.)

In Sec. 13.2 we present the chi-squared goodness of fit test, which can be used to determine how well a given data set fits a particular probability model. Its use enables us to test the validity of the probability model.

In Sec. 13.3 we consider populations in which each member is classified according to two distinct characteristics. We show how the goodness of fit test can be used to test the hypothesis that the two characteristics of a randomly chosen member of the population are independent.

The two characteristics of members of a population will be independent if knowledge of one of the characteristics of a randomly chosen member of the population does not affect the probabilities of the other characteristic of this member. Whereas in Sec. 13.3 we suppose that the data result from a random sample of the entire population, in Sec. 13.4 we consider a different type of sampling scheme. This new scheme starts by focusing attention on one of the characteristics. It determines the various possible values of this characteristic and then chooses random samples from the subpopulations of members having each of the possible values. For instance, if one of the characteristics is gender, then rather than choose a random sample from the entire population, as is done in Sec. 13.3, we now choose random samples from the subpopulations of men and women. A test for independence is presented when this type of sampling scheme is used. In addition, we show how the results of Sec. 13.4 can be used to test the hypothesis that an arbitrary number of population proportions are equal. In the special case of two populations, the test is identical to the one presented in Sec. 10.6.

13.2 CHI-SQUARED GOODNESS OF FIT TESTS

Consider a very large population, and suppose that each member of the population has a value that can be 1, or 2, or 3, . . . , or k. For a given set of probabilities p_i, $i = 1, . . . , k$, we will consider the problem of testing the null hypothesis that p_i represents, for each i, the proportion of the population that has value i. That is, if we let P_i denote the true proportion of the population that has value i, for $i = 1, . . . , k$, then we are interested in testing

$$H_0: P_1 = p_1, P_2 = p_2, \ldots, P_k = p_k$$

against the alternative hypothesis

$$H_1: P_i \neq p_i \qquad \text{for some } i, i = 1, \ldots, k$$

Example 13.1   It is known that 41 percent of the U.S. population has type A blood, 9 percent has type B, 4 percent has type AB, and 46 percent has type O. Suppose that we suspect that the blood type distribution of people suffering from stomach cancer is different from that of the overall population.

To verify that the blood type distribution is different for those suffering from stomach cancer, we could test the null hypothesis

$$H_0: P_1 = .41, P_2 = .09, P_3 = .04, P_4 = .46$$

where P_1 is the proportion of all those with stomach cancer who have type A blood, P_2 is the proportion of those who have type B blood, P_3 is the proportion who have type AB blood, and P_4 is the proportion who have type O blood. A rejection of H_0 would then enable us to conclude that the blood type distribution is indeed different for those suffering from stomach cancer.

In the preceding scenario, each member of the population of individuals who are suffering from stomach cancer is given one of four possible values according to his or her blood type. We are interested in testing the hypothesis that $P_1 = .41$, $P_2 = .09$, $P_3 = .04$, $P_4 = .46$ represent the proportions of this population having each of the different values.

To test the null hypothesis that $P_i = p_i$, $i = 1, \ldots, k$, first we need to draw a random sample of elements from the population. Suppose this sample is of size n. Let N_i denote the number of elements of the sample that have value i, for $i = 1, \ldots, k$. Now if the null hypothesis is true, then each element of the sample will have value i with probability p_i. Also, since the population is assumed to be very large, it follows that the successive values of the members of the sample will be independent. Thus, if the null hypothesis is true, then N_i will have the same distribution as the number of successes in n independent trials, when each trial is a success with probability p_i. That is, if H_0 is true, then N_i will be a binomial random variable with parameters n and p_i. Since the expected value of a binomial is the product of its parameters, we see that when H_0 is true,

$$E[N_i] = np_i \qquad i = 1, \ldots, k$$

For each i, let e_i denote this expected number of outcomes that equal i when H_0 is true. That is,

$$e_i = np_i$$

Thus, when H_0 is true, we expect that N_i would be relatively close to e_i. That is, when the null hypothesis is true, the quantity $(N_i - e_i)^2$ should not be too large, say, in relation to e_i. Since this is true for each value of i, a reasonable way of testing H_0 would be to compute the value of the test statistic

$$\text{TS} = \sum_{i=1}^{k} \frac{(N_i - e_i)^2}{e_i}$$

and then reject H_0 when TS is sufficiently large.

To determine how large TS need be to justify rejection of the null hypothesis, we use a result that was proved by Karl Pearson in 1900. This result states that for large values of n, TS will approximately have a chi-squared distribution with $k - 1$ degrees of freedom. Let $\chi^2_{k-1,\,\alpha}$ denote the $100(1 - \alpha)$ percentile of this distribution; that is, a chi-squared random variable having $k - 1$ degrees of freedom will exceed this value with probability α (see Fig. 13.3). Then the approximate significance level α test of the null hypothesis H_0 against the alternative H_1 is as follows:

Reject H_0 if TS $\geq \chi^2_{k-1,\,\alpha}$

Do not reject H_0 otherwise

The preceding is called the *chi-squared goodness of fit test*. For reasonably large values of n, it results in an hypothesis test of H_0 whose significance level is approximately equal to α. An accepted rule of thumb is that this approximation will be quite good provided n is large enough so that $e_i \geq 1$ for each i and at least 80 percent of the values e_i exceed 5.

Values of $\chi^2_{m,\,\alpha}$ for various values of m and α are given in App. Table D.3. A portion of this table is represented in Table 13.1.

Example 13.2   Suppose, in Example 13.1, that a random sample of 200 stomach cancer patients yielded 92 having blood type A, 20 having blood type B, 4 having blood type AB, and 84 having blood type O. Are these data significant enough, at the 5 percent level of significance, to enable us to reject the null hypothesis that the blood type distribution of stomach cancer sufferers is the same as that of the general population?

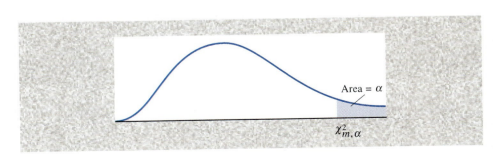

Figure 13.3
Chi-squared
percentile
$P\{\chi^2_m \geq \chi^2_{m,\alpha}\} = \alpha.$

Table 13.1	**Some Values of $\chi^2_{m,\alpha}$**			

m	$\alpha = .99$	$\alpha = .95$	$\alpha = .05$	$\alpha = .01$
1	.000157	.00393	3.841	6.635
2	.0201	.103	5.991	9.210
3	.115	.352	7.815	11.345
4	.297	.711	9.488	13.277
5	.554	1.145	11.070	15.086
6	.872	1.635	12.592	16.812
7	1.239	2.167	14.067	18.475

Solution The observed frequencies are

$$N_1 = 92 \qquad N_2 = 20 \qquad N_3 = 4 \qquad N_4 = 84$$

whereas the expected frequencies, when H_0 is true, are

$$e_1 = np_1 = 200 \times .41 = 82$$
$$e_2 = np_2 = 200 \times .09 = 18$$
$$e_3 = np_3 = 200 \times .04 = 8$$
$$e_4 = np_4 = 200 \times .46 = 92$$

Thus, the value of the test statistic is

$$\text{TS} = \frac{(92 - 82)^2}{82} + \frac{(20 - 18)^2}{18} + \frac{(4 - 8)^2}{8} + \frac{(84 - 92)^2}{92}$$
$$= 4.1374$$

Since this value is not as large as $\chi^2_{3,\,.05} = 7.815$ (obtained from Table 13.1), it follows that we cannot reject, at the 5 percent level of significance, the null hypothesis that the blood type distribution of people with stomach cancer is the same as that of the general public.

The chi-squared goodness of fit test can also be performed by determining the p value of the resulting data. If the data result in the test statistic's having value v, then the p value equals the probability that a value at least as large as v will have occurred if H_0 is true. Now, when H_0 is true, the distribution of the test statistic TS is approximately chi-squared with $k - 1$ degrees of freedom. Thus, it follows that the p value is approximately equal to the probability that a chi-squared random variable with $k - 1$ degrees of freedom is at least as large as v. The null hypothesis is then rejected at any significance level greater than or equal to the p value and is not rejected at all lower significance levels.

To determine the *p* value of the chi-squared test

1. Calculate the value of the test statistic TS.

2. If the value of TS is v, then the p value is

$$p \text{ value} = P\{\chi^2_{k-1} \geq v\}$$

where χ^2_{k-1} is a chi-squared random variable with $k - 1$ degrees of freedom.

Program 13-1 can be used to determine both the value of the test statistic TS and the resulting p value.

Example 13.3   To determine whether accidents are more likely to occur on certain days of the week, data have been collected on all the accidents requiring medical attention that occurred over the last 12 months at an automobile plant in northern California. The data yielded a total of 250 accidents, with the number occurring on each day of the week being as follows:

Monday	62
Tuesday	47
Wednesday	44
Thursday	45
Friday	52

Use the preceding data to test, at the 5 percent level of significance, the hypothesis that an accident is equally likely to occur on any day of the week.

Solution We want to test the null hypothesis that

$$P_i = \frac{1}{5} \qquad i = 1, 2, 3, 4, 5$$

The observed data are $N_1 = 62$, $N_2 = 47$, $N_3 = 44$, $N_4 = 45$, and $N_5 = 52$. Running Program 13-1 yields for the value of TS and the resulting p value

$$\text{TS} = 4.36 \qquad p \text{ value} = .359$$

Thus, a value of TS at least as large as the one obtained would be expected to occur 35.9 percent of the time when H_0 is true, and so the null hypothesis that accidents are equally likely to occur on any day of the week cannot be rejected.

Sometimes a data set is reported that is in such strong agreement with the expectations of the null hypothesis that one becomes suspicious about the possibility that the data may have been manipulated. One way of ascertaining the likelihood of this possibility is to calculate the value v of the test statistic TS and then determine how likely it is that a value as small as or smaller than v will have occurred when the null hypothesis is true. That is, one should determine $P\{\chi^2_{k-1} \leq v\}$. An extremely small value of this probability is then strong evidence for possible data manipulation.

Example 13.4   In the introduction to this chapter, we commented on an experiment performed by Gregor Mendel in which he reported that a cross of 8023 hybrid peas resulted in 6022 yellow and 2001 green peas. In theory, each cross should result in a yellow pea with probability 3/4 and a green one with probability 1/4. To determine if the data fit the model too well, we start by determining the value of the test statistic TS.

The parameters of this problem are

$$n = 8023 \qquad k = 2 \qquad p_1 = 3/4 \qquad p_2 = 1/4 \qquad N_1 = 6022 \qquad N_2 = 2001$$

Since

$$e_1 = 8023 \times \frac{3}{4} = 6017.25$$

$$e_2 = 8023 \times \frac{1}{4} = 2005.75$$

the value of the test statistic TS is

$$\text{TS} = \frac{(6022 - 6017.25)^2}{6017.25} + \frac{(2001 - 2005.75)^2}{2005.75}$$

$$= .015$$

Since .015 is greater than $\chi^2_{1, .95} = .004$, it follows that a value as small as or smaller than .015 would occur over 5 percent of the time (see Fig. 13.4). Thus, the data do not indicate any manipulation.

Indeed, it can be computed (say, from Program 13-2 which would give the p value $P\{\chi^2_1 \geq .015\}$ as output) that

$$P\{\chi^2_1 \leq .015\} = .0974$$

and so roughly 10 percent of the time TS would be as small as the value obtained when Mendel's data were used. While this by itself is not suggestive of any (conscious or unconscious) data manipulation, it turns out that almost all the data sets reported by Mendel fit his theoretical expectations as well as this one. Indeed, the probability that the sum of the values of all the chi-squared test statistics reported by Mendel, one

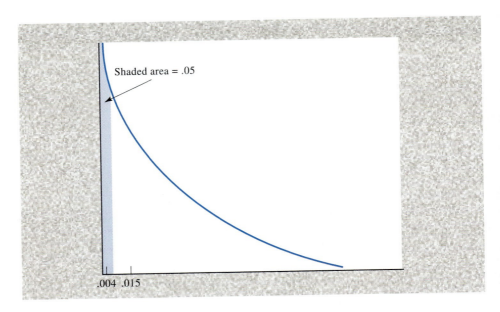

Figure 13.4
$\chi^2_{1, .95} = .004$
implies that
$P\{\chi^2_1 \leq .015\} \geq .05$.

for each experiment, would be as small as or smaller than the value obtained by using Mendel's data is .00004.

Table 13.2 summarizes the chi-squared goodness of fit test.

Problems

1. Determine the following chi-squared percentile values.
   (a) $\chi^2_{5, .01}$ (b) $\chi^2_{5, .05}$ (c) $\chi^2_{10, .01}$
   (d) $\chi^2_{10, .05}$ (e) $\chi^2_{20, .05}$

2. Consider a data set of 200 elements having the following frequency table.

Outcome	Frequency
1	44
2	38
3	57
4	61

Consider a test of the hypothesis that each of the 200 data values is equally likely to be any of the values 1 through 4.
(a) Express notationally the null and the alternative hypotheses.
(b) Compute the value of the test statistic.
(c) What conclusion is drawn at the 10 percent level of significance?
(d) Repeat part (c), using the 5 percent significance level.
(e) Repeat part (c), using the 1 percent significance level.

Table 13.2	**Chi-Squared Goodness of Fit Test**

Suppose that each member of a population has one of the values 1, 2, . . . , k. Let P_i be the proportion of the population that has value i, $i = 1, \ldots, k$. Let p_i, $i = 1, \ldots, k$, be a specified set of nonnegative numbers that sum to 1, and consider a test of

$$H_0: P_i = p_i \quad \text{for all } i = 1, \ldots, k$$

against

$$H_1: P_i \neq p_i \quad \text{for some } i = 1, \ldots, k$$

To test the above, draw a random sample of n members of the population. Let $e_i = np_i$, $i = 1, \ldots, k$, and make n large enough so that all the e_i are at least 1 and at least 80 percent of them are at least 5.

Let N_i equal the number of members of the sample that have value i. Use the test statistic

$$\text{TS} = \sum_{i=1}^{k} \frac{(N_i - e_i)^2}{e_i}$$

The significance level α test is to

Reject H_0	if	$\text{TS} \geq \chi^2_{k-1,\,\alpha}$
Not reject H_0	otherwise	

Equivalently, if the value of TS is t, then the p value is given by

$$p \text{ value} = P\{\chi^2_{k-1} \geq t\}$$

In the above χ^2_{k-1} is a chi-squared random variable with $k - 1$ degrees of freedom, and $\chi^2_{k-1,\,\alpha}$ is the $100(1 - \alpha)$ percentile of this distribution.

3. In a certain county, it has been historically accepted that 52 percent of the patients who go to hospital emergency rooms are in stable condition, 32 percent are in serious condition, and 16 percent are in critical condition. However, a particular county hospital feels that its percentages are different. To prove its claim, the hospital has randomly selected a sample of 300 patients who have visited its emergency room in the past 6 months. The numbers falling in each grouping are as follows:

Stable	148
Serious	92
Critical	60

Historical Perspective

(Science Photo Library/ Photo Researchers)

Karl Pearson

An epochal event in the history of statistics occurred in 1900, when Karl Pearson published a paper in *The London, Edinburgh, and Dublin Philosophical Magazine and Journal of Science*. In this paper he presented his chi-squared goodness of fit test. This was an event of great importance because it changed the way people viewed the subject of statistics. Whereas up to then most scientists thought of statistics as a discipline of data organization and presentation, many were led by this paper to view statistics as a discipline concerned with the testing of hypotheses.

Do the above data prove the claim of the hospital? Explain carefully what the null hypothesis is and use the 5 percent level of significance.

4. A random sample of 100 student absences yielded the following data on the days of the week on which the absences occurred.

Day	Monday	Tuesday	Wednesday	Thursday	Friday
Frequency	27	19	13	15	26

Test the hypothesis that an absence is equally likely to occur on any of the five days. What are your conclusions?

5. Consider an experiment having six possible outcomes whose probabilities are hypothesized to be .1, .1, .05, .4, .2, and .15. This is to be tested by performing 60 independent replications of the experiment. If the resultant number of times that each of the six outcomes occur is 4, 3, 7, 17, 16, and 13, should the hypothesis be rejected? Use the 5 percent level of significance.

6. In a certain region, 84 percent of drivers have no accidents in a year, 14 percent have exactly one accident, and 2 percent have two or more accidents. In a random sample of 400 lawyers, 308 had no accidents, 66 had one accident, and 26 had two or more. Could you conclude from this that lawyers do not exhibit the same accident profile as the rest of the drivers in the region?

7. The past output of a machine indicates that each unit it produces will be

Top grade	with probability .38
High grade	with probability .32
Medium grade	with probability .26
Low grade	with probability .04

A new machine, designed to perform the same job, has produced 500 items with the following results.

Top grade	222
High grade	171
Medium grade	98
Low grade	9

Can the difference in output be ascribed solely to chance? Explain!

8. Roll a die 100 times, keeping track of the frequency of each of the six possible outcomes. Use the resulting data to test the hypothesis that all six sides are equally likely to come up.

9. A marketing manager claims that mail-order sales are equally likely to come from each of four different regions. An employee does not agree and so has collected a random sample of 400 recent orders. They yielded the following numbers from each region.

Region 1	106
Region 2	138
Region 3	76
Region 4	80

Do the data disprove the manager's claim at the 5 percent level of significance? What about at the 1 percent level of significance?

10. A study was instigated to see if southern California earthquakes of at least moderate size (having values of at least 4.4 on the Richter scale) are more likely to occur on certain days of the week than on others. The catalogs yielded the following data on 1100 earthquakes.

Day	Sun	Mon	Tues	Wed	Thurs	Fri	Sat
Number of earthquakes	156	144	170	158	172	148	152

Test, at the 5 percent level, the hypothesis that an earthquake is equally likely to occur on any of the seven days of the week.

11. In certain state lotteries one buys a ticket and then chooses four different integers, between 0 and 36 inclusive. The lottery commission then randomly selects four numbers in this range in such a way that all possible choices are equally likely. After taking out its percentage—sometimes as high as 40 percent—the lottery commission then divides the remainder equally among all those players who had the correct choice of four numbers. Since all possible four-number choices are equally likely to be chosen by the commission, it is easy to see that it is best to select "unpopular" numbers so that if you do win, you will not

have to share the prize with too many others. That is, since your chances of being a winner and the amount of money returned to the winners' pool do not depend on your selection, it is best to choose numbers that others are unlikely to choose.

The above raises the question of whether there are indeed unpopular numbers, that is, numbers that are played less frequently than others. To answer this question, one could perform a chi-squared test of the hypothesis that all choices of lottery players are equally likely.

Consider a simplified lottery in which each player selects one of the integers between 1 and 10. Suppose that a random sample of 10,000 previously purchased lottery tickets yielded that each of the 10 numbers had been played with the following frequencies.

Number	Frequency
1	1122
2	1025
3	1247
4	818
5	1043
6	827
7	1149
8	946
9	801
10	1022

Do these data prove that the 10 numbers are not being played with equal frequency?

12. Data provided by the U.S. Bureau of Labor Statistics indicate that the age breakdown, by percentage, of all U.S. workers who are on flexible schedules is as follows:

Age range	Percentage
16–24	13.7
25–34	32.5
35–44	26.3
45–54	17.1
55 and up	10.4

Suppose that a random sample of 240 workers on flexible schedules in the city of Sacramento yielded 24 in the age range of 16 to 24, 94 in the age range of 25 to 34, 48 in the age range of 35 to 44, 35 in the age range of 45 to 54, and 39 in the age range of 55 and up. Can we conclude, at the 5 percent level of significance, that the age breakdown of Sacramento workers on flexible schedules is different from the national breakdown?

13. The 1987 annual report of the Girl Scouts of America indicates that 59.8 percent of members were 8 years old or younger, 32.4 percent were between 9 and 11 years old, and 7.8 percent were 12 years and older. In 1993, a random sample of 400 Girl Scouts contained 255 who were 8 years old or younger, 112 who were between 9 and 11 years old, and 33 who were 12 years old or over. Test the hypothesis that the 1993 percentages are the same as they were in 1987. Use the 5 percent level of significance.

14. Karl Pearson reported that he flipped a coin 24,000 times with 12,012 heads and 11,988 tails resulting. Is this believable? Explain the reasoning behind your answer.

15. The following gives the age breakdown, by percentages, of unmarried women having children in 1986.

Age range	Percentage
14 or less	1.1
15–19	32.0
20–24	36.0
25–29	18.9
30 and up	12.0

Source: U.S. National Center for Health Statistics, *Vital Statistics of the United States.*

A random sample of 1000 births to unmarried women in 1990 indicated that 42 of the mothers were age 14 or less, 403 were between 15 and 19 years old, 315 were between 20 and 24 years old, 150 were between 25 and 29 years old, and 90 were 30 years or older. Test the hypothesis that the 1990 percentages are the same as the 1986 percentages.

13.3 TESTING FOR INDEPENDENCE IN POPULATIONS CLASSIFIED ACCORDING TO TWO CHARACTERISTICS

Consider a large population in which each member is classified according to two distinct characteristics, which we shall designate as the X characteristic and the Y characteristic. Suppose that the possible values for the X characteristic are denoted as 1 or 2 or ... or r; similarly, the possible values of the Y characteristic are denoted as 1 or 2 or ... or s. Thus, there are r possible values for the X characteristic and s possible values for the Y characteristic.

Example 13.5   Consider a population of voting-age adults, and suppose that each adult is classified according to both gender—female or male—and political affiliation—Democrat, Republican, or Independent. Let the X characteristic represent gender and the Y characteristic represent political affiliation. Since there are two possible genders and three possible political affiliations, $r = 2$ and $s = 3$. Let us say that a person's X characteristic is 1 if the person is a woman and 2 if the person is a man. Also, say that a person's Y characteristic is 1 if the person is a Democrat, 2 if the person is a Republican, and 3 if he or she is an Independent. Thus, for instance, a woman who is a Republican would have X characteristic 1 and Y characteristic 2.

Let P_{ij} denote the proportion of the population that has both X characterization i and Y characterization j, for i being any of the values 1, 2, ... , r and j being any of the values 1, 2, ... , s. Also let P_i denote the proportion of the population who have X characteristic i, and let Q_j be the proportion who have Y characteristic j. Thus if X and Y denote the values of the X characteristic and Y characteristic of a randomly chosen member of the population, then

$$P\{X = i, Y = j\} = P_{ij}$$
$$P\{X = i\} = P_i$$
$$P\{Y = j\} = Q_j$$

Example 13.6   For the situation described in Example 13.5, P_{11} represents the proportion of the population consisting of women who classify themselves as Democrats, P_{12} is the proportion of the population consisting of women who classify themselves as Republicans, and P_{13} is the proportion of the population consisting of women who classify themselves as Independents. The proportions P_{21}, P_{22}, and P_{23} are defined similarly with *men* replacing *women* in the definitions. The quantities P_1 and P_2 are the proportions of the population that are, respectively, women and men; and Q_1, Q_2, and Q_3 are the proportions of the population that are, respectively, Democrats, Republicans, and Independents.

We will be interested in developing a test of the hypothesis that the X characteristic and Y characteristic of a randomly chosen member of the population are independent. Recalling that X and Y are independent if

$$P\{X = i, Y = j\} = P\{X = i\}P\{Y = j\}$$

it follows that we want to test the null hypothesis

$$H_0: P_{ij} = P_i Q_j \qquad \text{for all } i = 1, \ldots, r, j = 1, \ldots, s$$

against the alternative

$$H_1: P_{ij} \neq P_i Q_j \qquad \text{for some values of } i \text{ and } j$$

To test this hypothesis of independence, we start by choosing a random sample of size n of members of the population. Let N_{ij} denote the number of elements of the sample that have both X characteristic i and Y characteristic j.

Example 13.7   Consider Example 13.5, and suppose that a random sample of 300 people were chosen from the population, with the following data resulting.

	Democrat	Republican	Independent	Total
i		*j*		
Women	68	56	32	156
Men	52	72	20	144
Total	120	128	52	300

Thus, for instance, the random sample of size 300 contained 68 women who classified themselves as Democrats, 56 women who classified themselves as Republicans, and 32 women who classified themselves as Independents; that is, $N_{11} = 68$, $N_{12} = 56$, and $N_{13} = 32$. Similarly, $N_{21} = 52$, $N_{22} = 72$, and $N_{23} = 20$.

The above table, which specifies the number of members of the sample that fall in each of the *rs* cells, is called a *contingency table*. ◼ ◼

If the hypothesis is true that the *X* and *Y* characteristics of a randomly chosen member of the population are independent, then each element of the sample will have *X* characteristic *i* and *Y* characteristic *j* with probability $P_i Q_j$. Hence, if these probabilities were known then, from the results of Sec. 13.2, we could test H_0 by using the test statistic

$$\text{TS} = \sum_i \sum_j \frac{(N_{ij} - e_{ij})^2}{e_{ij}}$$

where

$$e_{ij} = nP_i Q_j$$

The quantity e_{ij} represents the expected number, when H_0 is true, of elements in the sample that have both *X* characteristic *i* and *Y* characteristic *j*. In computing TS we must calculate the sum of the terms for all *rs* possible values of the pair *i,j*. When H_0 is true, TS will approximately have a chi-squared distribution with $rs - 1$ degrees of freedom.

The trouble with directly using the above approach is that the $r + s$ quantities P_i and Q_j, $i = 1, \ldots, r$, $j = 1, \ldots, s$, are not specified by the null hypothesis. Thus, we need to first estimate them. To do so, let N_i and M_j denote the number of elements of the sample that have, respectively, *X* characteristic *i* and *Y* characteristic *j*. As N_i/n and M_j/n are the proportions of the sample having, respectively, *X* characteristic *i* and *Y* characteristic *j*, it is natural to use them as estimators of P_i and Q_j. That is, we estimate P_i and Q_j by

$$\hat{P}_i = \frac{N_i}{n} \qquad \hat{Q}_j = \frac{M_j}{n}$$

This leads to the following estimate of e_{ij}:

$$\hat{e}_{ij} = n\hat{P}_i\hat{Q}_j = \frac{N_i M_j}{n}$$

In words, $\hat{e}_{ij}$ is equal to the product of the number of members of the sample that have X characteristic i (that is, the sum of row i of the contingency table) and the number of members of the sample that have Y characteristic j (that is, the sum of column j of the contingency table) divided by the sample size n.

Thus, it seems that a reasonable test statistic to use in testing the independence of the X characteristic and the Y characteristic is the following:

$$TS = \sum_i \sum_j \frac{(N_{ij} - \hat{e}_{ij})^2}{\hat{e}_{ij}}$$

where $\hat{e}_{ij}$, $i = 1, \ldots, r$, $j = 1, \ldots, s$, are as given above.

To specify the set of values of TS which will result in rejection of the null hypothesis, we need to know the distribution of TS when the null hypothesis is true. It can be shown that when H_0 is true, the distribution of the test statistic TS is approximately a chi-squared distribution with $(r - 1)(s - 1)$ degrees of freedom. From this, it follows that the significance level α test of H_0 is as follows:

Reject H_0 if TS $\geq \chi^2_{(r-1)(s-1),\ \alpha}$

Do not reject H_0 otherwise

A technical remark: It is not difficult to see why the test statistic TS should have $(r - 1)(s - 1)$ degrees of freedom. Recall, from Sec. 13.2, that if all the values P_i and Q_j are specified in advance, then the test statistic has $rs - 1$ degrees of freedom. (This is so since k, the number of different types of elements in the population, is equal to rs.) Now, at first glance it may seem that we have to use the data to estimate $r + s$ parameters. However, since the P_i's and the Q_j's both sum to 1—that is, $\sum_i P_i = \sum_j Q_j = 1$—we really only need to estimate $r - 1$ of the P_i's and $s - 1$ of the Q_j's. (For instance, if r is equal to 2, then an estimate of P_1 will automatically provide an estimate of P_2 since $P_2 = 1 - P_1$.) Hence, we actually need to estimate $r - 1 + s - 1 = r + s - 2$ parameters. Since a degree of freedom is lost for each parameter estimated, it follows that the resulting test statistic has $rs - 1 - (r + s - 2) = rs - r - s + 1 = (r - 1)(s - 1)$ degrees of freedom.

Example 13.8 The data of Example 13.7 are as given below.

	j			
i	1	2	3	Total = N_i
1	68	56	32	156
2	52	72	20	144
Total = M_j	120	128	52	300

What conclusion can be drawn? Use the 5 percent level of significance.

Solution From the above data, the six values of

$$\hat{e}_{ij} = \frac{N_i M_j}{n}$$

are as follows:

$$\hat{e}_{11} = \frac{N_1 M_1}{n} = \frac{156 \times 120}{300} = 62.40$$

$$\hat{e}_{12} = \frac{N_1 M_2}{n} = \frac{156 \times 128}{300} = 66.56$$

$$\hat{e}_{13} = \frac{N_1 M_3}{n} = \frac{156 \times 52}{300} = 27.04$$

$$\hat{e}_{21} = \frac{N_2 M_1}{n} = \frac{144 \times 120}{300} = 57.60$$

$$\hat{e}_{22} = \frac{N_2 M_2}{n} = \frac{144 \times 128}{300} = 61.44$$

$$\hat{e}_{23} = \frac{N_2 M_3}{n} = \frac{144 \times 52}{300} = 24.96$$

The value of the test statistic is thus

$$\text{TS} = \frac{(68 - 62.40)^2}{62.40} + \frac{(56 - 66.56)^2}{66.56} + \frac{(32 - 27.04)^2}{27.04} + \frac{(52 - 57.60)^2}{57.60}$$

$$+ \frac{(72 - 61.44)^2}{61.44} + \frac{(20 - 24.96)^2}{24.96}$$

$$= 6.433$$

Since $r = 2$ and $s = 3$, $(r - 1)(s - 1) = 2$ and so we must compare the value of TS with the critical value $\chi^2_{2, .05}$. From Table 13.1,

$$\chi^2_{2, .05} = 5.991$$

Since TS ≥ 5.991, the null hypothesis is rejected at the 5 percent level of significance. That is, the hypothesis that gender and political affiliation of members of the population are independent is rejected at the 5 percent level of significance.   ◼ ◼

The test of the hypothesis that the X and Y characteristics of a randomly chosen member of the population are independent can also be performed by determining the p value of the data. This is accomplished by first calculating the value of the test statistic TS. If its value is v, then the p value is given by

| Table 13.3 | **Testing Independence of Two Characteristics of Members of a Population** |

Assume that each member of a population has both an X and a Y characteristic. Let r and s denote the number of possible X and Y characteristics, respectively. To test

H_0: characteristics of a randomly chosen member are independent

against

H_1: characteristics of a randomly chosen member are not independent

choose a sample of n members of the population. Let N_{ij} denote the number of these that have both X characteristic i and Y characteristic j. Also let

$$N_i = \sum_j N_{ij} \quad \text{and} \quad M_j = \sum_i N_{ij}$$

denote, respectively, the number of members of the sample that have X characteristic i and that have Y characteristic j. The test statistic is

$$TS = \sum_i \sum_j \frac{(N_{ij} - \hat{e}_{ij})^2}{\hat{e}_{ij}}$$

where $\hat{e}_{ij} = N_i M_j / n$. The significance level α test

Rejects H_0 if TS $\geq \chi^2_{(r-1)(s-1),\ \alpha}$
Does not reject H_0 otherwise

Equivalently, if the value of TS is t, then the p value is

$$p \text{ value} = P\{\chi^2_{(r-1)(s-1)} \geq v\}$$

$$p \text{ value} = P\{\chi^2_{(r-1)(s-1)} \geq v\}$$

where $\chi^2_{(r-1)(s-1)}$ is a chi-squared random variable with $(r-1)(s-1)$ degrees of freedom.

Program 13-2 will calculate the value of the test statistic and then determine the resulting p value. The program supposes that the data are arranged in the form of a contingency table and asks the user to input the successive rows of this table.

Example 13.9   A public health scientist wanted to learn about the relationship between the marital status of patients being treated for depression and the severity of their conditions. The scientist chose a random sample of 159 patients who had been treated for depression at a mental health clinic and had these patients classified according to the severity of their depression—severe, normal, or mild—and according to their marital status. The following data resulted.

Depressive state	Marital status			Totals
	Married	Single	Widowed or divorced	
Severe	22	16	19	52
Normal	33	29	14	79
Mild	14	9	3	28
Totals	69	54	36	159

Determine the p value of the test of the hypothesis that the depressive state of the clinic's patients is independent of their marital status.

Solution We run Program 13-2 to obtain that the value of the test statistic and the resulting p value are

$$\text{TS} = 6.828 \qquad p \text{ value} = .145$$

The test for independence is summarized in Table 13.3.

Problems

1. The following contingency table presents data from a sample of a population that is characterized in two different ways.

Y characteristic	X characteristic		
	A	B	C
1	32	12	40
2	56	48	60

(a) Determine the value of the test statistic when testing that the two characteristics are independent.
(b) Would the null hypothesis be rejected at the 5 percent level of significance?
(c) What about at the 1 percent level?

2. There is some evidence that ownership of a dog may have predictive value in determining whether an individual will survive a heart attack. The following data are from a random sample of 95 individuals who each suffered a severe heart attack. The data classify each of these individuals with respect to (1) whether they were still alive 1 year after their attack and (2) whether they had a dog as a pet.

Historical Perspective

(Photoworld/FPG)

Florence Nightingale

Her biographer called her the *passionate statistician.* It was an appropriate description of Florence Nightingale, the woman who almost single-handedly changed nursing into a science. During the Crimean war she searched out and collected data on sanitary conditions and mortality rates in military hospitals, and she used these data to statistically prove that the two were dependent. Her work was instrumental in improving the hygienic conditions in hospitals and resulted in the saving of untold lives.

Florence Nightingale was a follower of the Belgian statistician Adolphe Quetelet, and she believed as he did that "accidents occur with astonishing regularity when the same conditions exist." She held that successful administrators were ones who searched out data. She felt that the universe evolved in accordance with a divine plan and that it was every person's job to learn to live in harmony with it. But to understand this plan, she believed that one had to study statistics. In the words of Karl Pearson, "For Florence Nightingale, statistics were more than a study, they were her religion."

	Had pet	No pet	
Survived	28	44	72
Did not survive	8	15	23
	36	59	95

Does the above prove, at the 5 percent level of significance, that owning a pet and survival are dependent? Explain carefully what null hypothesis you are testing and what test statistic you are using.

3. A random sample of 187 voters were chosen, and the voters were asked to evaluate the performance of the first 100 days of President Clinton. Use the resulting data to test the hypothesis that the evaluation of an individual does not depend on whether that individual is a man or woman.

	Women	Men
Positive evaluation	54	47
Negative evaluation	20	32
Not sure	23	11

Use the 5 percent level of significance.

4. An insurance company is interested in determining whether there is a relationship between automobile accident frequency and cigarette smoking. It randomly sampled 597 policyholders and came up with the following data.

Number of accidents in last 2 years	Smokers	Nonsmokers
0	35	170
1	79	190
2 or more	57	66

Test the hypothesis, at the 5 percent level of significance, that the accident frequency of a randomly chosen policyholder is independent of his or her smoking habits.

5. The management of a certain hotel is interested in whether all its guests are treated the same regardless of the prices of their rooms. They randomly chose 155 recent guests and questioned them about the service they had received at the hotel. The following summary data resulted:

	Type of room		
Service ranking	Economy	Standard	Luxury
Excellent	30	21	9
Good	36	29	8
Fair	12	8	2

What conclusions would you draw?

6. The following data categorize a random selection of professors of a certain university according to their teaching performance (as measured by the students in their classes) in the most recent semester and the number of courses they were teaching at the time.

	Number of courses		
Student ranking	1	2	3 or more
Above average	12	10	4
Average	32	40	38
Below average	7	12	25

Test, at the 5 percent level, the hypothesis that a professor's teaching performance is independent of the number of courses she or he is teaching.

7. In Prob. 6, it is possible that only certain professors, usually ones who specialize in research, would teach only one course in a semester. These would then tend to be more advanced courses and to have fewer students than in most courses. Thus, to more directly learn whether teaching additional courses affects teaching performance, it might be reasonable to consider the data of Prob. 6 with the column pertaining to those teaching only one class deleted. Make this change and repeat Prob. 6.

8. The socioeconomic status of residents of a particular neighborhood can be classified as either lower or middle class. A sample of residents were questioned about their attitude toward a planned public health clinic for the neighborhood. The results are given below.

	Socioeconomic class	
Attitude	**Lower**	**Middle**
In favor	87	63
Against	46	55

Test the hypothesis, at the 5 percent level of significance, that lower- and middle-class residents of the neighborhood have the same attitude toward the new clinic.

9. A market research firm has distributed samples of a new shampoo to a variety of individuals. The following data summarize the comments of these individuals about the shampoo as well as provide the age group into which they fall.

	Age group (years)		
Rating	**15–20**	**21–30**	**Over 30**
1 Excellent	18	20	41	79
2 Good	25	27	43	95
3 Fair	17	15	26	58
4 Poor	3	2	8	13
	63	64	118	245

Do the above data prove that different age groups have different opinions about the shampoo? Use the 5 percent level of significance.

10. A random sample of 160 patients at a health maintenance organization yielded the following information about their smoking status and blood cholesterol counts.

	Blood cholesterol count		
Smoking status	**Low**	**Moderate**	**High**
Smoker			
Heavy	6	14	24
Light	12	23	15
Nonsmoker	23	32	11

(a) Would the hypothesis of independence between blood cholesterol count and smoking status be rejected at the 5 percent level of significance?
(b) Repeat part (a), but this time use the 1 percent level of significance.

(c) Do your results imply that a reduction in smoking will result in a lowered blood cholesterol level? Explain!

11. To see if there was any dependency between the type of professional job held and one's religious affiliation, a random sample of 638 individuals belonging to a national organization of doctors, lawyers, and engineers were chosen in a 1968 study. The results of the sample are given in the following contingency table.

	Doctors	**Lawyers**	**Engineers**
Protestant	64	110	152
Catholic	60	86	78
Jewish	57	21	10

Test the hypothesis, at the 5 percent level of significance, that the profession of individuals in this organization and their religious affiliation are independent. Repeat at the 1 percent level.

12. Look at the following contingency table and guess (without any computations) as to the result of a test, at the 5 percent level of significance, of the hypothesis that the two data characteristics are independent.

	A	*B*	*C*
1	26	44	30
2	14	30	25
3	30	45	33

Now perform the computations.

13. Repeat Prob. 12, but first double all the data values.

14. Repeat Prob. 13, but first double all the data values.

13.4   TESTING FOR INDEPENDENCE IN CONTINGENCY TABLES WITH FIXED MARGINAL TOTALS

In Example 13.5 we were interested in determining whether gender and political affiliation were dependent in a particular population. To test this hypothesis, we first chose a random sample of people from this population and then noted their characteristics. However, another way in which we could gather data is to fix in advance the numbers of men and women in the sample and then choose random samples of those sizes from the subpopulations of men and women. That is, rather than let the numbers of women and men in the sample be determined by chance, we might decide these numbers in advance. Because doing so would result in fixed specified values for the total numbers of men and women in the sample, the resulting contingency table is often said to have *fixed margins* (since the totals are given in the margins of the table).

It turns out that even when the data are collected in the manner prescribed above, the same hypothesis test as given in Sec. 13.3 can still be used to test for the independence of the two characteristics. The test statistic remains

$$\text{TS} = \sum_i \sum_j \frac{(N_{ij} - \hat{e}_{ij})^2}{\hat{e}_{ij}}$$

where

N_{ij} = number of members of sample who have both X characteristic i and Y characteristic j

N_i = number of members of sample who have X characteristic i

M_j = number of members of sample who have Y characteristic j

and

$$\hat{e}_{ij} = \frac{N_i M_j}{n}$$

where n is the total size of the sample.

In addition, it is still true that when H_0 is true, TS will approximately have a chi-squared distribution with $(r - 1)(s - 1)$ degrees of freedom. (The quantities r and s refer, of course, to the numbers of possible values of the X and Y characteristic, respectively.) In other words, the test of the independence hypothesis is unaffected by whether the marginal totals of one characteristic are fixed in advance or result from a random sample of the entire population.

Example 13.10   A randomly chosen group of 20,000 nonsmokers and one of 10,000 smokers were followed over a 10-year period. The following data relate the numbers developing lung cancer in that period.

	Smokers	Nonsmokers	Total
Lung cancer	62	14	76
No lung cancer	9,938	19,986	29,924
Total	10,000	20,000	30,000

Test the hypothesis that smoking and lung cancer are independent. Use the 1 percent level of significance.

Solution The estimated numbers expected to fall in each ij cell are

$$\hat{e}_{11} = \frac{(76)(10,000)}{30,000} = 25.33$$

$$\hat{e}_{12} = \frac{(76)(20,000)}{30,000} = 50.67$$

$$\hat{e}_{21} = \frac{(29,924)(10,000)}{30,000} = 9974.67$$

$$\hat{e}_{22} = \frac{(29,924)(20,000)}{30,000} = 19,949.33$$

Therefore, the value of the test statistic is

$$TS = \frac{(62 - 25.33)^2}{25.33} + \frac{(14 - 50.67)^2}{50.67} + \frac{(9938 - 9974.67)^2}{9974.67}$$

$$+ \frac{(19,986 - 19,949.33)^2}{19,949.33}$$

$$= 53.09 + 26.54 + .13 + .07 = 79.83$$

Since this is far larger than $\chi^2_{1, .01} = 6.635$, we reject the null hypothesis that whether a randomly chosen person develops lung cancer is independent of whether that person is a smoker.   ◼ ◼

We now show how we can use the framework of this section to test the hypothesis that m population proportions are equal. To begin, consider m separate populations of individuals. Suppose that the proportion of members of the ith population that are in favor of a certain proposition is p_i, and consider a test of the null hypothesis that all the p_i's are equal. That is, consider a test of

$$H_0: p_1 = p_2 = \cdots = p_m$$

against

$$H_1: \text{not all the } p_i\text{'s are equal}$$

To obtain a test of this null hypothesis, consider first the superpopulation consisting of all members of each of the m populations. Any member of this superpopulation can be classified according to two characteristics. The first characteristic specifies which of the m populations the member is from, and the second characteristic specifies whether the member is in favor of the proposition. Now the hypothesis that all the p_i's are equal is just the hypothesis that the same proportions of members of each population are in favor of the proposition. But this is exactly the same as stating that for members of the superpopulation the characteristic of being for or against the proposition is independent of the population that the member is from. That is, the null hypothesis H_0 is equivalent to the hypothesis of independence of the two characteristics of the superpopulation.

Therefore, we can test H_0 by first choosing independent random samples of fixed sizes from each of the m populations. If we let M_i be the sample size from population i, for $i = 1, \ldots, m$, we can test H_0 by testing for independence in the following contingency table.

	Population				
	1	2	$\cdots$	m	**Totals**
In favor	F_1	F_2	$\cdots$	F_m	N_1
Against	A_1	A_2	$\cdots$	A_m	N_2
Total	M_1	M_2	$\cdots$	M_m	

In the above, F_i refers to the number of members of population i who are in favor, and A_i refers to the number who are against the proposition.

Example 13.11 A recent study reported that 500 female office workers were randomly chosen and questioned in each of four different countries. One of the questions related to whether these women often received verbal or sexual abuse on the job. The following data resulted.

Country	Number reporting abuse
Australia	28
Germany	30
Japan	51
United States	55

Based on these data, is it plausible that the proportions of female office workers who often feel abused at work are the same for these countries?

Solution Putting the above data in the form of a contingency table gives the following:

	Country				
	1	2	3	4	**Totals**
Receive abuse	28	30	58	55	171
Do not receive abuse	472	470	442	445	1829
Totals	500	500	500	500	2000

We can now test the null hypothesis by testing for independence in the preceding contingency table. If we run Program 13-2, then the value of the test statistic and the resulting p value are

$$TS = 19.51 \qquad p \text{ value} = .0002$$

Therefore, the hypothesis that the percentages of women who feel they are being abused on the job are the same for these countries is rejected at the 1 percent level of significance (and, indeed, at any significance level above .02 percent).

When there are only two populations, the preceding test of the equality of population proportions is identical to the one presented in Sec. 10.6.

Problems

1. Can we conclude from the results of Example 13.10 that smoking causes lung cancer? What other explanations are possible?

2. A study of the relationship between school preferences and family income questioned 100 upper-income and 100 lower-income families in a certain city about the type of school they would most like their children to attend. These are the resulting data.

Preference	Upper income	Lower income	
Public	22	19	41
Private religious	31	39	70
Private nonreligious	47	42	89
	100	100	200

What conclusions can you draw?

3. A sample of 300 cars having cellular phones and one of 400 cars without phones were tracked for 1 year. The following table gives the number of these cars involved in accidents over that year.

	Accident	No accident
Cellular phone	22	278
No phone	26	374

Use the above to test the hypothesis that having a cellular phone in your car and being involved in an accident are independent. Use the 5 percent level of significance.

4. A newspaper chain sampled 100 readers of each of its three major newspapers to determine their economic class. The results were as follows:

	Newspaper		
Economic class	1	2	3
Lower middle	22	25	28
Middle	41	37	44
Upper middle	37	38	28

Test the hypothesis that the newspaper an individual reads and the economic class to which that person belongs are independent. Use the 5 percent level.

5. The following table shows the number of defective and acceptable items in samples taken both before and after the introduction of a modification in the manufacturing process.

	Defective	Nondefective
Before	22	404
After	18	422

Does the above prove that the modification results in a different percentage of defective items?

6. From a statistics class of 200 students, 100 were randomly chosen to watch the lectures on television rather than in person. The other 100 stayed in the lecture hall. The final grades of the students are given below.

	Final grades			
	A	B	C	Less than C
In-class students	22	38	35	5
Television students	18	32	40	10

Test the hypothesis that final grades are independent of whether the student watches on television or is present in the lecture hall. Can we reject at the 5 percent level of significance? What about at the 1 percent level?

7. To study the effect of fluoridated water supplies on tooth decay, two communities of roughly the same socioeconomic status were chosen. One of these communities had fluoridated water while the other did not. Random samples of 200 teenagers from both communities were chosen, and the numbers of cavities they had were determined. The following data resulted.

Cavities	Fluoridated town	Nonfluoridated town
0	154	133
1	20	18
2	14	21
3 or more	12	28

Do these data establish, at the 5 percent level of significance, that the number of dental cavities a person has is not independent of whether that person's water supply is fluoridated? What about at the 1 percent level?

8. An automobile dealership sent out postcards to 1000 potential customers, offering them a free test drive of one of its cars. Each postcard was colored red,

white, light blue, or green. Data relating the number of customers who responded and the colors of the postcards they had been sent are given below.

	Red	White	Blue	Green
Responded	108	106	105	127
No response	142	144	135	123

Test the hypothesis that the color of the postcard sent does not affect the recipient's chance of responding. Use the 5 percent level of significance.

9. Random samples of 50 college students, 40 college faculty, and 60 bankers yielded the following data relating to the numbers of smokers in these samples.

Group	Number who smoke
College students	18
College faculty	12
Bankers	24

   (a) Test the hypothesis, at the 10 percent level of significance, that the same percentages of college students, college faculty, and bankers are smokers.
   (b) Repeat part (a) at the 5 percent level of significance.
   (c) Repeat part (a) at the 1 percent level of significance.

10. To determine if a malpractice lawsuit is more likely to follow certain types of surgery, random samples of three different types of surgeries were studied, and the following data resulted.

Type of operation	Number sampled	Number leading to a lawsuit
Heart surgery	400	16
Brain surgery	300	19
Appendectomy	300	7

Test the hypothesis that the percentages of the surgical operations that lead to lawsuits are the same for each of the three types.
   (a) Use the 5 percent level of significance.
   (b) Use the 1 percent level of significance.

Key Terms

Goodness of fit test: A statistical test of the hypothesis that a specified set of k probabilities represents the proportion of members of a large population that fall into each of k distinct categories.

Contingency table: A table that classifies each element of a sample according to two distinct characteristics.

Summary

Goodness of Fit Tests

Consider a large population of elements, each of which has a value that is 1, or 2, ..., or k. Let P_i denote the proportion of the population that has value i, for $i = 1, \ldots, k$. For a given set of probabilities $p_1, \ldots, p_k$ ($p_i \geq 0, \sum_i p_i = 1$), consider a test of

$$H_0: P_i = p_i \quad \text{for all } i = 1, \ldots, k$$

against the alternative

$$H_1: P_i \neq p_i \quad \text{for some } i, i = 1, \ldots, k$$

To test this null hypothesis, first draw a random sample of n elements of the population. Let N_i denote the number of elements in the sample that have value i. The test statistic to be employed is

$$\text{TS} = \sum_{i=1}^{k} \frac{(N_i - e_i)^2}{e_i}$$

where

$$e_i = np_i$$

When H_0 is true, e_i is equal to the expected number of elements in the sample that have value i.

The hypothesis test is to reject H_0 when TS is sufficiently large. To determine how large, we use the fact that when H_0 is true, TS has a distribution that is approximately a chi-squared distribution with $k - 1$ degrees of freedom. This implies that the significance level α test is to

Reject H_0 if TS $\geq \chi^2_{k-1, \, \alpha}$

Not reject H_0 otherwise

The quantity $\chi^2_{k-1, \, \alpha}$ is defined by

$$P\{\chi^2_{k-1} \geq \chi^2_{k-1, \, \alpha}\} = \alpha$$

where χ^2_{k-1} is a chi-squared random variable with $k - 1$ degrees of freedom.

The above test is called the *chi-squared goodness of fit test*. It can also be implemented by determining the p value of the data set. If the observed value of TS is v, then the p value is

$$p \text{ value} = P\{\chi^2_{k-1} \geq v\}$$

Program 13-1 can be used to determine the value of the test statistic and the resulting p value.

Testing for Independence in Populations Whose Elements Are Classified according to Two Different Characteristics

Suppose now that each element of a population is classified according to two distinct characteristics, which we call the X characteristic and the Y characteristic. Suppose the possible values of the X characteristic are 1, or 2, . . . , or r, and the possible values for the Y characteristic are 1, or 2, . . . , or s. Let P_{ij} denote the proportion of the population that has X characteristic i and Y characteristic j. Let P_i denote the proportion of the population whose X characteristic is i, $i = 1, \ldots, r$; and let Q_j denote the proportion whose Y characteristic is j, $j = 1, \ldots, s$.

Consider a test of the null hypothesis that the X and Y characteristics of a randomly chosen member of the population are independent. That is, consider a test of

$$H_0: P_{ij} = P_i Q_j \qquad \text{for all } i, j$$

against

$$H_1: P_{ij} \neq P_i Q_j \qquad \text{for some } i, j$$

To test the above, draw a random sample of n elements of the population. Let N_{ij} denote the number of these having X characteristic i and Y characteristic j. Also let N_i denote the number that have X characteristic i; and let M_j denote the number that have Y characteristic j. The test statistic used to test H_0 is

$$\text{TS} = \sum_i \sum_j \frac{(N_{ij} - \hat{e}_{ij})^2}{\hat{e}_{ij}}$$

where

$$\hat{e}_{ij} = \frac{N_i M_j}{n}$$

The summation in the expression for TS is over all rs possible values of the pair i, j. The significance level α test is to

Reject H_0 if $\text{TS} \geq \chi^2_{(r-1)(s-1), \, \alpha}$

Not reject H_0 otherwise

Equivalently, the test can be implemented by computing the p value. If the value of TS is v, then the p value is given by

$$p \text{ value} = P\{\chi^2_{(r-1)(s-1)} \geq v\}$$

Program 13-2 can be used to determine both the value of TS and the resulting p value.

The same test as given above can also be employed when the sample chosen is not a random sample from the entire population but rather is a collection of random samples of fixed sizes from the r (or s) subpopulations whose X characteristic (or Y characteristic) is fixed.

The following tabular presentation of the data is called a *contingency table:*

X characteristic	1	2	$\cdots$	j	$\cdots$	s	Totals
			Y characteristic				
1	N_{11}	N_{12}	$\cdots$	N_{1j}	$\cdots$	N_{1s}	N_1
i	N_{i1}	N_{i2}	$\cdots$	N_{ij}	$\cdots$	N_{is}	N_i
r	N_{r1}	N_{r2}	$\cdots$	N_{rj}	$\cdots$	N_{rs}	N_r
Totals	M_1	M_2	$\cdots$	M_j	$\cdots$	M_s	n

Review Problems

1. The following data classify minor accidents over the past year at a certain industrial plant, according to the time periods in which these accidents occurred.

Time period	Number of accidents
8–10 a.m.	47
10–12 p.m.	52
1–3 p.m.	57
3–5 p.m.	63

Test the hypothesis that each accident was equally likely to occur in any of the four time periods. Use the 5 percent level of significance.

2. A movie distribution company often sets up sneak previews of new movies. In such a situation a movie whose title had not previously been announced is shown in addition to the regularly scheduled movie. When the audience leaves the theater, individuals are given questionnaires to be filled out at home and mailed back to the company. Such information is then used by the company in deciding how widely to distribute the movie. Of some interest is whether the popularity of a movie will be the same over different parts of the country. To

test this hypothesis, a sneak preview was scheduled in four theaters around the country. One theater was in New York, one in Chicago, one in Phoenix, and one in Seattle. The following are ratings given to the movie by audiences in these four places.

Rating	Location			
	New York	Chicago	Phoenix	Seattle
Excellent	234	141	108	142
Good	303	256	165	170
Poor	102	88	41	45

Test, at the 5 percent level, the hypothesis that the audience reaction is independent of the location. What about at the 1 percent level?

3. Suppose that a die is rolled 600 times. Consider a test of the hypothesis that each roll is equally likely to be any of the six faces. Make up data which you think will result in a p value approximately equal to
   (a) .50
   (b) .05
   (c) .95
   (d) Approximate the actual p values for the data you presented in parts (a), (b), and (c).

4. It has been claimed that the proportions of voters presently favoring the Democratic, Republican, or Independent candidate in an upcoming election are 40, 42, and 18 percent. To test this hypothesis, a random sample of 50 voters yielded the following results.

	Democrat	Republican	Independent
Number favoring	18	22	10

Is the claim consistent with the preceding data? Use the 5 percent level of significance.

5. The following data come from a study of randomly selected automobile accidents. It categorizes each accident by the weight of the car involved and the severity of injury suffered by the driver.

Injury	Weight of car (pounds)		
	Less than 2500	2500–3000	Greater than 3000
Very severe	34	22	8
Average	43	41	47
Moderate	51	60	50

Test, at the 5 percent level of significance, the hypothesis that the severity of injury and the weight of the car are independent.

6. A friend reported the following results when rolling a die 1000 times.

Outcome	Frequency
1	167
2	165
3	167
4	166
5	167
6	168

Do you believe these results? Explain!

7. A random sample of 527 earthquakes in western Japan yielded the following frequencies of occurrence at certain time periods in a day.

Time period	Frequency
12 a.m.–6 a.m.	123
6 a.m.–12 p.m.	135
12 p.m.–6 p.m.	141
6 p.m.–12 a.m.	128

Test the hypothesis that earthquakes are equally likely to occur in each of the four time periods.

8. The following data give the number of murders, by day of the week, in the state of Utah from 1978 through 1990.

Day	Sunday	Monday	Tuesday	Wednesday	Thursday	Friday	Saturday
Number	109	74	97	94	83	107	100

Test the hypothesis, at the 5 percent level, that a murder was equally likely to occur on any of the 7 days of the week.

9. The 1987 annual report of the Boy Scouts of America indicated that 7.0 percent of members were Tiger Cubs, 43.5 percent were Cub Scouts, 24.7 percent were Boy Scouts, and 24.8 percent were Explorers. In 1993, a random sample of 400 scouts included 34 Tiger Cubs, 150 Cub Scouts, 142 Boy Scouts, and 74 Explorers. Use these data to test the hypothesis that the 1993 percentages remain unchanged from those of 1987. Use the 5 percent level of significance.

10. One of Mendel's breeding experiments resulted in the following data.

Type of pea	Expected	Observed
Smooth yellow	313	315
Wrinkled yellow	104	101
Smooth green	104	108
Wrinkled green	35	32

Do you think such a good fit is "too good to be true"?

11. A public health clinic detailed the results of 260 elderly patients who had been advised to have a flu vaccine. A total of 184 agreed to have the vaccine, while the other 76 declined. The flu season results for this group were as follows.

	Vaccine	No vaccine
Flu	10	6
No flu	174	70

Does the above establish that those receiving the vaccine had a different chance of contracting the flu from those not receiving the vaccine? Use the 5 percent level of significance. If it does, check at the 1 percent level of significance; if not, check at the 10 percent level.

12. A random sample of 220 married men in their fifties were classified according to their education and number of children. The following contingency table describes the data.

	Number of children		
Education	0–1	2–3	More than 3
Elementary	10	28	22
Secondary	19	63	38
College	14	41	27

Test the hypothesis that the size of a family is independent of the educational level of the father. Use the 5 percent level of significance.

13. The following data relate a mother's age and the birthweight (in grams) of her child.

	Birthweight (grams)	
Mother's age (years)	Less than 2500	More than 2500
20 or less	12	50
Greater than 20	18	125

(a) Test the hypothesis, at the 5 percent level of significance, that the baby's birthweight is independent of the mother's age.

(b) What is the p value?

14. Repeat Prob. 13 when the four data values are all doubled.

15. On a course evaluation form, students are asked to rank the course as excellent, fair, or poor. In addition, students signify whether the course is required or not for them. A random sample of 121 such evaluations yielded the following data.

	Rating		
	Excellent	Average	Poor
Required	14	42	18
Not required	12	28	7

Test, at the 5 percent level, the hypothesis that course rating is independent of whether the course is required. What about at the 1 percent level?

16. A class of 154 students in statistics meets in a room that can hold 250 students. Out of curiousity, the instructor categorized each student by gender and seat location. Using the data below, test the hypothesis that these characterizations are independent.

	Front	Middle	Back
Females	22	40	18
Males	10	38	26

17. One might imagine that the first digits of numbers found in an almanac would be equally likely to be 1, or 2, or . . . , or 9. Make a random selection of numbers from an almanac, and note the first digit of each. Use the data to test the hypothesis that all nine digits are equally likely.

18. Repeat Prob. 17, this time using the second digit.

19. The following table gives the cumulative percentage distribution of the heights of U.S. women residents aged 18 to 24 in the years 1976 through 1980.

Height	Women's cumulative percentage
5 feet 0 inches	4.22
5 feet 3 inches	29.06
5 feet 5 inches	58.09
5 feet 7 inches	85.37
5 feet 8 inches	92.30

Source: U.S. National Center for Health Statistics, *Vital and Health Statistics,* series 11, no. 238.

Thus, for instance, 4.22 percent of women in those years were less than or equal to 5 feet 0 inches tall, 24.84 percent were larger than 5 feet 0 inches but less than or equal to 5 feet 3 inches in height, and so on.

Suppose that a random selection of 200 present-day women in the age bracket of 18 to 24 resulted in 6 of them being less than 5 feet 0 inches, 42 of them being between 5 feet 0 inches and 5 feet 3 inches, 48 of them being between 5 feet 3 inches and 5 feet 5 inches, 60 of them being between 5 feet 5 inches and 5 feet 7 inches, 21 being between 5 feet 7 inches and 5 feet 8 inches, and the rest being taller than 5 feet 8 inches. Would these data imply that the height distribution has changed? Use the 5 percent level of significance.

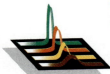

Chi-Squared Goodness of Fit Test

Purpose

Use Minitab to

1. Help perform chi-squared goodness of fit tests.

2. Perform tests for independence in populations classified according to two characteristics.

3. Help perform tests for independence in contingency tables with fixed marginal totals.

Procedures

First, load the Minitab (Windows version) software as in the Minitab lab for Chap. 1.

1. CHI-SQUARED GOODNESS OF FIT TEST

Example 1 Use Minitab to work Example 13.2 in Sec. 13.2 of the text.

Minitab does not currently have a listed macro to do the chi-squared goodness of fit test. However, with a few simple commands you can do most of the computations for the test in Minitab.

First you need to compute the expected frequencies from the information given. Save the observed frequencies in column C1, and rename as OF (for observed frequencies). Save the computed expected frequencies in column C2, and rename as EF (for expected frequencies). Select Calc→Mathematical Expressions, and the Mathematical expressions dialog box will be displayed. In the Variable text box type C3, and type in the expression shown in the Expression text box in Fig. M13.1. Observe that this expression represents the test statistic.

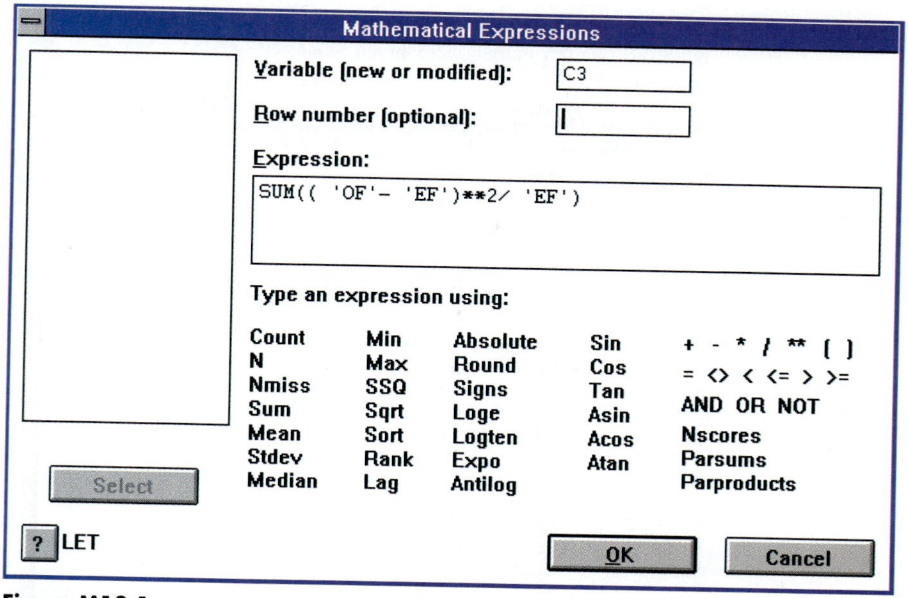

Figure M13.1

Click on the OK button, and a value of 4.13739 will be displayed in column C3 in the Data window. If you wanted to save the test statistic as a constant, you could have typed K1 (say) in the Variable text box, and the computed value would have been saved as the constant K1. However, in order to see the value of K1, you would need to type PRINT K1 at the MTB> prompt in the Session window. Next, to compute the p value for the test, select Calc→ Probability Distributions→Chisquare, and the Chi-square distribution dialog box will be displayed. Select the Cumulative probability check box, and type 3 in the Degrees of freedom text box. Select the Input column check box, and type C3 in the corresponding text box. You may save in a column if you wish by specifying where to save in the Optional storage text box.

Click on the OK button, and $P(X \leq x) = .7530$ will be displayed in the Session window since no storage was specified. Thus the p value $= 1 - .7530 = .2470$, and we will fail to reject the null hypothesis that the blood type distribution of people with stomach cancer is the same as that of the general public. Note that if you had selected Inverse cumulative probability, a critical (table) chi-squared value of 7.8147 would be displayed in the Session window at the 5 percent level of significance. You would need to select the Input constant check box and type .95 in the text box.

2. TESTS FOR INDEPENDENCE IN POPULATIONS CLASSIFIED ACCORDING TO TWO CHARACTERISTICS

Example 2   Use Minitab to work Example 13.9 in Sec. 13.3 of the text.

First, enter the table values in columns C1 through C3. Do not enter the row totals or column totals. Next select Stat→Tables→Chisquare Test, and the Chi-square test dialog box will be displayed. Select C1-C3 for the Columns containing the table text box.

Click on the OK button, and the computations will be displayed in the Session window. This output is shown in Fig. M13.2. Observe that the computed test statistic value is 6.828 with a p value of .146. Thus you would fail to reject the null hypothesis at the 5 percent level of significance that the depressive state of the clinic's patients is independent of their marital status. Note, no column names were given since we needed a two-way classification. Minitab automatically classified the rows as 1, 2, and 3.

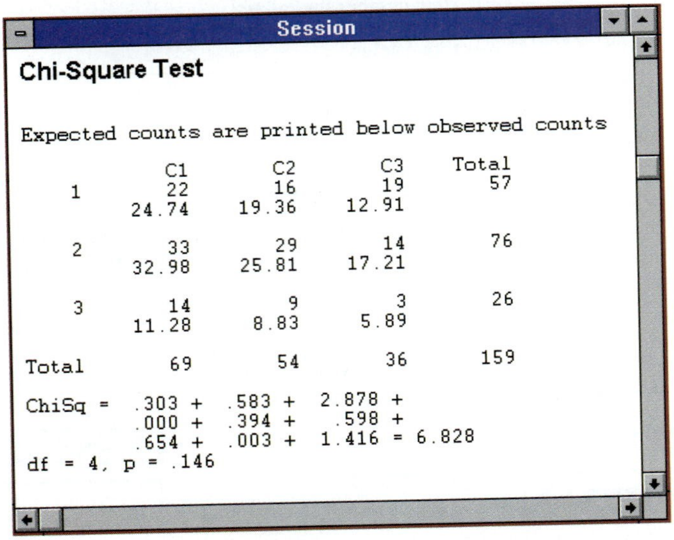

```
┌─────────────────────────────────────────────────────────────┐
│  ▼  ▲                    Session                        ▼  ▲ │
├─────────────────────────────────────────────────────────────┤
│ Chi-Square Test                                             │
│                                                             │
│ Expected counts are printed below observed counts          │
│                                                             │
│              C1        C2        C3     Total               │
│      1       22        16        19       57                │
│           24.74     19.36     12.91                         │
│                                                             │
│      2       33        29        14       76                │
│           32.98     25.81     17.21                         │
│                                                             │
│      3       14         9         3       26                │
│           11.28      8.83      5.89                         │
│                                                             │
│  Total       69        54        36      159               │
│                                                             │
│  ChiSq =  .303 +  .583 +  2.878 +                          │
│           .000 +  .394 +   .598 +                          │
│           .654 +  .003 +  1.416 = 6.828                    │
│  df = 4, p = .146                                          │
└─────────────────────────────────────────────────────────────┘
```

Figure M13.2

3. TESTING FOR INDEPENDENCE IN CONTINGENCY TABLES WITH FIXED MARGINAL TOTALS

Note The test for independence is unaffected by whether the marginal totals of one of the characteristics are fixed in advance or result from a random sample of the entire population. Thus, you can use the procedures in Sec. 2 to test for independence when the marginal totals are fixed.

Computer Exercises

1. Use Minitab or any other statistical software to help work Prob. 2 to 7 and 9 to 15 in Sec. 13.2.

2. Use Minitab or any other statistical software to work Probs. 1 to 14 in Sec. 13.3.

3. Use Minitab or any other statistical software to work Probs. 1 to 10 in Sec. 13.4.

4. From the Session window in Fig. M13.2, write up an appropriate hypothesis test for the example.

5. Use your library or any other resource to collect a set of data that could be analyzed with the chi-squared goodness of fit test. Use Minitab or any other statistical software to do a complete analysis of the data. Present any relevant analysis and discussion in a report. If you select data from published research, include relevant factors about the validity and reliability of the study.

6. Use your library or any other resource to collect a data set that could be classified according to two characteristics. Use Minitab or any other statistical software to do a complete analysis of the data. Present your analysis and discussion in a report. If you select data from published research, include relevant factors about the validity and reliability of the study.

7. Design your own experiment, and collect data that could be analyzed with the chi-squared goodness of fit test. Use Minitab or any other statistical software to present your findings in a report.

8. Design two experiments and collect data that could be classified according to two characteristics. Use both cases as discussed in your text for fixed and variable marginal totals. Use Minitab or any other statistical software to present your findings in a report.

Nonparametric
Hypotheses Tests

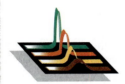

 We consider hypotheses tests in situations where the underlying population distribution is unknown and cannot be reasonably assumed to have any specified parametric form, such as normality. We show how to use the sign test to test hypotheses concerning the median of the distribution. The signed-rank test for testing that a population distribution is symmetric about 0 is introduced. We present the rank-sum test for testing the equality of two population distributions. Finally, we study the runs test that can be used to test the hypothesis that a sequence of 0s and 1s is a random sequence that does not follow any specified pattern.

14.1 INTRODUCTION

Are we making the earth warmer? More precisely, are humans' actions causing the earth's temperature to rise? Even though data appear to indicate that recent annual average temperatures are among the highest ever recorded, this question is surprisingly difficult to answer. One difficulty involves the change in the geographic locations where measurements are taken over time. For instance, past temperature measurements were usually taken in relatively secluded rural regions whereas present-day measurements are usually taken near cities having many paved roads (that tend to hold heat). This fact, in itself, will result in higher present-day temperature readings. Another difficulty results from uncertainty concerning the accuracy of measurements from long ago. In addition, there is the statistical question as to whether higher present-day temperatures are due to some real change, such as the burning of carbon-based products which might result in the trapping of the sun's energy in the earth's atmosphere, or whether these higher readings are just the chance fluctuations in random samples.

To get a handle on the statistical part of the problem, we want to be able to test whether a data set of temperatures over time represents a random sample from some fixed probability distribution, or whether the distribution of temperatures is itself changing over time.

In considering the above question—Is there a fixed underlying distribution of temperatures that is unchanging over time?—it is important to note that we are not specifying in advance the form of this distribution. In particular, since there is no à priori reason to believe that such an underlying distribution would necessarily be a normal distribution, we certainly do not want to make that supposition. Rather we need to develop a hypothesis test that is valid for any underlying type of distribution. Hypotheses tests which can be used in situations where the underlying distribution of the data is not required to have any particular form will be studied in this chapter. Because the validity of these tests does not rest on the assumption of any particular parametric form (such as normality) for the underlying distribution, these tests are called *nonparametric*.

14.2 SIGN TEST

Consider a large population of elements, each of which has a measurable value. Suppose that the distribution of population values is continuous and that we are interested in testing hypotheses concerning the median, or middle value, of this distribution. If the population distribution is normal, then the median is equal to the mean, and the methods of the previous chapters should be employed. However, we do not make the normality assumption here, but we present tests that can be used for any continuous distribution.

Let η denote the median value of the population. That is, exactly half of the members of the population have values less than η, and half have values greater than η. Equivalently, if X is a randomly chosen member of the population, then

$$P\{X < \eta\} = P\{X > \eta\} = \frac{1}{2}$$

Suppose now that we want to test the null hypothesis that the median is equal to some given value m. To obtain a test of

$$H_0: \eta = m$$

against

$$H_1: \eta \neq m$$

let p denote the proportion of the entire population whose value is less than m. That is,

$$p = P\{X < m\}$$

where X is a randomly chosen member of the population. Now if the null hypothesis is true and m is indeed the median, then p will equal 1/2. On the other hand, if m is not equal to the median, the p will not equal 1/2. Therefore, a test of the hypothesis that the median is equal to m is equivalent to a test of the null hypothesis

$$H_0: p = \frac{1}{2}$$

against the alternative

$$H_1: p \neq \frac{1}{2}$$

Thus we see that testing the hypothesis that the median is equal to m is equivalent to testing whether a population proportion is equal to 1/2. This proportion is, of course, equal to the proportion of the population whose value is less than m.

We can now make use of the results of Sec. 9.5.1 to obtain a test of the null hypothesis H_0 that the median of the population is equal to m. Namely, choose a random sample of n elements of the population, and let TS denote the number of them having values less than m. Note that when H_0 is true, TS will be a binomial random variable with parameters n and 1/2. The test is to reject the null hypothesis if the value of TS is too large or too small. Specifically, if the observed value of TS is i, then the significance level α test calls for rejecting H_0 if either

$$P\{N \geq i\} \leq \frac{\alpha}{2}$$

or

$$P\{N \leq i\} \leq \frac{\alpha}{2}$$

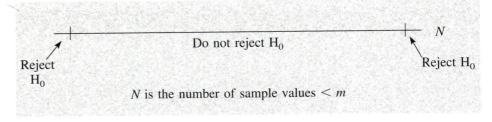

Figure 14.1
A test of H_0:
$\eta = m$ against
H_1: $\eta \neq m$.

where N is a binomial random variable with parameters $(n, 1/2)$. Figure 14.1 illustrates the test.

Because we have assumed that the population distribution is continuous there should not, in principle, be any data values exactly equal to m. However, since measurements are recorded to the accuracy of the instrumentation used, this may occur in practice. If there are values equal to m, they should be eliminated and the value of n reduced accordingly.

In terms of the p value, the above can be summed up as follows:

To test

$$H_0: \eta = m \quad \text{against} \quad H_1: \eta \neq m$$

choose a random sample. Discard any values equal to m. Let n be the number of values that remain. Let the test statistic be the number of values that are less than m. If there are i such values, then the p value is

$$p \text{ value} = 2 \, \text{Min}(P\{N \leq i\}, P\{N \geq i\})$$

where N is a binomial random variable with parameters n and $1/2$. The null hypothesis is then rejected at all significance levels greater than or equal to the p value and is not rejected otherwise.

To find the p value, it is not necessary to compute both $P\{N \leq i\}$ and $P\{N \geq i\}$. Rather we only need to compute the smaller of these two probabilities. Since $E[N] = n/2$, this will be $P\{N \leq i\}$ when i is small (compared to $n/2$) and $P\{N \geq i\}$ when i is large (compared to $n/2$). When i is near $n/2$, no computations are necessary since both probabilities are close to $1/2$ and so the p value is near 1. Thus, from a practical point of view, the p value can be expressed as

$$p \text{ value} = \begin{cases} 2P\{N \leq i\} & \text{if } i \leq \dfrac{n}{2} \\[2mm] 2P\{N \geq i\} & \text{if } i \geq \dfrac{n}{2} \end{cases}$$

where N is binomial with parameters n and 1/2.

Example 14.1 The inventory ordering policy of a particular shoe store is partly based on the belief that the median foot size of teenage boys is 10.25 inches. To test this hypothesis, the foot size of each of a random sample of 50 boys was determined. Suppose that 36 boys had sizes in excess of 10.25 inches. Does this disprove the hypothesis that the median size is 10.25?

Solution Let N be a binomial random variable with parameters (50, 1/2). Since 36 is larger than $50(1/2) = 25$, we see that the p value is

$$p \text{ value} = 2P\{N \geq 36\}$$

We can now use either the normal approximation or Program 5-1 to explicitly compute the above probability. Since

$$E[N] = 50 \times \frac{1}{2} = 25 \qquad \text{Var } (N) = 50 \times \frac{1}{2} \times \frac{1}{2} = 12.5$$

the normal approximation yields the following:

$$
\begin{aligned}
p \text{ value} &= 2P\{N \geq 36\} \\
&= 2P\{N \geq 35.5\} \qquad \text{(the continuity correction)} \\
&= 2P\left\{ \frac{N - 25}{\sqrt{12.5}} \geq \frac{35.5 - 25}{\sqrt{12.5}} \right\} \\
&\approx 2P\{Z \geq 2.97\} \\
&= .0030 \qquad \text{from Table D.1}
\end{aligned}
$$

(Program 5-1, which computes binomial probabilities, yields the exact value .0026.) Thus the belief that the median shoe size is 10.25 inches is rejected even at the 1 percent level of significance. There appears to be strong evidence that the median shoe size is greater than 10.25.

Suppose $X_1, \ldots, X_n$ are the n sample data values. Since the value of the test statistic depends on only the signs, either positive or negative, of the values $X_i - m$, the foregoing test is called the *sign test*.

14.2.1 Testing the Equality of Population Distributions When Samples Are Paired

The sign test can also be used to compare two populations when there is a natural pairing between the elements of their samples. We illustrate this by an example.

Example 14.2   An experiment was performed to see if two different sunscreen lotions, both having sun protection factor 15, are equally effective. A group of 12 volunteers exposed their backs to the sun for 1 hour in midday. Each volunteer had brand A sunscreen on one side of his or her spine and brand B on the other side. A measure of the amount of sunburn resulting on both sides of the spine was then determined for each volunteer. If 10 of the volunteers had less of a burn on the side receiving brand A sunscreen than on the side receiving brand B, can we conclude that the brands are not equally effective?

Solution   In this example we can imagine that we have two different populations, the population of backs receiving brand A sunscreen and the population receiving brand B. The "paired" members of the two samples are the two sides of each volunteer's back. Now if the two sunscreens were equally effective, then the median of the difference in sunburn of the two sides of a volunteer's back would equal 0. That is, just by chance, roughly half of the time brand A should perform better than brand B, and vice versa. Thus, we can test for equality of effectiveness by testing the hypothesis that the median of the difference between the brand A and the brand B sunburn of each volunteer is equal to 0.

Since the number of differences whose value is negative is 10, which is greater than $12(1/2) = 6$, we obtain from the sign test that the p value is

$$p \text{ value} = 2P\{N \geq 10\}$$

where N is binomial with parameters $(12, 1/2)$. Since

$$P\{N \geq 10\} = P\{N = 10\} + P\{N = 11\} + P\{N = 12\}$$

$$= \frac{12!}{10!\ 2!}\left(\frac{1}{2}\right)^{12} + \frac{12!}{11!\ 1!}\left(\frac{1}{2}\right)^{12} + \frac{12!}{12!\ 0!}\left(\frac{1}{2}\right)^{12}$$

$$= \left[\frac{12 \cdot 11}{2 \cdot 1} + 12 + 1\right]\left(\frac{1}{2}\right)^{12} = \frac{79}{4096}$$

we see that

$$p \text{ value} = \frac{158}{4096} = .0386$$

Thus, the null hypothesis of equal effectiveness is rejected at any significance level greater than or equal to 3.86 percent. (For instance, it is rejected at the 5, but not the 1, percent level of significance.)

14.2.2   One-Sided Tests

We can also use the sign test to test one-sided hypotheses about a population median. Suppose we want to test

$$H_0: \eta \leq m$$

against

$$H_1: \eta > m$$

where η is the population median and m is some specified value. Again, let p denote the proportion of the population whose values are less than m. Now if the null hypothesis is true and so m is at least as large as η, then the proportion of the population whose value is less than m is at least 1/2 (see Fig. 14.2). Similarly, if the alternative hypothesis is true and so m is less than η, then the proportion of the population whose value is less than m is less than 1/2 (see Fig. 14.2). Hence, the above is equivalent to testing

$$H_0: p \geq \frac{1}{2}$$

against

$$H_1: p < \frac{1}{2}$$

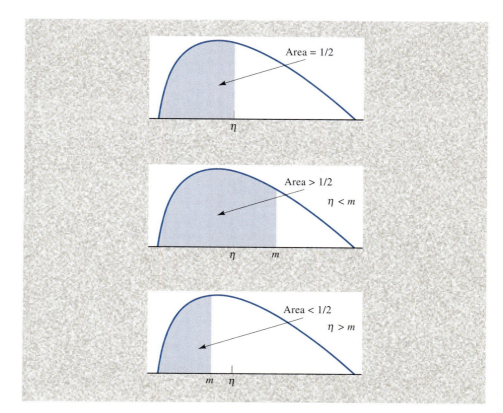

Figure 14.2
$P\{X < m\} < 1/2$
  if $\eta > m$
$P\{X < m\} = 1/2$
  if $\eta = m$
$P\{X < m\} > 1/2$
  if $\eta < m$

To use the sign test to test the above one-sided hypothesis, choose a random sample of n elements of the population. Suppose that i of them have values that are less than m. The resulting p value is the probability that a value as small as or smaller than i would have occurred by chance if each element had probability 1/2 of being less than or equal to m. That is, letting N be a binomial random variable with parameters $(n, 1/2)$, we have

$$p \text{ value} = P\{N \le i\}$$

Example 14.3   A bank has decided to build a branch office in a particular community if it can be established that the median annual income of residents of the community is greater than $40,000. To obtain information, a random sample of 80 families were chosen, and the families were questioned about their income. Of the 80 families, 52 had annual incomes above and 28 had annual incomes below $40,000. Is this information significant enough, at the 5 percent level of significance, to establish that the median income in the community is greater than $40,000?

Solution We need to see if the data are sufficiently strong to reject the null hypothesis when testing

$$H_0: \eta \le 40 \qquad \text{against} \qquad H_1: \eta > 40$$

If p is the proportion of families in the population with annual incomes below $40,000, then this is equivalent to testing

$$H_0: p \ge \frac{1}{2} \qquad \text{against} \qquad H_1: p < \frac{1}{2}$$

Since 28 of the 80 sampled families have annual incomes below $40,000, the p value of the data is

$$p \text{ value} = P\{N \le 28\}$$

where N is a binomial random variable with parameters $(80, 1/2)$. Using Program 5-1 (or the normal approximation) yields

$$p \text{ value} = .0048$$

For such a low p value, the null hypothesis that the median income is less than or equal to $40,000 is rejected, thus establishing that the median income almost certainly exceeds this value.

A test of the one-sided null hypothesis that the median is at least m is similar to the preceding. Thus the one-sided tests are as follows.

> ### One-sided hypotheses tests of the median
>
> To test either
>
> $$H_0: \eta \leq m \qquad \text{against} \qquad H_1: \eta > m \qquad\qquad (1)$$
>
> or
>
> $$H_0: \eta \geq m \qquad \text{against} \qquad H_1: \eta < m \qquad\qquad (2)$$
>
> choose a random sample from the population. Remove all values equal to m. Suppose n remain. Let TS denote the number of data values that are less than m. If TS is equal to i, then the p values are
>
> $$p \text{ value } = P\{N \leq i\} \qquad \text{in case (1)}$$
> $$p \text{ value } = P\{N \geq i\} \qquad \text{in case (2)}$$
>
> where N is a binomial random variable with parameters n and $1/2$.

Problems

1. The published figure for the median systolic blood pressure of middle-aged men is 128. To determine if there has been any change in this value, a random sample of 100 men have been selected. Test the hypothesis that the median is equal to 128 if
   (a) 60 men have readings above 128
   (b) 70 men have readings above 128
   (c) 80 men have readings above 128
   In each case, determine the p value.

2. In 1989, the median household income for the state of Connecticut was $41,721. A recent survey randomly sampled 250 households and discovered that 42 percent had incomes below the 1989 median and 58 percent had incomes above it. Does this establish that the median household income in Connecticut is no longer the same as in 1989? What is the p value?

3. Fifty students at the police academy took target practice, using two different types of guns. Each student took half of her or his shots with the less expensive gun and the other half with the more expensive gun. If 29 students had higher scores with the less expensive gun, does this establish that the two guns are not equally effective? Use the 5 percent level of significance.

4. A dermatology clinic wants to compare the effectiveness of a new hand cream and the one it presently recommends to patients suffering from eczema. To

gather information, half of its patients are told to put the new skin cream on their left hand and the old cream on their right hand each night for one week; and the other half are told to put the new cream on their right hand and the old one on their left. Each patient is examined after one week. Suppose that for 60 percent of the patients the hand receiving the new cream showed greater improvement than the one receiving the old cream.

When the number of patients involved is equal to

(a) 10   (b) 20   (c) 50   (d) 100   (e) 500

do these data prove that the two creams are not equally effective? Use the 5 percent level of significance. Also find the p value in each case.

5. A statistics instructor has made up an examination for a large class of students. She wants the median score on the examination to be at least 72 and thinks that this test will enable her to reach her goal. To be cautious, she has randomly chosen 13 students to take the examination early. If their scores are

65, 79, 77, 90, 56, 60, 65, 80, 70, 69, 83, 69, 65

should the hypothesis that the median score will be at least 72 be rejected? Use the 5 percent level of significance.

6. The median selling price of a home in a certain residential community has been steady at $122,000 for the past 2 years. To determine if the median price has increased, a random sample of 20 recently sold homes were chosen. The selling prices of these homes were (in units of $1000)

144, 116, 125, 128, 96, 92, 163, 130, 120, 142,
155, 133, 110, 105, 136, 140, 124, 130, 88, 146

Are these data strong enough to establish that the median price has increased? Use the 5 percent level of significance.

7. To test the hypothesis that the median weight of 16-year-old females from Los Angeles is at least 110 pounds, a random sample of 200 such females were chosen. If 120 females weighed less than 110 pounds, does this discredit the hypothesis? Use the 5 percent level of significance. What is the p value?

8. In an attempt to prove that fish oil lowers blood cholesterol levels, a nutritionist instructed 24 volunteers to take a certain fish oil supplement for 3 months. After this time each volunteer had his or her blood cholesterol level checked. Suppose that a comparison with their levels before the beginning of the experiment showed that 16 of the 24 volunteers experienced a reduction in cholesterol levels.
(a) What are the null and the alternative hypotheses?
(b) Is the null hypothesis rejected? Use the 5 percent level of significance.
(c) What is the p value?

9. In 1987, the national median salary of all U.S. physicians was $124,400. A recent random sample of 14 physicians showed 1990 incomes of (in units of $1000)

    125.5, 130.3, 133.0, 102.6, 198.0, 232.5, 106.8,
    114.5, 122.0, 100.0, 118.8, 108.6, 312.7, 125.5

Use these data to test the hypothesis that the median salary of physicians in 1990 was not greater than in 1987. What is the p value?

14.3 SIGNED-RANK TEST

In Sec. 14.2.1 we saw how the sign test could be used to test the null hypothesis that two populations have the same distribution of values, when the data consisted of paired samples. To test this hypothesis, we considered the differences of the paired-sample values. We noted that if the null hypothesis were true, then the median of these differences would be 0. The sign test was then used to test this latter hypothesis.

The only information needed for the sign test of the equality of two population distributions, when paired samples are used, is the number of times the first data value in a pair is larger than the second. That is, the sign test does not require the actual values of the data pairs, only knowledge of which is larger. However, although it is easy to use, the sign test is not a particularly efficient test of the null hypothesis that the population distributions are the same. For if this null hypothesis is indeed true, then not only will the distribution of paired differences have median zero but also it will have the stronger property of being symmetric about zero. That is, for any number x it will be just as likely for the first value in the pair to be larger than the second by the amount x as for the second value to be larger than the first by this amount (see Fig. 14.3). The sign test, however, does not check for symmetry of the distribution of differences, only that its median value is equal to zero.

Figure 14.3
A density symmetric about 0. Areas of shaded regions are equal.

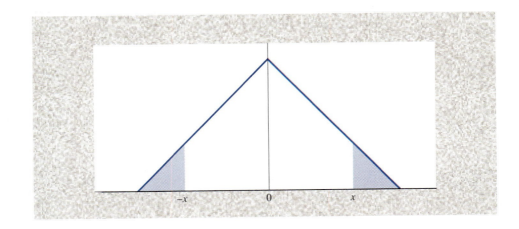

For instance, suppose that the data consist of 12 paired values whose differences are as follows:

$$2, 5, -.1, -.4, -.3, 9, 7, 8, 12, -.5, -1, -.6$$

Since six of the differences are positive and six are negative, this data set is perfectly consistent with the hypothesis that the median of the differences is 0. On the other hand, since all the large values are on the positive side, the data do not appear to be consistent with the hypothesis that their distribution is symmetric about 0. Thus, it seems highly unlikely that the population distributions are equal.

Suppose again that we want to use data consisting of paired samples to test the hypothesis that two population distributions are equal. We now present a test which is more sensitive than the sign test. It is called the *signed-rank test,* and it proceeds by testing whether the distribution of the differences of the paired values is symmetric about 0.

Suppose that paired samples of size n are chosen from the two populations. Let D_i denote the difference between the first population value and the second population value of the ith pair, for $i = 1, \ldots, n$. Now order these n differences according to their absolute values. That is, the first difference should be the value of D_i having smallest absolute value, and so on. The test statistic for the signed-rank test is the sum of the ranks (or positions) of the negative numbers in the resulting sequence.

Example 14.4   Suppose the data consist of the following four paired-sample values:

i	X_i	Y_i
1	4.6	6.2
2	3.8	1.5
3	6.6	11.7
4	6.0	2.1

The differences $X_i - Y_i$ are thus

$$-1.6, 2.3, -5.1, 3.9$$

Ordering these differences according to absolute value, from the smallest to the largest, gives the following:

$$-1.6, 2.3, 3.9, -5.1$$

Since the ranks of the negative values are 1 and 4, the value of the signed-rank test statistic is

$$\text{TS} = 1 + 4 = 5$$

In other words, since the ranked differences in positions 1 and 4 are negative, TS $=$ $1 + 4 = 5$.

The signed-rank test is like the sign test in that it considers those data pairs in which the first population value is less than the second. But whereas the sign test gives equal weight to each such pair, the signed-rank test gives larger weights to the pairs whose differences are farthest from zero.

The signed-rank test calls for the rejection of the null hypothesis when we are testing

H_0: two population distributions are equal

against

H_1: two population distributions are not equal

if the test statistic TS is either sufficiently large or sufficiently small. A large value of TS indicates that the majority of the larger values of the differences have negative signs; whereas a small value indicates that the majority have positive signs. Either situation would be evidence against the symmetry of the distribution of differences and thus evidence against H_0.

If the value of the test statistic is t, then the signed-rank test rejects H_0 if either

$$P\{\text{TS} \leq t\} \leq \frac{\alpha}{2}$$

or

$$P\{\text{TS} \geq t\} \leq \frac{\alpha}{2}$$

In the above, α is the level of significance, and the probabilities are to be computed under the assumption that H_0 is true. Equivalently, we have the following statement concerning the p value:

Suppose the value of TS is t. The p value of the signed-rank test is given by

$$p \text{ value} = 2 \, \text{Min}(P\{\text{TS} \leq t\}, P\{\text{TS} \geq t\})$$

where the above probabilities are to be determined under the assumption that H_0 is true.

To be able to implement the signed-rank test, we need to be able to compute the preceding probabilities. The key to accomplishing this is the fact than when H_0 is true, and so the distribution of differences is symmetric about zero, each of the differences is equally likely to be either positive or negative, independently of the others. Program 14-1 makes use of this fact to explicitly determine the necessary probabilities and the resulting p value. The inputs needed are the sample size n and the value of the test statistic TS.

Example 14.5   A psychology instructor wanted to see if students would perform equally well on two different examinations. He selected 12 students, who all agreed to take part in the experiment. Six of the students were given examination A, and the other six examination B. On the next day the students were tested on the examination they had not yet taken. Thus, each of the 12 students took both examinations. The following pairs of scores were obtained by the students on the two examinations.

						Student						
Examination	1	2	3	4	5	6	7	8	9	10	11	12
A	763	419	586	920	881	758	262	332	717	909	940	835
B	797	404	576	855	762	707	195	341	728	817	947	849

Thus, for instance, student 3 scored 586 on examination A and 576 on test B. The paired differences are as follows:

$$-34, \ 15, \ 10, \ 65, \ 119, \ 51, \ 67, \ -9, \ -11, \ 92, \ -7, \ -14$$

Ordering these in increasing order of their absolute values gives

$$-7, \ -9, \ 10, \ -11, \ -14, \ 15, \ -34, \ 51, \ 65, \ 67, \ 92, \ 119$$

Since the differences in positions 1, 2, 4, 5, and 7 are negative, the values of the test statistic is

$$\text{TS} = 1 + 2 + 4 + 5 + 7 = 19$$

To obtain the p value, we now run Program 14-1, which computes the p value for the signed-rank test that a population distribution is symmetric about 0. Our sample size is 12 and our observed value of the sum of signed ranks is 19. The p value as computed by Program 14-1 is .1293945.

Thus, the p value is .129, and so the null hypothesis that the distributions of student scores on the two examinations are identical cannot be rejected at the 10 percent level of significance.

By making use of the fact that the ordered differences are independent random variables which are each equally likely to be either positive or negative, it can be established that when H_0 is true, the mean and variance of TS are given by, respectively,

$$E[TS] = \frac{n(n + 1)}{4}$$

and

$$\text{Var (TS)} = \frac{n(n + 1)(2n + 1)}{24}$$

In addition, it can be shown that for moderately large values of n, TS will have a distribution, when H_0 is true, that is approximately normal with the above mean and variance. These facts enable us to approximate the p value when Program 14-1 is not available.

Example 14.6   Let us see how well the normal approximation of the p value works for the data of Example 14.5. Since $n = 12$, we obtain from the above formulas that, when H_0 is true,

$$E[TS] = \frac{12 \cdot 13}{4} = 39 \qquad \text{Var (TS)} = \frac{12 \cdot 13 \cdot 25}{24} = 162.5$$

The value of the test statistic is 19. Since this value is less than $E[TS]$, it is clear that $P\{TS \leq 19\}$ is smaller than $P\{TS \geq 19\}$. Therefore,

$$
\begin{aligned}
p \text{ value} &= 2P\{TS \leq 19\} \\
&= 2P\{TS \leq 19.5\} \qquad \text{(continuity correction)} \\
&= 2P\left(\frac{TS - 39}{\sqrt{162.5}} \leq \frac{19.5 - 39}{\sqrt{162.5}}\right) \\
&\approx 2P\{Z \leq -1.530\} \\
&= .126
\end{aligned}
$$

Thus, the normal approximation yields an approximate p value that is quite close to the actual p value of .129 given in Example 14.5.

A general rule of thumb is that the normal approximation to the p value will be quite good provided that n, the number of paired-data values, is at least 25. (In fact, this may be rather conservative, since in the above example the approximation was quite good and n was only equal to 12.)

Example 14.7   Suppose from a sample of 25 paired values the value of TS is 238. Assuming that the distribution of differences is symmetric about 0, we see from the preceding formulas that

$$E[\text{TS}] = \frac{25 \cdot 26}{4} = 162.5$$

$$\sqrt{\text{Var (TS)}} = \sqrt{\frac{25 \cdot 26 \cdot 51}{24}} = 37.165$$

The normal approximation to the p value is thus

$$
\begin{aligned}
p \text{ value} &= 2P\{\text{TS} \geq 238\} \\
&= 2P\{(\text{TS} \geq 237.5\} \\
&= 2P\left\{ \frac{\text{TS} - 162.5}{37.165} \geq \frac{237.5 - 162.5}{37.165} \right\} \\
&\approx 2P\{Z \geq 2.018\} \\
&= .0436
\end{aligned}
$$

On the other hand, using Program 14-1 yields the exact p value:

$$p \text{ value} = .0422$$

Thus, once again we see that the approximation is quite close to the exact value.

14.3.1   Zero Differences and Ties

If a difference has value 0 (because its paired values are equal), then that data value should be discarded and the value of n should be reduced by 1.

If some of the differences have the same absolute value, then the weight given to a negative difference should be the average of the ranks of all the differences with the same absolute value. For instance, if the differences are

$$-1, \ 3, \ 8, \ -3$$

then the ordered differences are

$$-1, \ 3, \ -3, \ 8$$

Since there is a tie between the second and third differences, the value of the test statistic is

Table 14.1	Signed-Rank Test

It tests the hypothesis that two population distributions are equal by using paired samples, where $X_1, \ldots, X_n$ are the sample from the first population; $Y_1, \ldots, Y_n$ are the sample from the second population; X_i and Y_i are paired; and $D_i = X_i - Y_i$, $i = 1, \ldots, n$.
To test

H_0: distribution of D_i is symmetric about 0

against

H_1: distribution of D_i is not symmetric about 0

first eliminate any D_i equal to 0. Change the value of n to let it reflect the number of non-zero differences. Take TS equal to the sum of the positions of the negative differences D_i, when the D_i are ranked in increasing order of their absolute values. If two or more of the D_i values are equal, then they are each given a rank equal to the average of their ranks.
If TS $= t$, then

$$p \text{ value} = 2 \text{ Min}(P\{TS \le t\}, P\{TS \ge t\})$$

where the probabilities are to be computed under the assumption that H_0 is true. The probability can be approximated by using the fact than when H_0 is true, TS is approximately a normal random variable with mean and variance, respectively, given by

$$E[TS] = \frac{n(n+1)}{4} \qquad \text{Var (TS)} = \frac{n(n+1)(2n+1)}{24}$$

If there are no ties, then the exact p value can be obtained by running Program 14-1.

$$TS = 1 + \frac{2+3}{2} = 3.5$$

Program 14-1 should not be used if there are any ties. Instead the normal approximation should be employed.
Table 14.1 sums up the signed-rank test.

Problems

1. Determine the value of the signed-rank test statistic if the differences of paired values are as follows.
   (a) $-17, 33, 22, -8, 55, -41, -18, 40, 39, 14, -88, 99, 102, -5, 7$
   (b) $44, 2, 1, -.4, -3, -13, 44, 50, 1.1, -2.2, .01, -4, -6.6$
   (c) $12, 15, 19, 8, -3, -7, -22, -55, 48, 31, 89, 92$

2. Assuming that the difference of paired values has a distribution that is symmetric about 0, determine the mean and variance for the signed-rank test statistic for each of the parts of Prob. 1.

3. For each part of Prob. 1, find the p value of the hypothesis that the distribution of the differences is symmetric about 0. Use the normal approximation.

4. Compare the answers obtained in Prob. 3 with the exact p values given by Program 14-1.

5. A history professor wondered whether the student graders for her course tended to take into account whether term papers were handwritten or typed. As an experiment, the professor divided up 28 students into 14 pairs. Each pair of students was regarded by the professor to have roughly the same abilities. The professor then assigned a project and asked one member of each pair to turn in a handwritten report and the other member to turn in a typed report. For each pair, the decision as to which student was asked for the handwritten report was based on the flip of a coin. The grades given to the projects were as follows.

Pair	Handwritten	Typed
1	83	88
2	75	91
3	75	72
4	60	70
5	72	80
6	55	65
7	94	90
8	85	89
9	78	85
10	96	93
11	80	86
12	75	79
13	66	64
14	55	68

(a) Would you conclude that how the paper is presented, either typed or handwritten, had an effect on the score given? Use the 5 percent level of significance.

(b) What is the p value?

6. To test the effectiveness of sealants on reducing cavities, half the teeth of 100 children were treated and the other half left untreated. After 6 months the difference between the number of cavities in the treated and untreated teeth of each child was determined. The signed-rank test statistic for these differences was 1830. Can we conclude, at the 5 percent level of significance, that sealants make a difference? What about at the 1 percent level of significance?

7. A consumer organization wanted to determine whether automobile repair shops were giving different estimates to women than to men. It selected two cars hav-

ing the identical defect and gave one to a man and the other to a woman. Randomly choosing eight repair shops, the organization had the man take his car to four of these shops and had the woman go to the other four. One week later they repeated the process, with the man going to the shops previously visited by the woman and vice versa. The dollar prices quoted were as follows.

Shop	Price quoted to man ($)	Price quoted to woman ($)
1	145	145
2	220	300
3	150	200
4	100	125
5	250	400
6	150	135
7	180	200
8	240	275

Test the hypothesis that the sex of the person bringing the car to the repair shop does not affect the quoted price, using the
(a) Sign test
(b) Signed-rank test

8. Eleven patients having high albumin content in their blood are treated with a medicine. The measured values of their albumin both before and after the medication are as follows:

Blood Content of Albumin (grams per 100 milliliters)

Patient	Before medication	After medication
1	5.04	4.82
2	5.16	5.20
3	4.75	4.30
4	5.25	5.06
5	4.80	5.38
6	5.10	4.89
7	6.05	5.22
8	5.27	4.69
9	4.77	4.52
10	4.86	4.72
11	6.14	6.26

(a) What is the value of the test statistic of the signed-rank test?
(b) What is the p value of the test that the treatment has no effect?

9. An engineer claims that painting the exterior of a particular aircraft will affect its cruising speed. To check this claim, 10 aircraft just off the assembly line

were flown to determine cruising speed prior to painting, and they were flown again after being painted. The following data resulted.

Aircraft	Cruising speed (miles per hour)	
	No paint	Paint
1	426.1	416.7
2	438.5	431.0
3	440.6	442.6
4	418.5	423.6
5	441.2	447.5
6	427.5	423.9
7	412.2	412.8
8	421.0	419.8
9	434.7	424.1
10	411.9	418.7

Do the data establish that the engineer is correct? Use the 5 percent level of significance.

10. Let $X_1, \ldots, X_n$ be a random sample of data from a certain population. Suppose we want to test the hypothesis that data from this population are symmetric about some value v. Explain how we could accomplish this by using the signed-rank test. (*Hint:* Let $D_i = X_i - v$.)

14.4 RANK-SUM TEST FOR COMPARING TWO POPULATIONS

Consider two populations having a certain measurable characteristic, and suppose we are interested in testing the hypothesis that the two population distributions of this characteristic are the same. To test this hypothesis, suppose that independent samples of sizes n and m are drawn from the two populations.

If we were willing to assume that the underlying probability distributions were both normal, then we would apply the two-sample tests developed in Chap. 10. However, instead we will develop a nonparametric test which does not require the assumption of normality.

To begin, rank the $n + m$ data values from the two samples from smallest to largest. That is, give rank 1 to the smallest data value, rank 2 to the second smallest, and so on. For the time being we will assume that the $n + m$ values are all distinct, so there are no ties. Designate one of the samples (it makes no difference which one) as the first sample. The test we will consider makes use of the test statistic TS, defined to equal the sum of the ranks of the first sample. That is,

TS = sum of ranks of data in first sample

Example 14.8   To determine if reflex reaction time is age-dependent, a sample of eight 20-year-old men and an independent sample of nine 50-year-old men were chosen. The following represents their reaction times (in seconds) to a given stimulus.

20-year-olds: 4.22, 5.13, 1.80, 3.34, 2.72, 2.80, 4.33, 3.60

50-year-olds: 5.42, 3.39, 2.55, 4.45, 5.55, 4.96, 5.88, 6.30, 5.10

Putting these 17 values in increasing order gives the following:

1.80,* 2.55, 2.72,* 2.80,* 3.34,* 3.39, 3.60,* 4.22,* 4.33,* 4.45, 4.96, 5.10,

5.13,* 5.42, 5.55, 5.88, 6.30

In the above we have put a star next to the data values that come from the 20-year-olds (which we are taking to be the first sample). Hence, the value of the sum of the ranks of the first sample is

$$TS = 1 + 3 + 4 + 5 + 7 + 8 + 9 + 13 = 50$$

Let H_0 be the hypothesis that the two population distributions are identical, and suppose that the value of the test statistic TS is t. Since we want to reject H_0 if the value of TS is either significantly large or significantly small, it follows that the significance level α test will call for rejection of H_0 if either

$$P\{TS \leq t\} \leq \frac{\alpha}{2}$$

or

$$P\{TS \geq t\} \leq \frac{\alpha}{2}$$

where both of the preceding probabilities are to be computed under the assumption that H_0 is true. In other words, the null hypothesis will be rejected if the sum of the ranks from the first sample is either too small or too large to be explained by chance. As a result, it follows that the significance level α test will call for rejection of H_0 if the p value of the data set, given by

$$p \text{ value} = 2 \text{ Min}(P\{TS \leq t\}, P\{TS \geq t\})$$

is less than or equal to α.

To determine the above probabilities, we need to know the distribution of TS when H_0 is true. To begin, suppose that the first sample is the one of size n. Now, when H_0 is true and so all the $n + m$ data values come from the same distribution, it follows that the set of ranks of the first sample will have the same distribution as a random selection of n of the values $1, 2, \ldots, n + m$. Using this, we can show that when H_0 is true, the mean and variance of TS are given by the following formulas.

When H_0 is true,

$$E[\text{TS}] = \frac{n(n + m + 1)}{2}$$

$$\text{Var (TS)} = \frac{nm(n + m + 1)}{12}$$

In addition, it can be shown that when n and m are both of at least moderate size (both being larger than 7 should suffice), TS will, when H_0 is true, approximately have a normal distribution. Hence, if the sample sizes are not too small, it follows that TS will be approximately normal with a mean and variance as stated in the preceding.

Example 14.9 In Example 14.8, $n = 8$, $m = 9$, and the value of the sum of the ranks of the first sample was TS $= 50$. Now

$$E[\text{TS}] = \frac{n(n + m + 1)}{2} = 72 \qquad \text{Var (TS)} = \frac{nm(n + m + 1)}{12} = 108$$

Since the observed value of TS was less than its mean, we have

$$
\begin{aligned}
p \text{ value} &= 2P\{\text{TS} \le 50\} \\
&= 2P\{\text{TS} \le 50.5\} \\
&= 2P\left\{ \frac{\text{TS} - 72}{\sqrt{108}} \le \frac{50.5 - 72}{\sqrt{108}} \right\} \\
&\approx 2P\{Z \le -2.069\} \\
&= .0385
\end{aligned}
$$

Therefore, the null hypothesis that the two population distributions are identical would be rejected at the 5 percent level of significance.

If there are any ties, then the rank of a data value should be the average of the ranks of all those with the same value. For instance, if the first-sample data are 2, 4, 4, 6 and the second-sample data are 5, 6, 7, then the sum of the ranks of the data of sample 1 is $1 + 2.5 + 2.5 + 5.5 = 11.5$. The test should then be run exactly as before. (It turns out that the possibility of ties has the effect of reducing the variance of TS when the null hypothesis is true. As a result, the p value previously given will be larger than the actual p value that takes into account the tied values. In consequence, the test presented will be conservative in that whenever ties are present and it calls

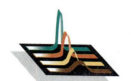

The preceding test is called the *two-sample rank-sum test*. Aside from the sign test, which goes back to Arbuthnot in 1710 (see Sec. 9.5), it was one of the first nonparametric tests to be developed. It was jointly and independently discovered in the mid-1940s by Wilcoxon and the team of Mann and Whitney. Because of this, the test is sometimes called the *Wilcoxon sum-of-ranks test* and is sometimes called the *Mann-Whitney test*. The publications of Wilcoxon and Mann-Whitney were the beginning of a wave of research on nonparametric tests, one which has not yet abated.

for rejection, then a more sophisticated test that takes the ties into account will also call for rejection of the null hypothesis.)

Example 14.10 In an attempt to determine if the vocabulary skills of two different students are similar, an English teacher had each of them write a short essay on the same topic. The teacher then counted the number of times each student used words having four or more letters. The following data resulted.

| | Number of words used having i letters | |
i	Student 1	Student 2
4	44	49
5	16	11
6	8	5
7	7	4
8	4	1
9	2	1
10	3	0

Thus, for instance, 8 out of the 84 words (having four or more letters) written by student 1 and 5 of the 71 words used by student 2 were six-letter words. Use these data to test the hypothesis that the word frequency distributions of the two students are the same.

Solution The data consist of one sample of 84 words and another sample of 71 words. Since in the combined samples of 155 words the data value 4 appears 93 times, each of these 93 data values is given a rank equal to the average of the rank numbers 1 through 93. That is, each is given rank

$$\frac{(1 + 2 + \cdots + 93)}{93} = \frac{1 + 93}{2} = 47$$

Also, since the next-smallest data value (the value 5) occurs 27 times, each of these values shares the ranks from 94 through 120. Therefore, each of the data values 5 is given rank

$$\frac{94 + 120}{2} = 107$$

Similarly, the data values 6, 7, 8, 9 and 10 are given rank values as shown below.

Data value	4	5	6	7	8	9	10
Rank	47	107	127	139	147	151	154

The sum of the ranks of the sample of 71 words has value

$$TS = 47 \times 49 + 107 \times 11 + 127 \times 5 + 139 \times 4 + 147 + 151 = 4969$$

Since $n = 71$, and $m = 84$, we see that

$$\frac{n(n + m + 1)}{2} = 5538 \qquad \frac{nm(n + m + 1)}{12} = 77{,}532$$

Thus, the approximate p value is

$$
\begin{aligned}
p \text{ value} &= 2P\{TS \le 4969\} \\
&= 2P\{TS \le 4969.5\} \\
&= 2P\left\{ \frac{TS - 5538}{\sqrt{77{,}532}} \le \frac{4969.5 - 5538}{\sqrt{77{,}532}} \right\} \\
&\approx 2P\{Z \le -2.04\} \\
&= .041
\end{aligned}
$$

and so the hypothesis that the word-length distributions of the two students are identical is rejected at the 5 percent level of significance.

In cases where the sample sizes are not large, say, when either is less than 8, we can no longer assume that the distribution of the sum of the ranks is approximately normal. However, we can still employ the above rank-sum test by directly computing the exact p value. To do so, we use the fact that when H_0 is true, the set of ranks of the first sample has the same distribution as a random selection of n of the values 1, 2, ..., $n + m$. By using this, it is possible (with the help of a computer) to explicitly determine the p value. Program 14-2 computes the exact p value for the rank-sum test. The inputs needed are the sizes of the first and second samples and the sum of the ranks of the elements of the first sample. Whereas either sample can be designated as the first sample, the program will run fastest if the first sample is the one whose sum of ranks is smaller. In addition, since this program implicitly assumes there are no ties, it can be used only when the value of TS is an integer.

Historical Perspective

Thomas Mendenhall
(Ohio State University
Photo Archive,
Columbus)

The use of statistical techniques to make literary comparisons goes back a long way. In 1901, Thomas Mendenhall, who had been a professor of physics at Ohio State University, published a comparison of the frequencies of the number of letters in the words used by Shakespeare and other authors. Mendenhall noted that nearly all Shakespeare's plays had approximately the same frequency distribution. He showed that Shakespeare used a higher proportion of words with one, two, four, or five letters and a lesser proportion of the others when compared with either Dickens or Thackeray. Francis Bacon's distribution was also found to be quite different from Shakespeare's. Excitement was stirred, however, as an analysis of the plays of Christopher Marlowe produced a word-size frequency distribution almost identical to that of Shakespeare.

More recently, statistical analysis has been employed to decide the author-ship of 12 *Federalist* papers. These papers, consisting of 77 letters, appeared anonymously in New York State newspapers between 1787 and 1788. The letters tried to persuade the citizens of New York to ratify the Constitution. Although it was generally known that the authors of the papers were Alexander Hamilton, John Jay, and James Madison, it was not known which of them was responsible for each specific paper. As of 1964, authorship of most of the papers had been determined. However, a long-standing dispute remained concerning the authorship of 12 of them. In a book published in 1964, Harvard statisticians Frederic Mosteller and David Wallace used a statistical analysis to conclude that all 12 papers had been written by Madison alone. Their analysis considered such things as the frequency distributions of each author's use of such words as *by, from, to,* and *upon.*

Example 14.11   Let us reconsider Example 14.9, this time using Program 14-2 to compute the p value. This program runs best if you designate as the first sample, the sample having the smaller sum of ranks. The size of the first sample is 8. The size of the second sample is 9. The sum of the ranks of the first sample is 50. Program 14-2 computes the p values as 3.595229E-02.

Thus the exact p value is .0359, which is reasonably close to the approximate value of .0385 obtained by using the normal approximation in Example 14.9.

14.4.1   Comparing Nonparametric Tests with Tests that Assume Normal Distributions

The strength of nonparametric tests is that they can be used without making any assumptions about the form of the underlying distributions. The price that one pays for using a nonparametric test is that it will not be as effective in cases where the distributions are normal or approximately normal as would a test that starts out by assuming normality. Somewhat surprisingly, the loss in effectiveness is relatively small. For instance, it can be shown that when sample sizes n and m are large, the efficiency of the nonparametric rank-sum test is approximately 95 percent of that of the two-sample t test when the distributions are indeed normal. By this, we loosely mean that when the population distributions are normal but unequal, then the chance of rejection with the nonparametric test with samples of size n is roughly the same as with the normal-based t test with samples of size $.95n$. This is an impressive result, and it might easily lead one to conclude that the nonparametric test is superior if one is not absolutely certain that the distributions are close to normal. For if the distributions are not normal, then the normal test is based on a false assumption; and even if the distributions are normal, the nonparametric one is almost as good. However, even when the underlying distributions are not normal, the normal test will be a good one when sample sizes n and m are large. This is so because this test is based on a test statistic which will be approximately normal even when the population distributions are not. We can thus conclude that, for large sample sizes, the normal-based t test will be an effective test.

Probably the best we can say is that if one is not certain that the underlying distribution is at least approximately normal, then for moderate sample sizes the rank-sum nonparametric test is preferred to the two-sample t test. On the other hand, in cases of large sample sizes, either test type can be used. A key difference, however, that can be useful in deciding which type of test to use is that the t test is designed to detect differences in the population means whereas the rank-sum test is designed to detect any difference in the population distributions.

Problems

1. The following data are from independent samples from two populations.
   Sample 1:   142, 155, 237, 244, 202, 111, 326, 334, 350, 247
   Sample 2:   212, 277, 175, 138, 341, 255, 303, 188
   (a) Determine the sum of the ranks of the data from sample 1.
   (b) Determine the sum of the ranks of the data from sample 2.

2. There is an algebraic identity stating that the sum of the first k positive integers is equal to $k(k+1)/2$. That is,

$$\sum_{i=1}^{k} i = \frac{k(k + 1)}{2}$$

Use this identity to determine the relationship between the sum of the ranks of the sample of size n and the sum of the ranks of the sample of size m. Use the results of Prob. 1 to check your result. Assume that all $n + m$ data values are different.

3. A study was carried out to determine if educational opportunities in rural and urban California counties are the same. Two counties of roughly the same socioeconomic makeup, one in an urban area and the other in a rural area, were chosen. The Scholastic Aptitude Test (SAT) scores of a random sample of high school graduates were obtained in both counties. The results were as follows:

Rural	Urban
544	610
567	498
475	505
658	711
590	545
602	613
571	509
502	514
578	609

Find the p value of the test of the hypothesis that the distributions of scores in both counties are identical. Use the normal approximation.

4. A group of 16 volunteers were randomly divided into two subgroups of 8 each. Members of the first subgroup were given daily tablets containing 5 grams of vitamin C, and members of the second subgroup were given a placebo. After 1 month the blood cholesterol levels of the 16 individuals were measured and compared with their levels at the beginning of the experiment. The reductions in blood cholesterol levels for the two subgroups were as follows.

Vitamin C	Placebo
6	9
12	-3
14	0
2	-1
7	5
7	3
1	-4
8	-1

Test the null hypothesis, at the 5 percent level of significance, that vitamin C and the placebo are equally effective in reducing cholesterol. Assume that the distribution of the test statistic is approximately normal when the null hypothesis is true. (A negative data value means that the blood cholesterol level increased. For instance, the data value -4 indicates an increase of 4 in the blood cholesterol reading.)

5. A study was conducted to test the hypothesis that the starting salary distribution of students graduating from Stanford University with a Masters in Business Administration (MBA) was the same as the one for MBA graduates of the University of California at Berkeley. A random sample of recently graduating students yielded the following yearly salaries (in units of $1000).

Stanford	Berkeley
57.8	52.6
60.4	56.6
71.2	61.0
52.5	47.9
68.0	55.0
69.6	62.5
70.0	66.4
54.0	57.5
48.8	56.5
57.6	49.8

What conclusion would you draw at the 5 percent level of significance?

6. An experiment designed to determine the effectiveness of vitamin B1 in stimulating the growth of mushrooms was performed. The vitamin was applied to 9 mushrooms, selected at random from a set of 17. The remaining 8 mushrooms were left untreated. The weights (in grams) of all 17 mushrooms at the end of the experiment were as follows:

Untreated mushrooms: 18, 12.4, 13.5, 14.6, 24, 21, 23, 17.5
Vitamin B1 mushrooms: 34, 27, 21.2, 29, 20.5, 19.6, 28, 33, 19
Test, at the 5 percent level, the hypothesis that the vitamin B1 treatment had no effect.

7. Twenty-four workers were randomly divided into two sets of 12 each. Each set of workers was put through a 2-week training program. However, the first set of workers spent an additional day on "motivational" material. At the end of the training session the workers were given a series of tests and then ranked according to their performances. If the sum of the ranks of the workers who went through the motivational material was 136, what is the p value of the test of the hypothesis that the motivational material has no effect?

8. Use Program 14-2 to find the exact p value in Prob. 4.

9. Use Program 14-2 to find the exact p value in Prob. 5.

10. Redo Prob. 1 in Sec. 10.4, this time using a nonparametric test.

14.5 RUNS TEST FOR RANDOMNESS

A basic assumption in much of data analysis is that a set of data constitutes a random sample from some population. However, sometimes the data set is not actually a random sample from a population but rather is one that has some internal pattern. For instance, the data might tend to be increasing or decreasing over time, or they may follow some cyclical pattern where they increase and then decrease in a cyclic manner (see Fig. 14.4). In this section we will develop a test of the hypothesis that a given data set constitutes a random sample.

To test the hypothesis that a given sequence of data values constitutes a random sample, suppose initially that each datum can take on only two possible values, which we designate 0 or 1. Consider any data set of 0s and 1s, and call any consecutive sequence of either 0s or 1s a run. For instance, the data set

0, 1, 1, 1, 0, 0, 1, 0, 0, 1, 1, 1, 1, 0, 1, 0, 0

contains a total of nine runs: 5 runs of 0s and four runs of 1s. The first run consists of the single value 0; the next run consists of the three values 1, 1, 1; the next one consists of the two values 0, 0; and so on.

Suppose that the data set consists of a total of $n + m$ values, of which n are equal to 1 and m are equal to 0. Let R denote the number of runs in the data set. Now if the data set were a random sample from some population, then all possible orderings of the $n + m$ values (consisting of n 1s and m 0s) would be equally likely. By using this result it is possible to determine the probability distribution of R and thus to test the null hypothesis H_0 that the data set is a random sample by rejecting H_0 if the value

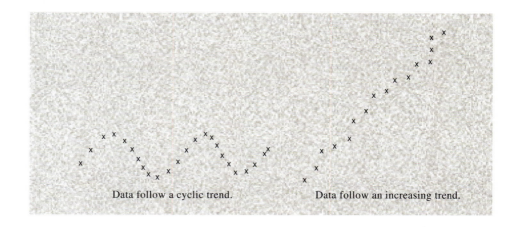

Figure 14.4
Nonrandom
data sets.

Data follow a cyclic trend.   Data follow an increasing trend.

of R is either too small or too large to be explained by chance. Specifically, if the value of R is r, then the significance level α test calls for rejecting H_0 if either

$$P\{R \leq r\} \leq \frac{\alpha}{2}$$

or

$$P\{R \geq r\} \leq \frac{\alpha}{2}$$

where the above probabilities are computed under the assumption that the null hypothesis is true. The resulting test is called the *runs test*.

We can also perform the runs test by observing the value of R, say it is r, and then calculating the resulting p value:

$$p \text{ value} = 2 \text{ Min}(P\{R \leq r\}, P\{R \geq r\})$$

Program 14-3 determines the above p value by calculating the relevant probabilities.

Example 14.12   The following are the outcomes of the last 24 games played by a local softball team. The letter W signifies a win and L a loss.

W, L, L, L, W, L, L, W, L, L, W, L, L, W, L, W, L, L, L, L, W, L, W, L

Is this data set consistent with randomness?

Solution To test the hypothesis of randomness, note that the data set of 8 W's and 16 L's contains a total of 16 runs. To see whether this justifies a rejection of the hypothesis of randomness, we run Program 14-3.

The number of ones is 8. The number of zeros is 16. The number of runs is 16. The p value as computed by Program 14-3 is 5.249703E-02.

Thus, since the p value is .0525, it follows that the hypothesis of randomness cannot be rejected at the 5 percent level of significance. That is, although the evidence of the data is against the hypothesis of randomness, it is not quite strong enough to cause us to reject that hypothesis at the 5 percent level of significance.

If Program 14-3 is not readily available to be run, then we can obtain an approximate p value by making use of a result which states that when the null hypothesis is true, R will have an approximately normal distribution with mean and variance given, respectively, by

$$\mu = \frac{2nm}{n + m} + 1$$

and

$$\sigma^2 = \frac{2nm(2nm - n - m)}{(n + m)^2 (n + m - 1)}$$

This will lead to a good approximation for the distribution of R provided that n and m, the numbers of 1s and 0s in the data set, are both of at least moderate size. (Both being at least 20 should suffice.)

If we observe a total of r runs, then the p value is given by

$$p \text{ value} = 2 \text{ Min}(P\{R \le r\}, P\{R \ge r\})$$

We can now use the preceding normal approximation to compute the relevant probability concerning R.

Example 14.13   Let us repeat Example 14.12, this time determining the approximate p value by using the above normal approximation. Since $n = 8$ and $m = 16$, we see that

$$\mu = \frac{2 \cdot 8 \cdot 16}{24} + 1 = 11.667$$

$$\sigma^2 = \frac{256(256 - 24)}{24 \cdot 24 \cdot 23} = 4.4831$$

Since there are a total of 16 runs, the p value is given by

$$p \text{ value} = 2 \text{ Min}(P\{R \le 16\}, P\{R \ge 16\})$$

Since $E[R] = 11.667$ is less than 16, it follows that $P\{R \ge 16\}$ is smaller than $P\{R \le 16\}$. Thus,

$$p \text{ value} = 2P\{R \geq 16\}$$
$$= 2P\{R \geq 15.5\} \quad \text{(continuity correction)}$$
$$= 2P\left\{\frac{R - 11.667}{\sqrt{4.4831}} \geq \frac{15.5 - 11.667}{\sqrt{4.4831}}\right\}$$
$$\approx 2P\{Z \geq 1.81\}$$
$$= .07$$

Since this approximate p value is greater than .05, the null hypothesis is not rejected at the 5 percent level of significance. Therefore, the approximate p value leads us to the same conclusion, at the 5 percent level of significance, as does the exact p value. However, the approximate value of .07 is not that close to the actual p value of .0525. Of course, in this example, the values of n and m (namely, 8 and 16) do not meet the rule of thumb that both should be at least 20 for the approximation to be accurate.

Example 14.14   Consider a sequence that contains 20 zeros and 20 ones. Suppose that there are 27 runs in this sequence. Compare the actual p value of the test of the hypothesis that the sequence is random with the approximate p value obtained by using the normal approximation.

Solution Let us start with the normal approximation. Since $n = m = 20$, the mean and standard deviation of the number of runs are, respectively,

$$\mu = \frac{2 \cdot 20 \cdot 20}{40} + 1 = 21 \qquad \sigma = \sqrt{\frac{2 \cdot 20 \cdot 20 \cdot 760}{40 \cdot 40 \cdot 39}} = 3.121$$

Therefore, since the observed number of runs is 27, the normal approximation gives the following:

$$p \text{ value} = P\{R \geq 27\}$$
$$= 2P\{R \geq 26.5\}$$
$$= 2P\left\{\frac{R - 21}{3.121} \geq \frac{26.5 - 21}{3.121}\right\}$$
$$\approx 2P\{Z \geq 1.762\}$$
$$= .078$$

On the other hand, running Program 14-3 gives the exact p value:

The number of ones is 20. The number of zeros is 20. The number of runs is 27. The p value is computed by program 14-3 as 7.599604E-02.

Therefore, in this example (where both n and m are equal to 20) the approximate p value of .078 is quite close to the actual p value of .076.

We can also use the runs test to test for randomness when the sequence of data values is not comprised of just 0s and 1s. To test whether a given sequence of data $X_1, X_2, \ldots, X_n$ constitutes a random sample from some population, let s_m denote the sample median of this data set. Now, for each data value determine whether it is less than or equal to s_m or whether it is greater than s_m. Put a 0 in position i if X_i is less than or equal to s_m, and put a 1 otherwise. If the original data set constituted a random sample from some distribution, then the sequence of 0s and 1s will also constitute a random sample. Therefore, we can test whether the original data set is a random sample by using the runs test on the resulting sequence of 0s and 1s.

Example 14.15   The average summer temperatures in degrees Fahrenheit for 20 successive years from 1971 to 1990 in a given west coast city are

72, 71, 70, 82, 80, 77, 71, 85, 75, 80, 82, 81, 83, 82, 85, 86, 83, 81, 82, 84

Test the hypothesis that the data constitute a random sample.

Solution The sample median m is the average of the 10th and the 11th smallest values. Therefore,

$$s_m = \frac{81 + 82}{2} = 81.5$$

The data of 0s and 1s that indicate whether each value is less than or equal to or greater than 81.5 are as follows.

0 0 0 1 0 0 0 1 0 0 1 0 1 1 1 1 1 0 1 1

Thus, the sequence of 0s and 1s consists of 10 of each and has a total of 10 runs. To determine if this value is significantly greater or larger than could be expected by chance if the data were truly random, we run Program 14-3.

The number of ones is 10. The number of zeros is 10. The number of runs is 10. The p value is computed as .8281409.

For such a large p value, the hypothesis of randomness is not rejected. That is, the data give no evidence of not being a random sample.

If we had used the normal approximation, then we would have first computed μ and σ, the null hypothesis mean and standard deviation of the total number of runs. Since $n = m = 10$

$$\mu = \frac{200}{20} + 1 = 11 \qquad \sigma = \sqrt{\frac{200(180)}{400 \cdot 19}} = 2.176$$

Since the observed number of runs is 10, the p value given by the normal approximation is

$$\begin{aligned}
p \text{ value} &= 2P\{R \le 10\} \\
&= 2P\{R \le 10.5\} \\
&= 2P\left\{\frac{R - 11}{2.176} \le \frac{10.5 - 11}{2.176}\right\} \\
&\approx 2P\{Z \le -.23\} \\
&= .818
\end{aligned}$$

Thus, the normal approximation is quite accurate in this example.

Example 14.16   The following are the successive numbers of points scored by a certain high school basketball team in the 23 games it played in the 1994–1995 season. Is it reasonable to suppose that the scores constitute a random sample?

77, 62, 58, 64, 66, 72, 59, 69, 80, 74, 72, 69
74, 83, 85, 87, 80, 88, 76, 77, 82, 85, 83

Solution   The sample median is the 12th smallest score, namely 76. The sequence of 0s and 1s indicating whether each value is less than or equal to or greater than 76 is as follows.

1 0 0 0 0 0 0 0 1 0 0 0 0 1 1 1 1 1 0 1 1 1 1

Thus, this sequence consists of twelve 0s and eleven 1s, and has seven runs. From Program 14-3, we see that p value $= .02997$, and thus the hypothesis that the data constitute a random sample is rejected at the 5 percent level of significance.

Problems

Unless otherwise stated, use either Program 14-3 or the normal approximation, whichever is more convenient, in answering the following questions.

1. Suppose a sequence of 0s and 1s contains twenty 0s and thirty 1s. Let R denote the total number of runs. What are the (a) largest and (b) smallest possible values of R?

2. Determine the number of runs for the following data sets of 0s and 1s.
   (a) 1 1 0 0 0 0 1 1 0 1 0 1 1 0 0 0 1 1
   (b) 0 1 1 0 0 0 0 1 1 1 0 1 1 0 1 1 1 1
   (c) 0 0 0 1 1 0 0 1 1 0 1 1 1 1 1 1 0 1 0

3. The following data relate to the acceptability of the 26 most recently produced watches at a Swiss watch factory. The value 1 signifies that the watch is acceptable and the value 0 that it is unacceptable.

   1 1 1 0 1 1 1 1 1 1 1 1 1 0 0 0 0 1 1 1 1 0 0 1 1

   Test the hypothesis, at the 5 percent level of significance, that the data constitute a random sample.

4. A production run of 60 items resulted in 12 defectives. The defectives are item numbers 9, 14, 15, 26, 30, 36, 37, 44, 45, 46, 59, and 60.
   (a) What is the value of R, the total number of runs?
   (b) Can we conclude, at the 5 percent level of significance, that the successive items do not constitute a random sample?

5. A total of 25 people, 10 of whom are women, are to be interviewed. The interviewer is told to interview them in a randomly chosen order. Suppose that the sequence of sexes of the successively interviewed people is as follows.

   F F M F F F F M M F F M F F M M M M M M M M M M M

   Did the interviewer follow instructions? Explain and give the relevant p value.

6. Over the last 50 days the Dow Jones industrial average increased on 32 days and decreased on the other 18 days. If the total number of runs (of increasing or decreasing Dow Jones average) was 22, what is the p value of the test of the hypothesis that the increases and decreases constituted a random sample?

7. The lifetimes in hours of 30 successively produced storage batteries are as follows:

   148, 152, 155, 147, 176, 170, 165, 149, 138, 155, 160, 153, 162,

   155, 159, 174, 168, 149, 182, 177, 191, 185, 178, 176, 182,

   184, 181, 177, 160, 154

   (a) What is the sample median?
   (b) What is the value of R, the number of runs in the corresponding data that details for each data value whether it is less than (or equal to) or greater than the sample median?
   (c) Do these data disprove the hypothesis that the sequence of values constitutes a random sample?

8. The following data represent end-of-year Dow Jones averages for a sequence of 10 consecutive years.

   910, 890, 1010, 1033, 1080, 1275, 1288, 1553, 1980, 2702

Test the hypothesis, at the 5 percent level of significance, that these data can be thought of as constituting a random sample.

Key Terms

Nonparametric hypotheses tests: A class of hypotheses tests about a population that do not assume that the population distribution is a specified type.

Sign test: A nonparametric test concerning the median of a population. The test statistic counts the number of data values less than the hypothesized median.

Signed-rank test: A nonparametric test of the null hypothesis that a population distribution is symmetric about a specified value.

Rank-sum test: A nonparametric test of the equality of two population distributions. It uses independent samples from the populations and then ranks the combined data from the two samples. The sum of the ranks of (either) one of the samples is the test statistic.

Runs test: A nonparametric test of the hypothesis that an ordered data sequence constitutes a random sample from some population.

Summary

In this chapter we learned how to test a statistical hypothesis without making any assumptions about the form of the underlying probability distributions. Such tests are called *nonparametric*.

Sign Test

The sign test can be used to test hypotheses concerning the median of a distribution. Suppose that for a specified value m we want to test

$$H_0: \eta = m$$

against

$$H_1: \eta \neq m$$

where η is the median of the population distribution. To obtain a test, choose a sample of elements of the population, discarding any data values exactly equal to m. Suppose n data values remain. The test statistic of the sign test is the number of remaining values that are less than m. If there are i such values, then the p value of the sign test is given by

$$p \text{ value} = \begin{cases} 2P\{N \le i\} & \text{if } i \le \dfrac{n}{2} \\ \\ 2P\{N \ge i\} & \text{if } i \ge \dfrac{n}{2} \end{cases}$$

where N is a binomial random variable with parameters n and $p = 1/2$. The computation of the binomial probability can be done either by running Program 5-1 or by using the normal approximation to the binomial.

The sign test can also be used to test the one-sided hypothesis

$$H_0: \eta \le m \qquad \text{against} \qquad H_1: \eta > m$$

It uses the same test statistic as above, namely, the number of data values that are less than m. If the value of the test statistic is i, then the p value is given by

$$p \text{ value} = P\{N \le i\}$$

where again N is binomial with parameters n and $p = 1/2$.

If the one-sided hypothesis to be tested is

$$H_0: \eta \ge m \qquad \text{against} \qquad H_1: \eta < m$$

then the p value, when there are i values less than m, is

$$p \text{ value} = P\{N \ge i\}$$

where N is binomial with parameters n and $p = 1/2$.

As in all hypothesis testing, the null hypothesis is rejected at any significance level greater than or equal to the p value.

Signed-Rank Test

The signed-rank test is used to test the hypothesis that a population distribution is symmetric about the value 0. In applications, the population often consists of the differences of paired data. The signed-rank test calls for choosing a random sample from the population, discarding any data values equal to 0. It then ranks the remaining nonzero values, say there are n of them, in increasing order of their absolute values. The test statistic is equal to the sum of the rankings of the negative data values. If the value of the test statistic TS is equal to t, then the p value is

$$p \text{ value} = 2 \text{ Min}(P\{TS \le t\}, P\{TS \ge t\})$$

where the probabilities are to be computed under the assumption that the null hypothesis is true. The p value can be found either by using Program 14-1 or by using the fact that TS will approximately have, when the null hypothesis is true and n is of least moderate size, a normal distribution with mean and variance, respectively, given by

$$E[\text{TS}] = \frac{n(n + 1)}{4} \qquad \text{Var (TS)} = \frac{n(n + 1)(2n + 1)}{24}$$

Rank-Sum Test

The rank-sum test can be used to test the null hypothesis that two population distributions are identical, when the data consist of independent samples from these populations. Arbitrarily designate one of the samples as the first sample. Suppose that the size of this sample is n and that of the other sample is m. Now rank the combined samples. The test statistic TS of the rank-sum test is the sum of the ranks of the first sample. The rank-sum test calls for rejecting the null hypothesis when the value of the test statistic is either significantly large or significantly small.

When n and m are both greater than 7, the test statistic TS will, when H_0 is true, approximately have a normal distribution with mean and variance given by, respectively,

$$E[\text{TS}] = \frac{n(n + m + 1)}{2} \qquad \text{Var (TS)} = \frac{nm(n + m + 1)}{12}$$

This enables us to approximate the p value, which when TS $= t$ is given by

$$p \text{ value} \approx \begin{cases} 2P\left\{Z \le \dfrac{t + .5 - n(n + m + 1)/2}{\sqrt{nm(n + m + 1)/12}}\right\} & \text{if } t < \dfrac{n(n + m + 1)}{2} \\[4mm] 2P\left\{Z \ge \dfrac{t - .5 - n(n + m + 1)/2}{\sqrt{nm(n + m + 1)/12}}\right\} & \text{if } t > \dfrac{n(n + m + 1)}{2} \end{cases}$$

For values of t near $n(n + m + 1)/2$, the p value is close to 1, and so the null hypothesis would not be rejected (and the preceding probability need not be calculated).

For small values of n and m the exact p value can be obtained by running Program 14-2.

Runs Test

The runs test can be used to test the null hypothesis that a given sequence of data constitutes a random sample from some population. It supposes that each datum is either a 0 or a 1. Any consecutive sequence of either 0s or 1s is called a *run*. The test statistic for the run test is R, the total number of runs. If the observed value of R is r, then the p value of the runs test is given by

$$p \text{ value} = 2 \text{ Min}(P\{R \leq r\}, P\{R \geq r\})$$

The probabilities in the above are to be computed under the assumption that the null hypothesis is true.

Program 14-3 can be used to determine the above p value. If Program 14-3 is not available, we can approximate the p value by making use of the fact that when the null hypothesis is true, R will approximately have a normal distribution. The mean and variance, respectively, of this distribution are

$$\mu = \frac{2nm}{n + m} + 1 \qquad \sigma^2 = \frac{2nm(2nm - n - m)}{(n + m)^2(n + m - 1)}$$

Review Problems

1. Use a nonparametric test to solve Prob. 2 in Sec. 10.4.

2. Use a nonparametric test to solve Prob. 3 in Sec. 10.3.

3. According to the *Federal Reserve Bulletin* of January 1992, in 1989 the sample median net worth of all 55-year-olds in the labor force was $104,500. (The sample mean was $438,300.) Suppose that a random sample of 1000 such workers today yielded the result that 421 had a family net worth (in 1989 dollars) of over $104,500. Can we conclude, at the 5 percent level of significance, that the median net worth has decreased?

4. An experiment was initiated to study the effect of a newly developed gasoline detergent on automobile mileage. The following data, representing mileage per gallon before and after the detergent was added for each of eight cars, resulted.

Car	Mileage without additive	Mileage with additive
1	24.2	23.5
2	30.4	29.6
3	32.7	32.3
4	19.8	17.6
5	25.0	25.3
6	24.9	25.4
7	22.2	20.6
8	21.5	20.7

Find the p value of the test of the hypothesis that mileage is not affected by the additive when
(a) The sign test
(b) The signed-rank test
is used. Compare your results with each other and with Example 10.8 of Sec. 10.5.

5. Test the hypothesis that the weights of the students given in App. A constitute a random sample. Use the first 40 data values to test this hypothesis. What is the p value?

6. Choose a random sample of 30 of the students in App. A. Use those data to test the hypothesis that the median weight of all the students listed is less than or equal to 130 pounds.

7. Choose a random sample of 40 students from App. A, and use this sample to test the hypothesis that the distribution of blood cholesterol readings is the same for both sexes.

8. A chemist tests a variety of blood samples for a certain virus. The successive results are as given below, with P meaning that the virus is present and A that it is absent.

$$A\ A\ P\ P\ P\ A\ A\ A\ A\ P\ P\ A\ A\ A\ A\ A\ A\ P\ P\ A\ A\ P\ P\ P\ P$$

Test the hypothesis that the chemist tested the samples in a random order. Use the 5 percent level of significance. Also determine the p value.

9. Repeat Example 10.9, this time using a nonparametric test.

10. Explain how we could have used a contingency table analysis to test the hypothesis in Example 14.10. Do this test, find the p value, and compare it with the one obtained in Example 14.10. Since the contingency table test is different from the one used in Example 14.10, the two p values need not be equal.

11. Consider Prob. 7 of Sec. 14.3. Suppose now that the same car had been brought to 16 different automobile repair shops, with the woman bringing it into 8 of them and the man to the other 8. Suppose the data on the quoted repair prices were as given in that problem. Test the hypothesis, at the 5 percent level of significance, that the distributions of price quotes received by the man and by the woman are the same.

MINITAB LAB

Nonparametric Hypotheses Tests

Purpose

Use Minitab to

1. Help perform the sign test.

2. Help perform the signed-rank test (Wilcoxon).

3. Help perform the rank-sum test for comparing two populations (Mann-Whitney).

4. Help perform the runs test for randomness.

Procedures

First, load the Minitab (Windows version) software as in the Minitab lab for Chap. 1.

1. ONE-SAMPLE SIGN TEST

Example 1 Use Minitab to work Example 14.1 in Sec. 14.2 of the text.

    To use the sign test in Minitab, you need a set (column) of data values. In this example, we are not given the data set, but we are told that 36 values exceeded 10.25 inches. Thus 14 values were less than 10.25 inches since the sample size was 50. We can use this information to generate a set of 50 values such that 14 values are below 10.25. Since the median value was 10.25, we can select the value of 10 (say, since $10 < 10.25$) and generate 14 values of it. Similarly we can select the value 11 (say, since $11 > 10.25$) and generate 36 values of it. To

727

generate the 14 values, select Calc→Set Patterned Data, and in the resulting dialog box store the generated values in C1. Select the Arbitrary list of constants check box and type 10 in the text box. In the Repeat each value text box type 14, and in the Repeat the whole list text box type 1.

Select the OK button, and the 14 values, each of size 10, will be displayed in column C1 in the Data window. Repeat the above procedure for the 36 values each of size 11, and store in column C2. Next, we need to stack these two columns so that we have a column with 50 values. To do this, select Manip→Stack, and the Stack dialog box will be displayed. Stack the two columns and store in column C3. Figure M14.1 shows the Stack dialog box with the appropriate entries. Click on the OK button, and the values in columns C1 and C2 will be stacked and displayed in column C3. Check to see that there are 50 values in column C3.

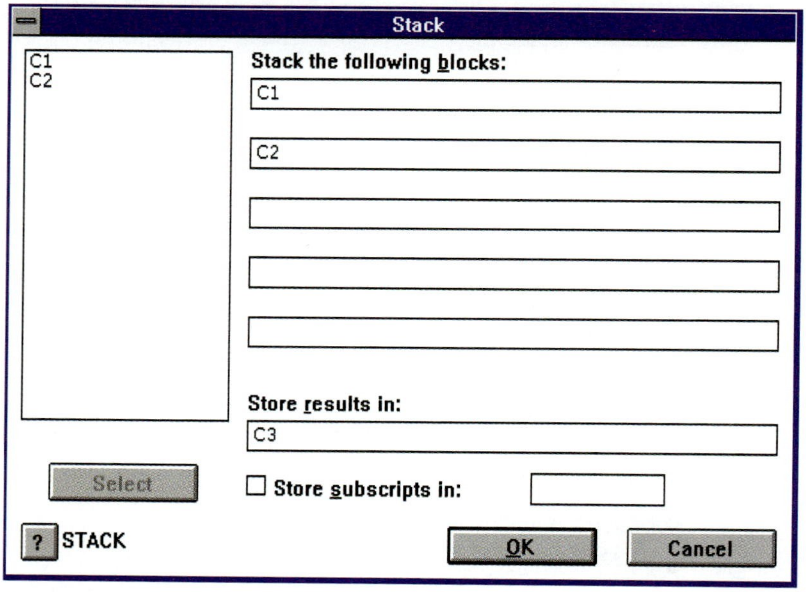

Figure M14.1

Now, to apply the sign test to column C3, select Stat→Nonparametrics→1-Sample Sign, and the 1-Sample sign dialog box will be displayed. Select the Test median check box, and type in 10.25 in the text box. Select Not equal to in the Alternative drop-down box. That is, we are testing whether the median is equal to 10.25 against the alternative of its not being equal to 10.25. The 1-sample sign dialog box is shown in Fig. M14.2 with the appropriate entries.

Click on the OK button, and the Session window as shown in Fig. M14.3 will display the results. Since the p value for the test is .0026 (rather small) we can reject the hypothesis of the median foot size of teenage boys being equal to 10.25 inches. Observe that the median value of 11 is meaningless.

Note

1. If the data set is given, just apply the sign test to the appropriate column.

2. We can also do one-sided test for the median by selecting less than or greater than in the Alternative drop-down box.

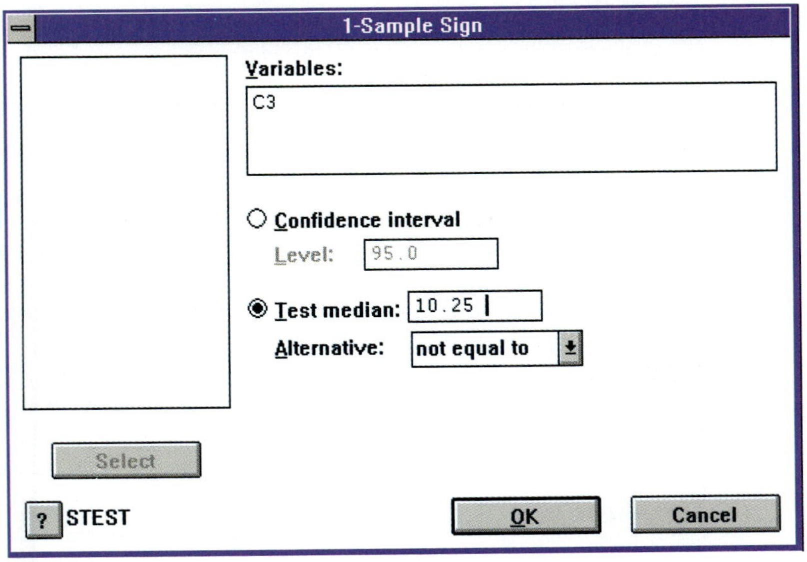

Figure M14.2

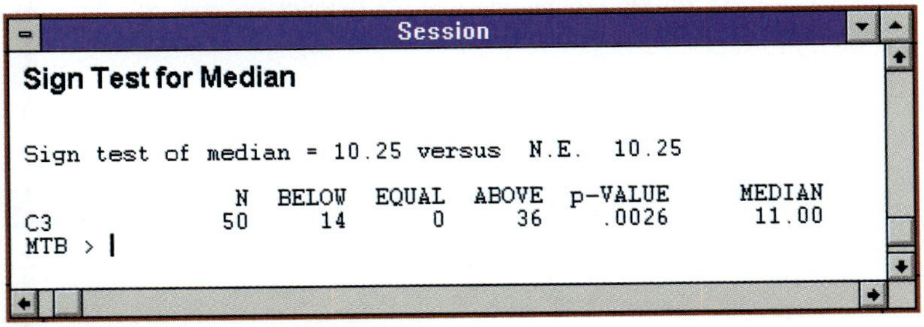

Figure M14.3

2. SIGNED-RANK TEST

Note   The signed-rank test as described in your text is also known as the Wilcoxon matched-pairs signed-ranks test. This test is used for dependent samples.

Example 2   Use Minitab to work Example 14.5 in Sec. 14.3 of the text.

This version of Minitab does not have a procedure to *directly* do the signed-ranks test (matched-pairs). However, we can manipulate things to enable Minitab to do an equivalent test. First, enter the two sets of values in columns C1 and C2, and observe that these are dependent samples since the individual pairs of values were observed for the same student. Find the difference of the values in C1 and C2, and save in column C3; then do a test for zero median for the differences in C3. To do this, select Stat→Nonparametrics→1-Sample Wilcoxon, and the 1-Sample Wilcoxon dialog box will be displayed. Select column C3 for the Variables text box, and select the check box for Test median. Make sure that a value of 0 is in the Test median text box and that Not equal to is in the Alternative drop-down box. Figure M14.4 shows the 1-Sample Wilcoxon dialog box with the appropriate entries.

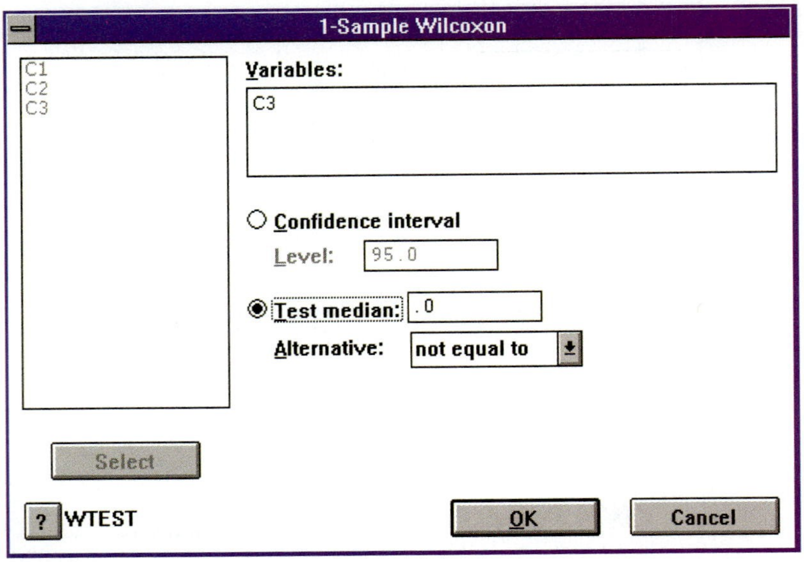

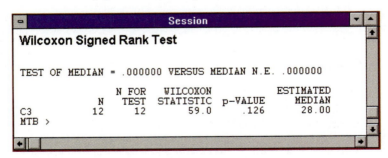

Figure M14.4

Click on the OK button, and the Session window will display the results. The Session window is shown in Fig. M14.5. Observe that the p value for this test is .126, which is the same value as in the text when the normal approximation is used. Thus we will fail to reject the null hypothesis that the distributions of the student scores on the two examinations are identical at the 10 percent level of significance.

Figure M14.5

3. RANK-SUM TEST FOR COMPARING TWO POPULATIONS

Note The rank-sum test for comparing two populations as described in your text is also known as the *Mann-Whitney test.* This test is used for *independent* samples.

Example 3 Use Minitab to work Example 14.8 in Sec. 14.4 of the text.

To apply the Mann-Whitney test to the set of independent data, enter the values in columns C1 and C2. Rename to column C1 as TWENTY for the 20-year-olds and C2 as FIFTY for the 50-year-olds. Select Nonparametrics→Mann-Whitney, and the Mann-Whitney dialog box will be displayed. Select the appropriate entries as shown in Fig. M14.6.

Click on the OK button, and the associated computations will be displayed in the Session window. This is shown in Fig. M14.7. The test statistic value is given by $W = 50$, and p value =

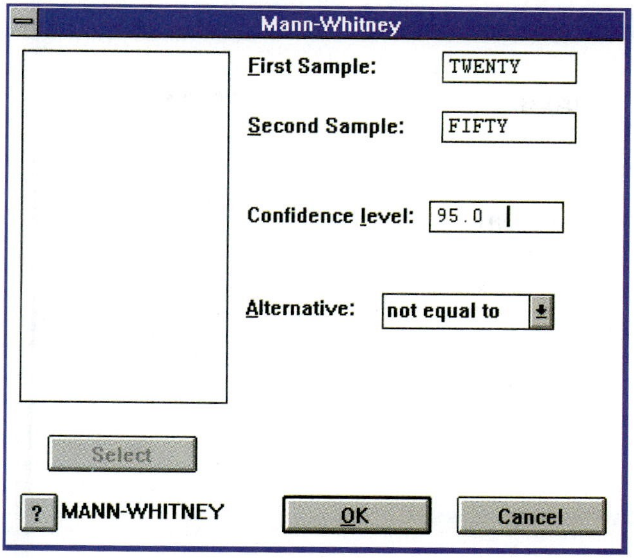

Figure M14.6

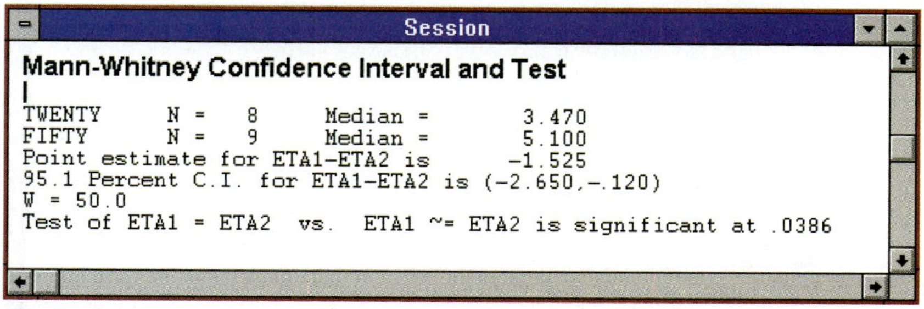

Figure M14.7

.0386. Thus at the 5 percent level of significance we will reject the null hypothesis that the two population distributions are identical since .0386 < .05. Also included in the output are the medians for the two samples, the point estimate for the difference of the two population medians, and the 95 percent confidence for this difference.

4. RUNS TEST FOR RANDOMNESS

Example 4 Use Minitab to work Example 14.12 in Sec. 14.5 of the text.

Enter W as a 1 and L as a 0 in column C1. Select Select→Nonparametrics→Runs Test, and the Runs test dialog box will be displayed. Select C1 for the Variables text box, and select the check box for Above and below the mean. By selecting this option you are telling Minitab to use the mean as the baseline to determine the number of runs.

Click on the OK button, and the Session window will display the results. This is shown in Fig. M14.8. Observe that the p value computed by Minitab is .0410, so we can reject the hypothesis of randomness at the 5 percent level of significance. The p value in the text is different (.0525) which leads us to fail to reject the null hypothesis of randomness. The difference in the p values could be due to rounding errors in the two programs.

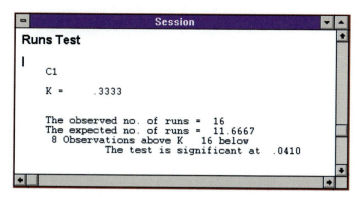

Figure M14.8

Note   You can work Example 14.15 directly with Minitab by entering the data values in column C1 and selecting and typing 81.5 in the Above and below text box of the Runs test dialog box.

Computer Exercises

1. Use Minitab or any other statistical software to help work Probs. 1 to 9 in Sec. 14.2.

2. Use Minitab or any other statistical software to work Probs. 5, 7, 8, and 9 in Sec. 14.3.

3. Use Minitab or any other statistical software to work Probs. 3 to 6 in Sec. 14.4.

4. Use Minitab or any other statistical software to work Probs. 3, 4, 5, 7, and 8 in Sec. 14.5.

5. Use your library or any other resource to collect a set of data that could be analyzed with the runs test. Use Minitab or any other statistical software to do a complete analysis of the data. Present any relevant analysis and discussion in a report. If you select data from published research, include relevant factors about the validity and reliability of the study.

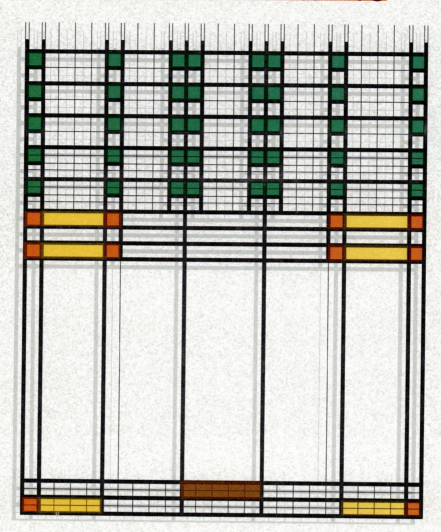

Quality Control

e introduce control charts, which are used to determine when a process that produces items has gone out of control. This is done both where the measurable value of an item is a continuous variable and where it is a discrete variable. In this latter case, each item has two possible values, with the value indicating whether the item is acceptable or not.

15.1   INTRODUCTION

Almost every industrial system—whether it involves the manufacturing of products or the servicing of customers—results in some random variation in the items it processes. That is, no matter how stringently the system is being controlled, there is always going to be some variation in the items processed. For instance, the successive items produced by a manufacturing process will not all be identical, and the successive times that it takes to service customers will often be different, even when the underlying operation is performing as required. This type of variation is called *chance variation* and is considered to be inherent to the system. However, there is another type of variation which sometimes appears. This variation, far from being inherent to the system, is due to some assignable cause, and it usually has an adverse effect on the quality of the industrial operation. For instance, in a manufacturing context, this latter variation may be caused by a faulty machine setting, or by poor quality of the raw materials being employed, or by incorrect software, or by human error, or by any of a large number of possibilities. When the only variation present is due to chance, and not to any assignable cause, we say that the process is *in control;* and a key problem in quality control is to determine when a process is in control and when it is out of control.

In this chapter we study control charts, which can be used to indicate when a process has gone out of control. The types of control charts we consider are determined by two numbers, called the *upper control limit (UCL)* and the *lower control limit (LCL).* To utilize these charts, first we divide the data generated by the industrial concern into subgroups. Then we compute the subgroup averages, and when one of these does not fall within the upper and lower control limits, we conclude that the process is out of control.

In Sec. 15.2 we suppose that the successive items processed have measurable characteristics—this could refer to their quality level in a manufacturing context or to their service time in a service industry—whose mean and variance are known when the process is operating in control. We show how to construct control charts that are useful for detecting a change in the mean of the in-control distribution. In Sec. 15.3 we construct a control chart for situations in which each item, rather than having some measurable characteristic, is classified as being either satisfactory or unsatisfactory.

15.2   THE $\bar{X}$ CONTROL CHART FOR DETECTING A SHIFT IN THE MEAN

Suppose that when an industrial system is in control, the successive items it processes have measurable values that are independent, normal random variables with mean μ and variance σ^2. However, due to unforeseeable circumstances, suppose that the system may go out of control and, as a result, begin to process items having values from a different distribution. We want to be able to recognize when this occurs so as to stop the system, learn what is wrong, and fix it.

Let the measurable values of the successive items processed by the system be denoted by $X_1, X_2, \ldots$. In our attempt to determine when the process goes out of control, we will find it convenient to first break up the data into subgroups of some fixed size—call this size n. Among other things, this value of n should be chosen so as to yield uniformity of data values within individual subgroups. That is, we should attempt to choose n so that it is reasonable, when a shift in distribution occurs, that it will occur between and not within subgroups. Thus, in practice n is often chosen so that all the data within a subgroup relate to items processed on the same day, or on the same shift, or with the same settings, etc.

Let $\bar{X}_i$, $i = 1, 2, \ldots$, denote the average of the ith subgroup. Since, when in control, all the data values are normal with mean μ and variance σ^2, it follows that $\bar{X}_i$, the sample mean of n of them, is normally distributed with mean and variance given, respectively, by

$$E[\bar{X}_i] = \mu$$

$$\text{Var}\,(\bar{X}_i) = \frac{\sigma^2}{n}$$

Hence, it follows that when the process is in control,

$$Z = \frac{\bar{X}_i - \mu}{\sqrt{\sigma^2/n}}$$

is a standard normal random variable. That is, if the process remains in control throughout the processing of subgroup i, then $\sqrt{n}(\bar{X}_i - \mu)/\sigma$ has a standard normal distribution. Now, a standard normal random variable Z will almost always be between -3 and $+3$. Indeed, from Table 6.1 we see that $P\{-3 < Z < 3\} = .9973$. Hence, if the process remains in control throughout the processing of subgroup i, then we would certainly expect that

$$-3 < \frac{\sqrt{n}(\bar{X}_i - \mu)}{\sigma} < 3$$

or, equivalently, that

$$\mu - \frac{3\sigma}{\sqrt{n}} < \bar{X}_i < \mu + \frac{3\sigma}{\sqrt{n}}$$

The values

$$\text{LCL} \equiv \mu - \frac{3\sigma}{\sqrt{n}}$$

$$\text{UCL} \equiv \mu + \frac{3\sigma}{\sqrt{n}}$$

are called, respectively, the *lower* and *upper control limit*.

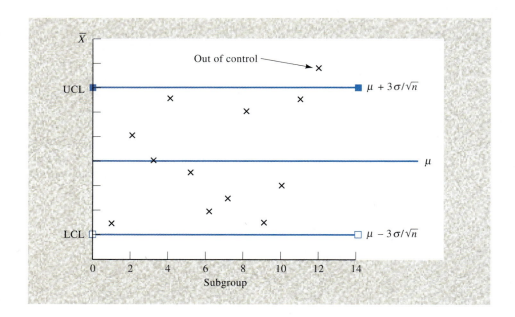

Figure 15.1
Control chart
for $\bar{X}$; n = size
of subgroup.

The $\bar{X}$ control chart, which is primarily designed to detect a change in the average value of an item processed, is obtained by plotting the successive subgroup averages $\bar{X}_i$ and declaring that the process is out of control the first time that $\bar{X}_i$ does not fall between LCL and UCL (see Fig. 15.1).

Since an $\bar{X}$ control chart will declare a process out of control only when a subgroup average falls outside the control limits, it is important that the subgroups are chosen so that it becomes highly likely that any shift in distribution that occurs is between subgroups. This is so because it is easier to detect a shift in a subgroup having all, rather than only some of, its values out of control.

Example 15.1   The time it takes a computer servicing firm to install a hard disk along with some sophisticated software for its use is a random variable having mean 25 minutes and standard deviation 6 minutes. The company has two employees who work on this operation. To monitor the efficiency of these employees, the company has plotted the successive average times that it takes them to complete four jobs. The even-number subgroups refer to the first employee, and the odd-number subgroups refer to the second. Suppose the first 20 of these successive subgroup averages are as follows (see Fig. 15.2 for a plot).

Subgroup	$\bar{X}$	Subgroup	$\bar{X}$	Subgroup	$\bar{X}$	Subgroup	$\bar{X}$
1	23.6	6	24.6	11	29.4	16	32.8
2	20.8	7	22.6	12	27.8	17	23.3
3	25.5	8	24.4	13	26.8	18	30.5
4	26.2	9	24.7	14	27.2	29	25.3
5	23.3	10	26.0	15	24.0	20	34.1

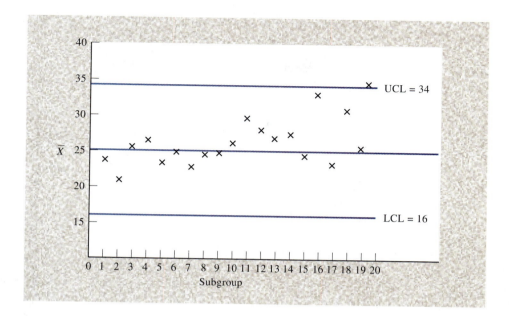

Figure 15.2
Control chart
for data of
Example 15.1.

What conclusions can be drawn?

Solution Since the subgroup size is 4 and since the successive data values have mean $\mu = 25$ and standard deviation $\sigma = 6$ when the process is in control, it follows that the control limits are given by

$$\text{LCL} = 25 - \frac{18}{\sqrt{4}} = 16 \qquad \text{UCL} = 25 + \frac{18}{\sqrt{4}} = 34$$

Since the average of subgroup 20 is greater than UCL, it appears that the system is no longer in control. Indeed, since all the last 6 even-number subgroup averages are larger than the in-control mean of 25 (with the final three being significantly larger), it seems probable that the second employee has been out of control for some time.

We have assumed above that when the process is in control, the underlying distribution of an item's measurable characteristic is a normal distribution; and, in fact, this is often the case in manufacturing processes. In addition, as long as this distribution is somewhat close to being normal, the subgroup averages will, because of the central limit theorem, be approximately normal and so would be unlikely to differ from their mean by more than 3 standard deviations. As a result, by utilizing subgroup averages it is not necessary for us to know the entire in-control distribution, only its mean and variance. This is the most important reason for utilizing subgroup averages rather than individual data points. Typical values of the subgroup size are 4, 5, or 6. The reason for this is that if one uses a smaller subgroup size, then the approximate normality of the subgroup average might come into question. On the other hand, since one cannot detect an out-of-control system without processing all the items of at least one subgroup, the subgroup size should not be too large.

Because the $\overline{X}$ control chart was first suggested by Walter Shewhart, it is often referred to as the *Shewhart control chart.* In essence, it works by using each subgroup to test the statistical hypothesis

$$H_0: \text{mean} = \mu \quad \text{against the alternative} \quad H_1: \text{mean} \neq \mu$$

at the significance level $\alpha = P\{|Z| > 3\} = .0027$. Whenever the null hypothesis is rejected, the process is declared to be out of control.

Although we usually talk about control charts in the context of a manufacturing process, they can be utilized in a variety of situations, as indicated by the next example.

Example 15.2   Consider a small video rental store in which the daily rentals on each of the weekdays Monday through Thursday have mean 52 and standard deviation 10. If the numbers of videos rented daily (Monday to Thursday) in the past week are

32, 38, 28, 30

can we conclude that a change in distribution has occurred?

Solution Let a subgroup consist of the number of rentals on the four specified week-days of each week. Since the in-control mean and standard deviation are $\mu = 52$ and $\sigma = 10$, respectively, it follows that the control limits are

$$\text{LCL} = 52 - \frac{3(10)}{\sqrt{4}} = 37$$

$$\text{UCL} = 52 + \frac{3(10)}{\sqrt{4}} = 67$$

Since the most recent subgroup average is $(32 + 38 + 28 + 30)/4 = 32$, which is outside these limits, we can declare that the average number of videos rented daily is no longer equal to 52.

At this juncture the store's manager should try to discover (1) the cause of the change in mean and (2) whether it appears to be a temporary or a more permanent change. For instance, he or she might discover that there were some particularly inter-esting television programs in the past week, such as a World Series, or the Olympics, or a political convention, which would lead to the belief that the change is of a short-term nature. Or else it might be discovered that the change in mean was caused by a new competitor in the neighborhood, and this might result in a more permanent change.

Sometimes not all the measurable values of the items produced are noted, only those of a randomly selected subset of items. When this is the situation, it is natural to let subgroups consist of items that are produced at roughly the same time.

Problems

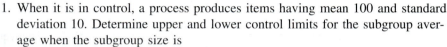

1. When it is in control, a process produces items having mean 100 and standard deviation 10. Determine upper and lower control limits for the subgroup average when the subgroup size is
   (a) 4   (b) 5   (c) 6   (d) 10

2. When a process is working properly, it produces items that have mean 35 and standard deviation 4. To monitor this process, subgroups of size 4 are sampled. If the following represents the averages of the first 20 subgroups, does it appear that the process was in control throughout?

Subgroup	$\bar{X}$	Subgroup	$\bar{X}$
1	31.2	11	36.4
2	38.4	12	31.1
3	35.0	13	32.3
4	33.3	14	37.8
5	34.7	15	36.6
6	31.1	16	40.4
7	35.8	17	41.2
8	34.4	18	35.9
9	37.1	19	40.4
10	34.2	20	32.5

3. When a manufacturing process is in control it produces wire whose diameter has mean 80 with standard deviation 10 (in units of 1/10,000 inch). The following data represent the sample mean of subgroups of size 5:

   85, 88, 90, 77, 79, 83, 90, 75, 94, 80, 84, 86, 88

   Does the process appear to have been in control?

4. A control chart is maintained on the time it takes workers to perform a certain task. The time that it should take is normal with mean 26 minutes and standard deviation 4.2 minutes. The following are $\bar{X}$ values for 10 subgroups of size 4.

   28.2, 28.4, 31.1, 27.3, 33.2, 31.4, 27.9, 30.4, 31.3, 30.4

   (a) Determine the upper and lower control limits.
   (b) Does it appear that the process was in control?

5. When a seam welding process is working correctly, the distance from the weld to the center of the seam is normally distributed with mean 0 and standard deviation .005 inch. If the following values are the average of the distances from the weld to the center in eight subgroups of size 5, does it appear that the process was in control during their processing?

Subgroup	Average distance
1	.0023
2	−.0012
3	−.0015
4	.0031
5	.0038
6	.0051
7	.0022
8	−.0033

6. Prior to 1982, the number of murders committed yearly in the United States per 100,000 population was normally distributed with mean 9.0 and standard deviation 1.2. The following table gives the murder rates in the years from 1982 to 1991.

Year	1982	1983	1984	1985	1986	1987	1988	1989	1990	1991
Murder rates per 100,000 population	9.1	8.3	7.9	7.9	8.6	8.3	8.4	8.7	9.4	9.8

Source: U.S. Federal Bureau of Investigation, *Crime in the United States,* annual.

During the years from 1982 to 1991, did the murder rate remain at its pre-1982 level? Use subgroups of size 2.

15.3 CONTROL CHARTS FOR FRACTION DEFECTIVE

The $\bar{X}$ control chart can be used when the data are measurements whose values can vary continuously over a region. However, in some situations the items processed have values that are classified as either acceptable or unacceptable. For instance, a manufactured item can be classified as defective or not; or the service of a customer could be rated (by the customer) as acceptable or not. In this section we show how to construct control charts for these situations.

Let us suppose that when the system is in control, each item processed will independently be defective with probability p. If we let X denote the number of defective items in a subgroup of n items, then, assuming the system has been in control, X will be a binomial random variable with parameters n and p; and thus

$$E[X] = np$$

$$\text{Var } (X) = np(1 - p)$$

Hence, when the system is in control, the number of defectives in a subgroup of size n should be, with high probability, between the lower and upper limits

$$\text{LCL} = np - 3\sqrt{np(1 - p)}$$

$$\text{UCL} = np + 3\sqrt{np(1 - p)}$$

The subgroup size n is usually much larger than the typical values of between 4 and 10 used in $\overline{X}$ control charts. The main reason is that if p is small (as is typically the case) and n is not of reasonable size, then most of the subgroups will have zero defects even when the process goes out of control; and thus it would take longer to detect this out-of-control situation than it would if n were chosen so that np was not too small. A secondary reason for using a larger value of n is that when np is of moderate size, X will approximately have a normal distribution, and so when in control, each subgroup statistic will fall within the control limits with probability approximately equal to $1 - .0027 = .9973$.

Example 15.3   Successive samples of 200 screws are drawn from the production of an automatic screw machine, with each screw being rated as acceptable or defective. Suppose that it is known from historical data that when the process is in control, each screw is independently defective with probability .07. If the following values represent the number of defective screws in each of 20 samples, would the process have been declared out of control at any time during the collection of these samples?

Subgroup	Defectives	Subgroup	Defectives
1	23	11	4
2	22	12	13
3	12	13	17
4	13	14	5
5	15	15	9
6	11	16	5
7	25	17	19
8	16	18	7
9	23	19	22
10	14	20	17

Solution Since $n = 200$ and $p = .07$, we have

$$np = 14 \qquad 3\sqrt{np(1 - p)} = 10.825$$

and so

$$LCL = 14 - 10.825 = 3.175$$

$$UCL = 14 + 10.825 = 24.825$$

Since the number of defectives in subgroup 7 falls outside the range from LCL to UCL, the process would have been declared out of control at that point.

Remark Note that we are attempting to detect any change in quality even when this change results in a quality improvement. That is, we regard the process as being out of control even when the probability of a defective item decreases. The reason for this is that it is important to recognize any change in quality, for either better or worse, so as to be able to evaluate the reason for the change. In other words, if an improvement in product quality occurs, then it is important to determine the reason for this improvement (what are we doing right?).

Problems

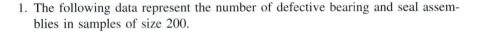

1. The following data represent the number of defective bearing and seal assemblies in samples of size 200.

Sample number	Number of defectives	Sample number	Number of defectives
1	7	11	4
2	3	12	10
3	2	13	0
4	6	14	8
5	9	15	3
6	4	16	6
7	3	17	2
8	3	18	1
9	2	19	6
10	5	20	10

Suppose that when the process is in control, each assembly is defective with probability .03. Does it appear that the process was in control throughout?

2. Suppose that when a process is in control, each item produced is defective with probability .04. If the control chart takes daily samples of size 500, compute the upper and lower control limits.

3. Historically, 4 percent of fiber containers are defective due to contamination from gluing. The following data represent the number of defective containers in successive samples of size 100.

3, 5, 1, 0, 4, 7, 8, 9, 5, 7, 1, 3, 0, 5, 3, 6, 4, 8, 3, 6

   (a) Does the process appear to have been in control?

   (b) What are the control limits?

Key Terms

Control chart: A graphical procedure for enabling one to detect when a production process has gone out of control.

Summary

Suppose that a production process produces items, each of which has a measurable value which, when the process is in control, has mean μ and standard deviation σ. To detect any change, the items are put into subgroups of size n, and the subgroup averages $\bar{X}$ are plotted. Whenever a subgroup average is either less than the lower control limit

$$\text{LCL} = \mu - \frac{3\sigma}{\sqrt{n}}$$

or greater than the upper control limit

$$\text{UCL} = \mu + \frac{3\sigma}{\sqrt{n}}$$

then the process is declared to be out of control.

Sometimes rather than having a continuous value each item is classified as being either acceptable or defective. Let p denote the probability that an item is defective when the process is in control. To determine when it goes out of control, items are again put into subgroups of size n. When the number of defectives in a subgroup falls outside the control limits

$$\text{LCL} = np - 3\sqrt{np(1-p)} \quad \text{and} \quad \text{UCL} = np + 3\sqrt{np(1-p)}$$

then the process is declared to be out of control.

Review Problems

1. The distance between two adjacent pins of a memory chip for enhanced graphics adapters has, when the production process is in control, a mean of 1.5 millimeters and standard deviation of .001 millimeter. Determine the upper and lower control limits for an $\bar{X}$ control chart, using subgroups of size 4.

2. The following table gives the yearly production of U.S. steel mill products, in units of millions of short tons, in the years from 1983 through 1991.

Year	1983	1984	1985	1986	1987	1988	1989	1990	1991
Production	83.5	98.9	96.4	90.0	95.9	102.7	96.8	97.8	88.3

Source: American Iron and Steel Institute, *Annual Statistical Report,* Washington.

If historical data through 1982 showed that the yearly production of U.S. steel mill products was normally distributed with mean 94.4 million and standard deviation 3.3 million, does it appear that a change has occurred since then? Take subgroups of size $n = 3$.

3. Prior to 1983, the number of burglaries committed yearly in the United States per 100,000 population was normally distributed with mean 1420 and standard deviation 120. The following table gives the rates in the years from 1983 to 1991.

Year	1983	1984	1985	1986	1987	1988	1989	1990	1991
Burglary rate per 100,000 people	1338	1264	1287	1345	1330	1309	1276	1236	1252

Source: U.S. Federal Bureau of Investigation, *Crime in the United States,* annual.

During the years from 1983 to 1991, did the burglary rate remain at its pre-1983 level? Use subgroups of size 3.

4. When a process is performing correctly, 1.5 percent of the items produced do not conform to specifications. If items are grouped into subgroups of size 300, determine the upper and lower control limits for this control chart.

5. The numbers of defective switches in 12 samples of size 200 are as follows:

   4, 7, 2, 5, 9, 5, 7, 10, 8, 3, 12, 9

Suppose that when the process is in control, each switch is defective with probability .03. Does the process appear to have been in control?

Quality Control

Purpose

Use Minitab to

1. Construct $\bar{X}$ control charts.

2. Construct control charts for fraction defective.

Procedures

First, load the Minitab (Windows version) software as in the Minitab lab for Chap. 1.

1. THE $\bar{X}$ CONTROL CHART

Example 1 Use Minitab to work Prob. 2 in Sec. 15.2 of the text.

Enter the $\bar{X}$ values in column C1. Next, select Stat→Control Charts→Xbar, and the Xbar chart dialog box will be displayed. In the Variable text box, type C1; in the Historical mu text box, type 35; and in the Sigma historical text box, type 4. Select the Subgroup size check box, and type 1 in the text box. We use a value of 1 here because the sample means—not the individual values for the subgroups—were given. So we assume that the subgroup size is 1. If the individual values had been given, then you would need to specify the subgroup size if they were all the same; or if they had been saved in a specified column, you would need to select the option Subgroups. You can click on Annotation, Frame, etc., to enter a title, as you did in the Minitab lab for Chap. 2. The Xbar chart dialog box with the appropriate entries is shown in Fig. M15.1.

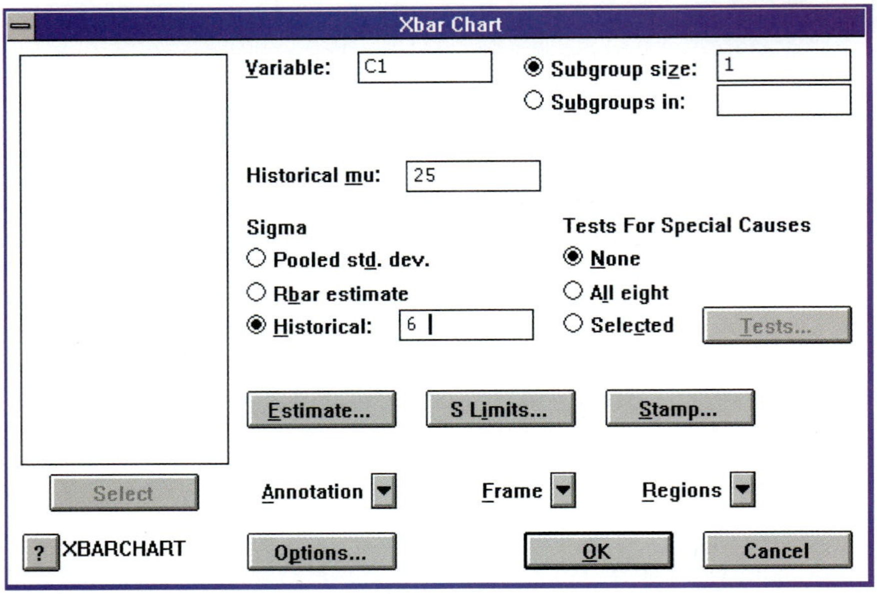

Figure M15.1

Click on the OK button, and the $\bar{X}$ chart will be displayed in the XbarChart window. This is shown in Fig. M15.2. Observe from this chart that the process is working properly since the $\bar{X}$ values are within the control limits. That is, the sample means are within the values UCL (upper control limit) = 47 and LCL (lower control limit) = 23.

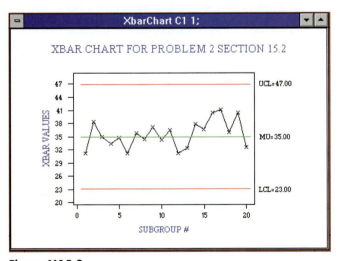

Figure M15.2

Note If we had selected Tests for special causes in Fig. M15.1, then we would have been testing for a special pattern in the data plotted on the chart. The occurrence of a pattern would suggest a special cause for variation that should be investigated.

2. CONTROL CHARTS FOR FRACTION DEFECTIVE

Example 2   Use Minitab to help work Example 15.3 in Sec. 15.3 of the text.

First, enter the observed number of defectives in column C1. This problem can be solved by using an *np* chart which draws a chart for the number of defectives. Select Stat→Control Charts→NP, and the NP chart dialog box will be displayed. Figure M15.3 shows this dialog box with the appropriate entries.

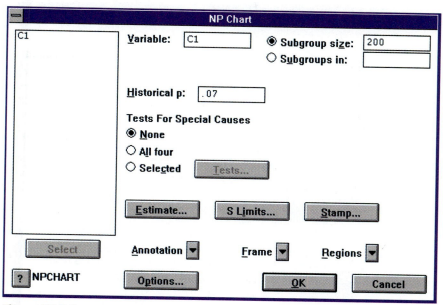

Figure M15.3

Click on the OK button, and the chart will be drawn as in Fig. M15.4. The Annotation, Frame, etc., options were used to present Fig. M15.4.

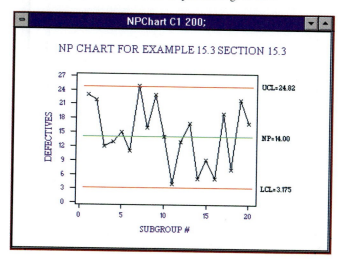

Figure M15.4

Observe that LCL = 3.175 and UCL = 24.82 and that subgroup 7 falls outside the UCL value. The process should be declared out of control at this point.

Computer Exercises

1. Use Minitab or any other statistical software to help work Probs. 3 to 6 in Sec. 15.2.

2. Use Minitab or any other statistical software to work Probs. 1 and 3 in Sec. 15.3.

3. Use your library or any other resource to collect a set of data that could be analyzed with the $\bar{X}$ chart (Fig. M15.2). Use Minitab or any other statistical software to do a complete analysis of the data. Present any relevant analysis and discussion in a report. If you select data from published research, include relevant factors about the validity and reliability of the study.

4. Use your library or any other resource to collect a set of data that could be analyzed with the *np* chart (Fig. M15.3). Use Minitab or any other statistical software to do a complete analysis of the data. Present any relevant analysis and discussion in a report. If you select data from published research, include relevant factors about the validity and reliability of the study.

5. Use Minitab or any other statistical software to simulate 30 subgroups of size 20 integer values between 0 and 9. Let the number of even numbers in each set represent the number of defectives of a process. Use the theory of this chapter to determine whether the process is out of control. Present your findings in a report. *Note:* Each set of values generated is equivalent to a binomial experiment with $n = 20$ and $p = 1/10$.

A P P E N D I X E S

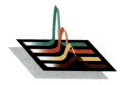

APPENDIX A

A Data Set

Student	Weight	Cholesterol	Pressure	Gender	Student	Weight	Cholesterol	Pressure	Gender
1	147	213	127	F	29	132	171	112	M
2	156	174	116	M	30	129	194	114	M
3	112	193	110	F	31	111	184	104	F
4	127	196	110	F	32	156	191	118	M
5	144	220	130	F	33	155	221	107	F
6	140	183	99	M	34	104	212	111	F
7	119	194	112	F	35	217	221	156	M
8	139	200	102	F	36	132	204	117	F
9	161	192	121	M	37	103	204	121	F
10	146	200	125	F	38	171	191	105	M
11	190	200	125	M	39	135	183	110	F
12	126	199	133	F	40	249	227	137	M
13	164	178	130	M	41	185	188	119	M
14	176	183	136	M	42	194	200	109	M
15	131	188	112	F	43	165	197	123	M
16	107	193	113	F	44	121	208	100	F
17	116	187	112	F	45	124	218	102	F
18	157	181	129	M	46	113	194	119	F
19	186	193	137	M	47	110	212	119	F
20	189	205	113	M	48	136	207	99	F
21	147	196	113	M	49	221	219	149	M
22	112	211	110	F	50	151	201	109	F
23	209	202	97	M	51	182	208	130	M
24	135	213	103	F	52	151	192	107	M
25	168	216	95	M	53	182	192	136	M
26	209	206	107	M	54	149	191	124	M
27	102	195	102	F	55	162	196	132	M
28	166	191	111	M	56	168	193	92	M

Student	Weight	Cholesterol	Pressure	Gender	Student	Weight	Cholesterol	Pressure	Gender
57	185	185	123	M	109	184	199	151	M
58	191	201	118	M	110	132	195	129	F
59	173	185	114	M	111	119	202	117	F
60	186	203	114	M	112	158	195	112	M
61	161	177	119	M	113	138	217	101	F
62	149	213	124	F	114	177	194	136	M
63	103	192	104	F	115	99	204	129	F
64	126	193	99	F	116	177	198	126	M
65	181	212	141	M	117	134	195	111	F
66	190	188	124	M	118	133	168	98	M
67	124	201	114	F	119	194	201	120	M
68	175	219	125	M	120	140	211	132	F
69	161	189	120	M	121	104	195	106	F
70	160	203	108	F	122	191	180	130	M
71	171	186	111	M	123	184	205	116	M
72	176	186	114	M	124	155	189	117	M
73	156	196	99	M	125	126	196	112	F
74	126	195	123	F	126	190	195	124	M
75	138	205	113	F	127	132	218	120	F
76	136	223	131	F	128	133	194	121	F
77	192	195	125	M	129	174	203	128	M
78	122	205	110	F	130	168	190	120	M
79	176	198	96	M	131	190	196	132	M
80	195	215	143	M	132	176	194	107	M
81	126	202	102	F	133	121	210	118	F
82	138	196	124	F	134	131	167	105	M
83	166	196	103	M	135	174	203	88	M
84	86	190	106	F	136	112	183	94	F
85	90	185	110	F	137	121	203	116	F
86	177	188	109	M	138	132	194	104	F
87	136	197	129	F	139	155	188	111	M
88	103	196	95	F	140	127	189	106	F
89	190	227	134	M	141	151	193	120	M
90	130	211	119	F	142	189	221	126	M
91	205	219	130	M	143	123	194	129	F
92	127	202	121	F	144	137	196	113	F
93	182	204	129	M	145	122	201	113	F
94	122	213	116	F	146	126	212	121	F
95	139	202	102	F	147	136	210	120	F
96	189	205	102	M	148	145	168	115	M
97	147	184	114	M	149	202	202	122	M
98	180	198	123	M	150	151	206	108	F
99	130	180	94	M	151	137	178	128	M
100	130	204	118	F	152	90	178	100	F
101	150	197	110	F	153	177	220	123	M
102	184	192	129	M	154	139	214	120	F
103	179	202	129	M	155	172	191	117	M
104	105	211	109	F	156	107	179	106	F
105	157	179	109	M	157	186	209	129	M
106	202	210	124	M	158	198	196	140	M
107	140	188	112	F	159	113	184	110	F
108	165	203	114	F	160	143	209	105	F

Student	Weight	Cholesterol	Pressure	Gender	Student	Weight	Cholesterol	Pressure	Gender
161	205	198	137	M	213	120	182	126	F
162	186	206	111	M	214	126	207	110	F
163	174	189	129	M	215	170	201	101	M
164	171	197	132	M	216	175	211	115	M
165	209	202	128	M	217	134	219	129	F
166	126	203	134	F	218	118	211	113	F
167	160	185	109	M	219	118	178	109	F
168	127	212	124	F	220	164	196	107	M
169	112	193	115	F	221	186	190	134	M
170	155	184	112	M	222	172	189	134	M
171	111	181	111	F	223	173	207	101	M
172	151	196	129	M	224	185	206	128	M
173	110	181	113	F	225	190	198	117	M
174	159	192	115	M	226	146	200	112	F
175	173	196	131	M	227	103	179	100	F
176	148	191	101	M	228	124	215	124	F
177	141	216	110	F	229	186	213	124	M
178	161	186	123	M	230	166	166	129	M
179	125	209	113	F	231	138	201	120	F
180	114	200	109	F	232	175	198	118	M
181	125	206	135	F	233	104	194	100	F
182	129	214	100	F	234	213	206	130	M
183	115	207	115	F	235	171	182	118	M
184	142	197	118	F	236	180	213	119	M
185	183	202	114	M	237	187	197	128	M
186	181	212	118	M	238	117	194	106	F
187	108	185	96	F	239	108	185	105	F
188	126	194	122	F	240	128	202	105	F
189	175	201	138	M	241	170	196	118	M
190	168	182	118	M	242	183	176	126	M
191	115	194	122	F	243	143	190	101	M
192	129	193	90	F	244	160	205	120	F
193	131	209	119	F	245	185	184	113	M
194	187	182	134	M	246	122	193	142	F
195	185	200	127	M	247	225	218	142	M
196	114	196	113	F	248	139	191	99	F
197	206	216	124	M	249	123	207	116	F
198	151	212	113	F	250	129	176	108	F
199	128	204	110	F	251	142	220	137	F
200	128	204	115	F	252	146	191	116	M
201	183	190	136	M	253	129	201	100	F
202	104	192	93	F	254	163	171	119	M
203	99	209	110	F	255	177	206	134	M
204	201	208	120	M	256	183	190	116	M
205	129	204	100	F	257	120	201	104	F
206	149	193	117	F	258	188	214	115	M
207	123	200	120	F	259	140	182	119	M
208	179	191	122	M	260	166	197	113	M
209	150	216	128	F	261	122	199	107	F
210	133	193	110	F	262	177	207	124	M
211	112	190	107	F	263	184	204	122	M
212	175	188	113	M	264	113	198	121	F

Student	Weight	Cholesterol	Pressure	Gender	Student	Weight	Cholesterol	Pressure	Gender
265	214	221	142	M	289	201	208	138	M
266	144	205	111	M	290	174	199	111	M
267	188	188	132	M	291	188	189	119	M
268	114	204	127	F	292	151	205	133	F
269	158	213	111	F	293	202	220	126	M
270	146	196	116	M	294	125	198	106	F
271	195	195	148	M	295	176	190	116	M
272	199	201	125	M	296	183	188	96	M
273	148	202	120	F	297	118	198	130	F
274	164	190	113	M	298	125	204	111	F
275	137	196	107	F	299	237	209	127	M
276	133	173	121	M	300	124	186	127	F
277	104	214	112	F	301	98	194	104	F
278	126	194	116	F	302	182	199	108	M
279	120	220	116	F	303	184	206	149	M
280	148	204	131	F	304	137	189	113	F
281	100	206	89	F	305	126	177	111	F
282	178	190	125	M	306	202	198	130	M
283	149	188	108	F	307	225	212	142	M
284	157	194	124	M	308	181	200	122	M
285	99	203	95	F	309	178	187	121	M
286	192	208	127	M	310	132	221	110	F
287	175	181	145	M	311	164	201	134	M
288	208	193	123	M	312	163	191	138	M

A P P E N D I X   B

Mathematical Preliminaries

B.1   SUMMATION

Consider four numbers which we will call x_1, x_2, x_3, and x_4. If s is equal to the sum of these numbers, then we can express this fact either by writing

$$s = x_1 + x_2 + x_3 + x_4$$

or by using the summation notation Σ. In this latter situation we write

$$s = \sum_{i=1}^{4} x_i$$

which means that s is equal to the sum of the x_i values as i ranges from 1 to 4.

The summation notation is quite useful when we want to sum a large number of quantities. For instance, suppose that we were given 100 numbers, designated as x_1, x_2, and so on, up to x_{100}. We could then compactly express s, the sum of these numbers, as

$$s = \sum_{i=1}^{100} x_i$$

If we want the sum to include only the 60 numbers starting at x_{20} and ending at x_{79}, then we could express this sum by the notation

$$\sum_{i=20}^{79} x_i$$

That is, $\displaystyle\sum_{i=20}^{79} x_i$ is the sum of the x_i values as i ranges from 20 to 79.

B.2   ABSOLUTE VALUE

The absolute value of a number is its magnitude regardless of its sign. For instance, the absolute value of 4 is 4, whereas the absolute value of -5 is 5. In general, the absolute value of a positive number is that number whereas the absolute value of a negative number is its negative. We use the symbol $|x|$ to denote the absolute value of the number x. Thus,

$$|x| = \begin{cases} x & \text{if } x \ge 0 \\ -x & \text{if } x < 0 \end{cases}$$

If we represent each real number by a point on a straight line, then $|x|$ is the distance from point x to the origin 0. This is illustrated by Fig. B.1.

Figure B.1
Distance from
-2 to 0 is
$|-2| = 2$.

If x and y are any two numbers, then $|x - y|$ is equal to the distance between x and y. For instance, if $x = 5$ and $y = 2$, then $|x - y| = |5 - 2| = |3| = 3$. On the other hand, if $x = 5$ and $y = -2$, then $|x - y| = |5 - (-2)| = |5 + 2| = 7$. That is, the distance between 5 and 2 is 3, whereas the distance between 5 and -2 is 7.

B.3   SET NOTATION

Consider a collection of numbers, for instance, all the real numbers. Sometimes we are interested in the subcollection of these numbers that satisfies a particular property. Let A designate a certain property; for instance, A could be the property that the number is positive, or that it is an even integer, or that it is a prime integer. We express the numbers in the collection that have the property A by the notation

$\{x: x \text{ has property } A\}$

which is read as "the set of all the values x in the collection that have the property A." For instance,

$\{x: x \text{ is an even integer between 1 and 7}\}$

is just the set consisting of the three values 2, 4, and 6. That is

$\{x: x \text{ is an even integer between 1 and 7}\} = \{2, 4, 6\}$

We are sometimes interested in the set of all numbers that are within some fixed distance of a specified number. For instance, consider the set of all numbers that are within 2 of the number 5. This set can be expressed as

$$\{x: |x - 5| \le 2\}$$

Because a number will be within 2 of the number 5 if and only if that number lies between 3 and 7, we have

$$\{x: |x - 5| \le 2\} = \{x: 3 \le x \le 7\}$$

APPENDIX C

How to Choose a Random Sample

As we have seen in this book, it is extremely important to be able to choose a random sample. Suppose that we want to choose a random sample of size n from a population of size N. How can we accomplish this?

The first step is to number the population from 1 to N in any arbitrary manner. Then we will choose a random sample by designating n elements of the population that are to be in the sample. To do this, we start by letting the first element of the sample be equally likely to be any of the N elements. The next element is then chosen so that it is equally likely to be any of the remaining $N - 1$ elements; the next so that it is equally likely to be any of the remaining $N - 2$ elements; and so on, until we have amassed a total of n elements, which constitute the random sample.

To implement the above scheme, it seems that we would always have to keep track of which elements had already been selected. However, by a neat trick, it turns out that this is not necessary. Indeed, we can arrange the N elements in an ordered list and then randomly choose not the elements themselves but rather the positions of the elements that are to be put in the random sample. Let us see how it works when $N = 7$ and $n = 3$. We start by numbering each of the 7 elements in the population and then arranging them in a list, say, the initial order is

1, 2, 3, 4, 5, 6, 7

We now choose a number that is equally likely to be 1, 2, 3, 4, 5, 6, or 7; say, 4 is chosen. This means that the element in position 4 (element number 4 in this case) is put in the random sample. To indicate that this element is in the random sample and to make certain that this element will not be chosen again, we interchange in the list the element in position 4 with the one in position 7. This results in the new list ordering

1, 2, 3, 7, 5, 6, 4

where we have underlined the element that is in the random sample. The next element to be put in the random sample should be equally likely to be any of the elements in the first 6 positions. Thus we select a value that is equally likely to be 1, 2, 3, 4, 5, or 6; the element in that position will become part of the random sample. And to indicate this and to leave the first 5 positions for the elements that have not yet been chosen, we interchange the element in the position chosen with the element in position 6. For instance, if the value chosen was 4, then the element in position 4 (that is, element number 7) becomes part of the random sample, and the new list ordering is

1, 2, 3, 6, 5, 7, 4

The final element of the random sample is equally likely to be any of the elements in positions 1 through 5, so we select a value that is equally likely to be 1, 2, 3, 4, or 5 and interchange the element in that position with the one in position 5. For instance, if the value is 2, then the new ordering is

1, 5, 3, 6, 2, 7, 4

Since there are now three elements in the random sample, namely, 2, 7, and 4, the process is complete.

To implement the above *algorithm* for generating a random sample, we need to know how to generate the value of a random quantity that is equally likely to be any of the numbers 1, 2, 3, . . . , k. The key to doing this is to make use of *random numbers* which are the values of random variables that are uniformly distributed over the interval $(0, 1)$. Most computers have a built-in random number generator that allows one to call for the value of such a quantity. If U designates a random number—that is, U is uniformly distributed over the interval $(0, 1)$—then it can be shown that

$$I = \text{Int } (kU) + 1$$

will be equally likely to be any of the values 1, 2, . . . , k, where Int (x) stands for the integer part of x. For instance,

Int $(4.3) = 4$
Int $(12.9) = 12$

and so on.

Program A-1 uses the above to generate a random sample of size n from the set of numbers 1, 2, . . . , N. When running this program, you will be asked to first enter the values of n and N and to then enter any four-digit number. For this last request, just type and enter any number that comes to mind. The output from this program is the subset of size n that constitutes the random sample.

Example C.1   Suppose we want to choose a random sample of size 12 from a population of 200 members. To do so, we start by arbitrarily numbering the 200 members of the population so that they now have numbers 1 to 200. We run Program A-1 to obtain the 12 members of the population that are to constitute the random sample.

THIS PROGRAM GENERATES A RANDOM SAMPLE OF K OF THE INTEGERS 1 THRU N
ENTER THE VALUE OF N
? 200
ENTER THE VALUE OF K
? 12
Random Number Seed (−32,768 to 32,767)? 355
THE RANDOM SAMPLE CONSISTS OF THE FOLLOWING 12 ELEMENTS
90 89 82 162 21 81 182 45 38 195 64 1

A P P E N D I X   D

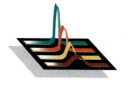

Tables

Table D.1 **Standard Normal Probabilities**

Table entries give $P\{Z \le x\}$.

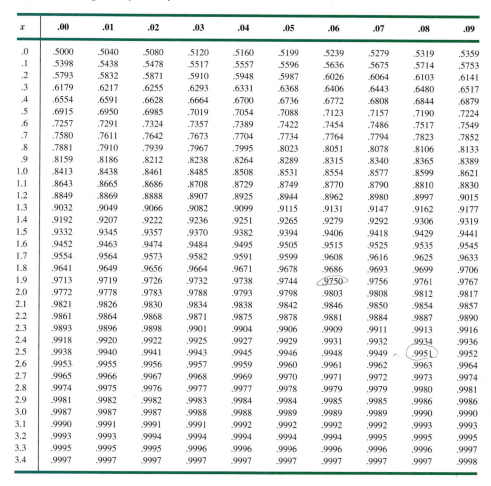

x	.00	.01	.02	.03	.04	.05	.06	.07	.08	.09
.0	.5000	.5040	.5080	.5120	.5160	.5199	.5239	.5279	.5319	.5359
.1	.5398	.5438	.5478	.5517	.5557	.5596	.5636	.5675	.5714	.5753
.2	.5793	.5832	.5871	.5910	.5948	.5987	.6026	.6064	.6103	.6141
.3	.6179	.6217	.6255	.6293	.6331	.6368	.6406	.6443	.6480	.6517
.4	.6554	.6591	.6628	.6664	.6700	.6736	.6772	.6808	.6844	.6879
.5	.6915	.6950	.6985	.7019	.7054	.7088	.7123	.7157	.7190	.7224
.6	.7257	.7291	.7324	.7357	.7389	.7422	.7454	.7486	.7517	.7549
.7	.7580	.7611	.7642	.7673	.7704	.7734	.7764	.7794	.7823	.7852
.8	.7881	.7910	.7939	.7967	.7995	.8023	.8051	.8078	.8106	.8133
.9	.8159	.8186	.8212	.8238	.8264	.8289	.8315	.8340	.8365	.8389
1.0	.8413	.8438	.8461	.8485	.8508	.8531	.8554	.8577	.8599	.8621
1.1	.8643	.8665	.8686	.8708	.8729	.8749	.8770	.8790	.8810	.8830
1.2	.8849	.8869	.8888	.8907	.8925	.8944	.8962	.8980	.8997	.9015
1.3	.9032	.9049	.9066	.9082	.9099	.9115	.9131	.9147	.9162	.9177
1.4	.9192	.9207	.9222	.9236	.9251	.9265	.9279	.9292	.9306	.9319
1.5	.9332	.9345	.9357	.9370	.9382	.9394	.9406	.9418	.9429	.9441
1.6	.9452	.9463	.9474	.9484	.9495	.9505	.9515	.9525	.9535	.9545
1.7	.9554	.9564	.9573	.9582	.9591	.9599	.9608	.9616	.9625	.9633
1.8	.9641	.9649	.9656	.9664	.9671	.9678	.9686	.9693	.9699	.9706
1.9	.9713	.9719	.9726	.9732	.9738	.9744	.9750	.9756	.9761	.9767
2.0	.9772	.9778	.9783	.9788	.9793	.9798	.9803	.9808	.9812	.9817
2.1	.9821	.9826	.9830	.9834	.9838	.9842	.9846	.9850	.9854	.9857
2.2	.9861	.9864	.9868	.9871	.9875	.9878	.9881	.9884	.9887	.9890
2.3	.9893	.9896	.9898	.9901	.9904	.9906	.9909	.9911	.9913	.9916
2.4	.9918	.9920	.9922	.9925	.9927	.9929	.9931	.9932	.9934	.9936
2.5	.9938	.9940	.9941	.9943	.9945	.9946	.9948	.9949	.9951	.9952
2.6	.9953	.9955	.9956	.9957	.9959	.9960	.9961	.9962	.9963	.9964
2.7	.9965	.9966	.9967	.9968	.9969	.9970	.9971	.9972	.9973	.9974
2.8	.9974	.9975	.9976	.9977	.9977	.9978	.9979	.9979	.9980	.9981
2.9	.9981	.9982	.9982	.9983	.9984	.9984	.9985	.9985	.9986	.9986
3.0	.9987	.9987	.9987	.9988	.9988	.9989	.9989	.9989	.9990	.9990
3.1	.9990	.9991	.9991	.9991	.9992	.9992	.9992	.9992	.9993	.9993
3.2	.9993	.9993	.9994	.9994	.9994	.9994	.9994	.9995	.9995	.9995
3.3	.9995	.9995	.9995	.9996	.9996	.9996	.9996	.9996	.9996	.9997
3.4	.9997	.9997	.9997	.9997	.9997	.9997	.9997	.9997	.9997	.9998

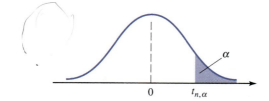

| Table D.2 | **Percentiles $t_{n,\alpha}$ of t Distributions** |

						α				
n	.40	.25	.10	.05	.025	.01	.005	.0025	.001	.0005
1	.325	1.000	3.078	6.314	12.706	31.821	63.657	127.32	318.31	636.62
2	.289	.816	1.886	2.920	4.303	6.965	9.925	14.089	23.326	31.598
3	.277	.765	1.638	2.353	3.182	4.541	5.841	7.453	10.213	12.924
4	.271	.741	1.533	2.132	2.776	3.747	4.604	5.598	7.173	8.610
5	.267	.727	1.476	2.015	2.571	3.365	4.032	4.773	5.893	6.869
6	.265	.718	1.440	1.943	2.447	3.143	3.707	4.317	5.208	5.959
7	.263	.711	1.415	1.895	2.365	2.998	3.499	4.029	4.785	5.408
8	.262	.706	1.397	1.860	2.306	2.896	3.355	3.833	4.501	5.041
9	.261	.703	1.383	1.833	2.262	2.821	3.250	3.690	4.297	4.781
10	.260	.700	1.372	1.812	2.228	2.764	3.169	3.581	4.144	4.587
11	.260	.697	1.363	1.796	2.201	2.718	3.106	3.497	4.025	4.437
12	.259	.695	1.356	1.782	2.179	2.681	3.055	3.428	3.930	4.318
13	.259	.694	1.350	1.771	2.160	2.650	3.012	3.372	3.852	4.221
14	.258	.692	1.345	1.761	2.145	2.624	2.977	3.326	3.787	4.140
15	.258	.691	1.341	1.753	2.131	2.602	2.947	3.286	3.733	4.073
16	.258	.690	1.337	1.746	2.120	2.583	2.921	3.252	3.686	4.015
17	.257	.689	1.333	1.740	2.110	2.567	2.898	3.222	3.646	3.965
18	.257	.688	1.330	1.734	2.101	2.552	2.878	3.197	3.610	3.922
19	.257	.688	1.328	1.729	2.093	2.539	2.861	3.174	3.579	3.883
20	.257	.687	1.325	1.725	2.086	2.528	2.845	3.153	3.552	3.850
21	.257	.686	1.323	1.721	2.080	2.518	2.831	3.135	3.527	3.819
22	.256	.686	1.321	1.717	2.074	2.508	2.819	3.119	3.505	3.792
23	.256	.685	1.319	1.714	2.069	2.500	2.807	3.104	3.485	3.767
24	.256	.685	1.318	1.711	2.064	2.492	2.797	3.091	3.467	3.745
25	.256	.684	1.316	1.708	2.060	2.485	2.787	3.078	3.450	3.725
26	.256	.684	1.315	1.706	2.056	2.479	2.779	3.067	3.435	3.707
27	.256	.684	1.314	1.703	2.052	2.473	2.771	3.057	3.421	3.690
28	.256	.683	1.313	1.701	2.048	2.467	2.763	3.047	3.408	3.674
29	.256	.683	1.311	1.699	2.045	2.462	2.756	3.038	3.396	3.659
30	.256	.683	1.310	1.697	2.042	2.457	2.750	3.030	3.385	3.646
40	.255	.681	1.303	1.684	2.021	2.423	2.704	2.971	3.307	3.551
60	.254	.679	1.296	1.671	2.000	2.390	2.660	2.915	3.232	3.460
120	.254	.677	1.289	1.658	1.980	2.358	2.617	2.860	3.160	3.373
∞	.253	.674	1.282	1.645	1.960	2.326	2.576	2.807	3.090	3.291

n = degrees of freedom.

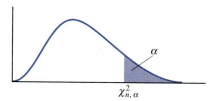

$$\chi^2_{n,\,\alpha}$$

Table D.3   Percentiles $\chi^2_{n,\,\alpha}$ of the Chi-Squared Distributions

n	.995	.990	.975	.950	.900	.500	.100	.050	.025	.010	.005
1	.00+	.00+	.00+	.00+	.02	.45	2.71	3.84	5.02	6.63	7.88
2	.01	.02	.05	.10	.21	1.39	4.61	5.99	7.38	9.21	10.60
3	.07	.11	.22	.35	.58	2.37	6.25	7.81	9.35	11.34	12.84
4	.21	.30	.48	.71	1.06	3.36	7.78	9.49	11.14	13.28	14.86
5	.41	.55	.83	1.15	1.61	4.35	9.24	11.07	12.83	15.09	16.75
6	.68	.87	1.24	1.64	2.20	5.35	10.65	12.59	14.45	16.81	18.55
7	.99	1.24	1.69	2.17	2.83	6.35	12.02	14.07	16.01	18.48	20.28
8	1.34	1.65	2.18	2.73	3.49	7.34	13.36	15.51	17.53	20.09	21.96
9	1.73	2.09	2.70	3.33	4.17	8.34	14.68	16.92	19.02	21.67	23.59
10	2.16	2.56	3.25	3.94	4.87	9.34	15.99	18.31	20.48	23.21	25.19
11	2.60	3.05	3.82	4.57	5.58	10.34	17.28	19.68	21.92	24.72	26.76
12	3.07	3.57	4.40	5.23	6.30	11.34	18.55	21.03	23.34	26.22	28.30
13	3.57	4.11	5.01	5.89	7.04	12.34	19.81	22.36	24.74	27.69	29.82
14	4.07	4.66	5.63	6.57	7.79	13.34	21.06	23.68	26.12	29.14	31.32
15	4.60	5.23	6.27	7.26	8.55	14.34	22.31	25.00	27.49	30.58	32.80
16	5.14	5.81	6.91	7.96	9.31	15.34	23.54	26.30	28.85	32.00	34.27
17	5.70	6.41	7.56	8.67	10.09	16.34	24.77	27.59	30.19	33.41	35.72
18	6.26	.7.01	8.23	9.39	10.87	17.34	25.99	28.87	31.53	34.81	37.16
19	6.84	7.63	8.91	10.12	11.65	18.34	27.20	30.14	32.85	36.19	38.58
20	7.43	8.26	9.59	10.85	12.44	19.34	28.41	31.41	34.17	37.57	40.00
21	8.03	8.90	10.28	11.59	13.24	20.34	29.62	32.67	35.48	38.93	41.40
22	8.64	9.54	10.98	12.34	14.04	21.34	30.81	33.92	36.78	40.29	42.80
23	9.26	10.20	11.69	13.09	14.85	22.34	32.01	35.17	38.08	41.64	44.18
24	9.89	10.86	12.40	13.85	15.66	23.34	33.20	36.42	39.36	42.98	45.56
25	10.52	11.52	13.12	14.61	16.47	24.34	34.28	37.65	40.65	44.31	46.93
26	11.16	12.20	13.84	15.38	17.29	25.34	35.56	38.89	41.92	45.64	48.29
27	11.81	12.88	14.57	16.15	18.11	26.34	36.74	40.11	43.19	46.96	49.65
28	12.46	13.57	15.31	16.93	18.94	27.34	37.92	41.34	44.46	48.28	50.99
29	13.12	14.26	16.05	17.71	19.77	28.34	39.09	42.56	45.72	49.59	52.34
30	13.79	14.95	16.79	18.49	20.60	29.34	40.26	43.77	46.98	50.89	53.67
40	20.71	22.16	24.43	26.51	29.05	39.34	51.81	55.76	59.34	63.69	66.77
50	27.99	29.71	32.36	34.76	37.69	49.33	63.17	67.50	71.42	76.15	79.49
60	35.53	37.48	40.48	43.19	46.46	59.33	74.40	79.08	83.30	88.38	91.95
70	43.28	45.44	48.76	51.74	55.33	69.33	85.53	90.53	95.02	100.42	104.22
80	51.17	53.54	57.15	60.39	64.28	79.33	96.58	101.88	106.63	112.33	116.32
90	59.20	61.75	65.65	69.13	73.29	89.33	107.57	113.14	118.14	124.12	128.30
100	67.33	70.06	74.22	77.93	82.36	99.33	118.50	124.34	129.56	135.81	140.17

n = degrees of freedom.

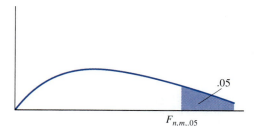

$F_{n,m,.05}$

.05

Cretical Values

| Table D.4 | **Percentiles of *F* Distributions** |

95th Percentiles of $F_{n,m}$ Distributions

									Degrees of freedom for the numerator *n*										
	1	**2**	**3**	**4**	**5**	**6**	**7**	**8**	**9**	**10**	**12**	**15**	**20**	**24**	**30**	**40**	**60**	**120**	**∞**
1	161.4	199.5	215.7	224.6	230.2	234.0	236.8	238.9	240.5	241.9	243.9	245.9	248.0	249.1	250.1	251.1	252.2	253.3	254.3
2	18.51	19.00	19.16	19.25	19.30	19.33	19.35	19.37	19.38	19.40	19.41	19.43	19.45	19.45	19.46	19.47	19.48	19.49	19.50
3	10.13	9.55	9.28	9.12	9.01	8.94	8.89	8.85	8.81	8.79	8.74	8.70	8.66	8.64	8.62	8.59	8.57	8.55	8.53
4	7.71	6.94	6.59	6.39	6.26	6.16	6.09	6.04	6.00	5.96	5.91	5.86	5.80	5.77	5.75	5.72	5.69	5.66	5.63
5	6.61	5.79	5.41	5.19	5.05	4.95	4.88	4.82	4.77	4.74	4.68	4.62	4.56	4.53	4.50	4.46	4.43	4.40	4.36
6	5.99	5.14	4.76	4.53	4.39	4.28	4.21	4.15	4.10	4.06	4.00	3.94	3.87	3.84	3.81	3.77	3.74	3.70	3.67
7	5.59	4.74	4.35	4.12	3.97	3.87	3.79	3.73	3.68	3.64	3.57	3.51	3.44	3.41	3.38	3.34	3.30	3.27	3.23
8	5.32	4.46	4.07	3.84	3.69	3.58	3.50	3.44	3.39	3.35	3.28	3.22	3.15	3.12	3.08	3.04	3.01	2.97	2.93
9	5.12	4.26	3.86	3.63	3.48	3.37	3.29	3.23	3.18	3.14	3.07	3.01	2.94	2.90	2.86	2.83	2.79	2.75	2.71
10	4.96	4.10	3.71	3.48	3.33	3.22	3.14	3.07	3.02	2.98	2.91	2.85	2.77	2.74	2.70	2.66	2.62	2.58	2.54
11	4.84	3.98	3.59	3.36	3.20	3.09	3.01	2.95	2.90	2.85	2.79	2.72	2.65	2.61	2.57	2.53	2.49	2.45	2.40
12	4.75	3.89	3.49	3.26	3.11	3.00	2.91	2.85	2.80	2.75	2.69	2.62	2.54	2.51	2.47	2.43	2.38	2.34	2.30
13	4.67	3.81	3.41	3.18	3.03	2.92	2.83	2.77	2.71	2.67	2.60	2.53	2.46	2.42	2.38	2.34	2.30	2.25	2.21
14	4.60	3.74	3.34	3.11	2.96	2.85	2.76	2.70	2.65	2.60	2.53	2.46	2.39	2.35	2.31	2.27	2.22	2.18	2.13
15	4.54	3.68	3.29	3.06	2.90	2.79	2.71	2.64	2.59	2.54	2.48	2.40	2.33	2.29	2.25	2.20	2.16	2.11	2.07
16	4.49	3.63	3.24	3.01	2.85	2.74	2.66	2.59	2.54	2.49	2.42	2.35	2.28	2.24	2.19	2.15	2.11	2.06	2.01
17	4.45	3.59	3.20	2.96	2.81	2.70	2.61	2.55	2.49	2.45	2.38	2.31	2.23	2.19	2.15	2.10	2.06	2.01	1.96
18	4.41	3.55	3.16	2.93	2.77	2.66	2.58	2.51	2.46	2.41	2.34	2.27	2.19	2.15	2.11	2.06	2.02	1.97	1.92
19	4.38	3.52	3.13	2.90	2.74	2.63	2.54	2.48	2.42	2.38	2.31	2.23	2.16	2.11	2.07	2.03	1.98	1.93	1.88
20	4.35	3.49	3.10	2.87	2.71	2.60	2.51	2.45	2.39	2.35	2.28	2.20	2.12	2.08	2.04	1.99	1.95	1.90	1.84
21	4.32	3.47	3.07	2.84	2.68	2.57	2.49	2.42	2.37	2.32	2.25	2.18	2.10	2.05	2.01	1.96	1.92	1.87	1.81
22	4.30	3.44	3.05	2.82	2.66	2.55	2.46	2.40	2.34	2.30	2.23	2.15	2.07	2.03	1.98	1.94	1.89	1.84	1.78
23	4.28	3.42	3.03	2.80	2.64	2.53	2.44	2.37	2.32	2.27	2.20	2.13	2.05	2.01	1.96	1.91	1.86	1.81	1.76
24	4.26	3.40	3.01	2.78	2.62	2.51	2.42	2.36	2.30	2.25	2.18	2.11	2.03	1.98	1.94	1.89	1.84	1.79	1.73
25	4.24	3.39	2.99	2.76	2.60	2.49	2.40	2.34	2.28	2.24	2.16	2.09	2.01	1.96	1.92	1.87	1.82	1.77	1.71
26	4.23	3.37	2.98	2.74	2.59	2.47	2.39	2.32	2.27	2.22	2.15	2.07	1.99	1.95	1.90	1.85	1.80	1.75	1.69
27	4.21	3.35	2.96	2.73	2.57	2.46	2.37	2.31	2.25	2.20	2.13	2.06	1.97	1.93	1.88	1.84	1.79	1.73	1.67
28	4.20	3.34	2.95	2.71	2.56	2.45	2.36	2.29	2.24	2.19	2.12	2.04	1.96	1.91	1.87	1.82	1.77	1.71	1.65
29	4.18	3.33	2.93	2.70	2.55	2.43	2.35	2.28	2.22	2.18	2.10	2.03	1.94	1.90	1.85	1.81	1.75	1.70	1.64
30	4.17	3.32	2.92	2.69	2.53	2.42	2.33	2.27	2.21	2.16	2.09	2.01	1.93	1.89	1.84	1.79	1.74	1.68	1.62
40	4.08	3.23	2.84	2.61	2.45	2.34	2.25	2.18	2.12	2.08	2.00	1.92	1.84	1.79	1.74	1.69	1.64	1.58	1.51
60	4.00	3.15	2.76	2.53	2.37	2.25	2.17	2.10	2.04	1.99	1.92	1.84	1.75	1.70	1.65	1.59	1.53	1.47	1.39
120	3.92	3.07	2.68	2.45	2.29	2.17	2.09	2.02	1.96	1.91	1.83	1.75	1.66	1.61	1.55	1.55	1.43	1.35	1.25
∞	3.84	3.00	2.60	2.37	2.21	2.10	2.01	1.94	1.88	1.83	1.75	1.67	1.57	1.52	1.46	1.39	1.32	1.22	1.00

Degrees of freedom for the denominator *m*

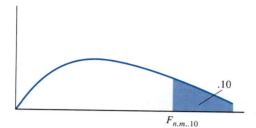

$F_{n,m,.10}$

| **Table D.4** | **Percentiles of _F_ Distributions (_Continued_)** |

90th Percentiles of F Distributions

							Degrees of freedom for the numerator _n_												
	1	**2**	**3**	**4**	**5**	**6**	**7**	**8**	**9**	**10**	**12**	**15**	**20**	**24**	**30**	**40**	**60**	**120**	**∞**
1	39.86	49.50	53.59	55.83	57.24	58.20	58.91	59.44	59.86	60.19	60.71	61.22	61.74	62.00	62.26	62.53	62.79	63.06	63.33
2	8.53	9.00	9.16	9.24	9.29	9.33	9.35	9.37	9.38	9.39	9.41	9.42	9.44	9.45	9.46	9.47	9.47	9.48	9.49
3	5.54	5.46	5.39	5.34	5.31	5.28	5.27	5.25	5.24	5.23	5.22	5.20	5.18	5.18	5.17	5.16	5.15	5.14	5.13
4	4.54	4.32	4.19	4.11	4.05	4.01	3.98	3.95	3.94	3.92	3.90	3.87	3.84	3.83	3.82	3.80	3.79	3.78	3.76
5	4.06	3.78	3.62	3.52	3.45	3.40	3.37	3.34	3.32	3.30	3.27	3.24	3.21	3.19	3.17	3.16	3.14	3.12	3.10
6	3.78	3.46	3.29	3.18	3.11	3.05	3.01	2.98	2.96	2.94	2.90	2.87	2.84	2.82	2.80	2.78	2.76	2.74	2.72
7	3.59	3.26	3.07	2.96	2.88	2.83	2.78	2.75	2.72	2.70	2.67	2.63	2.59	2.58	2.56	2.54	2.51	2.49	2.47
8	3.46	3.11	2.92	2.81	2.73	2.67	2.62	2.59	2.56	2.54	2.50	2.46	2.42	2.40	2.38	2.36	2.34	2.32	2.29
9	3.36	3.01	2.81	2.69	2.61	2.55	2.51	2.47	2.44	2.42	2.38	2.34	2.30	2.28	2.25	2.23	2.21	2.18	2.16
10	3.29	2.92	2.73	2.61	2.52	2.46	2.41	2.38	2.35	2.32	2.28	2.24	2.20	2.18	2.16	2.13	2.11	2.08	2.06
11	3.23	2.86	2.66	2.54	2.45	2.39	2.34	2.30	2.27	2.25	2.21	2.17	2.12	2.10	2.08	2.05	2.03	2.00	1.97
12	3.18	2.81	2.61	2.48	2.39	2.33	2.28	2.24	2.21	2.19	2.15	2.10	2.06	2.04	2.01	1.99	1.96	1.93	1.90
13	3.14	2.76	2.56	2.43	2.35	2.28	2.23	2.20	2.16	2.14	2.10	2.05	2.01	1.98	1.96	1.93	1.90	1.88	1.85
14	3.10	2.73	2.52	2.39	2.31	2.24	2.19	2.15	2.12	2.10	2.05	2.01	1.96	1.94	1.91	1.89	1.86	1.83	1.80
15	3.07	2.70	2.49	2.36	2.27	2.21	2.16	2.12	2.09	2.06	2.02	1.97	1.92	1.90	1.87	1.85	1.82	1.79	1.76
16	3.05	2.67	2.46	2.33	2.24	2.18	2.13	2.09	2.06	2.03	1.99	1.94	1.89	1.87	1.84	1.81	1.78	1.75	1.72
17	3.03	2.64	2.44	2.31	2.22	2.15	2.10	2.06	2.03	2.00	1.96	1.91	1.86	1.84	1.81	1.78	1.75	1.72	1.69
18	3.01	2.62	2.42	2.29	2.20	2.13	2.08	2.04	2.00	1.98	1.93	1.89	1.84	1.81	1.78	1.75	1.72	1.69	1.66
19	2.99	2.61	2.40	2.27	2.18	2.11	2.06	2.02	1.98	1.96	1.91	1.86	1.81	1.79	1.76	1.73	1.70	1.67	1.63
20	2.97	2.59	2.38	2.25	2.16	2.09	2.04	2.00	1.96	1.94	1.89	1.84	1.79	1.77	1.74	1.71	1.68	1.64	1.61
21	2.96	2.57	2.36	2.23	2.14	2.08	2.02	1.98	1.95	1.92	1.87	1.83	1.78	1.75	1.72	1.69	1.66	1.62	1.59
22	2.95	2.56	2.35	2.22	2.13	2.06	2.01	1.97	1.93	1.90	1.86	1.81	1.76	1.73	1.70	1.67	1.64	1.60	1.57
23	2.94	2.55	2.34	2.21	2.11	2.05	1.99	1.95	1.92	1.89	1.84	1.80	1.74	1.72	1.69	1.66	1.62	1.59	1.55
24	2.93	2.54	2.33	2.19	2.10	2.04	1.98	1.94	1.91	1.88	1.83	1.78	1.73	1.70	1.67	1.64	1.61	1.57	1.53
25	2.92	2.53	2.32	2.18	2.09	2.02	1.97	1.93	1.89	1.87	1.82	1.77	1.72	1.69	1.66	1.63	1.59	1.56	1.52
26	2.91	2.52	2.31	2.17	2.08	2.01	1.96	1.92	1.88	1.86	1.81	1.76	1.71	1.68	1.65	1.61	1.58	1.54	1.50
27	2.90	2.51	2.30	2.17	2.07	2.00	1.95	1.91	1.87	1.85	1.80	1.75	1.70	1.67	1.64	1.60	1.57	1.53	1.49
28	2.89	2.50	2.29	2.16	2.06	2.00	1.94	1.90	1.87	1.84	1.79	1.74	1.69	1.66	1.63	1.59	1.56	1.52	1.48
29	2.89	2.50	2.28	2.15	2.06	1.99	1.93	1.89	1.86	1.83	1.78	1.73	1.68	1.65	1.62	1.58	1.55	1.51	1.47
30	2.88	2.49	2.28	2.14	2.03	1.98	1.93	1.88	1.85	1.82	1.77	1.72	1.67	1.64	1.61	1.57	1.54	1.50	1.46
40	2.84	2.44	2.23	2.09	2.00	1.93	1.87	1.83	1.79	1.76	1.71	1.66	1.61	1.57	1.54	1.51	1.47	1.42	1.38
60	2.79	2.39	2.18	2.04	1.95	1.87	1.82	1.77	1.74	1.71	1.66	1.60	1.54	1.51	1.48	1.44	1.40	1.35	1.29
120	2.75	2.35	2.13	1.99	1.90	1.82	1.77	1.72	1.68	1.65	1.60	1.55	1.48	1.45	1.41	1.37	1.32	1.26	1.19
∞	2.71	2.30	2.08	1.94	1.85	1.77	1.72	1.67	1.63	1.60	1.55	1.49	1.42	1.38	1.34	1.30	1.24	1.17	1.00

Degrees of freedom for the denominator m

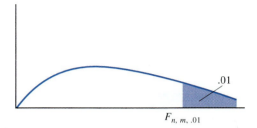

$F_{n, m, .01}$

Table D.4 Percentiles of *F* Distributions (*Continued*)

99th Percentiles of F Distributions

										Degrees of freedom for the numerator *n*										
		1	2	3	4	5	6	7	8	9	10	12	15	20	24	30	40	60	120	∞
	1	4052	4999.5	5403	5625	5764	5859	5928	5982	6022	6056	6106	6157	6209	6235	6261	6287	6313	6339	6366
	2	98.50	99.00	99.17	99.25	99.30	99.33	99.36	99.37	99.39	99.40	99.42	99.43	99.45	99.46	99.47	99.47	99.48	99.49	99.50
	3	34.12	30.82	29.46	28.71	28.24	27.91	27.67	27.49	27.35	27.23	27.05	26.87	26.69	26.60	26.50	26.41	26.32	26.22	26.13
	4	21.20	18.00	16.69	15.98	15.52	15.21	14.98	14.80	14.66	14.55	14.37	14.20	14.02	13.93	13.84	13.75	13.65	13.56	13.46
	5	16.26	13.27	12.06	11.39	10.97	10.67	10.46	10.29	10.16	10.05	9.89	9.72	9.55	9.47	9.38	9.29	9.20	9.11	9.02
	6	13.75	10.92	9.78	9.15	8.75	8.47	8.26	8.10	7.98	7.87	7.72	7.56	7.40	7.31	7.23	7.14	7.06	6.97	6.88
	7	12.25	9.55	8.45	7.85	7.46	7.19	6.99	6.84	6.72	6.62	6.47	6.31	6.16	6.07	5.99	5.91	5.82	5.74	5.65
	8	11.26	8.65	7.59	7.01	6.63	6.37	6.18	6.03	5.91	5.81	5.67	5.52	5.36	5.28	5.20	5.12	5.03	4.95	4.46
	9	10.56	8.02	6.99	6.42	6.06	5.80	5.61	5.47	5.35	5.26	5.11	4.96	4.81	4.73	4.65	4.57	4.48	4.40	4.31
	10	10.04	7.56	6.55	5.99	5.64	5.39	5.20	5.06	4.94	4.85	4.71	4.56	4.41	4.33	4.25	4.17	4.08	4.00	3.91
	11	9.65	7.21	6.22	5.67	5.32	5.07	4.89	4.74	4.63	4.54	4.40	4.25	4.10	4.02	3.94	3.86	3.78	3.69	3.60
	12	9.33	6.93	5.95	5.41	5.06	4.82	4.64	4.50	4.39	4.30	4.16	4.01	3.86	3.78	3.70	3.62	3.54	3.45	3.36
	13	9.07	6.70	5.74	5.21	4.86	4.62	4.44	4.30	4.19	4.10	3.96	3.82	3.66	3.59	3.51	3.43	3.34	3.25	3.17
	14	8.86	6.51	5.56	5.04	4.69	4.46	4.28	4.14	4.03	3.94	3.80	3.66	3.51	3.43	3.35	3.27	3.18	3.09	3.00
	15	8.68	6.36	5.42	4.89	4.36	4.32	4.14	4.00	3.89	3.80	3.67	3.52	3.37	3.29	3.21	3.13	3.05	2.96	2.87
	16	8.53	6.23	5.29	4.77	4.44	4.20	4.03	3.89	3.78	3.69	3.55	3.41	3.26	3.18	3.10	3.02	2.93	2.84	2.75
	17	8.40	6.11	5.18	4.67	4.34	4.10	3.93	3.79	3.68	3.59	3.46	3.31	3.16	3.08	3.00	2.92	2.83	2.75	2.65
	18	8.29	6.01	5.09	4.58	4.25	4.01	3.84	3.71	3.60	3.51	3.37	3.23	3.08	3.00	2.92	2.84	2.75	2.66	2.57
	19	8.18	5.93	5.01	4.50	4.17	3.94	3.77	3.63	3.52	3.43	3.30	3.15	3.00	2.92	2.84	2.76	2.67	2.58	2.59
	20	8.10	5.85	4.94	4.43	4.10	3.87	3.70	3.56	3.46	3.37	3.23	3.09	2.94	2.86	2.78	2.69	2.61	2.52	2.42
	21	8.02	5.78	4.87	4.37	4.04	3.81	3.64	3.51	3.40	3.31	3.17	3.03	2.88	2.80	2.72	2.64	2.55	2.46	2.36
	22	7.95	5.72	4.82	4.31	3.99	3.76	3.59	3.45	3.35	3.26	3.12	2.98	2.83	2.75	2.67	2.58	2.50	2.40	2.31
	23	7.88	5.66	4.76	4.26	3.94	3.71	3.54	3.41	3.30	3.21	3.07	2.93	2.78	2.70	2.62	2.54	2.45	2.35	2.26
	24	7.82	5.61	4.72	4.22	3.90	3.67	3.50	3.36	3.26	3.17	3.03	2.89	2.74	2.66	2.58	2.49	2.40	2.31	2.21
	25	7.77	5.57	4.68	4.18	3.85	3.63	3.46	3.32	3.22	3.13	2.99	2.85	2.70	2.62	2.54	2.45	2.36	2.27	2.17
	26	7.72	5.53	4.64	4.14	3.82	3.59	3.42	3.29	3.18	3.09	2.96	2.81	2.66	2.58	2.50	2.42	2.33	2.23	2.13
	27	7.68	5.49	4.60	4.11	3.78	3.56	3.39	3.26	3.15	3.06	2.93	2.78	2.63	2.55	2.47	2.38	2.29	2.20	2.10
	28	7.64	5.45	4.57	4.07	3.75	3.53	3.36	3.23	3.12	3.03	2.90	2.75	2.60	2.52	2.44	2.35	2.26	2.17	2.06
	29	7.60	5.42	4.54	4.04	3.73	3.50	3.33	3.20	3.09	3.00	2.87	2.73	2.57	2.49	2.41	2.33	2.23	2.14	2.03
	30	7.56	5.39	4.51	4.02	3.70	3.47	3.30	3.17	3.07	2.98	2.84	2.70	2.55	2.47	2.39	2.30	2.21	2.11	2.01
	40	7.31	5.18	4.31	3.83	3.51	3.29	3.12	2.99	2.89	2.80	2.66	2.52	2.37	2.29	2.20	2.11	2.02	1.92	1.80
	60	7.08	4.98	4.13	3.65	3.34	3.12	2.95	2.82	2.72	2.63	2.50	2.35	2.20	2.12	2.03	1.94	1.84	1.73	1.60
	120	6.85	4.79	3.95	3.48	3.17	2.96	2.79	2.66	2.56	2.47	2.34	2.19	2.03	1.95	1.86	1.76	1.66	1.53	1.38
	∞	6.63	4.61	3.78	3.32	3.02	2.80	2.64	2.51	2.41	2.32	2.18	2.04	1.88	1.79	1.70	1.59	1.47	1.32	1.00

Degrees of freedom for the denominator m

| Table D.5 | **Binomial Distribution Function** |

Data in the table are the values of $P\{\text{Bin }(n, p) \leq i\}$, where $\text{Bin }(n, p)$ is a binomial random variable with parameters n and p. For values of $p > .05$, use the identity $P\{\text{Bin }(n, p) \leq i\} = 1 - P\{\text{Bin }(n, 1 - p) \leq n - i - 1\}$.

							p				
n	i	.05	.10	.15	.20	.25	.30	.35	.40	.45	.50
2	0	.9025	.8100	.7225	.6400	.5625	.4900	.4225	.3600	.3025	.2500
	1	.9975	.9900	.9775	.9600	.9375	.9100	.8755	.8400	.7975	.7500
3	0	.8574	.7290	.6141	.5120	.4219	.3430	.2746	.2160	.1664	.1250
	1	.9928	.9720	.9392	.8960	.8438	.7840	.7182	.6480	.5748	.5000
	2	.9999	.9990	.9966	.9920	.9844	.9730	.9571	.9360	.9089	.8750
4	0	.8145	.6561	.5220	.4096	.3164	.2401	.1785	.1296	.0915	.0625
	1	.9860	.9477	.8905	.8192	.7383	.6517	.5630	.4752	.3910	.3125
	2	.9995	.9963	.9880	.9728	.9492	.9163	.8735	.8208	.7585	.6875
	3	1.0000	.9999	.9995	.9984	.9961	.9919	.9850	.9744	.9590	.9375
5	0	.7738	.5905	.4437	.3277	.2373	.1681	.1160	.0778	.0503	.0312
	1	.9774	.9185	.8352	.7373	.6328	.5282	.4284	.3370	.2562	.1875
	2	.9988	.9914	.9734	.9421	.8965	.8369	.7648	.6826	.5931	.5000
	3	1.0000	.9995	.9978	.9933	.9844	.9692	.9460	.9130	.8688	.8125
	4	1.0000	1.0000	.9999	.9997	.9990	.9976	.9947	.9898	.9815	.9688
6	0	.7351	.5314	.3771	.2621	.1780	.1176	.0754	.0467	.0277	.0156
	1	.9672	.8857	.7765	.6554	.5339	.4202	.3191	.2333	.1636	.1094
	2	.9978	.9842	.9527	.9011	.8306	.7443	.6471	.5443	.4415	.3438
	3	.9999	.9987	.9941	.9830	.9624	.9295	.8826	.8208	.7447	.6562
	4	1.0000	.9999	.9996	.9984	.9954	.9891	.9777	.9590	.9308	.8906
	5	1.0000	1.0000	1.0000	.9999	.9998	.9993	.9982	.9959	.9917	.9844
7	0	.6983	.4783	.3206	.2097	.1335	.0824	.0490	.0280	.0152	.0078
	1	.9556	.8503	.7166	.5767	.4449	.3294	.2338	.1586	.1024	.0625
	2	.9962	.9743	.9262	.8520	.7564	.6471	.5323	.4199	.3164	.2266
	3	.9998	.9973	.9879	.9667	.9294	.8740	.8002	.7102	.6083	.5000
	4	1.0000	.9998	.9988	.9953	.9871	.9712	.9444	.9037	.8471	.7734
	5	1.0000	1.0000	.9999	.9996	.9987	.9962	.9910	.9812	.9643	.9375
	6	1.0000	1.0000	1.0000	1.0000	.9999	.9998	.9994	.9984	.9963	.9922
8	0	.6634	.4305	.2725	.1678	.1001	.0576	.0319	.0168	.0084	.0039
	1	.9428	.8131	.6572	.5033	.3671	.2553	.1691	.1064	.0632	.0352
	2	.9942	.9619	.8948	.7969	.6785	.5518	.4278	.3154	.2201	.1445
	3	.9996	.9950	.9786	.9437	.8862	.8059	.7064	.5941	.4770	.3633
	4	1.0000	.9996	.9971	.9896	.9727	.9420	.8939	.8263	.7396	.6367
	5	1.0000	1.0000	.9998	.9988	.9958	.9887	.9747	.9502	.9115	.8555
	6	1.0000	1.0000	1.0000	.9999	.9996	.9987	.9964	.9915	.9819	.9648
	7	1.0000	1.0000	1.0000	1.0000	1.0000	.9999	.9998	.9993	.9983	.9961
9	0	.6302	.3874	.2316	.1342	.0751	.0404	.0207	.0101	.0046	.0020
	1	.9288	.7748	.5995	.4362	.3003	.1960	.1211	.0705	.0385	.0195

Appendix D: Tables

n	i	.05	.10	.15	.20	.25	.30	.35	.40	.45	.50
							p				
	2	.9916	.9470	.8591	.7382	.6007	.4628	.3373	.2318	.1495	.0898
	3	.9994	.9917	.9661	.9144	.8343	.7297	.6089	.4826	.3614	.2539
	4	1.0000	.9991	.9944	.9804	.9511	.9012	.8283	.7334	.6214	.5000
	5	1.0000	.9999	.9994	.9969	.9900	.9747	.9464	.9006	.8342	.7461
	6	1.0000	1.0000	1.0000	.9997	.9987	.9957	.9888	.9750	.9502	.9102
	7	1.000	1.0000	1.0000	1.0000	.9999	.9996	.9986	.9962	.9909	.9805
	8	1.0000	1.0000	1.0000	1.0000	1.0000	1.0000	.9999	.9997	.9992	.9980
10	0	.5987	.3487	.1969	.1074	.0563	.0282	.0135	.0060	.0025	.0010
	1	.9139	.7361	.5443	.3758	.2440	.1493	.0860	.0464	.0232	.0107
	2	.9885	.9298	.8202	.6778	.5256	.3828	.2616	.1673	.0996	.0547
	3	.9990	.9872	.9500	.8791	.7759	.6496	.5138	.3823	.2660	.1719
	4	.9999	.9984	.9901	.9672	.9219	.8497	.7515	.6331	.5044	.3770
	5	1.0000	.9999	.9986	.9936	.9803	.9527	.9051	.8338	.7384	.6230
	6	1.0000	1.0000	.9999	.9991	.9965	.9894	.9740	.9452	.8980	.8281
	7	1.0000	1.0000	1.0000	.9999	.9996	.9984	.9952	.9877	.9726	.9453
	8	1.0000	1.0000	1.0000	1.0000	1.0000	.9999	.9995	.9983	.9955	.9893
	9	1.0000	1.0000	1.0000	1.0000	1.0000	1.0000	1.0000	.9999	.9997	.9990
11	0	.5688	.3138	.1673	.0859	.0422	.0198	.0088	.0036	.0014	.0005
	1	.8981	.6974	.4922	.3221	.1971	.1130	.0606	.0302	.0139	.0059
	2	.9848	.9104	.7788	.6174	.4552	.3127	.2001	.1189	.0652	.0327
	3	.9984	.9815	.9306	.8389	.7133	.5696	.4256	.2963	.1911	.1133
	4	.9999	.9972	.9841	.9496	.8854	.7897	.6683	.5328	.3971	.2744
	5	1.0000	.9997	.9973	.9883	.9657	.9218	.8513	.7535	.6331	.5000
	6	1.0000	1.0000	.9997	.9980	.9924	.9784	.9499	.9006	.8262	.7256
	7	1.0000	1.0000	1.0000	.9998	.9988	.9957	.9878	.9707	.9390	.8867
	8	1.0000	1.0000	1.0000	1.0000	.9999	.9994	.9980	.9941	.9852	.9673
	9	1.0000	1.0000	1.0000	1.0000	1.0000	1.0000	.9998	.9993	.9978	.9941
	10	1.0000	1.0000	1.0000	1.0000	1.0000	1.0000	1.0000	1.0000	.9998	.9995
12	0	.5404	.2824	.1422	.0687	.0317	.0138	.0057	.0022	.0008	.0002
	1	.8816	.6590	.4435	.2749	.1584	.0850	.0424	.0196	.0083	.0032
	2	.9804	.8891	.7358	.5583	.3907	.2528	.1513	.0834	.0421	.0193
	3	.9978	.9744	.9078	.7946	.6488	.4925	.3467	.2253	.1345	.0730
	4	.9998	.9957	.9761	.9274	.8424	.7237	.5833	.4382	.3044	.1938
	5	1.000	.9995	.9954	.9806	.9456	.8822	.7873	.6652	.5269	.3872
	6	1.0000	.9999	.9993	.9961	.9857	.9614	.9154	.8418	.7393	.6128
	7	1.0000	1.0000	.9999	.9994	.9972	.9905	.9745	.9427	.8883	.8062
	8	1.0000	1.0000	1.0000	.9999	.9996	.9983	.9944	.9847	.9644	.9270
	9	1.0000	1.0000	1.0000	1.0000	1.0000	.9998	.9992	.9972	.9921	.9807
	10	1.0000	1.0000	1.0000	1.0000	1.0000	1.0000	.9999	.9997	.9989	.9968
	11	1.0000	1.0000	1.0000	1.0000	1.0000	1.0000	1.0000	1.0000	.9999	.9998
13	0	.5133	.2542	.1209	.0550	.0238	.0097	.0037	.0013	.0004	.0001
	1	.8646	.6213	.3983	.2336	.1267	.0637	.0296	.0126	.0049	.0017
	2	.9755	.8661	.6920	.5017	.3326	.2025	.1132	.0579	.0269	.0112
	3	.9969	.9658	.8820	.7437	.5843	.4206	.2783	.1686	.0929	.0461
	4	.9997	.9935	.9658	.9009	.7940	.6543	.5005	.3530	.2279	.1334

n	i	.05	.10	.15	.20	.25	.30	.35	.40	.45	.50
	5	1.0000	.9991	.9925	.9700	.9198	.8346	.7159	.5744	.4268	.2905
	6	1.0000	.9999	.9987	.9930	.9757	.9376	.8705	.7712	.6437	.5000
	7	1.0000	1.0000	.9998	.9988	.9944	.9818	.9538	.9023	.8212	.7095
	8	1.0000	1.0000	1.0000	.9998	.9990	.9960	.9874	.9679	.9302	.8666
	9	1.0000	1.0000	1.0000	1.0000	.9999	.9993	.9975	.9922	.9797	.9539
	10	1.0000	1.0000	1.0000	1.0000	1.0000	.9999	.9997	.9987	.9959	.9888
	11	1.0000	1.0000	1.0000	1.0000	1.0000	1.0000	1.0000	.9999	.9995	.9983
	12	1.0000	1.0000	1.0000	1.0000	1.0000	1.0000	1.0000	1.0000	1.0000	.9999
14	0	.4877	.2288	.1028	.0440	.0178	.0068	.0024	.0008	.0002	.0001
	1	.8470	.5846	.3567	.1979	.1010	.0475	.0205	.0081	.0029	.0009
	2	.9699	.8416	.6479	.4481	.2811	.1608	.0839	.0398	.0170	.0065
	3	.9958	.9559	.8535	.6982	.5213	.3552	.2205	.1243	.0632	.0287
		.9996	.9908	.9533	.8702	.7415	.5842	.4227	.2793	.1672	.0898
		1.0000	.9985	.9885	.9561	.8883	.7805	.6405	.4859	.3373	.2120
	6	1.0000	.9998	.9978	.9884	.9617	.9067	.8164	.6925	.5461	.3953
	7	1.0000	1.0000	.9997	.9976	.9897	.9685	.9247	.8499	.7414	.6074
	8	1.0000	1.0000	1.0000	.9996	.9978	.9917	.9757	.9417	.8811	.7880
	9	1.0000	1.0000	1.0000	1.0000	.9997	.9983	.9940	.9825	.9574	.9102
	10	1.0000	1.0000	1.0000	1.0000	1.0000	.9998	.9989	.9961	.9886	.9713
	11	1.0000	1.0000	1.0000	1.0000	1.0000	1.0000	.9999	.9994	.9978	.9935
	12	1.0000	1.0000	1.0000	1.0000	1.0000	1.0000	1.0000	.9999	.9997	.9991
	13	1.0000	1.0000	1.0000	1.0000	1.0000	1.0000	1.0000	1.0000	1.0000	.9999
15	0	.4633	.2059	.0874	.0352	.0134	.0047	.0016	.0005	.0001	.0000
	1	.8290	.5490	.3186	.1671	.0802	.0353	.0142	.0052	.0017	.0005
	2	.9638	.8159	.6042	.3980	.2361	.1268	.0617	.0271	.0107	.0037
	3	.9945	.9444	.8227	.6482	.4613	.2969	.1727	.0905	.0424	.0176
	4	.9994	.9873	.9383	.8358	.6865	.5155	.3519	.2173	.1204	.0592
	5	.9999	.9978	.9832	.9389	.8516	.7216	.5643	.4032	.2608	.1509
	6	1.0000	.9997	.9964	.9819	.9434	.8689	.7548	.6098	.4522	.3036
	7	1.0000	1.0000	.9996	.9958	.9827	.9500	.8868	.7869	.6535	.5000
	8	1.0000	1.0000	.9999	.9992	.9958	.9848	.9578	.9050	.8182	.6964
	9	1.0000	1.0000	1.0000	.9999	.9992	.9963	.9876	.9662	.9231	.8491
	10	1.0000	1.0000	1.0000	1.0000	.9999	.9993	.9972	.9907	.9745	.9408
	11	1.0000	1.0000	1.0000	1.0000	1.0000	.9999	.9995	.9981	.9937	.9824
	12	1.0000	1.0000	1.0000	1.0000	1.0000	1.0000	.9999	.9997	.9989	.9963
	13	1.0000	1.0000	1.9000	1.0000	1.0000	1.0000	1.0000	1.0000	.9999	.9995
	14	1.0000	1.0000	1.0000	1.0000	1.0000	1.0000	1.0000	1.0000	1.0000	1.0000
16	0	.4401	.1853	.0743	.0281	.0100	.0033	.0010	.0003	.0001	.0000
	1	.8108	.5147	.2839	.1407	.0635	.0261	.0098	.0033	.0010	.0003
	2	.9571	.7892	.5614	.3518	.1971	.0994	.0451	.0183	.0066	.0021
	3	.9930	.9316	.7899	.5981	.4050	.2459	.1339	.0651	.0281	.0106
	4	.9991	.9830	.9209	.7982	.6302	.4499	.2892	.1666	.0853	.0384
	5	.9999	.9967	.9765	.9183	.8103	.6598	.4900	.3288	.1976	.1051
	6	1.0000	.9995	.9944	.9733	.9204	.8247	.6881	.5272	.3660	.2272
	7	1.0000	.9999	.9989	.9930	.9729	.9256	.8406	.7161	.5629	.4018
	8	1.0000	1.0000	.9998	.9985	.9925	.9743	.9329	.8577	.7441	.5982
	9	1.0000	1.0000	1.0000	.9998	.9984	.9929	.9771	.9417	.8759	.7728

n	i					p					
		.05	**.10**	**.15**	**.20**	**.25**	**.30**	**.35**	**.40**	**.45**	**.50**
	10	1.0000	1.0000	1.0000	1.0000	.9997	.9984	.9938	.9809	.9514	.8949
	11	1.0000	1.0000	1.0000	1.0000	1.0000	.9997	.9987	.9951	.9851	.9616
	12	1.0000	1.0000	1.0000	1.0000	1.0000	1.0000	.9998	.9991	.9965	.9894
	13	1.0000	1.0000	1.0000	1.0000	1.0000	1.0000	1.0000	.9999	.9994	.9979
	14	1.0000	1.0000	1.0000	1.0000	1.0000	1.0000	1.0000	1.0000	1.0000	.9997
	15	1.0000	1.0000	1.0000	1.0000	1.0000	1.0000	1.0000	1.0000	1.0000	1.0000
17	0	.4181	.1668	.0631	.0225	.0075	.0023	.0007	.0002	.0000	.0000
	1	.7922	.4818	.2525	.1182	.0501	.0193	.0067	.0021	.0006	.0001
	2	.9497	.7618	.5198	.3096	.1637	.0774	.0327	.0123	.0041	.0012
	3	.9912	.9174	.7556	.5489	.3530	.2019	.1028	.0464	.0184	.0063
	4	.9988	.9779	.9013	.7582	.5739	.3887	.2348	.1260	.0596	.0245
	5	.9999	.9953	.9681	.8943	.7653	.5968	.4197	.2639	.1471	.0717
	6	1.0000	.9992	.9917	.9623	.8929	.7752	.6188	.4478	.2902	.1662
	7	1.0000	.9999	.9983	.9891	.9598	.8954	.7872	.6405	.4743	.3145
	8	1.0000	1.0000	.9997	.9974	.9876	.9597	.9006	.8011	.6626	.5000
	9	1.0000	1.0000	1.0000	.9995	.9969	.9873	.9617	.9081	.8166	.6855
	10	1.0000	1.0000	1.0000	.9999	.9994	.9968	.9880	.9652	.9174	.8338
	11	1.0000	1.0000	1.0000	1.0000	.9999	.9993	.9970	.9894	.9699	.9283
	12	1.0000	1.0000	1.0000	1.0000	1.0000	.9999	.9994	.9975	.9914	.9755
	13	1.0000	1.0000	1.0000	1.0000	1.0000	1.0000	.9999	.9995	.9981	.9936
	14	1.0000	1.0000	1.0000	1.0000	1.0000	1.0000	1.0000	.9999	.9997	.9988
	15	1.0000	1.0000	1.0000	1.0000	1.0000	1.0000	1.0000	1.0000	1.0000	.9999
	16	1.0000	1.0000	1.0000	1.0000	1.0000	1.0000	1.0000	1.0000	1.0000	1.0000
18	0	.3972	.1501	.0536	.0180	.0056	.0016	.0004	.0001	.0000	.0000
	1	.7735	.4503	.2241	.0991	.0395	.0142	.0046	.0013	.0003	.0001
	2	.9419	.7338	.4797	.2713	.1353	.0600	.0236	.0082	.0025	.0007
	3	.9891	.9018	.7202	.5010	.3057	.1646	.0783	.0328	.0120	.0038
	4	.9985	.9718	.8794	.7164	.5187	.3327	.1886	.0942	.0411	.0154
	5	.9998	.9936	.9581	.8671	.7175	.5344	.3550	.2088	.1077	.0481
	6	1.0000	.9988	.9882	.9487	.8610	.7217	.5491	.3743	.2258	.1189
	7	1.0000	.9998	.9973	.9837	.9431	.8593	.7283	.5634	.3915	.2403
	8	1.0000	1.0000	.9995	.9957	.9807	.9404	.8609	.7368	.5778	.4073
	9	1.0000	1.0000	.9999	.9991	.9946	.9790	.9403	.8653	.7473	.5927
	10	1.0000	1.0000	1.0000	.9998	.9988	.9939	.9788	.9424	.8720	.7597
	11	1.0000	1.0000	1.0000	1.0000	.9998	.9986	.9938	.9797	.9463	.8811
	12	1.0000	1.0000	1.0000	1.0000	1.0000	.9997	.9986	.9942	.9817	.9519
	13	1.0000	1.0000	1.0000	1.0000	1.0000	1.0000	.9997	.9987	.9951	.9846
	14	1.0000	1.0000	1.0000	1.0000	1.0000	1.0000	1.0000	.9998	.9990	.9962
	15	1.0000	1.0000	1.0000	1.0000	1.0000	1.0000	1.0000	1.0000	.9999	.9993
	16	1.0000	1.0000	1.0000	1.0000	1.0000	1.0000	1.0000	1.0000	1.0000	.9999
19	0	.3774	.1351	.0456	.0144	.0042	.0011	.0003	.0001	.0000	.0000
	1	.7547	.4203	.1985	.0829	.0310	.0104	.0031	.0008	.0002	.0000
	2	.9335	.7054	.4413	.2369	.1113	.0462	.0170	.0055	.0015	.0004
	3	.9868	.8850	.6841	.4551	.2630	.1332	.0591	.0230	.0077	.0022
	4	.9980	.9648	.8556	.6733	.4654	.2822	.1500	.0696	.0280	.0096

continued

n	i	p									
		.05	**.10**	**.15**	**.20**	**.25**	**.30**	**.35**	**.40**	**.45**	**.50**
	5	.9998	.9914	.9463	.8369	.6678	.4739	.2968	.1629	.0777	.0318
	6	1.0000	.9983	.9837	.9324	.8251	.6655	.4812	.3081	.1727	.0835
	7	1.0000	.9997	.9959	.9767	.9225	.8180	.6656	.4878	.3169	.1796
	8	1.0000	1.0000	.9992	.9933	.9713	.9161	.8145	.6675	.4940	.3238
	9	1.0000	1.0000	.9999	.9984	.9911	.9674	.9125	.8139	.6710	.5000
	10	1.0000	1.0000	1.0000	.9997	.9977	.9895	.9653	.9115	.8159	.6762
	11	1.0000	1.0000	1.0000	1.0000	.9995	.9972	.9886	.9648	.9129	.8204
	12	1.0000	1.0000	1.0000	1.0000	.9999	.9994	.9969	.9884	.9658	.9165
	13	1.0000	1.0000	1.0000	1.0000	1.0000	.9999	.9993	.9969	.9891	.9682
	14	1.0000	1.0000	1.0000	1.0000	1.0000	1.0000	.9999	.9994	.9972	.9904
	15	1.0000	1.0000	1.0000	1.0000	1.0000	1.0000	1.0000	.9999	.9995	.9978
	16	1.0000	1.0000	1.0000	1.0000	1.0000	1.0000	1.0000	1.0000	.9999	.9996
	17	1.0000	1.0000	1.0000	1.0000	1.0000	1.0000	1.0000	1.0000	1.0000	1.0000
20	0	.3585	.1216	.0388	.0115	.0032	.0008	.0002	.0000	.0000	.0000
	1	.7358	.3917	.1756	.0692	.0243	.0076	.0021	.0005	.0001	.0000
	2	.9245	.6769	.4049	.2061	.0913	.0355	.0121	.0036	.0009	.0002
	3	.9841	.8670	.6477	.4114	.2252	.1071	.0444	.0160	.0049	.0013
	4	.9974	.9568	.8298	.6296	.4148	.2375	.1182	.0510	.0189	.0059
	5	.9997	.9887	.9327	.8042	.6172	.4164	.2454	.1256	.0553	.0207
	6	1.0000	.9976	.9781	.9133	.7858	.6080	.4166	.2500	.1299	.0577
	7	1.0000	.9996	.9941	.9679	.8982	.7723	.6010	.4159	.2520	.1316
	8	1.0000	.9999	.9987	.9900	.9591	.8867	.7624	.5956	.4143	.2517
	9	1.0000	1.0000	.9998	.9974	.9861	.9520	.8782	.7553	.5914	.4119
	10	1.0000	1.0000	1.0000	.9994	.9961	.9829	.9468	.8725	.7507	.5881
	11	1.0000	1.0000	1.0000	.9999	.9991	.9949	.9804	.9435	.8692	.7483
	12	1.0000	1.0000	1.0000	1.0000	.9998	.9987	.9940	.9790	.9420	.8684
	13	1.0000	1.0000	1.0000	1.0000	1.0000	.9997	.9985	.9935	.9786	.9423
	14	1.0000	1.0000	1.0000	1.0000	1.0000	1.0000	.9997	.9984	.9936	.9793
	15	1.0000	1.0000	1.0000	1.0000	1.0000	1.0000	1.0000	.9997	.9985	.9941
	16	1.0000	1.0000	1.0000	1.0000	1.0000	1.0000	1.0000	1.0000	.9997	.9987
	17	1.0000	1.0000	1.0000	1.0000	1.0000	1.0000	1.0000	1.0000	1.0000	.9998
	18	1.0000	1.0000	1.0000	1.0000	1.0000	1.0000	1.0000	1.0000	1.0000	1.0000

A P P E N D I X E

ANSWERS TO ODD-NUMBERED EXERCISES

CHAPTER 1 PROBLEMS

1. (a) 1946
   (b) There were more years in which the average number of years completed by the older group exceeded that of the younger group.

3. (a) From 1985 to 1990 sales declined.
   (b) The total number of cars sold from 1985 to 1987 was 20,693,000 versus 18,120,000 from 1988 to 1990.
   (c) No

5. Researchers with such knowledge may be influenced by their own biases concerning the usefulness of the new drug.

7. (a) In 1936 automobile and telephone owners were probably not representative of the total voter population.
   (b) Yes. Automobile and telephone ownership is now more widespread and thus more representative of the total voter population.

9. The average age of death for U.S. citizens whose obituary is listed in *The New York Times* is about 82.4 years.

11. (a) No. Graduates who return the questionnaire may not be representative of the total population of graduates.
    (b) If the number of questionnaires returned were very close to 200—the number of questionnaires sent— then the approximation would be better.

13. Graunt implicitly assumed that the parishes he surveyed were representative of the total London population.

15. Data on the ages at which people were dying can be used to determine approximately how long on average the annuity payments will continue. This can be used to determine how much to charge for the annuity.

17. (a) 64%
    (b) 10%
    (c) 48%

SECTION 2.2

1. (a)

Family size	Frequency
4	1
5	1
6	3
7	5
8	5
9	3
10	5
11	2
12	3
13	1
14	0
15	1

(b)

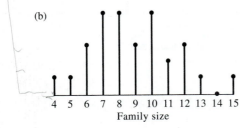

Family size

(c)

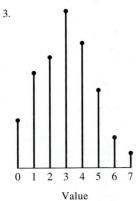

3.

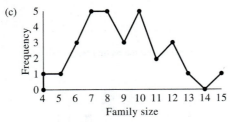

The data set is not symmetric. The data set is approximately symmetric.

5.

Value	Frequency
10	8
20	3
30	7
40	7
50	3
60	8

7.

Family size	Frequency	Relative frequency
4	1	.03
5	1	.03
6	3	.10
7	5	.17
8	5	.17
9	3	.10
10	5	.17
11	2	.07
12	3	.10
13	1	.03
14	0	.00
15	1	.03

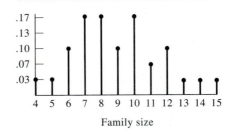

Family size

9. (a) .13
   (b) .25
   (c) No

11. (a) 16/37
    (b) 13/37
    (c) 22/37

13.

Average number of rainy days in Nov. or Dec.	Frequency
7	1
9	1
10	1
11	1
16	1
17	3
18	1
20	1
23	1
40	1

15.

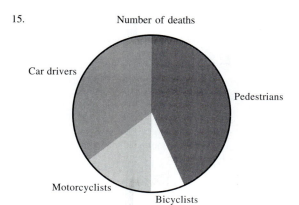

Number of deaths

(c) The chart in part (a) seems more informative since it shows a clearer pattern.

5. (a)

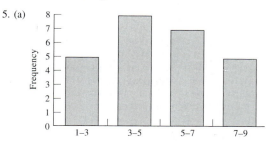

Ozone concentration (in parts per hundred million)

(b)

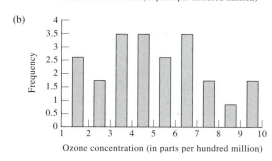

Ozone concentration (in parts per hundred million)

(c) The chart in part (b) seems more informative.

SECTION 2.3

1. (a)

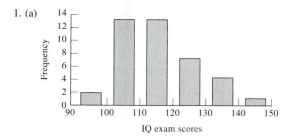

IQ exam scores

(b) Class intervals 100–110 and 110–120
(c) No
(d) No

3. (a)

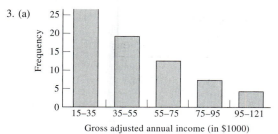

Gross adjusted annual income (in $1000)

(b)

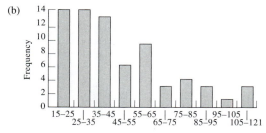

Gross adjusted annual income (in $1000)

7.

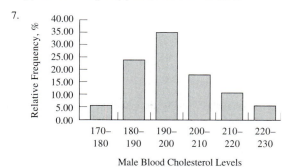

Male Blood Cholesterol Levels

Female Blood Cholesterol Levels

Female cholesterol	Frequency	Relative frequency
170–180	1	1/46 = .02
180–190	5	5/46 = .11
190–200	13	13/46 = .28
200–210	15	15/46 = .33
210–220	9	9/46 = .20
220–230	3	3/46 = .07

Male cholesterol	Frequency	Relative frequency
170–180	3	3/54 = .06
180–190	13	13/54 = .24
190–200	19	19/54 = .35
200–210	10	10/54 = .19
210–220	6	6/54 = .11
220–230	3	3/54 = .06

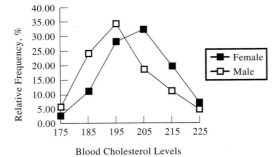

Blood Cholesterol Levels

Female students appear to have higher cholesterol levels.

9.

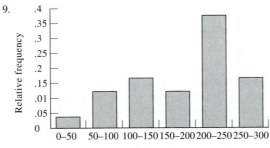

Death rate (per 100,000) by ischemic heart disease

11.

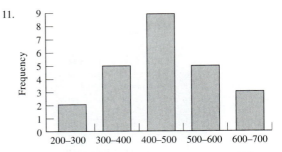

Total death rate by listed causes (per 100,000)

13.

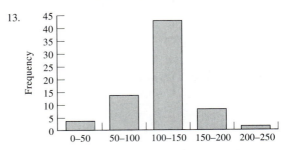

Average yearly number of rainy days

15. (a) It is the sum of the relative frequencies for all classes.

(b)

Blood pressure less than	Percentage of workers	
	Ages 30–40	Ages 50–60
90	.12	.14
100	.79	.41
110	5.43	3.56
120	23.54	11.35
130	53.78	28.04
140	80.35	48.43
150	92.64	71.27
160	97.36	81.26
170	99.13	89.74
180	99.84	94.53
190	99.96	97.26
200	100.00	98.50
210	100.00	98.91
220	100.00	99.59
230	100.00	99.86
240	100.00	100.00

(c) Ages 30 to 40 tend to have smaller values.

(d)

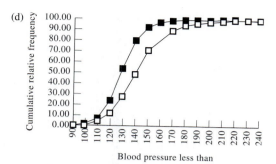

Blood pressure less than

■ Ages 30–40
□ Ages 50–60

SECTION 2.4

1. (a)

11	1, 4, 5, 6, 8, 8, 9, 9, 9
12	2, 2, 2, 2, 4, 5, 5, 6, 7, 7, 7, 8, 9
13	0, 2, 2, 3, 4, 5, 5, 7, 9
14	1, 1, 4, 6, 7

(b)

11	1, 4
11	5, 6, 8, 8, 9, 9, 9
12	2, 2, 2, 2, 4
12	5, 5, 6, 7, 7, 7, 8, 9
13	0, 2, 2, 3, 4
13	5, 5, 7, 9
14	1, 1, 4
14	6, 7

3.

1	4
1	5, 6, 6, 7, 7, 7, 7, 8, 8, 8, 9, 9, 9, 9
2	0, 0, 0, 0, 1, 2, 2, 2, 3, 4
2	5, 7, 7, 9
3	0, 1, 1, 2, 3
3	
4	0, 4, 4
4	5
5	1, 3
5	5
6	1
6	
7	
7	9

The interval 15–20 contains 14 data points.
The interval 16–21 contains 17 data points.

5. (a)

3	2
4	
5	2, 7, 8, 9
6	5, 8, 8
7	1, 4, 5, 5, 7, 8, 9
8	0, 1, 3, 3, 3, 4, 8, 8
9	0, 3, 4, 7
10	0, 4, 8

(b) Yes. The value 32 seems suspicious since it is so
much smaller than the others.

7. (a)

1	4, 6, 6, 6
2	0, 0, 1, 3, 4, 4, 6, 7, 7, 7
3	1, 2, 3, 5, 5, 8, 8, 9
4	2, 6
5	5

(b)

0	3, 6, 7, 7, 7, 7, 9
1	0, 0, 0, 0, 0, 0, 3, 4, 4, 6, 6, 7, 7, 9, 9
2	0, 1
3	1

(c)

0	1, 3, 4, 4, 4, 5, 7, 9
1	0, 0, 2, 6, 7, 7, 7, 8, 9, 9
2	1, 2, 5, 9
3	2, 6
4	5

9. (a) 6
   (b) 43.75%
   (c) 12.5%

11. (a) School B
    (b) School A
    (c) School A
    (d)

5	0, 3
5	5, 7
6	2
6	5, 5, 8, 8, 9, 9
7	0, 2, 3, 4
7	6, 7, 7, 8, 8, 9, 9
8	0, 2, 3, 3
8	5, 5, 6, 6, 6, 7, 7, 8, 8, 9
9	0, 0, 1, 3
9	5, 5, 5, 6, 6, 8, 8
10	0

SECTION 2.5

1. (a)

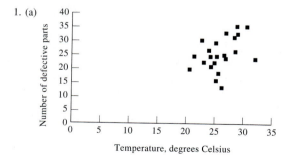

Temperature, degrees Celsius

(b) The number of defective parts tends to increase as the
temperature increases.
(c) About 23

3. (a)

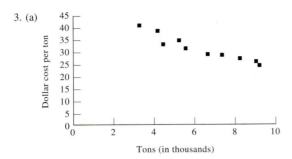

(b) The per-ton production cost decreases as production increases.

(c) $25

5. (a)

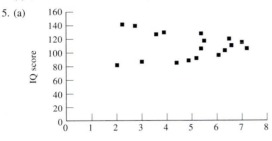

(b) Attention span and IQ are not related.

7. (a)

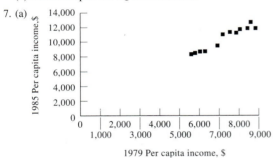

(b) $12,000

(c) $11,000

9. The two causes seem unrelated.

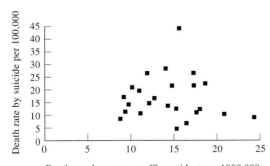

Death rate by mortor traffic accident per 1000,000

REVIEW PROBLEMS

1. (a)

Blood type	Frequency
A	19
B	8
O	19
AB	4

(b)

Blood type	Relative frequency
A	.38
B	.16
O	.38
AB	.08

(c)

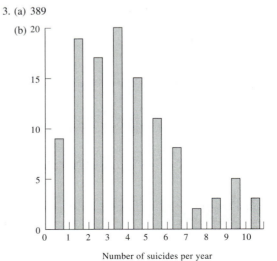

Blood types

3. (a) 389

(b)

Number of suicides per year

5. (a)

Value	Frequency
1	2
2	1
3	4
4	1
5	2

(b)

Value	Frequency
1	2
2	3
3	3
4	2

(c) 3, 2.5

7. (a)

1985	Stem	1990
	0	3.6, 6.7
6.3, 2.8	1	2.2, 2.8, 7.8, 9.3
8.8, 5.7, 4.7, 3.5	2	0.5, 5.5
3.9, 0.5	3	
	4	
	5	
	6	
	7	
0.1	8	8.6

(b)

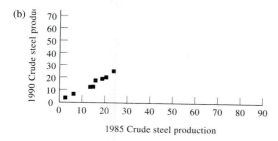

9. (a)

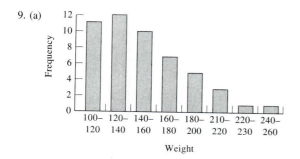

(b) There are relatively few weights near the upper end of the weight range.

11. Weight and blood pressure do not seem related.

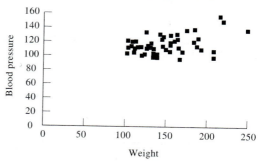

13. Yes, high scores on one examination tend to go along with high scores on the other.

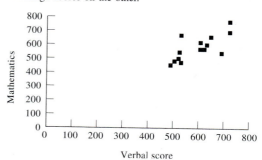

15. (a)

0	.27, .78, .93
1	.19, .31, .49, .53, .81
2	.30, .92, .93
3	.07, .21, .32, .39, .66, .68, .81
4	.02, .11, .43, .50
5	.35, .41

(b)

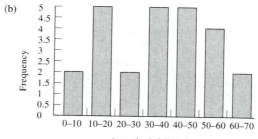

(c)

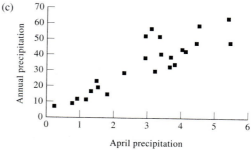

17. (a) Yes

(b)
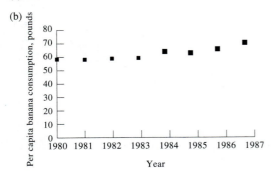

(c) 75 pounds

19.
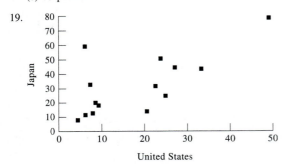

SECTION 3.2

1. 1196/15 = 79.73

3. 429.03/13 = 33.00 inches; 1331/13 = 102.38 days

5. 1807/12 = 150.58

7. 638/7 = 91.14 cases

9. 6; 18; 11

11. 2,533,000 fires

13. 15

15. $\frac{1}{2}(10) + \frac{1}{6}(20) + \frac{1}{3}(30) = 18.33$

17. $37,120

19. (a) −5, −4, −2, 1, 4, 6
    (b) −15, −12, −6, 3, 12, 18
    (c) same as (a)

SECTION 3.3

1. (a) 6580 yards
   (b) 6545 yards

3. 23

5. (a) 22.0
   (b) 8.1
   (c) 23.68
   (d) 9.68

7. 31.5 inches

9. (a) 99.4
   (b) 14.9
   (c) 204.55

11. (a) 20.74
    (b) 20.5
    (c) 19.74
    (d) 19.5
    (e) Mean = 20.21; median = 20.05

13. Median = 31.35, mean = 31.75

15. (a) 32.52
    (b) 24.25

17. (a) 253
    (b) 243.8

SECTION 3.3.1

1. (a) If the data are arranged in increasing order, then the sample 80 percentile is given by the average of the values in positions 60 and 61.
   (b) If the data are arranged in increasing order, then the sample 60 percentile is given by the average of the values in positions 45 and 46.
   (c) If the data are arranged in increasing order, then the sample 30 percentile is the value in position 23.

3. (a) 95.5
   (b) 96

5. (a) 70
   (b) 58
   (c) 52

7. 230c

9. (a) (i) 201.5, (ii) 217
   (b) (i) 218.5, (ii) 221
   (c) (i) 134.5, (ii) 137

11. (a) From 35 to 66
    (b) 47
    (c) 66

SECTION 3.4

1. 1B, 2C, 3A

3. (a) 126
   (b) 102, 110, 114
   (c) 196

5. 5, 6, 6, 6, 8, 10, 12, 14, 23 is one such data set.

7. (a) 8 loops
(b) 2 miles

SECTION 3.5

1. $s^2 = .037$; $\bar{x} = 26.22$

3. (a) 6.18
(b) 6.77

5. 17.90

7. 5932.57

9. 11,802,857.1

11. (a) $s^2 = 2.5$, $s = 1.58$
(b) $s^2 = 2.5$, $s = 1.58$
(c) $s^2 = 2.5$, $s = 1.58$
(d) $s^2 = 10$, $s = 3.16$
(e) $s^2 = 250$, $s = 15.81$

13. For the first 50 students, $s^2 = 172.24$ and $\bar{x} = 115.80$.
For the last 50 students, $s^2 = 178.96$ and $\bar{x} = 120.98$.
The values of the statistics for the two data sets are similar. This is not surprising.

15. 195,808.76

17. (a) .805
(b) 2.77
(c) 1.22

SECTION 3.6

1. (a)

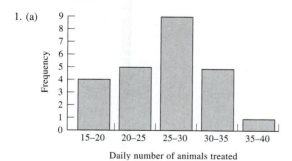

(b) 25.75
(c) 26.5
(d) No

3. (a)

11	.6
12	
13	.5, .7, .8
14	.3, .3, .4, .5, .5, .6, .6, .6, .6, .7, .8, .8, .9, .9, .9, .9
15	.0, .0, .0, .1, .1, .3, .5, .5, .5, .6, .7, .7, .7, .8, .9, .9
16	.0, .1, .2, .4, .7, .8
17	.1, .1, .3, .9
18	.0, .8, .8
19	
20	
21	.3, .4

(b) Yes
(c) 15.69
(d) 15.3

5. (a) 168,045
(b) 172,500

(c)

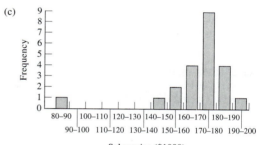

Sales price ($1000)

(d) Yes, if we ignore the data value 82. No, if we use all the data.

7. 95%, 94.85%

9. Sample mean

11. (a)

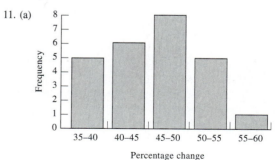

Percentage change

(b) 46.06
(c) 46
(d) 36.68
(e) No
(f) From 40.01 to 52.12
(g) From 33.95 to 58.18
(h) 68%
(i) 96%

13. (a)

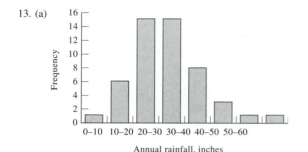

Annual rainfall. inches

(b) 46.98
(c) 46.1
(d) 46.69
(e) Yes
(f) From 40.15 to 53.81
(g) From 33.31 to 60.65
(h) 66%
(i) 96%

SECTION 3.7

1. Let (x_i, y_i), $i = 1, 2, 3$ be the middle set of data pairs. Then the first set is $(121x_i, 360 + y_i)$ and the third is $(x_i, \frac{1}{2}y_i)$, $i = 1, 2, 3$.

3. (a)

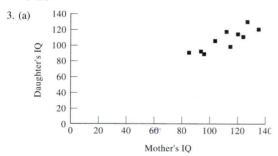

Mother's IQ

(b) Almost 1
(c) .86
(d) There is a relatively strong linear relationship between them.

5. .99; they have a strong positive correlation

7. $-.441202$; the linear relationship is relatively weak, but there is an indication that when one of the variables is high then the other tends to be low.

9. .95

11. All data $= -.33$; first seven countries $= -.046$

13. All data $= .25$; first seven countries $= -.303$

15. (d) Correlation is not causation.

17. No, correlation is not causation.

REVIEW PROBLEMS

1. (a) $-2, -1, 1, 2$
   (b) $-2, -1, 0, 1, 2$
   (c) Part (a): mean $= 0$, median $= 0$; part (b): mean $= 0$, median $= 0$

3. (a) 29.3
   (b) No
   (c) First quartile is 27.7; second quartile, 29.3; third quartile, 31.1.
   (d) 31.7

5. Yes

9. No

11. No, association is not causation.

SECTION 4.2

1. (a) $S = \{(R, R), (R, B), (R, Y), (B, R), (B, B), (B, Y), (Y, R), (Y, B), (Y, Y)\}$
   (b) $\{(Y, R), (Y, B), (Y, Y)\}$
   (c) $\{(R, R), (B, B), (Y, Y)\}$

3. (a) $\{(U of M, OSU), (U of M, SJSC), (RC, OSU), (RC, SJSC), (SJSC, OSU), (SJSC, SJSC), (Yale, OSU), (Yale, SJSC), (OSU, OSU), (OSU, SJSC)\}$
   (b) $\{(SJSC, SJSC), (OSU, OSU)\}$
   (c) $\{(U of M, OSU), (U of M, SJSC), (RC, OSU), (RC, SJSC), (SJSC, OSU), (Yale, OSU), (Yale, SJSC), (OSU, SJSC)\}$
   (d) $\{(RC, OSU), (OSU, OSU), (SJSC, SJSC)\}$

5. $S = \{(France, fly), (France, boat), (Canada, drive), (Canada, train), (Canada, fly)\}$
   $A = \{(France, fly), (Canada, fly)\}$

7. (a) $\varnothing$
   (b) $\{1, 4, 6\}$
   (c) $\{1, 3, 4, 5\}$
   (d) $\{2\}$

9. (a) $\{(1, g), (1, f), (1, s), (1, c), (0, g), (0, f), (0, s), (0, c)\}$
   (b) $\{(0, s), (0, c)\}$
   (c) $\{(1, g), (1, f), (0, g), (0, f)\}$
   (d) $\{(1, g), (1, f), (1, s), (1, c)\}$

11. (a) A^c is the event that a rolled die lands on an odd number.
    (b) $(A^c)^c$ is the event a rolled die lands on an even number.
    $(A^c)^c = A$.

13.

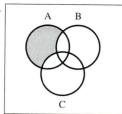

(a)

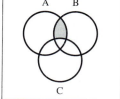

(b)

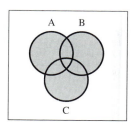

(c)

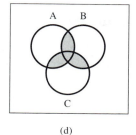

(d)

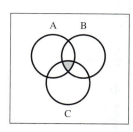

(e)

SECTION 4.3

1. (a) $P(E) = .35$; $P(E) = .65$; $P(G) = .55$
   (b) $P(E \cup F) = 1$
   (c) $P(E \cup G) = .8$
   (d) $P(F \cup G) = .75$
   (e) $P(E \cup F \cup G) = 1$
   (f) $P(E \cap F) = 0$
   (g) $P(F \cap G) = .45$
   (h) $P(E \cap G) = .1$
   (i) $P(E \cap F \cap G) = 0$

3. 1/10,000

5. If they are disjoint, it is impossible. If they are not disjoint, it is possible.

7. (a) 1
   (b) .8
   (c) .5
   (d) .1

9. (a) .95
   (b) .80
   (c) .20

11. .7

13. .31

15. .6

17. (a) $A \cap B^c$
    (b) $A \cap B$
    (c) $B \cap A^c$
    (d) $P(I) + P(II) + P(III)$
    (e) $P(I) + P(II)$
    (f) $P(II) + P(III)$
    (g) $P(II)$

SECTION 4.4

1. $88/216 \approx .41$

3. (a) $4/52 \approx .08$
   (b) $48/52 \approx .92$
   (c) $13/52 \approx .25$
   (d) $1/52 \approx .02$

5. (a) 78%
   (b) 69.9%
   (c) 24.3%
   (d) 61%
   (e) 87%

7. (a) .56
   (b) .1

9. (a) .4
   (b) .1

11. 56

13. 1/19

15. (a) .1
    (b) .1

17. (a) 10/31
    (b) 9/31
    (c) 1/3
    (d) 11/31
    (e) 7/31

SECTION 4.5

1. (a) .02/.3 ≈ .067
   (b) .02/.03 ≈ .667

3. (a) .245
   (b) .304

5. (a) .145
   (b) .176
   (c) .215
   (d) .152

7. (a) .46
   (b) .65

9. (a) .62
   (b) .50
   (c) .55
   (d) .49

11. $1/169 \approx .006$

13. .30

15. (a) $19/34 \approx .56$
    (b) $1 - 19/34 \approx .44$
    (c) $1/17 \approx .06$

17. Since $P(B|A) > P(B)$, $P(A \cap B) > P(B)P(A)$
    Hence, $P(A|B) = \dfrac{P(A \cap B)}{P(B)} > \dfrac{P(B)P(A)}{P(B)} = P(A)$

19. .24

21. .68

23. (a) $7/12 \approx .58$
    (b) 50
    (c) $13/119 \approx .11$
    (d) $35/204 \approx .17$
    (e) .338

25. (a) .79; .21
    (b) .81; .27

27. (a) 1/2
    (b) 3/8
    (c) 2/3

29. 1/16

31. No; the friends do not know each other.

33. $P(A) = 1/13$; $P(B) = 1/4$; $P(A \cap B) = 1/52$; thus
    $P(A \cap B) = P(A)P(B)$.

35. 1/365

37. (a) .64
    (b) .96
    (c) .8704

39. Yes, $P(A)P(B) = P(A \cap B)$.

41. (a) $32/4805 \approx .0067$
    (b) $729/1922 \approx .38$
    (c) .06
    (d) .04
    (e) .006
    (f) .016

43. (a) 1/4
    (b) 2/3

SECTION 4.6

1. (a) .55
   (b) 5/9

3. (a) .672
   (b) .893

5. .398

7. (a) .534
   (b) .402

REVIEW PROBLEMS

1. (a) 3/4
   (b) 3/4
   (c) 6/11
   (d) 1/22
   (e) 9/22

3. (a) .68
   (b) .06
   (c) .12

5. (a) 11/24
   (b) 13/23

7. (a) 1/64
   (b) 1/64
   (c) 1/64

9. (a) S = {(Chicken, rice, melon), (chicken, rice, ice cream), (chicken, rice, gelatin), (chicken, potatoes, melon), (chicken, potatoes, ice cream), (chicken, potatoes, gelatin), (roast beef, rice, melon), (roast beef, rice, ice cream), (roast beef, rice, gelatin), (roast beef, potatoes, melon), (roast beef, potatoes, ice cream), (roast beef, potatoes, gelatin)}
   (b) {(Chicken, potatoes, ice cream), (chicken, potatoes, gelatin), (roast beef, potatoes, ice cream), (roast beef, potatoes, gelatin)}
   (c) 1/3
   (d) 1/12

11. (a) 1/3
    (b) 1/3
    (c) 1/3
    (d) 1/2

13. $14/33 \approx .424$

15. (a) 1/52
    (b) 1/52
    (c) equally
    (d) 1/52

17. (a) .42
    (b) .18
    (c) .24
    (d) .58
    (e) .724

19. No

21. (a) .496
    (b) 54/252
    (c) 36/248
    (d) No

23. (a) 4
    (b) 4/86
    (c) 1/2
    (d) No

25. (a) .077
    (b) .0494
    (c) .0285

SECTION 5.2

1. $P\{Y = 0\} = 1/4$
   $P\{Y = 1\} = 3/4$

3. (a) 5/12
   (b) 5/12
   (c) 0
   (d) 1/4

5.

i	$P\{Y = i\}$
1	11/36
2	1/4
3	7/36
4	5/36
5	1/12
6	1/36

7.

i	$P\{X = i\}$
2	.58
3	.42

9.

i	$P\{X = i\}$
0	1199/1428
1	55/357
2	3/476

11.

i	$P\{X = i\}$
0	.075
1	.325
2	.6

13. No; $P(4)$ is negative.

15.

i	$P\{X = i\}$
0	38/223
1	82/223
2	57/223
3	34/223
4	10/223
5	2/223

17. (a) .1
    (b) .5

19.

i	$P\{X = i\}$
0	.57
1	.37
2	.06

21. (a) .0711
    (b) .0018

23.

i	$P\{X = i\}$
0	.855
100,000	.14
200,000	.005

SECTION 5.3

1. (a) 2
   (b) 5/3
   (c) 7/3

3. $8.40

5. 1.9

7. (a) 2.53
   (b) 4.47

9. $880

11. (a) 2/3
    (b) 4/3
    (c) 2

13. (a) Second location
    (b) First location

15. −$5

17. (a) No
    (b) No
    (c) Yes
    (d) $4/95 \approx .042$

19. −$0.40

21. 2.5

23. $150

25. 0

27. (a) $16,800
    (b) $18,000
    (c) $18,000

29. 3

31. (a) 7
    (b) 7

33. 23,789.25

35. 3.6

SECTION 5.4

1. Var $(U) = 0$, Var $(V) = 1$, Var $(W) = 100$

3. 0

5. .49

7. .25

9. (b) .8
   (c) .6

11. (a) .5
    (b) .5

13. (a) 0
    (b) $3666

15. (a) 4.06
    (b) 1.08

17. 3 SD$(X) = 6$

19. (a) 2
    (b) 2

SECTION 5.5

1. (a) 24
   (b) 120
   (c) 5040

3. 3,628,800

5. (a) .278692
   (b) .123863
   (c) .00786432

7. (a) .468559
   (b) .885735

9. (a) 3 or more
   (b) .00856

11. .144531

13. (a) .517747
    (b) .385802
    (c) .131944

15. (a) .421875
    (b) .421875
    (c) .140625
    (d) .015625

17. (a) 10/3
    (b) 20/3
    (c) 10
    (d) 50/3

19. (a) .430467
    (b) .382638
    (c) 7.2
    (d) .72

21. (a) .037481
    (b) .098345
    (c) .592571
    (d) 1.76
    (e) .992774

23. (a) .00604662
    (b) 0

25. (a) 50; 5
    (b) 40; 4.89898
    (c) 60; 4.89898
    (d) 25; 3.53553
    (e) 75; 6.12372
    (f) 50; 6.12372

SECTION 5.6

1. Hypergeometric, $n = 20$, $N = 200$, $p = .09$

3. Hypergeometric, $n = 6$, $N = 54$, $p = 6/54$

5. Hypergeometric, $n = 20$, $N = 100$, $p = .05$

7. Binomial, $n = 10$, $p = 1/13$

REVIEW PROBLEMS

1. (a) .4
   (b) .6

3. (a) 1, 2, 3, 4
   (b)

i	$P(X = i)$
1	.3
2	.21
3	.147
4	.343

   (c) .7599
   (d) 2.53
   (e) 1.53

5. (a) .723
   (b) No, because if she wins then she will win $1, whereas if she loses then she will lose $3.
   (c) −.108

7. (a)

i	$P(X = i)$
0	.7
4000	.15
6000	.15

   (b) 1500
   (c) 5,550,000
   (d) 2,355.84

9. The low bid will maximize their expected profit.

11. (a) 1/3
    (b) 1/4
    (c) 7/24
    (d) 1/12
    (e) 1/24
    (f) $625
    (g) $125

13. (a) 0
    (b) −6875
    (c) −6875

17. (a) .6
    (b) .648
    (c) .68256
    (d) .710208
    (e) .733432
    (f) .813908

19. (a) .064
    (b) .432
    (c) .820026

SECTION 6.2

1. (a) .29
   (b) .56
   (c) .33
   (d) .27

3. (a) 2/3
   (b) .7
   (c) .6
   (d) .6

5. (a) 2/3
   (b) 1/6
   (c) 1/3

7. (a) 1/2
   (b) 0
   (c) 3/4
   (d) 3/8

SECTION 6.3

1. (a) 108.8 to 148
   (b) 89.2 to 167.6
   (c) 69.6 to 187.2

3. (b)

5. (d)

7. (c)

9. (a)

11. (b)

13. (d)

15. (b)

17. (a) Y
    (b) X
    (c) X and Y are equally likely to exceed 100.

19. (a) No
    (b) No
    (c) No
    (d) Yes

SECTION 6.4

1. (a) .9861
   (b) .1357
   (c) .4772
   (d) .7007
   (e) .975
   (f) .2358
   (g) .899
   (h) .2302
   (i) .8710

3. 3

7. (a) 1.65
   (b) 1.96
   (c) 2.58
   (d) 0
   (e) .41
   (f) 2.58
   (g) 1.15
   (h) .13
   (i) .67

SECTION 6.6

1. Since $x > a$, $x - u > a - u$. It follows that $\dfrac{x - u}{\sigma} >$ $\dfrac{a - u}{\sigma}$ since σ is positive.

3. .3085

5. (a) .6179
   (b) .8289
   (c) .4468

7. .008

9. (a) .1587
   (b) .2514
   (c) .4772

11. .8664

13. (a) .0993
    (b) .011
    (c) .8996

15. (a) .2660
    (b) .9890

(c) .7230
(d) .9991
(e) .0384

SECTION 6.7

1. (a) 1.48
   (b) 1.17
   (c) .52
   (d) 1.88
   (e) −.39
   (f) 0
   (g) −1.64
   (h) 2.41

3. (a) 50
   (b) 57.68
   (c) 61.76
   (d) 40.16
   (e) 57.02

5. 464.22

7. 525.6

9. 746

11. (a) True
    (b) True

13. 99.28

REVIEW PROBLEMS

1. (a) .9236
   (b) .8515
   (c) .0324
   (d) .9676
   (e) .1423
   (f) .0007
   (g) 75.524
   (h) 73.592
   (i) 68.3

3. 4.969

5. (a) .1587
   (b) .1587
   (c) .1886
   (d) 576.8

7. (a) .881
   (b) .881
   (c) .762

9. (a) .4483
   (b) .201
   (c) .4247

11. (a) .6915
    (b) .3859
    (c) .1587

13. (a) 1/4
    (b) .28965

SECTION 7.3

1. (a) SD $(\bar{X}) = \dfrac{1/2}{\sqrt{3}} \approx .29$

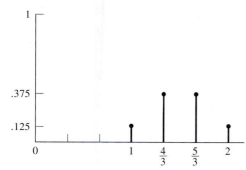

(b) SD $(\bar{X}) = \dfrac{1/2}{\sqrt{4}} = .25$

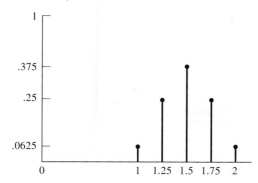

3. (a) 2
   (b) $\sqrt{2/3} \approx .82$
   (c)

i	$P\{\bar{X} = i\}$
1	1/9
1.5	2/9
2	3/9
2.5	2/9
3	1/9

(d) $E(\bar{X}) = 2$, SD$(\bar{X}) = 1/\sqrt{3} \approx .58$
(e) Yes

5. (a) $E(\bar{X}) = 2.4$, SD$(\bar{X}) = .2/\sqrt{36} \approx .033$
   (b) $E(\bar{X}) = 2.4$, SD$(\bar{X}) = .2/\sqrt{64} \approx .025$
   (c) $E(\bar{X}) = 2.4$, SD$(\bar{X}) = .2/\sqrt{100} \approx .02$
   (d) $E(\bar{X}) = 2.4$, SD$(\bar{X}) = .2/\sqrt{900} \approx .007$

7. Expected value = 15,500, standard deviation = 2800

SECTION 7.4

1. (a) .5468
   (b) .7888
   (c) .9876

3. .7888

5. (a) .0062
   (b) .7888

7. .9713

9. .1416

11. (a) .905
    (b) .5704

13. (a) 0
    (b) 0

15. (a) .6826
    (b) .9544
    (c) 1
    (d) 1
    (e) 1

SECTION 7.5

1. (a) $E(\bar{X}) = .6$, SD$(\bar{X}) = .15$
   (b) $E(\bar{X}) = .6$, SD$(\bar{X}) = .049$
   (c) $E(\bar{X}) = .6$, SD$(\bar{X}) = .015$
   (d) $E(\bar{X}) = .6$, SD$(\bar{X}) = .0049$

3. (a) .0122
   (b) 0.119
   (c) .5222

5. (a) .0764
   (b) .4681

7. (a) .1711
   (b) .0009

9. (a) .0125
   (b) .8508

11. .1949

13. .4602

15. (a) .9147
    (b) .0043
    (c) .5188

17. (a) .9599
    (b) .3121

19. (a) .9974
    (b) .0268

SECTION 7.6

1. (a) 5.7; 4 degrees of freedom
   (b) .018; 5 degrees of freedom
   (c) 1.13; 2 degrees of freedom

REVIEW PROBLEMS

1. (a) .8413
   (b) .5
   (c) .0228
   (d) .0005

3. $E(\bar{X}) = 3$; $SD(\bar{X}) = 1/\sqrt{2} \approx .71$

5. (a) Mean = 12, standard deviation = 3.25
   (b) .5588

7. (a) 300
   (b) $7\sqrt{20} \approx 31.3$
   (c) .5

9. .1003

11. (a) .3669
    (b) .9918
    (c) .9128

13. (a) .9984
    (b) 1
    (c) .0087
    (d) .0016
    (e) 0
    (f) .9744

SECTION 8.2

1. 145.5

3. $108

5. 165.6 hours

7. 12

9. 3.23

11. (a)

SECTION 8.3

1. .3849

3. .65; .107

5. .412; .05

7. (a) .122
   (b) .01

9. (a) .0233
   (b) .0375
   (c) .0867

11. (a) .245
    (b) .022

13. (c); accurate in terms of lowest standard error

SECTION 8.3.1

1. .28

3. (b) 3.32; 1.73; 1.45

SECTION 8.4

1. 18.36

3. 799.7; 193.12

5. 21.27

7. 30.5

9. 12.64

11. 1.35

13. .0474; .2386

SECTION 8.5

1. (a) (3.06, 3.24)
   (b) (3.03, 3.27)

3. (11.43, 11.53)

5. (a) (8852.87, 9147.13)
   (b) (8824.69, 9175.31)

7. (72.53, 76.67)

9. (a) (1337.35, 1362.65)
   (b) (1334.92, 1365.08)
   (c) (1330.18, 1369.82)

11. 13.716

13. 3176

15. (a) 72.99
    (b) 72.53
    (c) 76.67
    (d) 77.53

17. No

SECTION 8.6

1. (a) (5.15, 5.25)
   (b) (5.13, 5.27)

3. (a) (73.82, 93.91)
   (b) (71.63, 96.10)
   (c) (66.89, 100.84)

5. (a) (127.71, 163.29)
   (b) (119.18, 171.82)

7. (446.28, 482.01)

9. (280.04, 284.96)

11. (1849.4, 2550.6)

13. (a) (4.60, 4.80)
    (b) (4.58, 4.82)

15. (1124.95, 1315.05)

17. No

19. (a) (27.59, 38.64)
    (b) No

21. 68.897, 98.836

23. The average daily receipts exceeds $2857.

SECTION 8.7

1. (.548, .660)

3. (a) (.502, .519)
   (b) (.498, .523)

5. (.546, .734)

7. (.359, .411)

9. (0, .306)

11. (.801, .874)

13. (0, .45)

15. (a) (.060, .108)
    (b) (.020, .052)
    (c) (.448, .536)

17. (a) A 95% confidence interval is given by .75 $\pm$ .0346.
    (b) Rather than using $\hat{p}$ to estimate p in the standard error term they used the upper bound $p(1 - p) \leq 1/4$.

19. (a) 1692
    (b) Less than .04 but greater than .02
    (c) (.213, .247)

21. 6147

23. .868

25. (a) .139
    (b) .101

27. (a) No
    (b) No

REVIEW PROBLEMS

1. (a)

3. (22.35, 26.45)

5. (316.82, 323.18)

7. (a) (44.84, 54.36)
   (b) (45.66, 53.54)

9. (1527.47, 2152.53)

11. (a) 88.56
    (b) (83.05, 94.06)

13. (a) (34.02, 35.98)
    (b) (33.04, 36.96)
    (c) (31.08, 38.92)

15. (.487, .549)

17. .004

19. (a) (.373, .419)
    (b) (.353, .427)

21. Upper

SECTION 9.2

1. (a) Hypothesis B

3. (d) is most accurate; (b) is more accurate than not.

SECTION 9.3

1. TS = 1.55; $z_{\alpha/2}$ = 1.96; do not reject H_0.

3. (a) .0026
   (b) .1336
   (c) .3174
   At the 5% level of significance we reject H_0 in (a). At the 1% level of significance we reject H_0 in (a).

5. Yes

7. (a) No
   (b) 0

9. The data do not support a mean of 13,500 miles.

11. Yes; Yes

13. The p value is .281. Thus we reject this hypothesis at a level of significance of .281 or greater.

15. (a) .2616
    (b) .2616
    (c) .7549

SECTION 9.3.1

1. (a) No
   (b) No
   (c) .091

3. (a) 0
   (b) 0
   (c) .0085

5. (a) Yes
   (b) No, because the reduction in cavities is so small.

7. Yes, but increase the sample size.

9. The mean amount dispensed is less than 6 ounces; H_0: $\mu \geq 6$; H_1: $\mu < 6$; P value = 0

SECTION 9.4

1. The evidence is not strong enough to discredit the manufacturer's claim at the 5% level of significance.

3. (a) Yes
   (b) No

5. (a) No
   (b) No
   (c) No
   (d) The p value is .108.

7. Yes

11. H_0: $\mu \geq 23$ versus H_1: $\mu < 23$. The judge should rule for the bakery.

13. (a) H_0: $\mu \geq 31$
    (b) H_1: $\mu < 31$
    (c) No
    (d) No

15. No, the p value is .0068.

17. No; no

SECTION 9.5

1. p value = .0365; normal approximation is .0416

3. No

5. (a) H_0: $p \leq .5$; H_1: $p > .5$
   (b) .1356
   (c) .0519
   (d) .0042
   As n increases, the p value decreases, because we have more confidence in the estimate for larger n.

7. (a) No
   (b) No
   (c) No
   (d) Yes

9. No; no

11. No

13. (a) Yes
    (b) No
    (c) .2005

15. .0017

REVIEW PROBLEMS

1. (b)

5. (a) No
   (b) Yes
   (c) Yes

7. There is not sufficient evidence to support the claim at the 5% level of significance.

9. One would probably rule against Caputo since the p value of the test H_0: $p = 1/2$ against H_1: $p \neq 1/2$ is .000016.

SECTION 10.2

1. (a) No
   (b) 0

3. (a) There is evidence to support the hypothesis that the mean lengths of their cuttings are equal.
   (b) .8335

5. It suffices to relabel the data sets and use the given test.

7. No

SECTION 10.3

1. Yes; $H_0: \mu_x = \mu_y$; $H_1: \mu_x \neq \mu_y$; p value $= .0206$

3. p value $= .5664$

5. Yes; 0

7. No; $H_0: \mu_x \leq \mu_y$; $H_1: \mu_x > \mu_y$ where x corresponds to rural students and y corresponds to urban students.

9. $H_0: \mu_B \leq \mu_A$; $H_1: \mu_B > \mu_A$; p value $= .0838$. At the 5% level of significance supplier B should be used.

11. (a) $H_0: \mu_m \leq \mu_f$; $H_1: \mu_f < \mu_m$
    (b) .0004
    (c) It indicates that the female average wage is less than the male average wage.

13. (a) The null hypothesis should be rejected for $\alpha = .01$.
    (b) .0066
    (c) Reduction in mean score

SECTION 10.4

1. No; yes

3. (a) No
   (b) No

5. Yes

7. Reject $H_0: \mu_x = \mu_y$ for $\alpha = .05$. p value $= .0003$

9. (a) Reject $H_0: \mu_x = \mu_y$.
   (b) Reject $H_0: \mu_x = \mu_y$.
   (c) Do not reject $H_0: \mu_x = \mu_y$.

SECTION 10.5

1. (a) Reject the hypothesis at $\alpha = .05$.
   (b) p value $= .0015$

3. Do not reject H_0.

5. (a) Do not reject the hypothesis.
   (b) There is not evidence to reject the hypothesis at the 5% level of significance.

7. Reject the hypothesis at $\alpha = .05$.

9. (a) $H_0: \mu_{before} \leq \mu_{after}$; $H_1: \mu_{before} > \mu_{after}$
   (b) No

11. The null hypothesis is not rejected.

SECTION 10.6

1. (a) No
   (b) No

3. (a) Yes
   (b) .0178

5. (a) No
   (b) .0856

7. Reject the hypothesis that the proportions were the same in 1983 and 1990; p value $= .0017$.

9. Reject the hypothesis for $\alpha = .05$; p value $= 0$.

11. (a) Yes
    (b) 0

13. No

15. Yes; $H_0: \hat{p}_{placebo} \leq \hat{p}_{aspirin}$ (where $\hat{p}$ is the proportion that suffered heart attacks); $\hat{p}$ value $= 0$.

REVIEW PROBLEMS

1. (a) Reject $H_0: \mu_x = \mu_y$.
   (b) 0

3. (a) Do not reject the hypothesis that the probabilities are the same.
   (b) .5289
   (c) No
   (d) $\alpha \geq .26446$

5. (a) Reject $H_0: \mu_x = \mu_y$.

7. Do not reject the hypothesis that the probabilities are the same.

9. Do not reject the hypothesis (p value $= .79$).

11. Do not reject the hypothesis that the proportions are the same in both sports.

SECTION 11.2

1. (a) $\bar{X}_1 = 8$, $\bar{X}_2 = 14$, $\bar{X}_3 = 11$
   (b) $\bar{\bar{X}} = 11$

3. Yes

5. No

7. Do not reject the hypothesis for $\alpha = .05$.

9. Reject the hypothesis that death rates do not depend on season for $\alpha = .05$.

11. No

SECTION 11.3

1. $\hat{\alpha} = 68.8$, $\alpha_1 = 14.2$, $\alpha_2 = 6.53$, $\alpha_3 = -3.47$, $\alpha_4 = -3.47$, $\alpha_5 = -13.8$, $\beta_1 = .8$, $\beta_2 = -2.4$, $\beta_3 = 1.6$

3. $\hat{\alpha} = 28.33$, $\alpha_1 = 1$, $\alpha_2 = -2$, $\alpha_3 = 1$, $\beta_1 = 3.67$, $\beta_2 = -.67$, $\beta_3 = -3$

5. (a) 14.625
   (b) .85
   (c) −.975

7. $\hat{\alpha} = 9.585$, $\alpha_1 = 1.74$, $\alpha_2 = -1.96$, $\alpha_3 = 4.915$, $\alpha_4 = -1.36$, $\alpha_5 = -3.335$, $\beta_1 = .495$, $\beta_2 = -.405$, $\beta_3 = .795$, $\beta_4 = -.885$

9. (a) 44
   (b) 48
   (c) 52
   (d) 144

SECTION 11.4

1. (a) Yes
   (b) No

3. (a) No
   (b) No

5. (a) No
   (b) Yes

7. (a) Reject the hypothesis for $\alpha = .05$.
   (b) Reject the hypothesis for $\alpha = .05$.
   (c) Again reject both for $\alpha = .05$.

9. (a) Reject the hypothesis for $\alpha = .05$.
   (b) Do not reject the hypothesis for $\alpha = .05$.

REVIEW PROBLEMS

1. Reject the hypothesis for $\alpha = .05$.

3. Yes for $\alpha = .05$, no for $\alpha = .01$.

5. Do not reject the hypothesis for $\alpha = .05$.

7. (a) Do not reject the hypothesis for $\alpha = .05$.
   (b) 30.6
   (c) Reject the hypothesis for $\alpha = .05$.

9. (a) Do not reject the hypothesis for $\alpha = .05$.
   (b) Reject the hypothesis for $\alpha = .05$.

SECTION 12.2

1. (a)

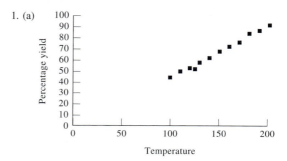

   (b) Yes

3. (a) Density; speed

   (b)

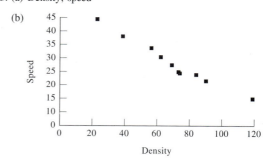

   (c) Yes

5. (a)

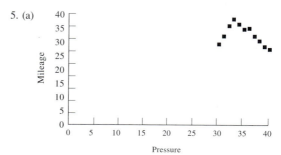

   (b) No

SECTION 12.3

1. (a)

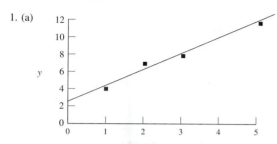

(b)

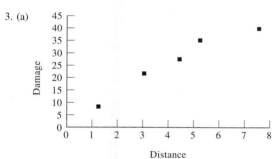

3. (a)

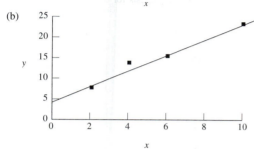

(c) $y = 14.79 + 2.43x$

5. (a)

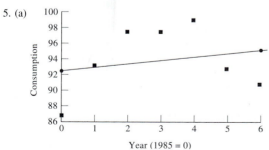

Year (1985 = 0)

(b) $y = 92.575 + .418x$
(d) 93.83

7. $y = 655.36 - 26.64x$

9. $y = 7.45 - .035x$

11. (a) $y = -8.31 + .27x$
    (b) 31.66
    (c) $y = 31.66 + 3.61x$
    (d) 147.12

13. At random

15. (a) $y = 67.56 + .23x$
    (b) 204.62
    (c) 261.73
    (d) 296.00

17. 121.85

19. (a) 433.87
    (b) 448.92
    (c) 463.98

SECTION 12.4

1. 2.32

3. (a) 6
    (b) 6
    (c) 76

5. .000156

7. 6970.21

SECTION 12.5

1. Do not reject H_0: $\beta = 0$.

3. Reject the hypothesis.

5. (a)

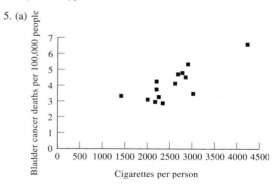

(b) $y = .75 + .0013x$
(c) Reject the hypothesis.
(d) Reject the hypothesis.

7. (a)

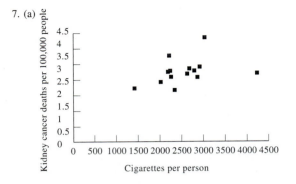

Cigarettes per person

(b) $y = 2.12 + .0003x$
(c) Do not reject the hypothesis.
(d) Do not reject the hypothesis.

9. Reject the hypothesis.

11. (a) Do not reject the hypothesis for $\alpha = .05$.
   (b) Do not reject the hypothesis for $\alpha = .05$.
   (c) Do not reject the hypothesis for $\alpha = .05$.

SECTION 12.6

1. (a) $\alpha = 10.48$, $\beta = .325$
   (b) Yes

5. (a) $y = 2.57 + .82x$
   (b) Yes

7. Not as well as the heights.

SECTION 12.7

1. (a) 12.6
   (b) (6.4, 18.8)

3. (a) $y = 45.42 - .66x$
   (b) $7773.49
   (c) (6659.78, 8887.21)
   (d) (6026.89, 9520.09)

5. (a) 2.501
   (b) (2.493, 2.510)

7. (a) $33,266
   (b) (27,263, 39,268)
   (c) $42,074; (35,608, 48,541)

SECTION 12.8

1. (a)

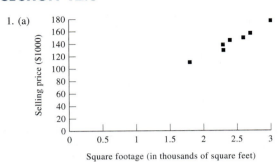

Square footage (in thousands of square feet)

(b) $y = 8.885 + 56.32x$
(c) 97%
(d) (144,628, 165,929)

3. (a) .9996
   (b) Yes
   (c) 41.975
   (d) (40.692, 43.258)

5. .149

7. .059

SECTION 12.9

1. (a) .9796; .9897
   (b) .9796; .9897
      This indicates that the value of the sample correlation coefficient does not depend on which variable is considered the independent variable.

3. (a) .8
   (b) .8
   (c) −.8
   (d) −.8

5. (a) $y = -3.16 + 1.24x$
   (b) $y = 7.25 + .66x$
   (c) .818; .904
   (d) .818; .904

SECTION 12.10

1. (a) $.985 + .0329x_1 + 43.65x_2 + 10.39x_3$
   (b) $121,594
   (c) $131,989

3. $y = -153.51 + 51.75x_1 + .077x_2 + 20.92x_3 + 13.10x_4$; 183.62

5. 69.99

REVIEW PROBLEMS

1. (a)

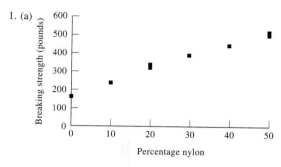

(b) $y = 177.93 + 6.89x$
(c) 522.61
(d) (480.53, 564.68)

3. (a) $\alpha = 94.13$; $\beta = .155$
(b) (93.17, 132.34)
(c) 100%

5. Not necessarily, doing well (or poorly) might just be a chance phenomenon which will tend to regress to the mean on the next attempt.

7. (a) $y = 71.975 + 1.196x$ (1980 = year 0)
(b) Reject H_0: $\beta = 0$.

9. (a) 34.9
(b) (4.34, 23.40)

11. (a) $y = 177.41 + 1.07x_1 + 11.7x_2$
(b) 241.90

13. (a) $\beta_0 = 2.84$, $\beta_1 = .16$, $\beta_2 = -.11$
(b) It tends to decrease.
(c) 7.3

15. Alcohol consumption, which is associated with both cigarette consumption and bladder cancer incidence, might be the primary cause. A multiple linear regression would be useful.

SECTION 13.2

1. (a) 15.086
(b) 11.070
(c) 23.209
(d) 18.307
(e) 31.410

3. H_0: $P_1 = .52$, $P_2 = .32$, $P_3 = .16$. No, H_0 is not rejected.

5. Yes, the null hypothesis is rejected.

7. No, p value = .0002.

9. Yes; yes

11. Yes

13. Do not reject the hypothesis.

15. Reject the hypothesis.

SECTION 13.3

1. (a) 7.08
(b) Yes
(c) No

3. Reject the hypothesis.

5. The characteristics are independent.

7. Reject the hypothesis.

9. No

11. Reject the hypothesis; reject the hypothesis.

13. Do not reject the hypothesis.

SECTION 13.4

1. No, we cannot conclude that smoking causes lung cancer, but we can conclude that the per capita lung cancer rate is higher for smokers than for nonsmokers.

3. Do not reject the hypothesis.

5. No

7. Yes; no

9. Do not reject in each case.

REVIEW PROBLEMS

1. Do not reject the hypothesis.

5. Reject the hypothesis.

7. Do not reject the hypothesis for $\alpha = .05$.

9. Reject the hypothesis.

11. No; no

13. (a) Do not reject the hypothesis.
(b) .208

15. Do not reject the hypothesis; do not reject the hypothesis.

19. Yes

SECTION 14.2

1. (a) p value $= 0.057$. Reject the null hypothesis at any significance level greater than or equal to .057.
   (b) p value ≈ 0. Reject the null hypothesis at any significance level.
   (c) p value ≈ 0. Reject the null hypothesis at any significance level.

3. We cannot reject the null hypothesis that the two guns are equally effective.

5. Since n is small we use the binomial distribution to calculate the p value $= 0.291$. Thus we cannot reject the hypothesis that the median score will be at least 72.

7. Yes, this discredits the hypothesis. p value $= .0029$.

9. Here we have 7 scores below and 7 scores above the 1988 median value. Thus the p value is 0.5 and we cannot reject the null hypothesis.

SECTION 14.3

1. (a) TS $= 39$
   (b) TS $= 42$
   (c) TS $= 20$

3. (a) p value $= 0.2460$
   (b) p value $= 0.8336$
   (c) p value $= 0.1470$

5. (a) Yes, how the paper is presented had an effect on the score given.
   (b) p value $= 0.0102$

7. (a) The null hypothesis is rejected at any significance level greater than or equal to 0.1250.
   (b) The null hypothesis is rejected at any significance level greater than or equal to 0.0348.

9. No, we cannot reject the null hypothesis. Painting does not affect an aircraft's cruising speed.

SECTION 14.4

1. (a) 94
   (b) 77

3. p value $= 0.8572$

5. Since the p value $= 0.2112$, we cannot reject the null hypothesis that the starting salary distribution for MBAs from the two schools are the same.

7. p value $= 0.4418$

SECTION 14.5

1. (a) 41
   (b) 2

3. Since the p value $= 0.0648$, we cannot reject the hypothesis that the data constitutes a random smaple.

5. Since the p value $= 0.0548$, we cannot reject the null hypothesis that the interviewer interviewed them in a randomly chosen order.

7. (a) Median $= 163.5$
   (b) Seven runs
   (c) Since the p value $= 0.0016$, we must reject the null hypothesis at any significance level greater than or equal to 0.0016. The sequence of values do not constitute a random sample.

CHAPTER 14 REVIEW

1. Using the rank-sum test with TS $= 113$, we obtain a p value of 0.0348. So we cannot reject the null hypothesis at the 1% level of significance, but we must reject the null hypothesis at the 5% level.

3. Since p value ≈ 0, reject the null hypothesis, the median net worth has decreased.

5. We do a runs test, with median $= 145$ and $n = m = 20$, and $r = 21$. Since $5 = 21$, the p value is 1.0.

9. Using the signed rank test with TS $= 0$. The p value $= 0.0444$. Thus we reject the null hypothesis that there is no difference in the shoe sales at any level of significance above 4.44%.

11. Since the p value $= 0.5620$, we cannot reject the null hypothesis.

SECTION 15.2

1. (a) LCL $= 85$, UCL $= 115$
   (b) LCL $= 86.58$, UCL $= 113.42$
   (c) LCL $= 87.75$, UCL $= 112.25$
   (d) LCL $= 90.51$, UCL $= 109.49$

3. LCL $= 66.58$, UCL $= 93.42$. Since subgroup 9 falls outside this range, the process would have been declared out of control at that point.

5. LCL $= -0.00671$, UCL $= 0.00671$. Since all the subgroups are within these control limits, the process is in control.

SECTION 15.3

1. LCL = 0, UCL = 13.23. Since all the subgroups are within the control limits, the process is in control.

3. (a) Since all the subgroups are within the control limits, the process is in control.
   (b) LCL = 0, UCL = 9.88

CHAPTER 15 REVIEW

1. LCL = 1.4985, UCL = 1.5015.

3. LCL = 1212.2, UCL = 1627.8 Since all the subgroups are within these control limits, the burglary rate remained at its pre-1983 rate.

5. LCL = 0, UCL = 13.23. Since all the subgroups are within these control limits, the process is in control.

Index